HANDBOOK ON BIODIVERSITY AND ECOSYSTEM SERVICES IN IMPACT ASSESSMENT

RESEARCH HANDBOOKS ON IMPACT ASSESSMENT

Series Editor: Frank Vanclay, *University of Groningen, the Netherlands*

The objective of this series of *Research Handbooks on Impact Assessment* is to provide a critical assessment of the state-of-the-art in the research and thinking in the various fields of impact assessment, including environmental impact assessment, strategic environmental assessment, social impact assessment, health impact assessment, biodiversity assessment, cumulative impact assessment, and so on. The series will also consider the use of impact assessment across a variety of sectors, including agriculture, energy, forestry, mining, transport. The aim is to produce prestigious high quality works of lasting significance, offering a comprehensive overview of the research fields in question. With oversight from the Series Editor, Professor Frank Vanclay, a noted specialist in the field of impact assessment, the *Research Handbooks* comprise carefully commissioned chapters from leading academics and/or reflexive practitioners selected by editors who are recognized leaders in their field. Taking a genuinely international and often transdisciplinary approach, the books in the series address current and sometimes controversial issues. Through clear analysis and lucid writing, these *Handbooks* are designed to shed light on contemporary issues and contribute to current debates. Offering unrivalled analysis and discussion, each *Handbook* will be an invaluable source of reference for an international audience of scholars, researchers and practitioners involved with the study and utilization of impact assessment.

Titles in the series include:

Handbook of Sustainability Assessment
Edited by Angus Morrison-Saunders, Jenny Pope and Alan Bond

Handbook of Regulatory Impact Assessment
Edited by Claire A. Dunlop and Claudio M. Radaelli

Handbook on Biodiversity and Ecosystem Services in Impact Assessment
Edited by Davide Geneletti

Handbook on Biodiversity and Ecosystem Services in Impact Assessment

Edited by

Davide Geneletti

Associate Professor, University of Trento, Italy

RESEARCH HANDBOOKS ON IMPACT ASSESSMENT

Edward Elgar
PUBLISHING

Cheltenham, UK • Northampton, MA, USA

Published by
Edward Elgar Publishing Limited
The Lypiatts
15 Lansdown Road
Cheltenham
Glos GL50 2JA
UK

Edward Elgar Publishing, Inc.
William Pratt House
9 Dewey Court
Northampton
Massachusetts 01060
USA

A catalogue record for this book
is available from the British Library

Library of Congress Control Number: 2016931484

This book is available electronically in the **Elgar**online
Economics subject collection
DOI 10.4337/9781783478996

MIX
Paper from responsible sources
FSC® C013604

ISBN 978 1 78347 898 9 (cased)
ISBN 978 1 78347 899 6 (eBook)

Typeset by Servis Filmsetting Ltd, Stockport, Cheshire

Printed and bound by CPI Group (UK) Ltd, Croydon, CR0 4YY

Contents

PART II APPLICATIONS IN DIFFERENT SECTORS

PART III CURRENT ISSUES AND CHALLENGES

Contributors

Christian Albert, Leibniz Universität and Helmholtz Centre for Environmental Research (UFZ), Germany.

Assumpció Antón, Institute of Research and Technology of Food and Agriculture (IRTA), Spain.

Matthew Ashley, Plymouth Marine Laboratory, UK.

Juan Azcárate, Royal Institute of Technology (KTH), Sweden.

Berit Balfors, Royal Institute of Technology (KTH), Sweden.

Susie Brownlie, deVilliers Brownlie Associates, South Africa.

Lea Bulling, Environmental Assessment and Planning Research Group, Berlin Institute of Technology (TU Berlin), Germany.

Chiara Cortinovis, Department of Civil, Environmental and Mechanical Engineering, University of Trento, Italy.

Richard T.T. Forman, Graduate School of Design, Harvard University, USA.

Susanne Frank, Karlsruhe Institute of Technology, Germany.

Christine Fürst, Karlsruhe Institute of Technology, Germany.

Davide Geneletti, Department of Civil, Environmental and Mechanical Engineering, University of Trento, Italy.

Joshua Goldstein, The Nature Conservancy, USA.

Tara Hooper, Plymouth Marine Laboratory, UK.

Pierre Horwitz, School of Natural Sciences, Edith Cowan University, Australia.

Michael Hughes, Environment and Conservation Sciences Group, School of Veterinary and Life Sciences, Murdoch University, Australia.

Pekka Itkonen, Helsinki City Rescue Department, Finland.

Marcos Jiménez, University of Bonn, Germany.

Mårten Karlson, Royal Institute of Technology (KTH), Sweden.

Leena Karrasch, COAST – Centre for Environment and Sustainability Research, University of Oldenburg, Germany.

Christina M. Kennedy, The Nature Conservancy, USA.

Joseph M. Kiesecker, The Nature Conservancy, USA.

Johann Köppel, Environmental Assessment and Planning Research Group, Berlin Institute of Technology (TU Berlin), Germany.

Leena Kopperoinen, Finnish Environment Institute (SYKE), Finland.

Olivia Langmead, Marine Biological Association, UK.

Danielle Maia de Souza, Swedish University of Agricultural Sciences, Sweden; University of Alberta and Lethbridge Research and Development Centre (AAFC), Canada.

Lisa Mandle, The Natural Capital Project, Stanford University, USA.

Llorenç Milà i Canals, United Nations Environment Programme, Division of Technology, Industry and Economics, France.

Ulla Mörtberg, Royal Institute of Technology (KTH), Sweden.

David Newsome, Environment and Conservation Sciences Group, School of Veterinary and Life Sciences, Murdoch University, Australia.

Sara Odelius Gordon, Royal Institute of Technology (KTH), Sweden.

Margot W. Parkes, School of Health Sciences, University of Northern British Columbia, Canada.

Frank Pietzsch, PiSolution, Germany.

Katrin Pietzsch, PiSolution, Germany.

Asha Rajvanshi, Wildlife Institute of India, India.

Dilys Roe, International Institute for Environment and Development, UK.

Daniel Alejandro Rozas Vásquez, Laboratorio de Planificación Territorial, Universidad Católica de Temuco, Chile.

Mary Ruckelshaus, The Natural Capital Project, Stanford University, USA.

Heather Tallis, The Nature Conservancy, USA and University of California, Santa Cruz, USA.

Léa Tardieu, INRA, AgroParisTech, UMR 356 Laboratoire d'Economie Forestière, France.

Félix Teillard, Food and Agriculture Organization of the United Nations, Animal Production and Health Division, Italy & INRA-AgroParisTech, UMR 1048 SAD APT, France.

Jo Treweek, eCountability Ltd., UK.

Jianguo (Jingle) Wu, School of Life Sciences and School of Sustainability, Arizona State University, USA.

Linda Zardo, Department of Civil, Environmental and Mechanical Engineering, University of Trento, Italy.

1. Introduction

Davide Geneletti

1.1 SETTING THE CONTEXT

This *Handbook* addresses the consideration of biodiversity and ecosystem services (ES) in impact assessment, by providing a critical analysis of some of the latest research and practice in this field. The book was written to support researchers and practitioners in the conceptual development and operational implementation of biodiversity and ES-inclusive impact assessment processes.

Impact assessment is the 'process of identifying the future consequences of a current or proposed action' (IAIA, 2009, p. 1). At its heart, impact assessment aims to provide information for decision-making. With this intention, the National Environmental Policy Act (NEPA) – signed into law in 1970 – introduced the process of Environmental Impact Assessment (EIA) in the United States of America. This Act represented the first legislation that formally established an impact assessment process to state the potential effects of actions prior to their approval (O'Riordan and Sewell, 1981). NEPA inspired the institutionalization of EIA as a decision-support tool all over the world (Canter, 1996). Currently, EIA is firmly embedded in the legislation of nearly all UN countries, as well as in international law (Morgan, 2012).

Even though the scope of EIA varies in different contexts, its focus (especially in early applications) has been on the biophysical aspects of the environment and on the project level of decision-making (Wood, 2003; Morgan, 2012). This has led to the development of other forms of impact assessment that expanded the scope, both thematically and across the decision spectrum. For example, Social Impact Assessment (SIA) (Esteves et al., 2012) and Health Impact Assessment (HIA) (Harris-Roxas et al., 2012) address in a more comprehensive way social aspects, and community and individual health issues respectively. Strategic Environmental Assessment (SEA) (Therivel, 2010) reflects the need to consider the higher levels of decision-making, such as policies, plans and programmes. More recently, sustainability assessment (Bond et al., 2012) has emerged to emphasize the analysis of the sustainability gains associated with any kind of decision.

The content of impact assessments is constantly evolving to reflect new perspectives and emerging issues and concerns. The treatment of biodiversity is a case in point. Early legislation (e.g., Directive 85/337/EEC, the first European Union directive on EIA) described the biotic aspects to be included in impact assessment studies simply as 'flora and fauna'. This narrow scope is reflected in the findings of reviews of EIA practice in the 1990s and early 2000s (Treweek et al., 1993; Thompson et al., 1997; Atkinson et al., 2000; Byron et al., 2000). These reviews report a tendency to limit biodiversity and ecological analysis to individual species (e.g., iconic, endangered or protected species) and sites already designated for nature conservation, concluding that the majority of EIA did not discuss impacts on the different levels at which biodiversity occurs. Meanwhile, the Convention on Biological Diversity endorsed and popularized a broader definition of biodiversity, seen as 'the variability among living organisms from all sources including, inter alia, terrestrial, marine and other aquatic ecosystems and the ecological complexes of which they are part; this includes diversity within species, between species and of ecosystems' (CBD, 1992, Art. 2).

More comprehensive approaches to address biodiversity in impact assessment were developed in the years that followed. For example, Treweek (1999) and Treweek et al. (2005) identified important principles to ensure that biodiversity considerations are appropriately addressed in EIA and SEA, respectively. These studies also illustrate key analysis to be performed in different stages of the EIA and SEA processes. Geneletti (2003, 2004) proposed methods to consider the spatial configuration of ecosystems in impact assessment, and to assess and compare their value using decision-support techniques. Geneletti et al. (2003) advanced the treatment of uncertainty factors in Biodiversity Impact Assessment (BIA). Balfors et al. (2005) set the foundations for landscape- and regional-level analysis of biodiversity in SEA. Gontier et al. (2006, 2010) applied innovative modelling tools to predict and assess impacts on biodiversity of projects and plans. Even though all these studies demonstrate a more systematic understanding of biodiversity and its complexity, they look at biodiversity from a nature conservation perspective, with limited (or only indirect) consideration of human development and well-being.

Two seminal works in 1997 reinforced the recognition that biodiversity underpins the delivery of ES that are fundamental for our well-being. Daily (1997) provided the first comprehensive account of the many services that nature offers to people, and the way in which we all depend on them. Costanza et al. (1997) estimated the overall economic value of the ES of the entire biosphere, stimulating a surge of interest in this topic. The Millennium Ecosystem Assessment (MA, 2005) and many subsequent studies documented the rate of degradation of many services, and the associated negative consequences for human well-being (Nelson et al., 2013; Barnosky et al., 2014). The MA defines ES as the benefits human populations derive from ecosystems, such as goods and products (e.g., fresh water, fuel), regulation of natural processes (e.g., climate, flooding, erosion), and non-material benefits (e.g., recreation, aesthetic enjoyment). Over the last decade, the concept of ES has been widely taken up in both science and policy (Polasky et al., 2015). Publications in this field have grown exponentially (Braat and De Groot, 2012), and a number of international initiatives gained media and policy attention, such as The Economics of Ecosystems and Biodiversity (TEEB, 2010) and the Intergovernmental Platform on Biodiversity and Ecosystem Services (IPBES) (Díaz et al., 2015).

The field of impact assessment has been receptive to this widening in perspective. EIA and SEA guidance published after the MA (Slootweg et al., 2006; OECD, 2008), and subsequent academic work (Slootweg et al., 2010), proposed integrative impact frameworks that include the effects of development on ES and human well-being, along with more traditional biodiversity conservation issues. Geneletti (2013) has collected recent experiences and viewpoints in this area, and Rosa and Sánchez (2015) and Geneletti et al. (2015) have reviewed practice. Methods have been advanced to include ES in impact assessment for higher-level policies (Helming et al., 2013; Diehl et al., 2016), plans (Geneletti, 2011, 2015), projects (Landsberg et al., 2013), as well as for specific sectors (e.g., infrastructure development; Mandle et al., 2015). All these efforts show the emerging interest in this area, but also the many open challenges.

This *Handbook* presents state-of-the-art methodological guidance and discussion of international practice related to the incorporation of biodiversity and ES in impact assessment. As one of the series of *Research Handbooks on Impact Assessment*, the book provides a critical assessment of the research and thinking in this field, emerging from different parts of the world. The case studies presented in the chapters span the five continents, and a broad range of sectors and biomes. The *Handbook* is divided into four parts. Part I, described in Section 1.2, looks at how biodiversity and ES information can be mainstreamed in different impact assessment types to improve their salience and effectiveness. Part II (Section 1.3) presents a range of applications in key policy and planning sectors. Part III (Section 1.4) addresses selected issues and challenges in contemporary practice and research. Part IV (Section 1.5) concludes with a final chapter that summarizes the key messages and provides directions on the way forward.

1.2 PART I: MAINSTREAMING BIODIVERSITY AND ECOSYSTEM SERVICES IN IMPACT ASSESSMENT TYPES

Over the last decades, a number of impact assessment types have emerged in the literature and practice, raising concerns about their efficiency and effectiveness. Morrison-Saunders et al. (2014), after having identified more than 40 different specialist types of impact assessment, argue that this proliferation creates separate silos of expertise and undermines interdisciplinary practice. This *Handbook* endorses the integration of biodiversity and ES within existing impact assessment types, rather than via separate stand-alone processes. It acknowledges that these issues are inherently interdisciplinary and better addressed from broader perspectives, as already demonstrated by Geneletti (2011, 2015) and Partidário and Gomes (2013) in relation to SEA, and by Karjalainen et al. (2013) in relation to EIA. The chapters of Part I present innovative conceptual frameworks, and practical methodologies, to strengthen the consideration of biodiversity and ES in different types of impact assessment.

Lisa Mandle and Heather Tallis (Chapter 2) address EIA of projects, by proposing an approach for carrying out spatially explicit assessments of development impacts on ES. They argue that assessments of environmental impacts of projects have traditionally focused primarily on quantifying change on the basis of area, habitat quality or ecological processes, without linking clearly to people and their well-being. Hence, an accounting of impacts on ES – not just ecosystems – is needed to make the impacts on people transparent. With their approach, environmental impacts are not only quantified, but importantly also tracked between ecosystems and people to account for where and to whom the costs and benefits accrue. The flexibility of the approach is demonstrated with two case studies, in which free, open-source, science-based software tools and commonly available data are used to assess the impacts of mine and road development on ES in Latin America.

Chapter 3, by Davide Geneletti, shifts the focus from project-level to strategic-level decisions, by presenting a sequence of key analyses to mainstream ES information in SEA. These analyses include: building a conceptual framework for ES production and use in the study region; reviewing relevant regulations, plans and policies; determining priority ES and assessing their baseline conditions and trends; developing possible alternatives that enhance opportunities and reduce risks for ES, and assessing their impact. The analyses are illustrated through a variety of examples and short case studies in different contexts. The chapter demonstrates how information of ES represents a valuable addition to the content of SEA by providing a means to translate unattended and unintended consequences of plans and policies implementation on human well-being, as argued by Kumar et al. (2013).

Pierre Horwitz and Margot W. Parkes (Chapter 4) examine the potential value of the ES concept as a framing device for Health Impact Assessment (HIA). Particularly, the chapter addresses the scoping stage of HIA. The authors identify overarching health issues, and associate them with relevant ES, which then enables estimation of health benefits when these ES are maintained or enhanced, and the consequences for health where these ES are disrupted. This provides a framework for the scoping phase in HIA, which can be used to promptly identify and negotiate trade-offs. A case study involving a resource development proposal (coal seam gas extraction in eastern Australia) is used to demonstrate the applicability of the proposed approach. The chapter concludes by discussing some of the remaining challenges and unanswered questions associated with the consideration of ES in the context of HIA.

In Chapter 5, Leena Karrasch discusses ways to match the ES concept with Social Impact Assessment (SIA). The author presents a framework to identify the links between ES and social impacts, accounting for the social preferences and needs of local decision-makers. The framework consists of three steps, starting with scientific-based input and moving towards intensive stakeholder engagement. The case study in regional development that is presented in the chapter shows that the framework could be used to assist planners and researchers in establishing a more integrated, participatory development process, particularly in relation to the consideration of complex interactions between nature and society.

The economic dimension is the focus of Chapter 6. In this contribution, Léa Tardieu presents the theory and practice of performing monetary evaluation of ES loss induced by transportation infrastructures. The chapter begins by reviewing the legal framework of transportation infrastructure projects in France, with the purpose of illustrating how the legal context influences the most appropriate way of incorporating ES. An application to a real case study (a high-speed rail project) is then presented. Finally, the advantages and challenges of integrating monetary evaluation into more traditional decision-support tools, such as EIA and cost–benefit analysis, are discussed.

In Chapter 7, Assumpció Antón et al. address the integration of biodiversity and ES in Life Cycle Assessment (LCA). The aim of LCA is to provide a quantification of the potential environmental impacts of goods and processes for the cycle of production from 'cradle to grave'. The first section of this chapter describes the environmental mechanisms affecting biodiversity and ecosystems services and their quantification in LCA. The second section introduces the state of the art of the different impact indicator proposals addressing biodiversity loss and change in ES in the frame of LCA. The review focuses on the effects of land use and land use change as the key drivers of biodiversity loss. Finally, the last section addresses the main challenges in the quest for good and harmonized indicators to account properly for biodiversity and ES in LCA.

1.3 PART II: APPLICATIONS IN DIFFERENT SECTORS

Part II of the *Handbook* provides a collection of applications in a broad swathe of sectors. The purpose is to illustrate experiences and approaches that are relevant for different decision-making areas and levels, and discuss the associated challenges, solutions and opportunities. The following sectors are featured: urban development, spatial planning (both for terrestrial and marine areas), agriculture and forestry, hydropower, wind energy, and tourism. The different chapters in this part of the book were selected also to cover different key impact assessment issues (e.g., cumulative effects, impact trade-offs, consideration of stakeholders' perception), and a variety of ecosystem types, including agro-ecosystems and grasslands, urban green infrastructures, forests, marine and fluvial ecosystems, and coastal zones.

Berit Balfors et al. (Chapter 8) address the impacts of urbanization on biodiversity and the capacity of urban green areas to provide ES. A description of the role of biodiversity and ES is provided as a framework for a landscape approach in biodiversity assessments and for presenting the practical examples. Three different case studies in urban planning from the Stockholm region are presented and discussed, together with tools to predict and assess biodiversity impacts at a landscape level and to plan and manage urban green areas. The chapter concludes with lessons learned and key recommendations for best practice.

In Chapter 9, Christine Fürst et al. consider how to assess the impact of agricultural and forest management on biodiversity and ES. Agricultural and forest land management practices play a key role in affecting biodiversity shifts and sustaining natural capacities to provide a multitude of ES. This contribution presents a modelling and assessment platform – GISCAME – that was developed to connect planning and decision-making at different scales. The approach particularly supports the comparative analyses of combined land use and land management change scenarios to guide spatial planning processes. To demonstrate how the platform can be applied, two applications in agriculture and forestry at regional level (in Germany and Chile) are presented.

The purpose of Chapter 10, by Leena Kopperoinen et al., is to present key issues related to including biodiversity and ES in impact assessment of spatial planning. First, a set of useful tools to support impact assessment is presented. This includes tools to assess ES at the local scale, to improve participatory mapping and to analyse the spatial configuration and connectivity of green infrastructures. The usefulness of these tools is then illustrated through three real-life spatial planning processes in Finland and Germany. The chapter concludes by providing a checklist for implementing impact assessment in practice.

Tara Hooper et al. examine marine spatial planning and impact assessment in Chapter 11. They argue that ES approaches to marine EIA have tended to consider only a small number of services, which is of limited use in project or strategic planning. The chapter then explores the process required for adopting an ES approach to EIA in marine systems, including the role of stakeholders in identifying ES and selecting assessment priorities. The need for appropriate indicators that permit the manageable measurement and reporting of ES is discussed. Modelling and spatial mapping are key tools for evaluating and communicating potential change, and methods for these are demonstrated using two case studies in the context of regional marine energy planning and marine protected areas, respectively.

In Chapter 12, David Newsome and Michael Hughes discuss the impacts of ecotourism on biodiversity, by providing a multi-scale perspective that considers the influence of people's perceptions and politics. They explore the complex spectrum within which ecotourism exists, ranging from mass tourism at one end to highly specialized niche tourism at the other. Positive and negative impacts are identified, acknowledging that the nature of such impacts is influenced by political and socio-economic factors that characterize the areas in which the biodiversity occurs. The chapter concludes with an exploration of the implications for impact assessment concerning the complex characteristics of ecotourism and their interaction with biodiversity. This is achieved via examining the case of the ecotourism–conservation nexus associated with iconic species such as gorillas in East Africa.

The last two chapters of Part II address the energy sector. Lea Bulling and Johann Köppel (Chapter 13) explore the trade-offs between wind energy and biodiversity conservation. They draw on the argument that the assessment of these trade-offs is bound to become more and more critical due to the expected increase in wind energy development, posing a great challenge to impact assessment processes. They provide an overview of current conflicts between biodiversity conservation and energy generation, looking specifically at the species groups that are affected the most by wind energy development (birds and selected mammals). Then they consider potential solutions and present exemplary case studies that can pave the way for future best practices.

Finally, in Chapter 14 Asha Rajvanshi examines the hydropower sector. She first reviews the potential impacts of dams on biodiversity, and then presents the merits and challenges of conducting Cumulative Impact Assessment of dams for the identification of basin-wide effects that may cause aggregate, multiple and altogether new impacts on ecological systems. The chapter provides examples that reflect the contributions of Cumulative Impact Assessment to SEA for planning hydropower projects in locations that are least disruptive to key ecological processes, and operating them in ways that protect biodiversity and maintain key ecological services. Operational recommendations for successful assessment of the cumulative effects of dams are provided at the end of the chapter.

1.4 PART III: CURRENT ISSUES AND CHALLENGES

Part III of the *Handbook* addresses selected issues and challenges in contemporary practice and research related to the consideration of biodiversity and ES in impact assessment. The purpose is to advance new ideas and discuss possible solutions, as well as to identify priority directions for future research.

In Chapter 15, Dilys Roe and Davide Geneletti consider the interactions between biodiversity conservation and poverty alleviation in impact assessment. They start by presenting a conceptual framework to guide thinking about the interactions between different components of biodiversity and different dimensions of poverty. The framework recognizes that both biodiversity and poverty are complex, multi-dimensional concepts and, furthermore, that the interactions between the two are mediated by a wide range of factors. In particular, cross-cutting determinants, such as governance and policies on poverty and biodiversity, are critical in determining whether biodiversity leads to actual poverty reduction. They then review the existing evidence base on biodiversity–poverty linkages, drawing on a systematic mapping of the published and grey literature. The findings of the review are used to identify key issues that should be addressed in Biodiversity Impact Assessment of any planned development interventions, if the subsequent implications for poverty are to be taken into account.

The next two chapters discuss open challenges, and make innovative proposals, in relation to biodiversity and ES mitigation and offset measures. Susie Brownlie and Jo Treweek (Chapter 16) explore biodiversity offsets in the context of impact assessment processes. Particularly, they examine the role of offsets in achieving 'no net loss' or 'net gain' of biodiversity when development takes place. The chapter looks at offset practice around the world and considers how issues in offset design and implementation have been addressed. It then goes on to consider some of the aspects that need further consideration to allay fears that use of offsets might reduce the effectiveness of impact assessment and accelerate loss of biodiversity due to development. Challenges to effective offset design, and responses to these challenges, are discussed. Finally, some emerging trends in offset practice that might help to address the challenges are presented.

In Chapter 17, Heather Tallis et al. draw on the argument that we lack a unified conceptual framework and analytical precedent to guide the integration of ES into more commonly practised biodiversity mitigation contexts. They present a four-step framework that addresses key deficiencies in current biodiversity mitigation practice and recommend how ES can be included in the context of existing regulatory approaches. Within this framework, they address the conceptual and analytical advances needed to establish ES targets, delineate a spatial extent that captures ES supply and delivery, establish avoidance thresholds for services, quantitatively estimate impacts on services, consistently construct mitigation replacement ratios, and identify and design potential ES offsets. In each of these areas, they identify opportunities to embed ES alongside biodiversity in a single integrated framework.

Climate change is expected to be an increasingly important source of impacts on biodiversity and ES, both directly and in combination with other activities and phenomena. The potential of nature-based solutions is more and more acknowledged in the climate adaptation debate, because they offer the advantage of promoting 'no regrets' interventions, and potentially delivering multiple economic, social and environmental co-benefits (Jones et al., 2012; Munang et al., 2013). In Chapter 18, Davide Geneletti et al. address the role that impact assessment can play to promote nature-based solutions for climate adaptation in urban areas. They first analyse the current trends and gaps, and then illustrate a methodology to increase the evidence base that planners have at their disposal to develop and compare nature-based solutions. The application of the method is demonstrated for a case study related to impact assessment of urban planning in Italy. Finally, some conclusions and recommendations for directing future practice and research in this area are provided.

Richard T.T. Forman and Jianguo (Jingle) Wu round up Part III of the

Handbook with an analysis of the best places for the next billion people, based on environmental constraints (Chapter 19). The analysis is first conducted globally, and then downscaled to urban regions. The purpose is to identify the types of urban places where major population growth could be concentrated, in order to reduce the impacts on farmland and natural land. An array of urban-region solutions are then suggested to improve food-producing land, nature and built communities despite a new billion people. The authors acknowledge that no single overarching solution emerges. However, packages of specific solutions seem promising to address a number of problems and mould a better land ahead – a very relevant message for impact assessment practitioners and researchers.

1.5 PART IV: CONCLUSIONS

The overall conclusions of the *Handbook* are provided in Chapter 20. This final chapter summarizes the key messages that emerged from the contributions, and that are relevant across impact assessment types and planning and policy sectors. These messages are articulated into two parts, addressing impact assessment frameworks and impact analyses respectively.

REFERENCES

Atkinson, S.F., S. Bathia and F. Schoolmaster et al. (2000), 'Treatment of biodiversity impacts in a sample of US environmental impact statements', *Impact Assessment and Project Appraisal*, **18**(4), 271–82.

Balfors, B., U. Mörtberg and M. Gontier et al. (2005), 'Impacts of region-wide urban development on biodiversity in strategic environmental assessment', *Journal of Environmental Assessment Policy and Management*, **7**(2), 229–46.

Barnosky, A.D., J.H. Brown and G.C. Daily et al. (2014), 'Introducing the scientific consensus on maintaining humanity's life support systems in the 21st century: information for policy makers', *The Anthropocene Review*, **1**(1), 78–109.

Bond, A., A. Morrison-Saunders and J. Pope (2012), 'Sustainability impact assessment: the state of the art', *Impact Assessment and Project Appraisal*, **30**(1), 53–62.

Braat, L. and R. de Groot (2012), 'The ecosystem services agenda: bridging the worlds of natural science and economics, conservation and development, and public and private policy', *Ecosystem Services*, **1**(1), 4–15.

Byron, H.J., J.R. Treweek and W.R. Sheate et al. (2000), 'Road developments in the UK: an analysis of ecological assessment in environmental impact statements produced between 1993 and 1997', *Journal of Environmental Planning and Management*, **43**(1), 71–97.

Canter, L.W. (1996), *Environmental Impact Assessment*, 2nd edition, New York: McGraw-Hill.

Convention on Biological Diversity (CBD) (1992), *Convention on Biological Diversity*, accessed 11 January 2016 at https://www.cbd.int/doc/legal/cbd-en.pdf.

Costanza, R., R. d'Arge and R. de Groot et al. (1997), 'The value of the world's ecosystem services and natural capital', *Nature*, **387**(6230), 253–60.

Daily, G. (1997), *Nature's Services: Societal Dependence on Natural Ecosystems*, Washington, DC: Island Press.
Díaz, S., S. Demissew and J. Carabias et al. (2015), 'The IPBES conceptual framework – connecting nature and people', *Current Opinion in Environmental Sustainability*, **14**, 1–16.
Diehl, K.E., D. Burkhard and K. Jacob (2016), 'Should the ecosystem services concept be used in European Commission impact assessment?', *Ecological Indicators*, **61**(Part 1), 6–17.
Esteves, A.M., D. Franks and F. Vanclay (2012), 'Social impact assessment: the state of the art', *Impact Assessment and Project Appraisal*, **30**(1), 34–42.
Geneletti, D. (2003), 'Biodiversity impact assessment of roads: an approach based on ecosystem rarity', *Environmental Impact Assessment Review*, **23**(3), 343–65.
Geneletti, D. (2004), 'Using spatial indicators and value functions to assess ecosystem fragmentation caused by linear infrastructures', *International Journal of Applied Earth Observations and Geoinformation*, **5**(1), 1–15.
Geneletti, D. (2011), 'Reasons and options for integrating ecosystem services in strategic environmental assessment of spatial planning', *International Journal of Biodiversity Science, Ecosystem Services & Management*, **7**(3), 143–9.
Geneletti, D. (2013), 'Ecosystem services in environmental impact assessment and strategic environmental assessment. Editorial', *Environmental Impact Assessment Review*, **40**, 1–2.
Geneletti, D. (2015), 'A conceptual approach to promote the integration of ecosystem services in strategic environmental assessment', *Journal of Environmental Assessment Policy and Management*, **17**(4), 1550035.
Geneletti, D., E. Beinat and C.J. Chung et al. (2003), 'Accounting for uncertainty factors in biodiversity impact assessment: lessons from a case study', *Environmental Impact Assessment Review*, **23**(4), 471–87.
Geneletti, D., A. Bond and D. Russel et al. (2015), 'Ecosystem services and sustainability assessment: theory and practice', in A. Morrison-Saunders, J. Pope and A. Bond (eds), *Handbook of Sustainability Assessment*, Cheltenham, UK and Northampton, MA, USA: Edward Elgar Publishing.
Gontier, M., U. Mörtberg and B. Balfors (2006), 'Biodiversity in environmental assessment – current practice and tools for prediction', *Environmental Impact Assessment Review*, **26**(3), 268–86.
Gontier, M., U. Mörtberg and B. Balfors (2010), 'Comparing GIS-based habitat models for applications in EIA and SEA', *Environmental Impact Assessment Review*, **30**(1), 8–18.
Harris-Roxas, B., F. Viliani and A. Bond et al. (2012), 'Health impact assessment: the state of the art', *Impact Assessment and Project Appraisal*, **30**(1), 43–52.
Helming, K., K. Diehl and D. Geneletti et al. (2013), 'Mainstreaming ecosystem services in European policy impact assessment', *Environmental Impact Assessment Review*, **40**, 82–7.
International Association for Impact Assessment (IAIA) (2009), 'What is impact assessment?', accessed 11 January 2016 at http://www.iaia.org/publicdocuments/special-publications/What%20is%20IA_web.pdf.
Jones, H.P., D.G. Hole and E.S. Zavaleta (2012), 'Harnessing nature to help people adapt to climate change', *Nature Climate Change*, **2**(7), 504–9.
Karjalainen, T.P., M. Marttunen and S. Sarkki et al. (2013), 'Integrating ecosystem services into environmental impact assessment: an analytic–deliberative approach', *Environmental Impact Assessment Review*, **40**, 54–64.
Kumar, P., S.E. Esen and M. Yashiro (2013), 'Linking ecosystem services to strategic environmental assessment in development policies', *Environmental Impact Assessment Review*, **40**, 75–81.
Landsberg, F., J. Treweek and N.M. Stickler et al. (2013), *Weaving Ecosystem Services Into Impact Assessment*, Washington, DC: World Resource Institute.
Mandle, L., B.P. Bryant and M. Ruckelshaus et al. (2015), 'Entry points for considering ecosystem services within infrastructure planning: how to integrate conservation with development in order to aid them both', *Conservation Letters* [online], DOI: 10.1111/conl.12201.

Millennium Ecosystem Assessment (MA) (2005), *Ecosystems and Human Well Being: Synthesis*, Washington, DC: Island Press.
Morgan, R.K. (2012), 'Environmental impact assessment: the state of the art', *Impact Assessment and Project Appraisal*, **30**(1), 5–14.
Morrison-Saunders, A., J. Pope and J.A.E. Gunn et al. (2014), 'Strengthening impact assessment: a call for integration and focus', *Impact Assessment and Project Appraisal*, **32**(1), 2–8.
Munang, R., I. Thiaw and K. Alverson et al. (2013), 'Climate change and ecosystem-based adaptation: a new pragmatic approach to buffering climate change impacts', *Current Opinion in Environmental Sustainability*, **5**, 67–71.
Nelson, E.J., P. Kareiva and M. Ruckelshaus et al. (2013), 'Climate change's impact on key ecosystem services and the human well-being they support in the US', *Frontiers in Ecology and the Environment*, **11**(9), 483–93.
Organisation for Economic Co-operation and Development (OECD) (2008), *Strategic Environmental Assessment and Ecosystem Services*, Paris: OECD.
O'Riordan, T. and W.R.D. Sewell (1981), 'From project appraisal to policy review', in *Project Appraisal and Policy Review*, Chichester, UK: Wiley, pp. 1–28.
Partidário, M.R. and R.C. Gomes (2013), 'Ecosystem services inclusive strategic environmental assessment', *Environmental Impact Assessment Review*, **40**, 36–46.
Polasky, S., H. Tallis and B. Reyers (2015), 'Setting the bar: standards for ecosystem services', *Proceedings of the National Academy of Sciences*, **112**(24), 7356–61.
Rosa, J.C.S. and L.E. Sánchez (2015), 'Is the ecosystem service concept improving impact assessment? Evidence from recent international practice', *Environmental Impact Assessment Review*, **50**, 134–42.
Slootweg, R., A. Kolhoff and R. Verheem et al. (2006), *Biodiversity in EIA and SEA. Background Document to CBD Decision VIII/28: Voluntary Guidelines on Biodiversity-Inclusive Impact Assessment*, Amsterdam: Commission for Environmental Assessment.
Slootweg, R., A. Rajvanshi and V.B. Mathur et al. (2010), *Biodiversity in Environmental Assessment: Enhancing Ecosystem Services for Human Well-being (Ecology, Biodiversity and Conservation Series)*, Cambridge, UK: Cambridge University Press.
The Economics of Ecosystems and Biodiversity (TEEB) (2010), *The Economics of Ecosystems and Biodiversity. Mainstreaming the Economics of Nature: A Synthesis of the Approach, Conclusions and Recommendations of TEEB*, Nairobi: United Nations Environment Programme.
Therivel, R. (2010), *Strategic Environmental Assessment in Action*, 2nd edition, London: Earthscan.
Thompson, S., J.R. Treweek and N. Veicht et al. (1997), 'The ecological component of environmental impact assessment: a critical review of British environmental statements', *Journal of Environmental Planning and Management*, **40**(2), 157–71.
Treweek, J. (1999), *Ecological Impact Assessment*, Oxford, UK: Blackwell Science Ltd.
Treweek, J., R. Therivel and S. Thompson et al. (2005), 'Principles for the use of strategic environmental assessment as a tool for promoting the conservation and sustainable use of biodiversity', *Journal of Environmental Assessment Policy and Management*, **7**(2), 173–99.
Treweek, J.R., S. Thompson and N. Veicht et al. (1993), 'Ecological assessment of proposed road developments: a review of environmental statements', *Journal of Environmental Planning and Management*, **36**(3), 295–307.
Wood, C. (2003), *Environmental Impact Assessment. A Comparative Review*, 2nd edition, Harlow, UK: Pearson.

PART I

MAINSTREAMING BIODIVERSITY AND ECOSYSTEM SERVICES IN IMPACT ASSESSMENT TYPES

2. Spatial ecosystem service analysis for Environmental Impact Assessment of projects

Lisa Mandle and Heather Tallis

2.1 INTRODUCTION

Governments, multi-lateral lending institutions and investors increasingly require project developers to complete impact assessments, reflecting the growing recognition that avoiding social harm through environmental impact is critical to maximizing the economic and social benefits of development projects such as roads, mines, dams and energy infrastructure (Morgan, 2012; The Biodiversity Consultancy, 2013). However, assessments of environmental impacts have largely focused on quantifying impacts on the basis of area, habitat quality or ecological processes, without linking clearly to people and their well-being (e.g., Quétier and Lavorel, 2011; Robertson et al., 2014). An accounting of impacts on ecosystem services (ES), or the benefits that flow from ecosystems to people – and not just ecosystems – is needed to make the impacts on people transparent and facilitate equitable distribution of benefits from development.

Recognition of this need can be seen in the growing requirements to explicitly account for ES in impact assessments for planning, permitting and safeguards processes. In 2012, the International Finance Corporation (IFC) issued new performance standards requiring IFC-financed projects to address ES in impact assessments (IFC, 2012). These performance standards have been adopted widely, by the World Bank for private lending (World Bank, 2013) and by Equator Principles Financial Institutions, which provide over 70 percent of international project finance debt in emerging markets (Equator Principles Association, 2013). A growing body of guidance on how to incorporate ES in different forms of environmental assessment has accompanied these new requirements (IPIECA/OGP, 2011; Landsberg et al., 2013; Geneletti, 2014; The Biodiversity Consultancy, 2014). Nevertheless, the perception remains that spatially explicit analysis of impacts on ES – especially if modeled – is challenging and resource intensive, and so while encouraged is not usually required (e.g., Landsberg et al., 2014). However, with recently developed approaches and tools alongside growing, global open data availability, spatially explicit accounting of impacts on ES is increasingly practical. In this chapter, we show that spatially explicit ES analysis is both feasible and can provide valuable information not captured by a qualitative, descriptive approach.

We begin with a brief overview of the benefits that a spatial ES analysis approach can contribute to the impact assessment process. Next, we offer a framework for integrating development impacts to the environment and to people through spatial analysis of ES. We illustrate this framework with two case studies: in the first case study, from Colombia, we use this approach to evaluate the impacts of proposed coal mine development on a subset of ES; in the second case study, we assess the direct and indirect impacts of a proposed highway in Peru and consider the equity implications for indigenous and non-indigenous communities in the area. We synthesize lessons learned from the case studies and discuss additional ways that this approach can be extended to other contexts and aspects of the impact assessment process.

2.2 WHY DO WE NEED SPATIALLY EXPLICIT ECOSYSTEM SERVICE ANALYSIS IN IMPACT ASSESSMENT?

Some have questioned whether the concept of ES provides anything useful and new to the impact assessment process, or whether it might be 'a methodology looking for a purpose' (Baker et al., 2013, p. 10). Indeed, impact assessment carried out with an ES perspective will contain many components already familiar to impact assessment practitioners, even those who have never explicitly framed their work in terms of ES or natural capital. However, evidence from existing impact and offset programs makes clear that biodiversity mitigation, as practiced, does not accurately and comprehensively account for services. Wetland mitigation programs in the United States, implemented under provisions of the Clean Water Act, frequently result in the relocation of wetlands away from densely populated urban areas towards rural areas with few people (King and Herbert, 1997; Ruhl and Salzman, 2006; BenDor et al., 2007). This transfer often leads to a loss of wetlands – and the many services they provide – for poor and minority communities.

First, we should make clear what we mean when we talk about 'ecosystem services'. Varying definitions and their inconsistent use has led to confusion as to how ES-based concepts can effectively be put into practice (Nahlik et al., 2012). Most broadly, ES are the diverse benefits provided by nature that sustain and fulfill human life. The Millennium

Ecosystem Assessment classification, which has also been adopted by the IFC performance standards, divides ES into four categories: provisioning, regulating, cultural and supporting services (Millennium Ecosystem Assessment, 2005). Provisioning services are the tangible goods obtained from ecosystems, including food, fuel and fiber. Regulating services are the benefits that result from the regulation of ecosystem processes, such as climate regulation, maintenance of air and water quality, pollination and pest control. Cultural ES are the non-material benefits from nature, and include spiritual, esthetic and recreational enjoyment of nature. Finally, supporting services are the ecological processes on which all other services ultimately depend. These include soil formation, primary production, and nutrient cycling. Some have suggested that 'supporting services' such as nutrient cycling and primary production, should be more accurately termed 'supporting functions' as they do not provide benefits directly, but rather support the production of other services. However, the same applies to many regulating services as well. For example, pollination – classified as a regulating service – is an ecosystem function that contributes to food production – a provisioning service.

The distinction between ecosystem functions and services becomes clearer if the production of ES is thought of as a supply chain, with ecological systems producing the supply, the location and demands of people converting the supply into a service, and people's preferences for that level of service expressing its value (Tallis, Mooney and Andelman et al., 2012). Ecosystem structure and function determine the potential supply of an ES that a particular area is capable of producing. For example, the full amount of wave dampening that mangroves provide captures the potential for mangroves to protect shorelines from storm surge and erosion. While many ES assessments stop at assessing the level of ecosystem function, supply alone does not provide a good indication of the amount of ES delivered. An ecosystem function becomes a service only when and where it provides benefits to people. Therefore, the location and demands of people determine whether the *supply* provided in a particular place produces an ecosystem *service*. In the case of mangroves, for example, mangroves provide the ES of coastal protection only when they occur in a place that protects coastline of value to people – for cultural, economic, esthetic or other reasons. This means that mangroves protecting coastal homes are providing hazard mitigation services, while mangroves bordering uninhabited coastlines without interest to any social group are not (though these mangroves may nevertheless be providing other valuable services, such as carbon sequestration for climate regulation and nursery habitat for commercial fish species). The value of a particular service can be quantified using monetary or non-monetary metrics. For example, the private property protection value provided by coastal habitats can be estimated in terms of the monetary value of avoided damage to coastal properties, or in terms of the number of people protected (Arkema et al., 2013). Given the nature of the ES supply chain and the intersection between the ecological and socio-economic dimensions that is at its core, activities aimed at protecting or enhancing biodiversity are not guaranteed to maintain or improve ES benefits to people.

In other words, ES provision depends not just on how much of a particular ecosystem is present, or the level of biodiversity or ecosystem functions within it, but also where that ecosystem is located on the landscape and, critically, where relative to people who have the potential to benefit from its functions. The concept of 'servicesheds' provides a practical way to account for this spatial aspect of ES provision. Servicesheds are defined as those areas that provide a specific ES to a specific beneficiary, which may be an individual, a group of people, a government or a corporation (Tallis, Polasky and Lozano et al., 2012). Servicesheds link the places on the landscape where ES are supplied to the people and places to which the benefits of those services accrue. They make it possible to identify who will be affected by impacts to the environment from development, and which ecosystems services will be affected. With this information, impacts between alternative project designs then can be directly compared or mitigation activities can be designed to return benefits back to the same people who lost services with development.

Defining and spatially demarcating servicesheds requires consideration of both the supply and flows of ES, as well as the physical and institutional ability of people to access those benefits. These factors and their relative importance vary depending on the particular service under consideration. For water-related services such as drinking water provision or purification, the serviceshed is often the upstream watershed from the point of extraction. For water users who rely on piped water, their serviceshed may be located a great distance from the point of use. This means that environmental impacts in one location can have far-reaching consequences for ES, depending on the spatial configuration of servicesheds and beneficiaries.

In contrast, the serviceshed for benefits from carbon sequestration for climate regulation is global. Because the Earth's atmosphere is well mixed, changes in carbon sequestration anywhere in the world will affect climate regulation for all people. However, the global serviceshed for climate regulation is an exception among ES. Most services operate at smaller spatial scales and therefore require tracking via explicitly demarcated servicesheds. For example, the serviceshed for crop pollination by wild insects for a particular farm is the area around the farm's agricultural fields within pollinators' flight ranges, which is likely to extend only hundreds of meters. Or in the marine realm, the serviceshed for coastal protection from storm surge is the area offshore of property of concern (for economic, social or cultural reasons), in the direction of possible storms, where natural habitats (e.g., mangroves, coral reefs) can dampen waves and provide protection.

Because ES benefits depend so critically on the spatial configuration of both ecosystems and people, a *spatially explicit* approach to ES impact assessment is essential to achieving the benefits of an ES framework. In particular, spatial analysis enables the environmental impacts of development to be traced to specific beneficiaries. It provides more explicit information that can be incorporated into environmental and social management plans, as compared to qualitative and non-spatial approaches, by illuminating where and how environmental changes are affecting benefits to people. In this way, it also enables identification of more efficient mitigation options by bringing together environmental and social aspects. By allowing tracking of benefits to specific people or groups of people, spatially explicit analysis enables consideration of equity and provides the opportunity to ensure that development and any associated mitigation actions do not lead to the creation or extension of inequality in service provision. All these aspects mean that a spatially explicit ES-based approach to impact assessment could reduce the likelihood of project delays due to unforeseen impacts, reduce reputational risk to developers from unintended social impacts and improve overall outcomes of development and mitigation.

With the promise of a spatial approach, there also comes a need for practical methods for implementation. In the next section, we provide an introduction to best practices for spatial analysis of ES for impact assessment processes.

2.3 AN APPROACH FOR SPATIALLY EXPLICIT ECOSYSTEM SERVICE ANALYSIS FOR IMPACT ASSESSMENT

A comprehensive, integrated framework for biodiversity and ES mitigation has been proposed (Tallis et al., 2015). Here we summarize the key impact assessment steps for ES.

First, the scope of analysis must be defined, including the spatial extent of analysis and focal services and beneficiaries. Biodiversity Impact Assessment efforts have traditionally focused on project-scale assessments, but recent calls have emphasized the need to expand to a landscape scale approach that allows for consideration of cumulative impacts and intentional consideration of efficient and effective siting and offset design (Margules and Pressey, 2000; Geneletti, 2002; Groves et al., 2002; Landis, 2003; Chan et al., 2006; Landis and Wiegers, 2007). These same considerations are salient in an ES impact assessment, and serviceshed boundaries can help identify the relevant extent. Serviceshed boundaries can be delineated for any prioritized service and related beneficiaries, just as boundaries for biodiversity can be proxied by ranges, occurrences, densities, numbers and interactions of species and habitats (Tallis, Polasky and Lozano et al., 2012; Arkema et al., 2013). Ultimately, the spatial extent of impact assessment and mitigation should be defined as the fullest extent needed to capture both biodiversity (and associated ecological processes), and the servicesheds (and associated beneficiaries) of all services targeted.

In most cases, resource, data and capacity constraints make it impossible to estimate impacts on all ecosystem goods and services. As in Biodiversity Impact Assessments, a subset of ES should be chosen that reasonably represents the full set of services relevant to the system. Existing methods can be used to narrow in on key services (e.g., Landsberg et al., 2013), and we recommend ensuring representation by considering a suite of target services that fall under the broad categories established by the Millennium Ecosystem Assessment: specifically, provisioning, regulating and cultural services. As supporting services largely underpin the other categories, and by definition are not directly connected to benefit flows, they will be well represented by selection of services from the other categories. Not all categories of services will be relevant in all contexts, but considering them will ensure major services are not overlooked.

Within major ES categories, further selection of representative services will be necessary. Services may be targeted if they provide benefits to a large population and/or a particularly vulnerable population (such as the elderly, poor, indigenous), if they are expected to suffer large impacts or are unlikely to recover through mitigation, if social or economic values associated with benefits are particularly high, or if alternatives for replacing the service are not available either through mitigation or other solutions such as built infrastructure.

In the selection of any services as proxies for others, close attention must be paid to the relationship among services in the analysis region. Unlike 'umbrella species' in the biodiversity context that have consistent ability to reflect trends in other associated species, ES relationships often vary from place to place. This means that a service that might be a good surrogate for other services in one place is not guaranteed to play that role well in another. For example, water quality regulation and carbon sequestration were positively correlated (and so good proxies for each other) in southwestern Australia but held the opposite relationship in Argentina

(Jackson et al., 2005). For this reason, the relationship among services must be explored in each case before selecting proxy services.

Once target services are selected, impact assessment will need to estimate how service delivery, and possibly value, is impacted by proposed development activities. Estimating impact on service supply is not sufficient, as it misses important connections to people and their well-being. Ideally, ES impact assessment would focus on value impacts as they reflect the full context of how much an ecosystem is providing to people and how much people prefer to receive those benefits. However, limited data on values may restrict the set of services that can be assessed, so calculation of impacts on service delivery should be considered the minimal step required in impact assessment. This step captures both the environmental changes *and* their resulting impacts on people.

For ES that are provided by a single, identifiable species or habitat type, the same methods used for Biodiversity Impact Assessment can serve as a starting point to estimate changes in supply. However, even for species-associated services, additional methods must be applied to assess impact on service delivery and value. For example, a fishery stock assessment may indicate impacts of proposed activities on fish populations, but without indication of how many of those fish are landed, by whom and for what purposes (e.g., recreation, consumption, commercial sale), we do not yet have an indication of ES impact.

ES impact assessment for any kind of service should focus on the marginal change in service delivery to each beneficiary (Tallis and Polasky, 2009). Many common ES assessment methods ask how much service is provided under a certain context, rather than focusing on marginal change, or how much service delivery or value will change as a result of proposed action. It is unlikely that all of a current service's value will be lost under proposed actions, so marginal change analyses are needed to identify specifically how much will be lost.

Multiple methods have been advanced for ES assessment, but only some align with the needs for spatially explicit, marginal change methods necessary for impact assessment. For example, habitat area-based assessments are of limited relevance to ES impact assessment because they seldom allow for estimation of impacts on specific services, connection of environmental impact to specific beneficiaries or marginal change estimation. Habitat area-based methods commonly focus on total ecosystem value estimates that require the lumping of services, and as such may not accurately reflect the set of services prioritized for a given impact assessment. These methods also commonly assign a consistent value of service to a unit area of habitat, overlooking known variation in such services caused by differences in ecological function, demand for services, institutions that regulate access to and value of services, or physical features that limit access to services (Barbier et al., 2008; Ricketts et al., 2008; Koch et al., 2009). Given these limitations, tools and datasets focused on simple benefits transfer have limited applicability to ES impact assessment.

Production function approaches provide a much better methodological fit to the requirements of ES impact assessment. Production functions estimate ES flows based on environmental conditions and the location, activities and preferences of beneficiaries (Boyd and Wainger, 2003; National Research Council, 2005). Tools have been built to capture production functions and make the approach more easily accessible to analysts for use in this and other ES assessment contexts (e.g., ARIES, InVEST, MEASURES, MIMES). These tools have been reviewed and discussed extensively elsewhere (e.g., Vigerstol and Aukema, 2011; Bagstad et al., 2013; Waage and Kester, 2014). Other tools that take a biophysical perspective (such as modeling water yield or agricultural production) can serve the same purpose when coupled with an ES framing, the use of servicesheds and economic or other social methods.

The most appropriate model or tool for evaluating impacts to ES depends on a variety of factors, including the services being considered, the relevant spatial and temporal scales, the level of precision needed in the results, and the resources available for carrying out the analysis. Whatever tools are chosen, our experiences with ES analysis have highlighted the importance of transparency and credibility when incorporating ES into decision-making processes (Ruckelshaus et al., 2015).

In the following sections we demonstrate how this approach can be applied to impact assessment with two case studies. In each case study, we have used free, open-source, science-based software tools and commonly available data to quantify the impacts of development on ES. Here, we provide only an overview of the more technical aspects of the analysis but provide references to more comprehensive documentation of the methods used for those interested in pursuing these details further.

In the first case study, from the Cesar Department in Colombia, we evaluated the impacts of existing and proposed coal mining permits on water quality for downstream communities. We identified which requested mining permits would have the greatest impacts on water quality regulation services, as well as which cities would experience losses in drinking water quality. In the second case study, an analysis of a proposed road through the Peruvian Amazon, we illustrate how this approach can be used to trace distributional effects, tackling equity concerns head on. This case study focuses on the distribution of impacts between indigenous and non-indigenous populations and the implications for returning benefits to each group through efficient, spatially targeted mitigation plans.

2.4 CASE STUDY 1: MINING IMPACTS ON WATER QUALITY REGULATION SERVICES, CESAR DEPARTMENT, COLOMBIA

The Cesar Department, in the northern part of Colombia, harbors some of the country's largest remaining coal reserves. It is also the home of the Cesar River, which flows into the Zapatosa wetland before joining the Magdalena River, one of the principal large river basins in the country. Remaining native habitats are few, as the state is largely converted to agriculture and rangelands. The Colombian Ministry of Environment is considering further permitting of coal mining activities in the Cesar Department, and we conducted partial ES impact assessments on two scenarios of proposed mine permit allocations in the region (Tallis and Wolny, 2010). The moderate mine expansion scenario represented full development of currently permitted mines with no new mines, whereas the full exploitation scenario represented full development of all currently permitted and proposed future mines.

2.4.1 Scope of Assessment: Selecting Target Ecosystem Services and Delineating Servicesheds

We designed the assessment to ask two key questions: which proposed mine permits will have the greatest impacts on target ES, and which cities have the potential to suffer ES impacts even after efficiently targeted mitigation offsets?

The removal of vegetation and topsoil in the creation of open pit mines likely leads to a suite of ES impacts including the loss of carbon sequestration, erosion control (which impacts drinking water quality and the lifetime of downstream infrastructure), surface water flow (for drinking, irrigation, recreation, etc.), flood regulation, and groundwater recharge (for drinking), among others (Carroll et al., 2000; Palmer et al., 2010). The provision of clean drinking water is of primary interest in the Cesar region. Most population centers in the watershed get their drinking water from groundwater wells. However, observed and further expected drawdown of groundwater from mine-related pumping will make groundwater use impractical and shift reliance to surface water (Juan Carlos González, TNC, personal communication). While many countries regulate direct pollution from mines, they do not account for the lost ES of water quality regulation provided by vegetation and soils that are removed in the mining process. Given the importance of drinking water in the region, the complexity of groundwater–surface water interactions, the lack of available information on aquifer boundaries, and the likelihood that local water

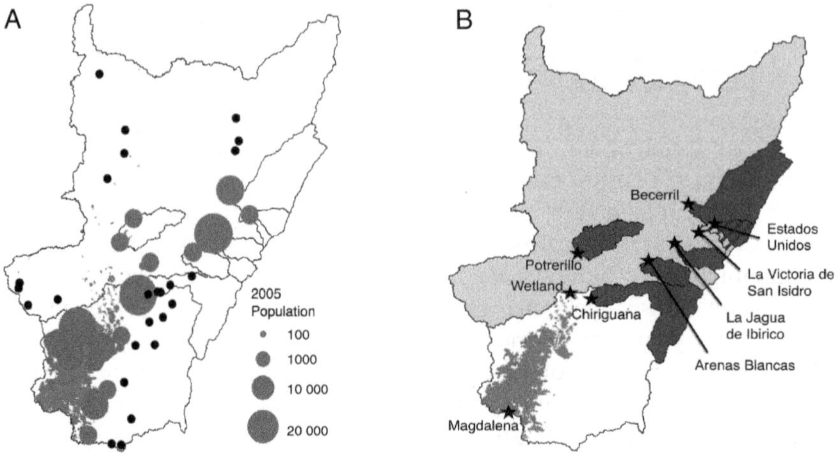

Note: Beneficiaries were defined as towns and cities (black and gray circles in A). Those likely to be impacted by at least one proposed mining scenario are shown with gray circles (A). Servicesheds were defined as the watershed area upstream of each potentially impacted beneficiary. Only cities with potential for mitigation within their serviceshed were analyzed (B, stars). Most serviceshed areas are shown in dark gray. All cities bordering the wetland are considered as one beneficiary (B, light gray serviceshed), and all cities downstream of the Cesar River watershed on the Magdalena River are represented by the point named Magdalena (B), with the entire Cesar River watershed as their serviceshed.

Figure 2.1 Cesar Department with beneficiaries of nitrogen and sediment retention for drinking water quality and associated servicesheds

sources will switch from groundwater to surface water in the near future, we focused on two components of surface water quality: total nitrogen retention and sediment retention (or erosion control).

Towns and cities were identified as the major beneficiaries of interest in the region (Figure 2.1A). As we focused on two water-related ES, we used watershed boundaries to define servicesheds (Figure 2.1B). We considered towns at risk of impact if they fell directly within the boundaries of a proposed mine, or if a proposed mine fell upstream of the town in the same watershed. There are 24 towns or cities that could be impacted (49 percent of cities in the region), with a total population of over 69 000 (Figure 2.1). These are the populations that will lose water quality regulation benefits unless ES mitigation is implemented. There are two servicesheds that reflected benefits to more than one population center. We used the inlet of the Zapatosa wetland to represent the towns surrounding it, including Antequera, Belén, Candelaria, Chimichagua, El Trebol, Hojancha, La Mata, Saloa, San José, Sempegua, Soledad and Zapatosa.

Here, the assumption is that the wetland is relatively well mixed and that the impacts of higher nitrogen and sediment loads entering the wetland as a result of mining will be felt equally by all cities and towns bordering the wetland. Similarly, we considered the outflow of the wetland as indicative of impacts to all cities downstream on the Magdalena River.

2.4.2 Quantifying Impacts to Beneficiaries

Colombian law currently requires the avoidance of impacts on biodiversity, with subsequent mitigation of any allowed impacts, following the mitigation hierarchy. Given this legal requirement, priority biodiversity areas are avoided when possible inside proposed permit areas. To reflect this reality, we excluded the priority biodiversity areas, identified by the Humboldt Institute and The Nature Conservancy (Cabrera et al., 2009) through a biodiversity prioritization process, from the proposed coal mining permit areas. We then assessed the likely impact of converting the remaining proposed permit areas to open pit coal mines.

To estimate impacts, we used InVEST Version 2.3 (Tallis, Ricketts and Guerry et al., 2012). We used the water purification model to assess impacts on total nitrogen regulation for drinking water quality and the avoided sedimentation model to assess impacts on sediment retention for drinking water quality. These models were developed to use the simplest possible approach to modeling complex processes as a way to provide sound scientific estimates while requiring minimal data, time and resources to apply, and are detailed in the *InVEST User's Guide*.

2.4.3 Assessing Outcomes

Using InVEST models, we were able to estimate marginal impacts of mine permits on target ES (Figure 2.2). Total nitrogen (TN) retention was reduced in all servicesheds under both the moderate (leading to a 3–46 percent increase in nitrogen loads at drinking water intake points per serviceshed) and full exploitation (1–46 percent increase in nitrogen loads per serviceshed) scenarios. Erosion control was also reduced under both scenarios (moderate: 7–1020 percent increase in sediment loads at drinking water intake points per serviceshed; full exploitation: 5–1038 percent increase in sediment loads per serviceshed). Overall, people living in the city of Potrerillo will receive the highest proportional loss of nitrogen retention and sediment retention services for drinking water quality because the majority of their serviceshed falls within proposed mine areas in both scenarios.

From these impact estimates, we were able to identify which mines are

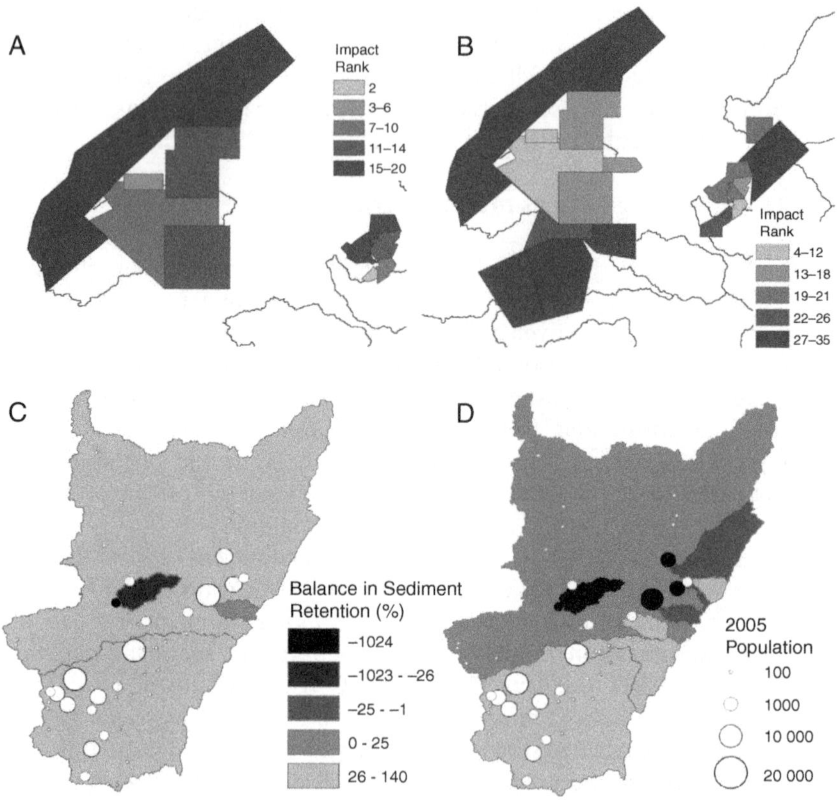

Note: The moderate scenario (A) represents full development of currently permitted mines with no new mines, whereas the full exploitation scenario (B) represents full development of all currently permitted and proposed future mines. Mine permits with the largest estimated population-level impacts on both nitrogen retention and erosion received the highest ranks, and are therefore the darkest shade of gray. Ranks start at numbers greater than 1 because the mine permits were ranked for each service (nutrient and sediment retention) separately, then the ranks for the two services were added. Protection and restoration in priority conservation areas provide enough recovery of sediment retention and nitrogen retention for drinking water quality to offset impacts (neutral or positive balance) for all but one relatively small city in the moderate mining scenario (C, E). The same is true for the nitrogen retention balance under the full exploitation scenario (F), but more beneficiaries in the eastern mountains will lose sediment retention benefits for drinking water quality even after mitigation (D). Cities impacted after mitigation are shown by black symbols. Fully mitigated cities are shown by white symbols.

Figure 2.2 Ranked impacts of proposed mining areas on nutrient retention and erosion control scaled to downstream beneficiary population size for the moderate and full exploitation scenarios

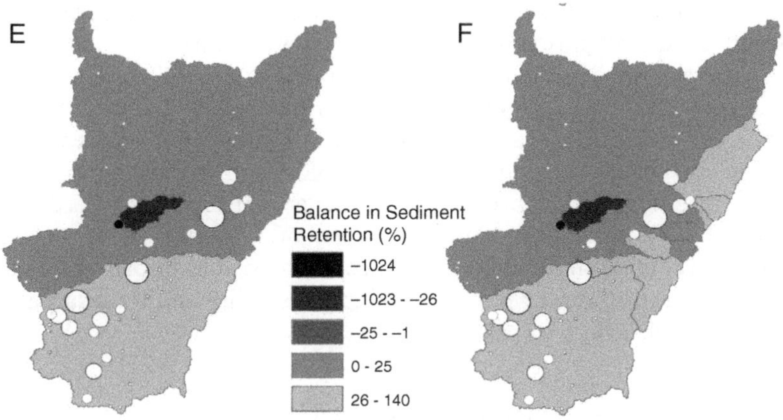

Figure 2.2 (continued)

likely to contribute the most to downstream population-level impacts. Larger mines and those in the eastern part of the region are estimated to have higher total impacts. The eastern region has steep mountains and some of the little remaining native vegetation in the area, so loss of this vegetation leads to larger marginal losses of ES. In this ranking exercise, we considered the two services to be equally important based on local stakeholder information, but unequal weights could be applied among services in other contexts. These findings emphasize the spatial heterogeneity of service impact and the utility of considering this spatial variation in impact assessment. Our results indicate that not all mine impacts will be equal, and avoiding or repositioning some mines will have much larger ES and beneficiary benefits than others.

Taking this impact assessment further, we asked where and how much mitigation (through protection and restoration) could offset expected impacts on target ES. Habitat protection and restoration focused in priority conservation areas outside the proposed mining areas but within each serviceshed could offset both nitrogen pollution and erosion impacts in three out of four servicesheds impacted under the moderate mining scenario (Figure 2.2C and E), representing full mitigation (with a 1:1 mitigation ratio) for 98 percent of the impacted population.

In the full exploitation scenario, the same actions could offset nitrogen pollution for eight out of nine impacted servicesheds (Figure 2.2F), representing 99 percent of the population, but erosion impacts could be offset for only five out of nine impacted servicesheds (Figure 2.2D), representing only 24 percent of the impacted population. In this case, the landscape holds much greater potential for offsetting nitrogen pollution impacts than erosion impacts, and under the full exploitation scenario there is no possibility of full mitigation for the impacted population.

The scenarios we analyzed were a small subset of the full suite of serviceshed analyses the Colombian Ministry of Environment could consider for permit granting. For example, explorations of lower-impact scenarios could include options for splitting permit blocks to avoid high-impact areas, greater investment in on-site treatments that lessen the need for landscape-based mitigation, and avoidance of a subset of proposed permits. The change analyses conducted could be used to inform the consideration of each of these options, providing flexibility and transparency in the decision-making process.

2.5 CASE STUDY 2: TRACING THE DISTRIBUTIONAL EFFECTS OF ROAD DEVELOPMENT TO INDIGENOUS AND NON-INDIGENOUS COMMUNITIES IN THE PERUVIAN AMAZON

In our second case study, we expand on the previous example from Colombia in two ways. First, we use servicesheds to separately track impacts to different social groups in the project area. Second, we use the approach to estimate the project's indirect impacts in addition to its direct impacts.

The Pucallpa–Cruzeiro do Sul road is planned as part of the Initiative for Integration of Regional Infrastructure in South America (IIRSA). First appearing in the IIRSA portfolio in 2003 (Salisbury et al., 2013), the road would link the two Amazon cities of Pucallpa in Peru with Cruzeiro do Sul in Brazil and is of particular interest because it is expected to spur development of agriculture, forestry and tourism sectors within the region (IIRSA, 2009).

Anticipation of the economic benefits to local, regional and national economies has been countered by the recognition that the proposed project occurs in an environmentally sensitive region (ibid.). Local and international non-governmental organizations have raised concerns about the project's potential to have detrimental social and environmental impacts (Salisbury et al., 2013). The route originally proposed for the road would cut through transboundary protected areas and the lands of indigenous communities, as well as encroach on indigenous groups in voluntary isolation. These concerns center on not only the direct impacts of the paved road but also the likely indirect impacts that have been prevalent in previous road development projects in the Amazon, where new roads have caused rapid deforestation in the surrounding landscape (e.g., Laurance et al., 2009). Construction of a rail line has also been proposed as a less environmentally damaging option (Salisbury et al., 2013; Hopkins et al., 2014).

The studies needed for project approval and acquisition of an environmental license are underway, but as of early 2016 have not yet been completed (IIRSA, 2013). Here, we show how spatial analysis of ES can be used to evaluate impacts on the environment, the services provided to people and the equity implications based on the distribution of impacts on indigenous and non-indigenous communities. We considered impacts on the nearly 250 000 people in the study area, including more than 15 000 members of indigenous communities from three ethnic groups. We estimated both the direct impacts likely to result from road construction, as well as the indirect impacts likely to occur as the increased access a new road provides has often led to conversion of natural vegetation to agriculture or other uses.

2.5.1 Scope of Assessment: Selecting Target Ecosystem Services and Delineating Servicesheds

We focused on five ES: erosion control, nitrogen regulation and phosphorus regulation for surface drinking water quality, carbon sequestration for climate regulation, and wild-harvested products, which provide medicine, food and fiber. We selected this particular set of services because they are likely to be affected by road development, are important to human well-being at local and global scales and because of the availability of data and models needed to estimate changes in these services.

We included in this study the 107 cities, towns and settlements whose water quality regulation servicesheds fell within 5 km of the proposed road, the likely epicenter of direct and indirect impacts of road development (Figure 2.3). The well-being of the residents of these population centers depends on locally provided ES.

We made the decision to bound our area of analysis based on water quality regulation services for practical reasons. For the three water quality-related services (erosion control, nitrogen and phosphorus regulation), the serviceshed is the upstream catchment area that delivers water to the river or stream nearest the settlement. For carbon sequestration, the whole world is the serviceshed because the atmosphere is well mixed, so changes in carbon sequestration have a global effect through climate change. Here, we restrict our impact estimates to reflect only the local population impacted, rather than the population of the entire planet. For wild-harvested products, the serviceshed is the area where these products

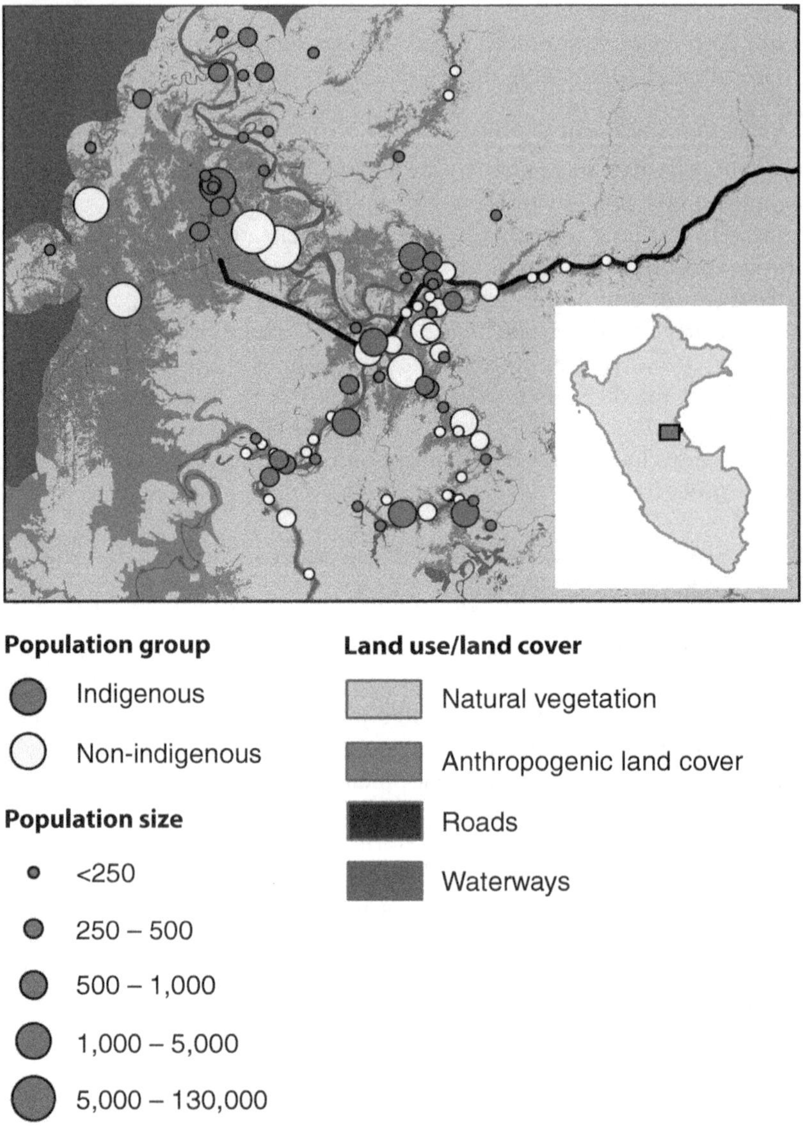

Note: The thick black line is the proposed route for the new road that would provide a connection between Peru and Brazil.

Figure 2.3 Map of the Pucallpa–Cruzeiro do Sul road case study area in the Ucayali region of Peru

are present and for which members of a population center have physical access and access rights. However, we were unable to delineate serviceshed for wild-harvested products in the project area because data on access and harvest practices were not available. Existing maps provided the location of important sites but did not provide information on who used each site (Instituto del Bien Común, unpublished data), so we could not trace the impacts on sites back to particular people in this case.

We evaluated the impact of the most likely route of the road, asking two main questions: how much ES provision is likely to be lost due to road development and which local beneficiaries will be affected, and how are losses distributed between indigenous and non-indigenous communities?

2.5.2 Quantifying Impacts on Beneficiaries

We quantified two forms of impact from the road. First we estimated the direct impacts from the creation of a new paved road, in which we assumed no other changes to the surrounding landscape. In addition, we estimated the indirect impacts of road construction that are likely to result from increases in deforestation in the surrounding landscape as a result of improved access to those areas. Therefore, we also modeled the combined impacts of the road and road-associated deforestation.

We accounted for the impact of road-associated deforestation by creating scenarios of future deforestation both without the road present (road + baseline deforestation) and with the road present (road + increased deforestation). We accomplished this by projecting rates of deforestation to 2030 using Terra-i land cover change data (Touval et al., 2011) and the IDRISI Land Change Modeler, which provides spatially explicit estimates of deforestation rates accounting for drivers of deforestation such as distance from cities and roads. As expected, these models predicted that with a new road, deforestation rates in the study area would increase overall, especially in areas close to the road. Based on recent trends in land use change in the area, we assumed that forest lost within 5 km of the road would be converted to oil palm and that forest lost beyond 5 km from the road would be converted to pasture.

We estimated the impacts of the road and associated deforestation on carbon storage and water-related ES using InVEST Version 2.4.4 (Tallis et al., 2011). We determined the impact of road development on sites used for the collection of wild-harvested products based on 2010 resource use maps from Instituto del Bien Común (IBC, unpublished data). Additional details about the methods and models used can be found in Mandle et al. (2013).

We summed the change in service provision within each population center's serviceshed to understand how impacts accrued to different beneficiary groups across the study area. In addition, we summed the total impacts to indigenous and non-indigenous population centers to assess whether negative impacts were distributed evenly between these two groups.

2.5.3 Assessing Outcomes

Our analysis shows that road development is likely to impact and cause losses of all five ES considered, affecting both indigenous and non-indigenous populations in the project area. Overall, the impact of road-associated deforestation is substantially greater than the impact of the road alone (Figure 2.4). This large additional impact of road-associated deforestation highlights the importance of accounting for indirect impacts in environmental assessments in order to comprehensively capture environmental change and its consequences for people.

Although the overall impact in biophysical terms is relatively small (< 2.5 percent loss), this landscape-level metric masks the relatively large consequences for some population centers in the region. Certain population centers were estimated to experience up to a 50 percent loss of sediment retention, 20 percent loss of phosphorous regulation and 15 percent loss of nitrogen regulation services.

These losses of services are not distributed equally across population centers, or between indigenous and non-indigenous communities (Figure 2.4, pie charts). In general, indigenous communities would be expected to suffer greater losses of ES. Less than 40 percent of the non-indigenous focal population would experience loss of at least one water quality regulation service as compared to more than 75 percent of the indigenous population, when considering the combined impacts of the road and associated deforestation. A greater proportion of the indigenous population is likely to experience losses of nitrogen and phosphorous regulation services (30 and 66 percent for indigenous, versus 3 and 36 percent for non-indigenous, for nitrogen and phosphorous respectively), while a slightly greater proportion of the non-indigenous population is expected to experience losses of erosion control services (38 percent non-indigenous vs 30 percent indigenous). Losses of carbon sequestration will affect everyone in the project area (and indeed the globe) because the atmospheric pool of carbon is mixed and therefore will affect climate globally. However, carbon sequestration is the exception. We estimated the proportion of wild-harvested product sites that would be affected either directly by the road, or indirectly by deforestation. However, because of the lack of data on who accesses particular sites, at this point we are unable to

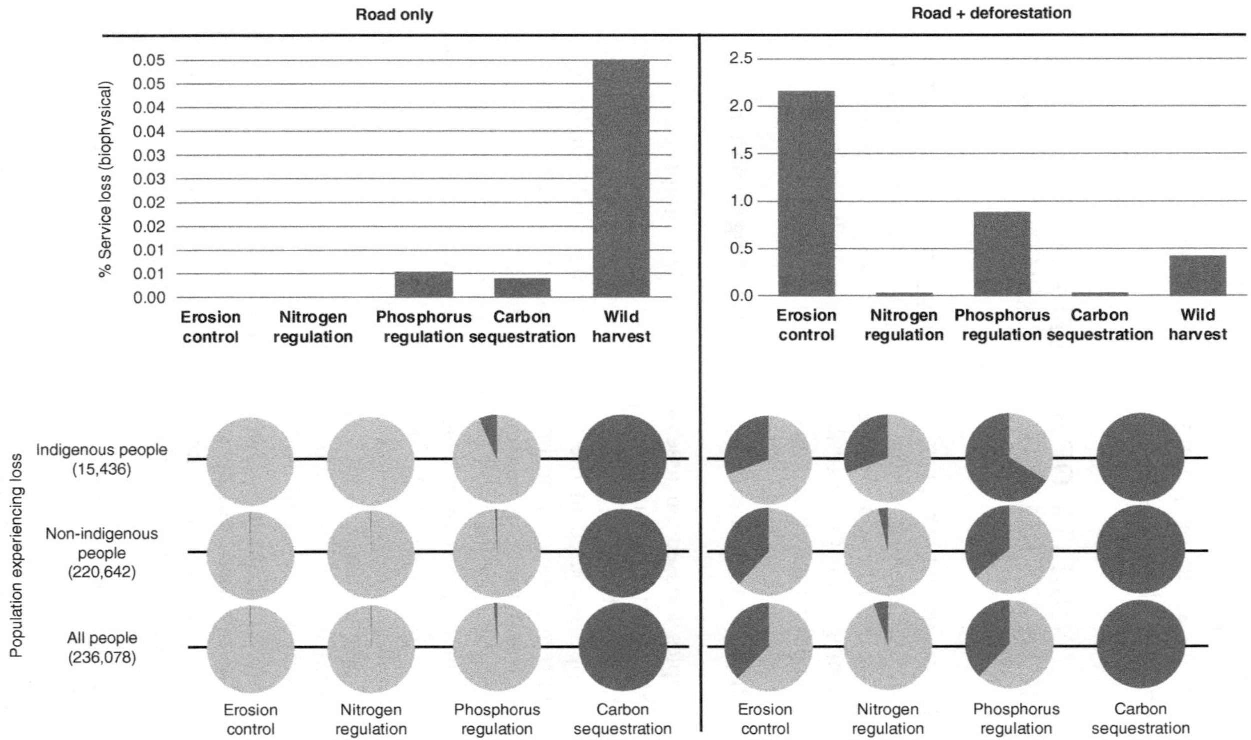

Note: Figures on the left show the direct impacts of road, while figures on the right show direct impacts plus indirect impacts of road-associated deforestation.

Figure 2.4 ES losses associated with road development in terms of percentage of biophysical service losses (upper bar charts) and the local population experiencing these losses (lower pie charts)

know which communities would be affected by these losses, though the effects are again likely to disproportionately affect indigenous communities. Access information would be an important addition. Overall, our results highlight the ability of spatially explicit ES analysis to distinguish where and to whom impacts accrue.

2.6 THE FLEXIBILITY OF A SPATIAL ECOSYSTEM SERVICES APPROACH: EXTENSIONS FROM CASE STUDIES

The case studies we present here provide a starting point from which many variations on or extensions from the basic approach can be envisioned. In this section, we describe some of these most promising extensions and opportunities, emphasizing the flexibility of the approach and its ability to be customized to particular needs and contexts.

2.6.1 From Strategic Assessments to Project-level Mitigation Planning

Both our examples focused on assessing project-level impacts, from a single proposed road and its indirect impacts in the case study from Peru, to the cumulative effects of two scenarios of mine expansion in Colombia. However, the approach used in the two case studies could similarly be incorporated into any stage of an impact assessment process for which policies, plans or activities can be translated into spatially explicit maps of changes in land use or access. Possibilities range from strategic-level impact assessment of plans or policies involving multiple sectors (e.g., Clarke et al., 2013; Bhagabati et al., 2014), to alternative analysis of potential designs for a single project. For example, to return to the Pucallpa case study, rail is being considered as an alternative to a roadway. The advantage of rail is that it provides much more limited access to surrounding areas and thus would be expected to result in less associated deforestation, leading to lower indirect impacts. The same analysis we conducted for the roadway could be repeated with the proposed rail route to quantitatively evaluate differences in the level of impact, and the consequences of those impacts for local communities and the ES they rely on.

2.6.2 Spatial Ecosystem Service Analysis Across the Mitigation Hierarchy

While our two case studies focused on quantification of impacts, the same approach can also be used to address multiple levels of the mitigation hierarchy. The avoidance and offset stages are particularly promising (see Chapter 17). Spatial analysis of ES can highlight areas that are especially critical to ES provision, either because of the high level of service they provide or because of the irreplaceability of the services produced. Avoiding impacts to both kinds of locations should be a high priority to minimize impacts and subsequent mitigation needs. Using servicesheds to design and evaluate offset options can make transparent the degree to which mitigation returns services back to the same people negatively impacted by development (Tallis, Wolny and Lozano et al., 2012; Mandle et al., 2013). This approach directly addresses concerns that an ES approach could undermine the mitigation hierarchy by allowing losses of vital services to one community to be offset in a different location (ibid.). Though the use of servicesheds as part of a spatially explicit ES analysis does not guarantee that it will be possible to completely offset ES losses, it does make clear who wins and who loses as a result of development and accompanying mitigation and allows project planners to allocate additional compensation activities in a transparent and equitable way.

2.6.3 Evaluating Project Dependence on Ecosystem Services

While environmental assessments may traditionally be limited to evaluating a project's impacts, consideration of a project's dependence on ES can be equally informative to project planning and implementation. The 2012 IFC performance standards require an analysis of project dependence (IFC, 2012), and a growing body of guidance is available to help companies and project developers recognize and manage risks and opportunities based on their dependence on ES (Hanson et al., 2012; Landsberg et al., 2013). Once again, the spatial component of ES provision is key to evaluating and managing project dependencies. For example, hydropower projects often depend on erosion control services to minimize sediment levels in dam reservoirs and intake water in order to maintain projected levels of energy production. By delineating the sediment retention serviceshed for a hydropower facility and identifying where vegetation within the serviceshed is most important for keeping sediment levels low, project developers can target protection and restoration activities to locations where the benefits will be greatest (Vogl et al., 2014). By considering a project's dependence on ES, and particularly by determining where provision of those services is most critical, project developers can target management activities to the most effective locations to minimize risk and improve project performance.

2.6.4 Integrating Stakeholder Participation and Ecosystem Service Modeling

Finally, and related to the importance of credibility and transparency, our case studies point to the important role for stakeholder participation in any spatially explicit ES assessment. ES modeling and direct stakeholder consultation should be seen as complementary, with the results provided by each process serving to enhance – rather than substitute for – the information produced from the other.

Stakeholder engagement is particularly important for prioritizing key services to include in analysis and for ensuring that model inputs and results are realistic and credible. Stakeholder knowledge can be especially valuable for delineating servicesheds, as this process requires information about both physical access ability and access rights. For example, in the case of the Pucallpa road, we estimated communities' water extraction points along rivers by assuming communities acquired water from the closest stream, based on combining a stream network map and a map of population centers. Obtaining information directly from communities on the location of their water sources would improve the delineation of servicesheds and thus improve estimates of changes in drinking water quality. Similar information on access to and management of wild-harvested products would have allowed delineation of the servicesheds for these services as well.

In the other direction, ES modeling can inform stakeholder engagement by guiding what questions are useful to ask stakeholders, and can reveal changes in services that may not be easily anticipated by stakeholders. While understanding how the loss of a forest area would reduce provisioning services such as fuel wood can be relatively straightforward, the impacts of that loss on regulating services such as erosion control, water purification or pollination services can be more difficult to foresee. Spatially explicit ES models based on ecological production functions can provide helpful information on likely changes to regulating services, which may be no less important than the more direct impacts to provisioning services but are often overlooked in stakeholder-driven assessment processes (Tallis, Polasky and Lozano et al., 2012; Honrado et al., 2013).

2.7 CONCLUSIONS

Inclusion of ES in environmental and social impact assessment processes is not only important for meeting the intent of impact assessment policy, but also explicitly required by a growing number of development project funders. Recent advances in methods, tools, data and guidance make spatial assessment of ES increasingly feasible as part of impact assessment practices. As demonstrated by the case studies presented in this chapter, spatial analysis of ES provides the ability to trace environmental impacts to specific beneficiaries, making the consequences of development more transparent and enabling more effective and efficient design of mitigation activities. Inclusion of spatial ES analysis holds promise for improving the impact assessment process, facilitating the development of policies and projects that deliver benefits to people while preserving the environment and services it provides.

REFERENCES

Arkema, K.K., G. Guannel and G. Verutes et al. (2013), 'Coastal habitats shield people and property from sea-level rise and storms', *Nature Climate Change*, **3**(10), 913–18.

Bagstad, K.J., D.J. Semmens and S. Waage et al. (2013), 'A comparative assessment of decision-support tools for ecosystem services quantification and valuation', *Ecosystem Services*, **5**, 27–39.

Baker, J., W.R. Sheate and P. Phillips et al. (2013), 'Ecosystem services in environmental assessment – help or hindrance?', *Environmental Impact Assessment Review*, **40**, 3–13.

Barbier, E.B., E.W. Koch and B.R. Silliman et al. (2008), 'Coastal ecosystem-based management with nonlinear ecological functions and values', *Science*, **319**(5861), 321–3.

BenDor, T., N. Brozović and V.G. Pallathucheril (2007), 'Assessing the socioeconomic impacts of wetland mitigation in the Chicago region', *Journal of the American Planning Association*, **73**(3), 263–82.

Bhagabati, N.K., T. Ricketts and T.B.S. Sulistyawan et al. (2014), 'Ecosystem services reinforce Sumatran tiger conservation in land use plans', *Biological Conservation*, **169**, 147–56.

Boyd, J.M. and L. Wainger (2003), *Measuring Ecosystem Service Benefits: The Use of Landscape Analysis to Evaluate Environmental Trades and Compensation*, Washington, DC: Resources for the Future.

Cabrera, E., J. Rodriguez and J. Otero et al. (2009), *Portafolio de áreas prioritarias de conservación en la zona central del Cesar; producto 16, convenio de asociación No. 09 de 2008* [Portfolio of Priority Conservation Areas in the Central Region of Cesar; Product 16, Partnership Agreement No. 09 2008], Bogotá, Colombia: Instituto de Investigación de Recursos Biológicos Alexander von Humboldt, The Nature Conservancy, Ministerio de Ambiente, Vivienda y Desarrollo Territorial, World Wildlife Fund and Conservación Internacional.

Carroll, C., L. Merton and P. Burger (2000), 'Impact of vegetative cover and slope on runoff, erosion, and water quality for field plots on a range of soil and spoil materials on Central Queensland coal mines', *Soil Research*, **38**(2), 313–28.

Chan, K.M.A., M.R. Shaw and D.R. Cameron et al. (2006), 'Conservation planning for ecosystem services', *PLoS Biology*, **4**(11), 2138–52.

Clarke, C., M. Canto and S. Rosado (2013), *Belize Integrated Coastal Zone Management Plan*, Belize City: Coastal Zone Management Authority and Institute.

Equator Principles Association (2013), *Equator Principles III*, London: Equator Principles Association.

Geneletti, D. (2002), *Ecological Evaluation for Environmental Impact Assessment*, Utrecht: Koninklijk Nederlands Aardrijkskundig Genootschap.

Geneletti, D. (2014), *Integrating Ecosystem Services in Strategic Environmental Assessment: A Guide for Practitioners*, Nairobi: UNEP.

Groves, C.R., D.B. Jensen and L.L. Valutis et al. (2002), 'Planning for biodiversity conservation: putting conservation science into practice', *Bioscience*, **52**(6), 499–512.

Hanson, C., J. Ranganathan and C. Iceland et al. (2012), *The Corporate Ecosystem Services Review: Guidelines for Identifying Business Risks and Opportunities Arising from Ecosystem Change Version 2.0*, Washington, DC, World Resources Institute.

Honrado, J.O.P., C. Vieira and C. Soares et al. (2013), 'Can we infer about ecosystem services from EIA and SEA practice? A framework for analysis and examples from Portugal', *Environmental Impact Assessment Review*, **40**(1), 14–24.

Hopkins, A., A. Malky and M. Glave et al. (2014), *Infraestructura y conservación: El caso Pucallpa-Cruzeiro do Sul* [Infrastructure and Conservation: The Case of Pucallpa-Cruzeiro do Sul], Conservation Strategy Fund (CSF), Consorcio Paisajes Indígenas de la Iniciativa para la Conservación en la Amazonía Andina (ICAA) and The Nature Conservancy (TNC).

Initiative for the Integration of Regional Infrastructure in South America (IIRSA) (2009), *Indicative Territorial Planning, Project Portfolio*, accessed 14 January 2016 at http://www10.iadb.org/intal/intalcdi/PE/2009/04494en.pdf.

Initiative for the Integration of Regional Infrastructure in South America (IIRSA) (2013), 'Project sheet: Pucallpa–Cruzeiro do Sul land interconnection', accessed 23 January 2015 at http://www.iirsa.org/proyectos/detalle_proyecto.aspx?h=29.

International Finance Corporation (IFC) (2012), *Performance Standard 6: Biodiversity Conservation and Sustainable Management of Living Natural Resources*, Washington, DC: IFC, World Bank Group.

IPIECA/OGP (2011), *Ecosystem Services Guidance: Biodiversity and Ecosystem Services Guide and Checklist*, London: International Petroleum Industry Environmental Conservation Association and International Association of Oil and Gas Producers.

Jackson, R.B., E.G. Jobbágy and R. Avissar et al. (2005), 'Trading water for carbon with biological carbon sequestration', *Science*, **310**(5756), 1944–7.

King, D.M. and L.W. Herbert (1997), 'The fungibility of wetlands', *National Wetlands Newsletter*, **19**(5), 10–13.

Koch, E.W., E.B. Barbier and B.R. Silliman et al. (2009), 'Non-linearity in ecosystem services: temporal and spatial variability in coastal protection', *Frontiers in Ecology and the Environment*, **7**(1), 29–37.

Landis, W.G. (2003), 'Twenty years before and hence: ecological risk assessment at multiple scales with multiple stressors and multiple endpoints', *Human and Ecological Risk Assessment: An International Journal*, **9**(5), 1317–26.

Landis, W.G. and J.K. Wiegers (2007), 'Ten years of the relative risk model and regional scale ecological risk assessment', *Human and Ecological Risk Assessment: An International Journal*, **13**(1), 25–38.

Landsberg, F., J. Treweek and M.M. Stickler et al. (2013), *Weaving Ecosystem Services into Impact Assessment: A Step-by-step Method (Version 1.0)*, Washington, DC: World Resources Institute.

Landsberg, F., J. Treweek and M.M. Stickler et al. (2014), *Weaving Ecosystem Services into Impact Assessment: Technical Appendix (Version 1.0)*, Washington, DC: World Resources Institute.

Laurance, W.F., M. Goosem and S.G.W. Laurance (2009), 'Impacts of roads and linear clearings on tropical forests', *Trends in Ecology & Evolution*, **24**(12), 659–69.

Mandle, L., H. Tallis and A.L. Vogl et al. (2013), *Can the Pucallpa-Cruzeiro do Sul Road be Developed With No Net Loss of Natural Capital in Peru? A Framework for Including Natural Capital in Mitigation*, Stanford, CA: Natural Capital Project.

Margules, C.R. and R.L. Pressey (2000), 'Systematic conservation planning', *Nature*, **405**(6783), 243–53.

Millennium Ecosystem Assessment (MA) (2005), *Ecosystems and Human Well-being: Synthesis*, Washington, DC: Island Press.

Morgan, R.K. (2012), 'Environmental impact assessment: the state of the art', *Impact Assessment and Project Appraisal*, **30**(1), 5–14.

Nahlik, A.M., M.E. Kentula and M.S. Fennessy et al. (2012), 'Where is the consensus? A proposed foundation for moving ecosystem service concepts into practice', *Ecological Economics*, **77**(C), 27–35.

National Research Council (2005), *Valuing Ecosystem Services: Toward Better Environmental Decision Making*, Washington, DC: The National Academies Press.

Palmer, M.A., E.S. Bernhardt and W.H. Schlesinger et al. (2010), 'Mountaintop mining consequences', *Science*, **327**(5962), 148–9.

Quétier, F. and S. Lavorel (2011), 'Assessing ecological equivalence in biodiversity offset schemes: key issues and solutions', *Biological Conservation*, **144**(12), 2991–9.

Ricketts, T.H., J. Regetz and I. Steffan-Dewenter et al. (2008), 'Landscape effects on crop pollination services: are there general patterns?', *Ecology Letters*, **11**(5), 499–515.

Robertson, M., T.K. BenDor and R. Lave et al. (2014), 'Stacking ecosystem services', *Frontiers in Ecology and the Environment*, **12**(3), 186–93.

Ruckelshaus, M., E. McKenzie and H. Tallis et al. (2015), 'Notes from the field: lessons learned from using ecosystem service approaches to inform real-world decisions', *Ecological* Economics, **115**, 11–21.

Ruhl, J.B. and J. Salzman (2006), 'The effects of wetland mitigation banking on people', *National Wetlands Newsletter*, **28**(2), 1.

Salisbury, D.S., M.C. Sanchez-Moreno and L.D. Torres et al. (2013), 'Border integrations: the fusion of political ecology and land change science to inform and contest transboundary integration in Amazonia', in C. Brannstrom and J.M. Vadjune (eds), *Land Change Science, Political Ecology, and Sustainability: Synergies and Divergences*, New York: Routledge.

Tallis, H. and S. Polasky (2009), 'Mapping and valuing ecosystem services as an approach for conservation and natural-resource management', *Annals of the New York Academy of Sciences*, **1162**(1), 265–83.

Tallis, H. and S. Wolny (2010), *Including Ecosystem Services in Mitigation: Report to the Colombian Ministry of the Environment, Mines and Territorial Development*, Stanford, CA: The Natural Capital Project.

Tallis, H., C.M. Kennedy and M. Ruckelshaus et al. (2015), 'Mitigation for one and all: an integrated framework for mitigation of impacts to biodiversity and ecosystem services', *Environmental Impact Assessment Review*, **55**, 21–34.

Tallis, H., H. Mooney and S. Andelman et al. (2012), 'A global system for monitoring ecosystem service change', *Bioscience*, **62**(11), 977–86.

Tallis, H., S. Polasky and J.S. Lozano et al. (2012), 'Inclusive wealth accounting for regulating ecosystem services', in United Nations University International Human Dimensions Program and United Nations Environment Program (eds), *Inclusive Wealth Report 2012. Measuring Progress Toward Sustainability*, Cambridge, UK: Cambridge University Press.

Tallis, H.T., T. Ricketts and A.D. Guerry et al. (2011), *InVEST 2.4.4. User's Guide*, Stanford, CA: Natural Capital Project.

Tallis, H.T., T. Ricketts and A.D. Guerry et al. (2012), *InVEST 2.3 Beta User's Guide*, Stanford, CA: Natural Capital Project.

Tallis, H., S. Wolny and J.S. Lozano et al. (2012), *Servicesheds Enable Mitigation of Development Impacts on Ecosystem Services*, Stanford, CA: Natural Capital Project.

The Biodiversity Consultancy (2013), *Government Policies on Biodiversity Offsets*, Cambridge, UK: The Biodiversity Consultancy.

The Biodiversity Consultancy (2014), *Ecosystem Services and IFC Performance Standard 6*, Cambridge, UK: The Biodiversity Consultancy.

Touval, J., A. Jarvis and L. Reymondin et al. (2011), *Terra-i: An Eye on Habitat Change*, Cali, Colombia: International Center for Tropical Agriculture (CIAT).

Vigerstol, K.L. and J.E. Aukema (2011), 'A comparison of tools for modeling freshwater ecosystem services', *Journal of Environmental Management*, **92**(10), 2403–9.

Vogl, A.L., S. Wolny and P. Hamel et al. (2014), *Valuation of Hydropower Services. Final Report Submitted to the World Bank for Contract #7167223*, Stanford, CA: Natural Capital Project.

Waage, S. and C. Kester (2014), *Making the Invisible Visible: Analytical Tools for Assessing Business Impacts and Dependencies Upon Ecosystem Services*, San Francisco, CA: Business for Social Responsibility.

World Bank (2013), *OP 4.03 – Performance Standards for Private Sector Activities*, Washington, DC: World Bank.

3. Ecosystem services analysis for Strategic Environmental Assessment: concepts and examples*

Davide Geneletti

3.1 INTRODUCTION

The ultimate objective of Strategic Environmental Assessment (SEA) is to help to protect the environment and promote sustainability, by ensuring that environmental considerations inform 'strategic actions', that is, policies, plans and programmes. SEA is considered one of the most promising tools with which to mainstream the environment into strategic decision-making, and is facing development challenges (World Bank, 2009). The Millennium Ecosystem Assessment (MA, 2005), as well as many other studies that followed, highlighted the importance of integrating ecosystem services (ES) into strategic decision-making, so that the effects of certain development options on ecosystems, and the services they provide, can be considered at the earliest appropriate stage. SEA is potentially a very suitable tool to achieve this purpose (Geneletti, 2011). Information on ES may represent a valuable addition to the content of SEA, as it provides a means to translate unattended and unintended consequences of plans and policies implementation on human well-being (Kumar et al., 2013).

There is a growing interest in the potential of SEA, and impact assessment in general, to mainstream ES concerns in decision-making, as shown by recent publications (Geneletti, 2013a; Karjalainen et al., 2013; Geneletti et al., 2015; Mandle et al., 2015) and reviews of practices (Honrado et al., 2013; Rosa and Sánchez, 2015). Experiences in this field have begun to emerge in the last few years (Partidário and Gomes, 2013), showing the need for comprehensive guidance (Helming et al., 2013). This chapter addresses this need by proposing a methodological approach to integrating ES analysis into SEA, and providing some examples. The analyses are illustrated and exemplified with a specific focus on the planning level of strategic decision-making. The next two sub-sections present an overview of SEA, and of the analyses required to mainstream ES into it, respectively.

3.1.1 Strategic Environmental Assessment

The field of SEA has developed rapidly over the last 15 years, and SEA is undertaken, both formally and informally, in an increasing number of countries and international organizations (Sadler, 2011). Two EU-led initiatives promoted the application of SEA: first, the European Union 'SEA Directive' (2001/42/EC), which requires an environmental assessment for plans and programmes that are likely to have significant effects on the environment; second, the SEA Protocol to the Espoo Convention (UNECE Convention on EIA in a Transboundary Context), agreed in 2003, which encourages the use of SEA in the context of policies and legislation (OECD, 2006). Currently, several dozen countries around the world have either national legislative or other provisions for SEA – for example, statutory instruments, cabinet and ministerial decisions, circulars and advice notes. Increasingly, developing countries are introducing legislation or regulations to undertake SEA, or are applying SEA-type processes (Sadler, 2011).

Even though SEA approaches vary in different countries, for different sectors and for different levels of decision-making, there is broad agreement on certain defining principles (Therivel, 2004):

- SEA is a tool for improving strategic actions. Hence, SEA should start early, and be undertaken as an integral part of the decision-making process.
- SEA should promote stakeholders' participation and ensure transparency in the decision-making process.
- SEA should focus on key environmental and sustainability concerns that are appropriate for the specific strategic action, considering the timescale and resources of the decision-making process.
- SEA should include the analysis and comparison of possible options for the strategic action, and the identification of the most suitable one(s).
- SEA should aim at minimizing negative effects, enhancing positive ones, compensating for the loss of valuable features and benefits, and ensuring that irreversible damage is not caused.

In short, a good-quality SEA process informs planners, decision-makers and affected public on the sustainability of strategic decisions, facilitates the search for the best alternative and ensures a democratic decision-making process (IAIA, 2002).

A number of methodological approaches have been proposed over the years to tailor SEA to different decision-making contexts, and to

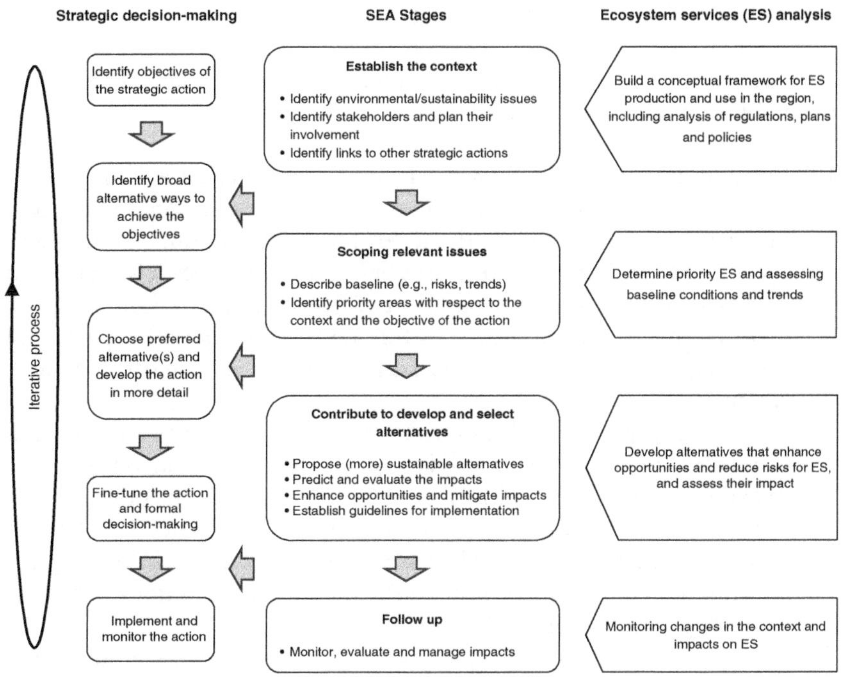

Source: The left-hand diagram is based on Therivel (2004).

Figure 3.1 Strategic decision-making and SEA stages and the analyses proposed to mainstream ES information

show the broad range of possible SEA forms (Abaza et al., 2004; OECD, 2006; Fischer, 2007; Ahmed and Sánchez-Triana, 2008; Partidário, 2012). However, even though SEA cannot be represented by a standard sequence of activities, the principles described earlier allow us to identify four typical stages through which SEA can feed decision-making, as shown in Figure 3.1. In reality, decision-making seldom follows this idealized model, hence SEA needs to be flexible and able to adapt to the specific context (including legal, institutional, procedural and political factors), which may be very different among countries, decision tiers (national, regional, etc.), and sectors (land use, agriculture, water, energy, etc.). Additionally, the specific circumstances of the strategic action under consideration (in terms of content, level of definition, availability of data, timing, consultation with stakeholders, etc.) will determine the way in which SEA is undertaken.

3.1.2 Mainstream Ecosystem Services into SEA

For each of the four SEA stages presented in Figure 3.1, key analyses are proposed to mainstream ES information, as follows:

1. Building a conceptual framework for ES production and use in the study region, including analysis of relevant regulations, plans and policies. In the first stage, SEA needs to provide an understanding of the context within which the strategic action will be developed and implemented. This analysis aims at establishing the 'ES context', by identifying ES and beneficiaries for the strategic action region, and reviewing relevant regulations, plans and policies concerning these services.
2. Determining priority ES and assessing their baseline conditions and trends. In this SEA stage, scoping is performed to focus on the most relevant issues in the light of the context and the objectives of the strategic action. The analysis aims at identifying a limited set of priority ES (which are considered relevant for shaping and informing the development of the strategic action), and generating detailed information about these services.
3. Developing possible alternatives that enhance opportunities and reduce risks for ES, and assessing their impact. In this stage, the strategic action is taking shape and specific alternatives are proposed and compared to achieve the objectives of the action. The analysis aims at contributing to the development and assessment of the alternatives in order to promote the conservation of ES, or at least to minimize the negative impacts and to propose mitigation and offset measures.
4. Monitoring changes in the context and impacts on ES. This SEA stage begins when the strategic action has been approved. The analysis aims at understanding the effective impacts on ES associated with the implementation of the action, as well as relevant contextual changes. The ultimate purpose is to enable timely interventions and adjustments to the action in order to address detected problems.

Each analysis builds on the previous ones, even though the sequence is not intended to be followed strictly. SEA is an iterative process, and many analyses may take place in parallel or in a different order, or may need to be iterated during the process. The overall purpose of the proposed analyses is to ensure that all relevant information on ES is collected, processed and used to support decision-making effectively. This information is intended to supplement the more traditional content of SEA, which focuses on issues such as environmental safety, air and water pollution, waste, and so on. Stakeholder consultation, including public participation and gender mainstreaming, is a vital component of SEA, and it is relevant in all the analyses. The following sections present an overview of the analyses and application examples. Further details on the approach and its link with the SEA process can be found in Geneletti (2015) and UNEP (2014).

3.2 BUILDING A CONCEPTUAL FRAMEWORK FOR ECOSYSTEM SERVICES IN THE STUDY REGION

In order to incorporate information on ES into SEA, a general understanding of how ES are produced and used in the planning region needs to be achieved from the very beginning of the process. This requires building a conceptual framework to link the socio-economic system with the biophysical environment via the flow of ES. Many frameworks can be used for this purpose, including the Millennium Ecosystem Assessment (MA, 2005), The Economics of Ecosystems and Biodiversity (TEEB, 2011), and Intergovernmental Platform on Biodiversity and Ecosystem Services (IPBES, 2013) frameworks, the ES cascade model (Haines-Young and Potschin, 2010) and the EU framework for ecosystem assessments (Maes et al., 2013). These frameworks share similarities, but each of them has its peculiar aspects, for example in the description of the well-being components or in the definition of the relationships between ecosystems and the values provided to people. Figure 3.2 presents an example of application of the MA framework to inform the early stage of SEA for a municipal development plan. This analysis enabled the identification of the main direct and indirect drivers of ecosystem change and their effects on ES and ultimately on human well-being.

Whenever possible during this analysis, details should be added concerning the relevance of ES for the well-being of different groups of beneficiaries, with specific attention to the most vulnerable groups in terms of geographical location, as well as socio-economic conditions (e.g., by considering the level of dependence of different livelihoods on a given ES and the 'substitutability' of that service). Information on the spatial patterns of vulnerability allows us to target the subsequent design and implementation of suitable planning and policy measures (Geneletti and Zardo, 2016). Stakeholder consultation is essential for this analysis. Stakeholders' opinions can help to simplify the problem and to get the essentials right. For example, participatory mapping approaches can be undertaken to gain a clearer view on what matters for people's well-being, and how this is associated with ecosystems and their services (see example in Chong, 2005).

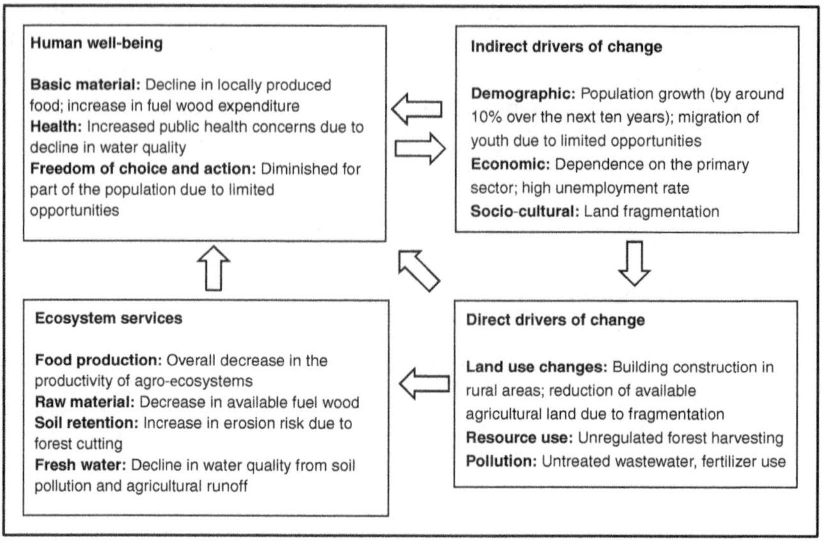

Source: Own representation based on the SEA report. The SEA process has been supported by the Municipal Spatial Planning Support Programme (MuSPP) implemented by UN-HABITAT.

Figure 3.2 *Analysis of the interaction between socio-economic and ecological systems in the municipality of Partesh/Partes (south-eastern Kosovo) to support the SEA of the Municipal Development Plan*

Another important analysis that helps to understand the ES context in the study region consists of reviewing existing regulations, plans and policies that establish conditions for the use or protection of ES. These may range from plans that regulate forestry and fishery to legislative acts related to indigenous people (e.g., regulating issues such as access to cultural sites, traditional remedies, etc.), and from policies about urban green areas to Payment for Ecosystem Services (PES) schemes. For each of these documents, the implications for the development of the strategic action under consideration should be analysed, for example by providing an answer to the following questions (see example in Box 3.1):

- What ecosystem types are addressed by the regulation/plan/policy?
- What stakeholders and beneficiary groups, disaggregated by gender and other sensitive groups, are mainly concerned?
- Does the regulation/plan/policy set constraints to decision-making? How?

> **BOX 3.1 EXPLORING THE INTERACTION WITH EXISTING REGULATIONS CONCERNING ES**
>
> This analysis is exemplified by considering the possible interaction between a hypothetical Coastal Zone Management Plan (hereafter 'the Plan') under development and an existing Act on coastal ecosystem conservation. The Act regulates trimming and alteration of mangroves, bans the use of herbicides and other chemicals, and identifies replenishment initiatives. Answers to the questions below are useful to identify possible synergies and constraints between the Plan and the Act:
>
> - What geographical areas/ecosystem types are addressed by the regulation?
>   *Mangrove ecosystems and other coastal ecosystems that play a key role in terms of protection from storms.*
> - What stakeholders and beneficiary groups are mainly concerned?
>   *Tourism operators, developers, fishing villages, inhabitants of risk-prone areas.*
> - Does the regulation set constraints to the Plan's decision-making? How?
>   *By imposing no-development areas and compensation measures for* ecosystem services *loss; by restricting permitted activities and land uses.*
> - Does the regulation offer opportunities for synergy with the Plan? How can the Plan contribute to the regulation objectives and vice versa?
>   *The policies and actions of the Plan can be directed at promoting ecotourism activities and other recreational uses that require well-preserved coastal ecosystems. In this way their preservation is also instrumental to the economic development of the area.*
> - What specific elements of the Plan concern the regulation most?
>   *The Zoning Map, which identifies permitted, prohibited and preferred land use conversion in the coastal areas.*

- Does the regulation/plan/policy offer opportunities for synergy with the strategic action? How can the strategic action contribute to the regulations objectives and vice versa?
- What specific elements of the strategic action are concerned the most by the regulation/plan/policy?

The overall purpose of this analysis is to help to harmonize the strategic action with the context, in order to exploit synergies and reduce risks in terms of ES conservation and use. The analysis needs to be repeated throughout the planning/policy-making process. In the preliminary stages, it is typically conducted by considering the objectives proposed in the strategic action. Later on, it can be performed by analysing the specific policies and activities proposed to achieve such objectives.

3.3 IDENTIFYING AND ASSESSING PRIORITY ECOSYSTEM SERVICES

In order for SEA to be effective, the number of services included in the analysis should be kept to a minimum, by considering only those that are relevant for the specific context, and content of the strategic action. In these initial stages, the content of the action is typically in the form of a draft set of problems that the action wishes to solve, and objectives that it wishes to achieve. By analysing this content, a preliminary screening of ES can be performed to identify the services upon which the strategic action depends, and the services that the strategic action may affect (positively or negatively).

A strategic action depends on an ES if the service is an input or if it enables, enhances or regulates the conditions necessary for a successful outcome of the action (OECD, 2008; Landsberg et al., 2013). For example, an agricultural development plan may depend upon water regulation services provided by watersheds. A strategic action affects an ES if it triggers drivers that decrease (negative impact) or enhance (positive impact) the quantity or quality of that service (OECD, 2008). For example, a tourism development plan may promote land-use changes in coastal areas that negatively affect the provision of fish. Box 3.2 presents an example of analysis of dependencies and impacts to identify priority ES for the SEA of a spatial plan. This analysis too requires wide stakeholder consultation to obtain existing information and to confirm values, interests and dependencies with the people who need and use the ES. The selection of priority services may change throughout the SEA process, according to changes in the content of the strategic action, and changes in the available knowledge about the biophysical and socio-economic context.

An assessment of the baseline condition of the identified priority ES should be performed in order to understand their current distribution, the benefits that they provide to different stakeholders, as well as the likely future trends, threats and opportunities. Information about future trends provides the basis to support the development of the strategic action in a way that it can adequately fit these trends. The assessment of ES can be conducted in a qualitative or quantitative way. Quantitative assessments can be based on monetary or non-monetary (e.g., biophysical) measures. Monetary valuation offers many advantages (e.g., in terms of communication), but it may not be always appropriate or even possible (TEEB, 2013). In many situations, the assessment of ES in their own terms may be more meaningful for stakeholders than a monetary value (e.g., the recreational or spiritual value of a landscape feature). Different types of assessment can be chosen for the different ES. The objective and scope of the SEA

BOX 3.2 ANALYSIS OF DEPENDENCIES AND IMPACTS TO IDENTIFY PRIORITY ECOSYSTEM SERVICES FOR THE SEA OF A SPATIAL PLAN

The analysis of the possible relationships between some of the objectives of a hypothetical spatial plan for the development of a mountain region in Trinidad and Tobago and ecosystem services is presented in Table 3.1 below. For each of the three plan objectives, the first column indicates the ecosystem services required for its achievement. The table also indicates when such dependence may extend beyond the boundary of the area being planned, hence requiring a broader-scale analysis (e.g., the regulation of water may depend upon decisions taken outside the region). The second column identifies situations where the achievement of the objective will have a positive/negative effect on the ecosystem services. For instance, the protection of natural areas is bound to contribute to soil formation and retention, but it may have both positive and negative effects on recreation opportunities. Similar analyses are useful also to test the 'internal consistency' of the strategic action. Potential inconsistencies exist whenever the achievement of one objective relies on a given service, which can be affected by a different objective. These situations can be detected by looking at each row of the table (see, for instance, the case of water regulation). The results of these types of analyses can suggest, for example, revisions of the objectives, but also additional stakeholders to be consulted (i.e., beneficiaries of the services affected).

Table 3.1 Analysis of the relationships between the objectives of a plan and the ecosystem services

	To Protect Water-sheds		To Ensure Suitable Development of Quarrying, Housing, Transportation and Agriculture Activities								To Develop a Non-invasive Eco-tourism Industry	
			Quarrying		Housing		Transport		Agriculture			
	D	I	D	I	D	I	D	I	D	I	D	I
Soil conservation	●	++		– –		–		–	○	+/–		–
Water regulation	●	++		–	●	– –			●	– –	●	
Hunting		+		–		–		–		–		
Water purification		++		– –	●	–				– –		

Table 3.1 (continued)

	To Protect Water-sheds		To Ensure Suitable Development of Quarrying, Housing, Transportation and Agriculture Activities								To Develop a Non-invasive Eco-tourism Industry	
			Quarrying		Housing		Transport		Agriculture			
	D	I	D	I	D	I	D	I	D	I	D	I
Carbon sequestration		++		–		– –						
Recreation		+/–		–				+/–		+	o	++
Aesthetic appreciation		+/–		–	●					+	o	+/–
Food and medicine		–/+		–		–		+		++		

Notes:
The table has been adapted from the table produced by the participants of the workshop 'Integrating ecosystem services in Strategic Environmental Assessment for policy support', held in Port of Spain on 3–5 December 2014. All elements of this table are purely hypothetical and intended for illustrative purposes only.
(+)+ (Very) Positive impact.
(–)– (Very) Negative impact.
● Dependence between objective and ecosystem service.
o Dependence that extends beyond the planning region.
D: dependence; I: impact.

(including the foreseen interactions with stakeholders throughout the process), as well as the availability of data, time and resources, will play a key role in selecting the appropriate way to assess ES, as well as the specific methods. Guidance on selecting and applying ES assessment methods can be found in UNEP-WCMC (2011); Crossman et al. (2013); Alkemade et al. (2013) and Burkhard et al. (2013). Examples of tools that can be used to support ES assessment are presented in Table 3.2.

Table 3.2 Examples of existing tools to support ES assessment

Tool	Description	URL
Integrated Valuation of Environmental Services and Tradeoffs (InVEST)	Open source ecosystem service mapping and valuation models	http://www.naturalcapitalproject.org
Artificial Intelligence for Ecosystem Services (ARIES)	Open source modelling framework to map ecosystem service flows	http://www.ariesonline.org
Multiscale Integrated Model of Ecosystem Services (MIMES)	Open source dynamic modelling system for mapping and valuing ecosystem services	http://www.ebmtools.org/mimes.html
Co$ting Nature	Web-accessible tool to map ecosystem services and conservation priority areas	http://www.policysupport.org/costingnature
Social Values for Ecosystem Services (SolVES)	A GIS application for mapping social values for ecosystem services based on survey data or value transfer	http://solves.cr.usgs.gov
Ecosystem Valuation Toolkit	An appraisal tool to estimate the value of a specific area's ecosystem services	http://www.esvaluation.org

Source: Modified after Bagstad et al. (2013).

3.4 DEVELOPING ALTERNATIVES AND ASSESSING IMPACTS ON ECOSYSTEM SERVICES

This analysis aims to contribute to the identification of possible alternatives to enhance priority ES, or at least to minimize negative effects on them. Alternatives can be generated as a reaction to proposals formulated by planners and policy-makers, or as a response to issues that emerged during the previous SEA stages, and that need to be adequately addressed by the strategic action. Ideally, the proposed alternatives should contribute to:

- promoting synergy (or at least avoiding conflict) with existing regulations, plans and policies relevant for ES;

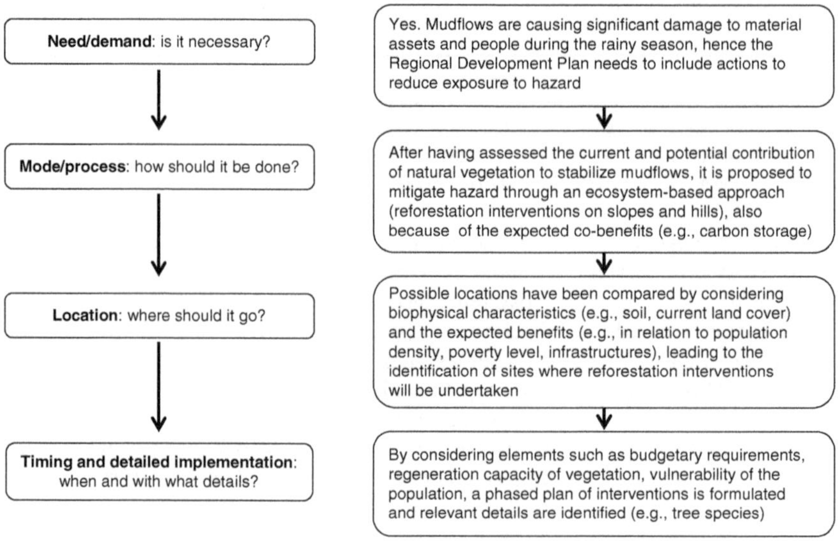

Note: The example illustrates the sequence of alternatives developed to address one of the Plan's objective, namely the exposure of population exposure to natural hazards.

Source: The diagram on the left is modified after ODPM (2005).

Figure 3.3 Hierarchy of alternatives (left) and application example for the SEA of a Regional Development Plan (right).

- minimizing the dependency of the strategic action on priority ES;
- enhancing positive (and minimizing negative) effects on priority ES;
- having desirable effects on drivers of ES change;
- generating a more equitable distribution of cost and benefits, particularly considering vulnerable stakeholders.

In SEA, different types of alternatives might need to be considered. To illustrate this concept, a hierarchy of alternatives has been proposed (Figure 3.3). The first two levels of alternatives are of a more strategic nature, hence typically applicable to policies at the national or regional level. The other two levels are more operational, and in general more suitable for plans and programmes, especially at the local level.

After having identified possible alternatives, the next analysis assesses their effects on ES, both in terms of changes in the supply of the services (e.g., quality/quantity of yield of crop, denitrification capability within a watershed, area of landscape in attractive condition), and in terms of importance that the changes have for the well-being of

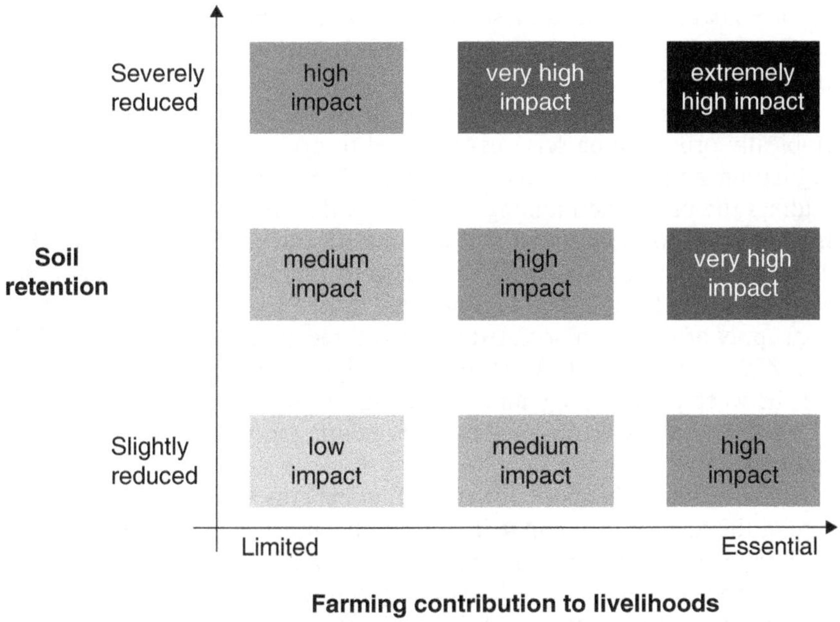

Note: The impact significance is estimated by combining the expected magnitude of change in the capability of ecosystems to prevent erosion (y-axis: soil retention), and the importance of well-preserved soil to the livelihoods of local villages (x-axis: contribution of farming to livelihoods). The highest impacts correspond to severe reductions in soil retention in areas where people rely on subsistence farming.

Figure 3.4 Conceptual diagram to assess the significance of ES impact

beneficiaries (e.g., considering vulnerability, dependency, replaceability, access) (Figure 3.4). The baseline and trends analyses performed earlier provide the reference against which the performance of different alternatives can be measured (including the 'do-nothing' alternative, if applicable), and the basis for suggesting the alternative(s) that better fit the desirable future conditions that the strategic action is pursuing.

For example, a case study conducted in the Araucanía (one of Chile's Administrative Regions) empirically assessed how the implementation of different land-use zoning policies will affect the future provision of ES (Geneletti, 2013b). Zoning policies give spatial representation to regional development strategies, by specifying where the strategies' objectives are to be achieved and with which uses of land. Land-use scenarios associated with the different policies were generated, by considering different trends for key land-use change processes. For each scenario, the provision of a set of ES (water purification, soil conservation, habitat for species, carbon sequestration and timber production) was modelled, and the results compared (Figure 3.5).

The MA demonstrated how actions to enhance the supply of some ES (typically provisioning services) have led to declines in other ES (mainly regulating and cultural services) (MA, 2005). SEA needs to explicitly address the interaction among ES, that is, the fact that multiple services may respond to the same driver, causing ES trade-offs or synergies. For example, in the case study in the Araucanía region mentioned above, the analysis of the trade-offs was conducted at two levels. First, trade-offs in the supply of services associated with the three policy scenarios (ZP0, ZP1 and ZP2) were assessed. As an example, Figure 3.6 (top) shows the trade-offs between provisioning, and regulating and supporting services (panels a, b and c), and the trade-off between regulating services at different scales (panel d).

Second, the implications for the actual benefits were assessed by combining the analysis of change in ES supply with spatially resolved socio-economic variables that estimate the appropriation of services by people (e.g., population density, livelihood systems, poverty indicators). This allowed representing trade-offs in the actual benefits from ES enjoyed by people under the different policy scenarios. As an example, Figure 3.6 (bottom) presents the trade-off between benefits associated with soil retention and water purification for one of the policy scenarios. The map shows areas where synergy occurs and people will be better off (i.e., increased benefits from at least one ES and no decrease from the other), areas where people will be worse off (vice versa), and areas where people will experience a trade-off in the benefits they receive (e.g., big loss from one ES, and small gain from the other).

3.5 MONITORING CHANGES IN THE CONTEXT AND IMPACTS ON ECOSYSTEM SERVICES

These analyses begin when all options have been closed, and the strategic action formally approved. They aim to understand the effective impacts on ES associated with the implementation of the action, as well as relevant contextual changes. Concerning the first objective, evidence needs to be collected about the actual impacts of the strategic actions on ES, in order to evaluate to what extent the observed impacts differ from the predictions performed during the previous stages. The ultimate purpose is to enable timely intervention and correction of detected problems (e.g., unforeseen impacts, mitigation and enhancement measures not carried out).

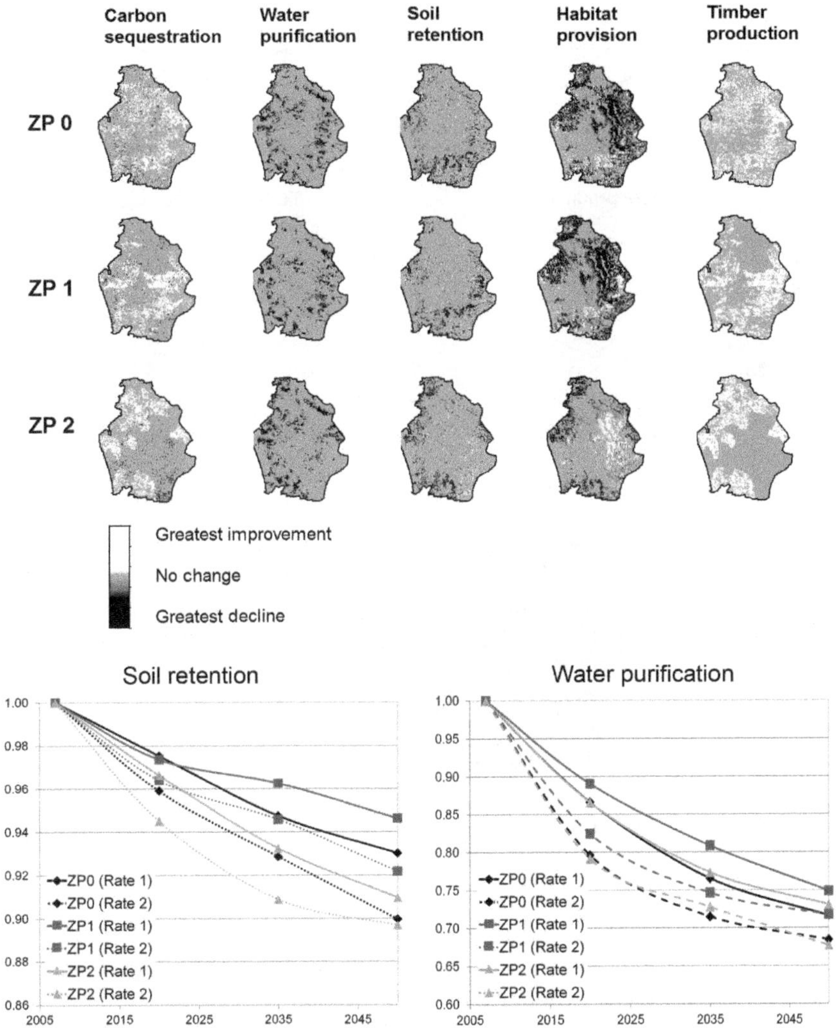

Note: ZP0 represents the 'business-as-usual' conditions; ZP1 and ZP2 are alternative land-use policies; Rate 1 and 2 refer to different rates of land-use change processes (see text for further explanation).

Source: Geneletti (2013b).

Figure 3.5 Comparing the effects of alternative land-use policies on ES in a spatially explicit way for a given time horizon (top), and in a non-spatial way through time (bottom)

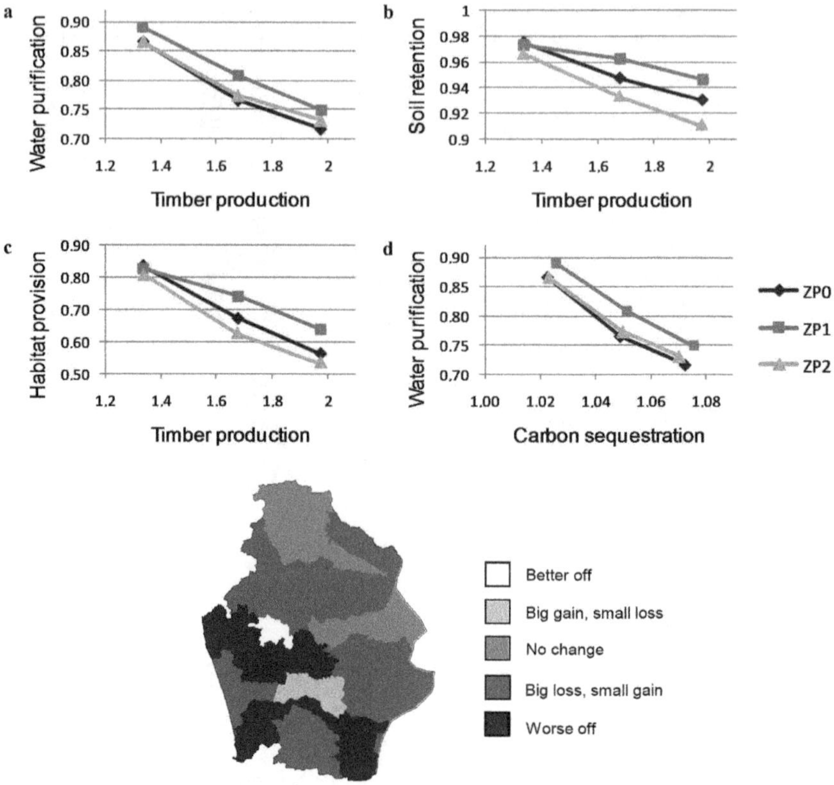

Note: See text for further explanation.

Source: Geneletti (2013b).

Figure 3.6 *Curves showing the trade-off between pairs of ES under different policy scenarios (top) and spatial representation of trade-off between benefits associated with soil retention and water purification for one of the policy scenarios (bottom)*

Important elements of an efficient monitoring system include (Morrison-Saunders and Arts, 2004):

- identification of indicators and description of the methods, frequency and responsibility for data collection (responsibility may vary according to the nature of the indicators);
- guidance on data evaluation (appraisal of the conformity with predictions or expectations);

- guidance on management (how to take appropriate response to issues arising from evaluation);
- communication strategies (informing stakeholders about the results or, where appropriate, involving stakeholders in the evaluation and/or management stages).

The selected indicators should be consistent with the ones used to inform the previous SEA analyses, limited in number (to ensure viability of the monitoring system), and possibly easy to measure, interpret and communicate. Besides contributing to a better implementation of the strategic action under consideration, this task has an important learning objective for future SEA: it helps us to understand what went wrong with impact prediction and how to make better predictions, hence improving the practice of mainstreaming ES into SEA and decision-making.

Concerning the second objective, insights need to be gained into the environmental and socio-economic changes that occur in the area during the implementation of the strategic action, and that may affect the supply, demand or use of ES. Relevant changes that have direct or indirect effects on ES can be related to the state of the environment (e.g., climate trends, natural disasters), the social profile (e.g., migration patterns), the socio-economic situation (e.g., shift in livelihood systems), and the legislative and regulatory framework (e.g., designation of new protected areas; land reform policy). Existing monitoring and reporting programmes (e.g., state of the environment reports) represent a useful starting point for this activity. Early detection of relevant contextual changes enables adjustments to the strategic action, thus ensuring the continuing facilitating role of SEA (Partidário, 2012). This task is also instrumental to filling the gap in knowledge (e.g., on the quantification of ES) that arose during the SEA, but could not be addressed for lack of time or resources.

The following guiding questions can be used to periodically monitor the context and detect changes that need to be brought to the attention of the people in charge of implementing or revising the strategic action:

1. Have new formal regulations linked to ES been approved in the study region (e.g., designated areas, Payment for Ecosystem Services schemes)?
2. Have other strategic actions or projects been approved or implemented that could affect the supply, demand or use of ecosystems services (e.g., energy policy, agricultural reform, urban plan)?
3. Has the demand for a particular priority ecosystems service by stakeholders changed (e.g., due to droughts, change in trade policy, change in access to specific locations)?

4. Has the supply of a particular priority ecosystems service changed (e.g., following a natural disaster)?
5. Is there new evidence available concerning ecosystems services (e.g., ecosystem maps, economic valuations, surveys on users' needs)?
6. Does this new evidence or knowledge suggest changes in the analysis carried out during the SEA (e.g., assessment of baseline conditions and future trends, impact prediction), hence in the final recommendations?

3.6 CONCLUSIONS

This chapter provided guidance for SEA to perform analyses that can clarify the potential impacts of strategic decisions on ES, in order to avoid unintended negative consequences, and seize opportunities for improvement. The proposed analyses can be used in different contexts and for different types of SEA processes. Practitioners and decision-makers are faced with the challenge of including ES and showing their added value to decision-making, within the time and resource constraints of real-life planning and policy-making.

The integration of ES in SEA has various benefits in terms of contributing to better design of policies and plans, but there are also critical issues that need to be recognized. These include the complexity of appropriately evaluating ES, and the lack of well-established indicators and assessing methods. These issues can be addressed by learning from the pilot applications and case studies that are being carried out around the world (as shown by the other chapters of this book), as well as by taking stock of the data, tools, and methods for ES analysis that are becoming increasingly available in the scientific (and grey) literature (Guerry et al., 2015).

NOTE

* The text and figures in this chapter draw on a guidance manual produced by the author for the UNEP's GEF-funded 'Project for Ecosystem Services' (ProEcoServ) implemented by the Ecosystem Services Economics Unit, DEPI-UNEP. The author is grateful for permission to draw on the results of this work. A substantially different version of this chapter, albeit based on the same guidance manual, was published in the *Journal of Environmental Assessment, Policy and Management* (Geneletti, 2015).

REFERENCES

Abaza, H., R. Bisset and B. Sadler (2004), *Environmental Impact Assessment and Strategic Environmental Assessment: Towards an Integrated Approach*, Geneva: UNEP.

Ahmed, K. and E. Sánchez-Triana (eds) (2008), *Strategic Environmental Assessment for Policies*, Washington, DC: World Bank.

Alkemade, R., B. Burkhard and N.D. Crossman et al. (2013), 'Quantifying ecosystem services and indicators for science, policy and practice. Special section', *Ecological Indicators*, **37**(Part A), 161–265.

Bagstad, K.J., D.J. Semmens and S. Waage et al. (2013), 'A comparative assessment of decision-support tools for ecosystem services quantification and valuation', *Ecosystem Services*, **5**, 27–39.

Burkhard, B., N.D. Crossman and S. Nedkov et al. (eds) (2013), 'Mapping and modelling ecosystem services for science, policy and practice. Special issue', *Ecosystem Services*, **4**, 1–146.

Chong, J. (2005), *Valuing the Role of Aquatic Resources in Livelihoods: Economic Aspects of Community Wetland Management in Stoeng Treng Ramsar site, Cambodia*, IUCN Water, Nature and Economics Technical Paper No. 3, accessed 29 January 2016 at https://portals.iucn.org/library/efiles/documents/2005-099.pdf.

Crossman, N.D., B. Burkhard and S. Nedkov et al. (2013), 'A blueprint for mapping and modelling ecosystem services', *Ecosystem Services*, **4**, 4–14.

Fischer, T.B. (2007), *Theory and Practice of Strategic Environmental Assessment: Towards a More Systematic Approach*, London: Earthscan.

Geneletti, D. (2011), 'Reasons and options for integrating ecosystem services in strategic environmental assessment of spatial planning', *International Journal of Biodiversity Science, Ecosystem Services & Management*, **7**(3), 143–9.

Geneletti, D. (2013a), 'Ecosystem services in environmental impact assessment and strategic environmental assessment. Special issue', *Environmental Impact Assessment Review*, **40**, 1–87.

Geneletti, D. (2013b), 'Assessing the impact of alternative land-use zoning policies on future ecosystem services', *Environmental Impact Assessment Review*, **40**, 25–35.

Geneletti, D. (2015), 'A conceptual approach to promote the integration of ecosystem services in strategic environmental assessment', *Journal of Environmental Assessment Policy and Management*, **17**(4), 1550035.

Geneletti, D. and L. Zardo (2016), 'Ecosystem-based adaptation in cities: an analysis of European climate adaptation plans', *Land Use Policy*, **50**, 38–47.

Geneletti, D., A. Bond and D. Russel et al. (2015), 'Ecosystem services and sustainability assessment: theory and practice', in A. Morrison-Saunders, J. Pope and A. Bond (eds), *Handbook of Sustainability Assessment*, Cheltenham, UK and Northampton, MA, USA: Edward Elgar Publishing.

Guerry, A.D., S. Polasky and J. Lubchenco et al. (2015), 'Natural capital and ecosystem services informing decisions: from promise to practice', *Proceedings of the National Academy of Sciences*, **112**(24), 7348–55.

Haines-Young, R.H. and M.P. Potschin (2010), 'The links between biodiversity, ecosystem services and human well-being', in D.G. Raffaelli and C.L.J. Frid (eds), *Ecosystem Ecology: A New Synthesis*, Cambridge, UK: Cambridge University Press, pp. 110–39.

Helming, K., K. Diehl and D. Geneletti et al. (2013), 'Mainstreaming ecosystem services in European policy impact assessment', *Environmental Impact Assessment Review*, **40**, 82–7.

Honrado, J.P., C. Vieira and C. Soares et al. (2013), 'Can we infer about ecosystem services from EIA and SEA practice? A framework for analysis and examples from Portugal', *Environmental Impact Assessment Review*, **40**, 14–24.

Intergovernmental Platform on Biodiversity and Ecosystem Services (IPBES) (2013), 'Recommended conceptual framework of the Intergovernmental Science-Policy Platform on Biodiversity and Ecosystem Services', *Working Document No. IPBES/2/*, accessed 14 January 2016 at http://www.ipbes.net/images/decisions/Decision%20IPBES_2_4.pdf.

International Association for Impact Assessment (IAIA) (2002), *Strategic Environmental Assessment. Performance Criteria, International Association for Impact Assessment Special Publication Series No. 1*, accessed 12 January 2016 at http://www.iaia.org/publicdocuments/special-publications/sp1.pdf.

Karjalainen, T.P., M. Marttunen and S. Sarkki et al. (2013), 'Integrating ecosystem services into environmental impact assessment: an analytic–deliberative approach', *Environmental Impact Assessment Review*, **40**, 54–64.

Kumar, P., S.E. Esen and M. Yashiro (2013), 'Linking ecosystem services to strategic environmental assessment in development policies', *Environmental Impact Assessment Review*, **40**, 75–81.

Landsberg, F., J. Treweek and N.M. Stickler et al. (2013), *Weaving Ecosystem Services into Impact Assessment*, Washington, DC: World Resource Institute.

Maes, J., A. Teller and M. Erhard et al. (2013), *Mapping and Assessment of Ecosystems and their Services. An Analytical Framework for Ecosystem Assessments Under Action 5 of the EU Biodiversity Strategy to 2020*, Luxembourg: Publications Office of the European Union.

Mandle, L., B.P. Bryant and M. Ruckelshaus et al. (2015), 'Entry points for considering ecosystem services within infrastructure planning: how to integrate conservation with development in order to aid them both', *Conservation Letters* [online], DOI: 10.1111/conl.12201.

Millennium Ecosystem Assessment (MA) (2005), *Ecosystems and Human Well-being: The Assessment Series* (4 vols and summary), Washington, DC: Island Press.

Morrison-Saunders, A. and J. Arts (2004), 'Introduction to EIA follow-up', in A. Morrison-Saunders and J. Arts (eds), *Assessing Impact: Handbook of EIA and SEA Follow-up*, London: Earthscan.

ODPM (2005), *A Practical Guide to the Strategic Environmental Assessment Directive*, London: Office of the Deputy Prime Minister.

Organisation for Economic Co-operation and Development (OECD) (2006), *Applying Strategic Environmental Assessment. Good Practice Guidance for Development Co-operation (DAC Guidelines and Reference Series)*, Paris: Organisation for Economic Co-operation and Development.

Organisation for Economic Co-operation and Development (OECD) (2008), *Strategic Environmental Assessment and Ecosystem Services. DAC Network on Environment and Development Cooperation (ENVIRONET)*, Paris: Organisation for Economic Co-operation and Development.

Partidário, M. (2012), *Strategic Environmental Assessment Better Practice Guide. Methodological Guidance for Strategic Thinking in SEA*, Lisbon: Portuguese Environment Agency and Redes Energéticas Nacionais.

Partidário, M.R. and R.C. Gomes (2013), 'Ecosystem services inclusive strategic environmental assessment', *Environmental Impact Assessment Review*, **40**, 36–46.

Rosa, J.C.S. and L.E. Sánchez (2015), 'Is the ecosystem service concept improving impact assessment? Evidence from recent international practice', *Environmental Impact Assessment Review*, **50**, 134–42.

Sadler, B. (2011), 'Taking stock of SEA', in B. Sadler, R. Aschemann and J. Dusikw et al. (eds), *Handbook of Strategic Environmental Assessment*, London: Earthscan.

The Economics of Ecosystems and Biodiversity (TEEB) (2011), *The Economics of Ecosystems and Biodiversity in National and International Policy Making*, edited by P. ten Brink, London and Washington, DC: Earthscan.

The Economics of Ecosystems and Biodiversity (TEEB) (2013), *Guidance Manual for TEEB Country Studies – The Economics of Ecosystems and Biodiversity. Version 1.0*, accessed 8 September 2015 at http://www.unep.org/pdf/TEEB_GuidanceManual_2013.pdf.

Therivel, R. (2004), *Strategic Environmental Assessment in Action*, London: Earthscan.

UNEP (2014), *Integrating Ecosystem Services in Strategic Environmental Assessment: A Guide for Practitioners*, ProEcoServ report by D. Geneletti for UNEP, accessed 14 January 2016 at http://www.ing.unitn.it/~genelab/documents/GuidelineESintoSEA.pdf.

UNEP-WCMC (2011), *Developing Ecosystem Service Indicators: Experiences and Lessons Learned from Sub-global Assessments and Other Initiatives (Technical Series No. 58)*, Montreal: Secretariat of the Convention on Biological Diversity, accessed 12 January 2016 at www.cbd.int/doc/publications/cbd-ts-58-en.pdf.

World Bank (2009), *Strategic Environmental Assessment in Policy and Sector Reform. Conceptual Models and Operational Guidance*, Washington, DC: World Bank.

4. Scoping Health Impact Assessment: ecosystem services as a framing device

Pierre Horwitz and Margot W. Parkes

4.1 INTRODUCTION

Many, indeed probably most, decisions made in societies have intended and unintended health outcomes; this alone provides sufficient reasoning to ensure that health professionals are engaged in these decisions rather than deal with the consequences at a later date. Health Impact Assessment (HIA) provides one approach to ensuring that health outcomes of decisions are taken into account (Harris-Roxas et al., 2012).

Harris-Roxas et al. (2012) show that historically HIA has emerged from three strands: environmental health concerns (where the environment has an effect on human health); the social determinants of health (where the concern is the 'upstream' causes of 'downstream' health conditions); and a focus on health equity (where interventions can address or promote health equity through planning). When any of these strands are performed effectively, or even more, when the three might be combined, the capacity to produce a powerful analysis of health consequences can be envisaged. HIA has considerable value as a proactive and systematic approach to identifying the health impacts of projects, programmes, plans and policies, including those associated with resource development (Utzinger et al., 2005; Winkler et al., 2012; Parkes, 2016).

An overarching theme in the development of HIA has been the engagement by the public health sector in non-health sector activities (Harris-Roxas et al., 2012; Harris-Roxas and Harris, 2013; Lee et al., 2013), and this intrinsically demands interdisciplinary and multidisciplinary approaches capable of seeking both to minimize negative health effects and to enhance positive health effects (Winkler et al., 2010). In their work on sustainability assessment, Bond et al. (2012) expand on this latter point, outlining a series of reasons why more integrative, deliberative and positive approaches to 'impacts' are desirable. They cite the shortfalls of focusing on mitigation and minimizing impacts (as compared to a systems approach), and emphasize the value of going beyond the pillars approach of sustainability (which encourages trade-offs), and working to deliver net sustainability gains (through greater system health and resilience over the long term).

To encourage an integrative and systems-informed approach to HIA we should therefore look for the following different strands, approaches and themes:

- interdisciplinarity as a capacity to arch across environmental, social and economic concerns in a health context;
- multi-causality to draw on understandings of the non-linear nature of social-ecological systems (*sensu* Walker et al., 2004);
- adopting a 'beyond duality', where need is recognized to include both mitigation that focuses on the negative impacts, and the positive impacts or beneficial roles provided by natural capital (see Briggs, 2008); and
- contextuality, taking account of adaptive and behavioural changes that characterize the systems concerned.

Most authors of HIA suggest that a broad consensus exists on its procedural aspects, which adopt a stepwise process involving: (I) Screening; (II) Scoping; (III) Risk Assessment; (IV) Appraisal and Mitigation; (V) Implementation and Monitoring; and (VI) Evaluation and Verification of Performance and Effectiveness (see Winkler et al., 2010). There is a recognition that improving the practice standards around each step of HIA will lead to an overall improvement in the quality of HIAs (Harris-Roxas et al., 2012).

The scoping phase (planning the HIA, and determining the required information and resources) is integral, not least since it will determine the orientation of all subsequent phases. It must be unconstrained by: the health issues under consideration, their environmental context, the groups of people who are chosen for attention (e.g., gender, age, socioeconomic status, ethnicity etc.), and the need for baseline population health data and information about existing health, social and environmental vulnerabilities (ibid.).

The challenge for the scoping phase in particular, and the HIA in general, is to find a framing device that will embrace the above imperatives. Here we examine whether this framing can be achieved by bringing together two practical concepts. From the Ottawa Charter for Health Promotion and recent derivations of it we recognize that health outcomes can be best enhanced by foregrounding the *settings* in which individuals and populations are experiencing health effects (WHO, 1986; Parkes and Horwitz, 2009). A useful way of accounting for these health outcomes is provided by the ecosystem services (ES) concept, where ecosystems can be shown to provide a characteristic set of benefits (as goods and services) (see, e.g., the Millennium Ecosystem Assessment, 2005).

The idea that the ES concept can be used to inform impact assessment is one that has already received attention in the context of Environmental Impact Assessment and Strategic Environmental Assessment, as described in the previous two chapters. The novelty explored in this chapter is to examine the potential value of ES as a framing device for HIA, operationalized via a reinvigorated sense of the socio-ecological[1] and settings-based approaches of health promotion. The pin that holds these concepts together is the realization that the foundational settings for human health, indeed human societies, are ecosystems (Parkes and Horwitz, 2009). A core challenge for HIA is to adopt an ecosystem approach to health (*sensu* Charron, 2012).

This chapter therefore seeks to explore this further, starting with a consideration of the ways in which ecosystems are represented in the settings approach from health promotion. We then examine ES, and highlight the characteristic ways in which health issues are linked to ecosystems, thereby providing a framework for the scoping phase in HIA. With this framework we consider how to identify and negotiate trade-offs, before applying these considerations to a case study involving a resource development proposal (coal seam gas extraction in an agricultural zone).

4.2 ECOSYSTEMS AS SETTINGS FOR HEALTH

The linkage between ES and human health is consistent with the Ottawa Charter for Health Promotion that recognized as prerequisites for health: peace, shelter, education, food, income, a stable ecosystem, sustainable resources, social justice, and equity (WHO, 1986). The Ottawa Charter (and more recently the Bangkok Charter for Health Promotion in a Globalized World, 2006) identified five major strategies for promoting health: (1) building healthy public policy; (2) creating supportive environments; (3) strengthening community action; (4) developing personal skills; and (5) re-orienting health services. The linkage is most clearly expressed in the text about 'creating supportive environments':

> Our societies are complex and interrelated. Health cannot be separated from other goals. The inextricable links between people and their environment constitutes the basis for a socioecological approach to health. The overall guiding principle for the world, nations, regions and communities alike, is the need to encourage reciprocal maintenance – to take care of each other, our communities and our natural environment. The conservation of natural resources throughout the world should be emphasised as a global responsibility.

A central tenet of the Ottawa Charter was that 'health is created and lived by people within the settings of their everyday life: where they learn, work, play, and love' (ibid.). This perspective has been central to the healthy-settings approach to health promotion, where people actively use and shape the environment and thus create or solve problems relating to health. Together these statements provide for an ecological model of health promotion, a systems perspective and a focus on whole system organization development and change (Dooris, 2006; Poland and Dooris, 2010).

Recent applications of the settings approach (see, e.g., Parkes and Horwitz, 2009; Horwitz and Finlayson, 2011) make the case for a re-orientation of public health in general, and health promotion in particular, towards ecosystems we depend on (watersheds, wetlands etc.), posing an explicit challenge to an 'ecological model' for public health that may be systems oriented, but is often abstracted from the living systems we depend on. A related challenge is to keep ecosystems and their ecological, contextual complexities embedded in policy, in ways that influence the institutional and governmental processes that deliver health services, address health inequalities, and intervene to improve public health.

4.3 LINKING THE SETTINGS APPROACH WITH ECOSYSTEM SERVICES

Health issues and the environment run much broader than just the negative impact of exposures. ES (as described earlier in this volume) are provided in settings in which the determinants of human health are expressed across short, medium and long time frames (Figure 4.1).

The characteristic ways that ecosystems, and the services they provide, determine human health and well-being can be outlined in relation to nine areas that are potentially influential in any context or setting. The following nine areas are adapted from Horwitz and Finlayson (2011) and references cited therein unless otherwise stated, and examined further in relation to Table 4.1.

1. *Contributors to hydration and safe water:* The principal supply of renewable fresh water for human use comes from an array of inland ecosystems, including lakes, rivers, swamps, and shallow groundwater aquifers, and they play a major role in treating and detoxifying a variety of waste products.
2. *Contributors to nutrition and microbiomes:* The world's major food items, core requirements for human health, come from terrestrial or aquatic ecosystems, and/or are irrigated by water from wetland ecosystems. Staple food items for almost all of the world's population are grown in a wide range of environments. Ecosystems also have

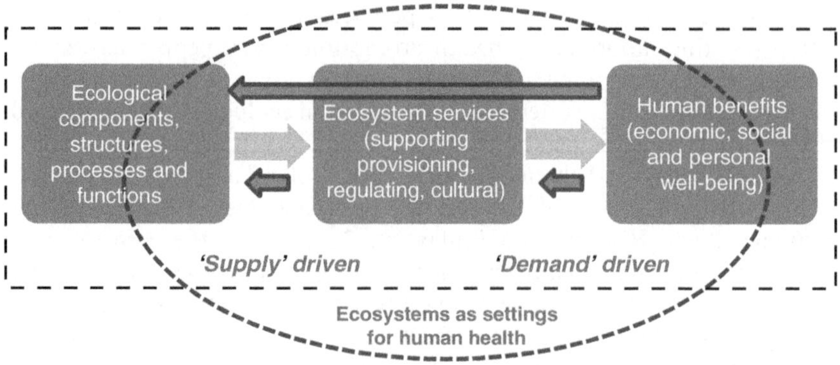

Note: This is a broad conceptual model showing the relationship between ecological integrity, ES that are provided, and the well-being benefits that humans derive (and demand). At first these appear like a linear chain of relationships as shown by the lighter arrows; but feedback loops related to the ways that humans draw upon ES have consequences for ecological components, structures and processes that constitute a broader understanding of biodiversity.

Figure 4.1 Settings for ecosystem services and human health

a role in maintaining dietary diversity, contributing to a multi-dimensional agenda focused on nutritional and health status, socio-cultural traditions, income generation and biodiversity conservation. A requirement for microbial input from the environment to drive immunoregulation is also an ecosystem contribution (through both food and water, and other exposures); Rook (2013) and others have argued that this is an essential ecosystem service for our well-being.

3. *Sites of exposure to infectious diseases:* Ecosystems are the loci for communicable disease, where microorganisms (the pathogens) are transmitted through water, people, animals as vectors, surfaces, foods, sediments or air. Poor-quality water (as unsafe water), inadequate sanitation and insufficient hygiene are the major risk factors for diarrhoeal disease, which is the second leading contributor to the global burden of disease. An important share of the total burden of disease worldwide could be prevented by improvements related to drinking water, sanitation, hygiene and water resource management. Infectious diseases associated with wetland ecosystems have profoundly influenced the discipline of public health, and this is probably the source of the erroneous oversimplification that wetlands or the environment in general are bad for human health.
4. *Sites of exposure to pollution or toxicants:* Human health can be

affected by acute or chronic exposure to toxicants, through the media of water, soil/sediments, or air. The nature of these exposures is exacerbated by human behaviours and activities. With respect to wetland ecosystems, Horwitz and Roiko (2015, p. 75) argue that:

> [. . .]the risks can increase (sometimes dramatically) if disruption to ecosystems, and the services they provide, is profound. Two principal forms of human exposures in wetland settings are distinguished: where the type or form of exposure is determined by the service that is provided (e.g., when drinking water contains a pollutant), and where services are eroded, creating the conditions for exposure (e.g., where water purification capacities of wetlands are overwhelmed by an oversupply of nutrients, resulting in an exposure to a microbial toxin). In both cases, addressing an imbalance of ecosystem services is required to ensure any necessary interventions are effective.

5. *Settings for mental health and psychosocial well-being:* Ecosystems, in their myriad forms, become embedded in the human psyche in formulations of 'sense of place'. Changes to ecosystems, to their products, their ability to deliver a livelihood, or to become a source of toxic exposure or disease, can influence a person's mental health. These potentialities are increasingly recognized as being part of the natural resource manager's and public health practitioner's spheres of prevention and intervention.
6. *Sites of exposure to physical hazards:* Climate externalities like floods, severe storms, drought, and fire, and geological externalities like earthquakes and tsunamis, can magnify any exposures; in fact because most people live in, on or near waterways, the conditions of the wetland and its ability to absorb external forces, will determine to a large extent the degree to which human health is affected. The disease burden following major disaster events ranges from psycho-social issues, infectious diseases, to physical injury and systemic chronic illness. The pathways to such disease events may be direct or indirect, and include a spectrum of community members including those directly injured, rescuers, people who have lost property, belongings or capacity to sustain a livelihood, families of those injured, and from there the more general population (Cook et al., 2008).
7. *Places where people derive their livelihood:* Livelihoods comprise the capabilities, assets and activities to make a living. Family and community livelihoods are sustained in the context of communities and economies they live in, which in turn are also dependent on the ecosystems in which these are embedded. Some livelihoods can be seen as tightly coupled to specific places, where it may be possible to understand the social and ecological systems of the family and community livelihoods

by listening to their stories, hopes and wishes, and by acting according to them, recognizing the combined context of local and traditional knowledge, government requirements, and market forces. Other livelihoods (and economies) are more usefully seen as embedded in much larger social-ecological systems whereby people may be earning a living from ES across local and regional through to global scales. This is exemplified by processes of natural resource development that may create multiple, often disconnected, sites of social and ecological impact – and benefit – along the full 'life cycle' of resource extraction and conversion to raw materials, through to production, transportation, use, end-of-life treatment, recycling and final disposal (Dewulf et al., 2015).

8. *Places for lifestyles, personal behaviours and community engagement (places to enrich people's lives, enable them to cope, and to help others)*: For ecosystem settings, the different behaviours and activities of people will be a proximal determinant of their health. Lifestyle factors are related to the ES, particularly leisure, recreation, sporting activities, education, and cultural heritage (including, e.g., a spiritual significance of water), providing for both physical and mental health given human affinities for ecosystem attributes.
9. *Sites where medicinal and other products can be derived or accessed:* Health benefits will accrue to societies in general and individuals in particular as biodiversity products of ecosystems can be used for pharmaceutical or other medicinal purposes. Animals, fungi, bacteria and lower plants (algae), some of them living in extreme conditions, provide productive sources of new natural products. The medicinal qualities of these are a good example of the continued value of traditional knowledge to health care today (see, e.g., Cunningham, 2015).

These nine areas play out in complex ways in the settings shown diagrammatically in Figure 4.1. The point here about biodiversity, ecosystems, and ES is their multi-dimensional nature – this is not a linear exposure pathway, as is often inferred in the field of environmental health. Instead, ES are derived and deliver health and well-being benefits for some people, but in the process may disrupt or diminish the availability of other ES that might have poor health outcomes for other people. We revisit this theme under 'trade-offs', below.

Horwitz and Finlayson (2011) also propose that these health areas can be clustered into three overarching clusters of health issues, presented in Table 4.1 in relation to (1) core requirements; (2) health risks and (3) determinants of health. Each of these health areas can also be associated with principally relevant ES, which then enables estimation of health benefits when these ES are maintained or enhanced, and the consequences for

Table 4.1 Major classes of health issues are associated with specific ES, leading to health benefits from the maintenance of these services, and the health consequences when these services are diminished or disrupted

Health Issues	Relevant Ecosystem Services	Health Effects, Health Outcomes from Ecosystem Services	
		Benefits if services are maintained or enhanced	Consequences of disruption to the services
1. *Core requirements:* Contributions to hydration, safe water and air	Drinking water for humans/livestock Groundwater replenishment Water purification/waste treatment or dilution Flood control, flood storage Air purification	Adequate water quantity and availability to meet human needs, including sanitation and hygiene Good water quality facilitates access to drinking water, appropriate hygiene and sanitation Good air quality	Inadequate water quantity leading to dehydration, access inequities, conflict, immigration Contaminated water (chemical and/or microbiological) complicating hygiene and sanitation Physical hazards (poor air quality, land subsidence, flood damage etc.)
2. *Core requirements:* Contributions to nutrition and microbiomes	Water for irrigated agriculture Food for humans Food for livestock Biological control agents for pests/vectors/pathogens Other products and resources, including genetic material Soil, sediment and nutrient retention Microbial input from the environment to drive immunoregulation	Availability of adequate food for good nutrition, growth and development Enables appropriate diet and digestion (including gut microbiome, essential vitamins and trace elements) and behaviour associated with a healthy diet (i.e., preparation of food and physical exercise) Microbiomes (gut, skin, airways, etc.) appropriately stimulate regulatory immune system	Decreased food security, leading to malnutrition, starvation, stunting Lack of nutritious or accessible food increases the likelihood of inappropriate diet, leading to vitamin deficiencies, obesity, diabetes, eating disorders Insufficient development of the immune system (contributing to allergies, autoimmune and inflammatory diseases)

Table 4.1 (continued)

Health Issues	Relevant Ecosystem Services	Health Effects, Health Outcomes from Ecosystem Services	
		Benefits if services are maintained or enhanced	Consequences of disruption to the services
3. *Health risk:* (Sites of) Exposure to infectious diseases	Drinking water for humans and/or livestock Water purification/waste treatment or dilution Biological control agents for pests/vectors/pathogens Soil, sediment and nutrient retention	Confidence in the quality and safety of interacting with local ecosystem Enhanced abilities to interact with ecosystems to derive other benefits, like those that accrue from provisioning or cultural services, or to derive an income	Exposure to health risks including: Water-borne diseases Vector-borne diseases Emerging infectious diseases Antimicrobial resistance
4. *Health risk:* (Sites of) Exposure to pollution or toxicants	Water purification/waste treatment or dilution Other hydrological services: hydrological maintenance of biogeochemical processes Soil, sediment and nutrient retention Air purification	Prevention of exposure to environmental contaminants Enhanced abilities to interact with ecosystems to derive other benefits, like those that accrue from provisioning and cultural services, or to derive an income	Exposure to: Soil or water-borne inorganic chemicals Microbial toxins in soil or water Atmospheric particles or chemicals
5. *Health risk:* Settings for mental health and psychosocial well-being	Contemporary cultural significance, including for arts, creative inspiration, and existence values Aesthetic and 'sense of place' values Spiritual and religious values Important knowledge systems, and importance for research	Meaningful interactions with ecosystems as places Enhanced abilities to derive benefits from cultural services and traditional practices	Depression, suicide (associated with hopelessness and helplessness of ecosystem degradation and other environmental change; and their social consequences; these are 'exposures') Grieving over loss of place ('solastalgia', *sensu* Albrecht et al., 2007)

6. *Health risk:* (Sites of) Exposure to physical hazards	Climate regulation Flood control, flood storage Soil, sediment and nutrient retention Coastal shoreline and river bank stabilization and storm protection Local climate regulation/buffering of change	Stable environment without imperative to migrate Investment in protection from temperature extremes or physical hazards is not essential Any of the benefits described in this table, associated with enhanced livelihoods from ecosystems	Forced migration Exposure to extremes of temperature; increased demand for investment and infrastructure to protect against temperature extremes Exposure to floods and droughts, cyclones, hurricanes, tsunamis etc.; increased demands for protection Any risk of exposure where change to an ecosystem has been implicated
7. *Determinants of health:* Places where people derive their livelihood	Most regulating services: local climate and air quality, moderation of extreme events, water purification/ waste treatment or dilution Most provisioning services: food, raw materials, water, medicinal resources Supporting services that enable livelihoods: primary production and soil formation Cultural services associated with livelihoods: spiritual, recreation, education	Socio-economic conditions to obtain sufficient water and food Sustained living made from ecosystem assets, including resource extraction and development Capacity to manage ecosystem assets to continue to receive these benefits Safe work place, occupational exposures	Insufficient water, insufficient nutrition, exposure to toxics, diseases and psychological stresses associated with loss of livelihoods from change to an ecosystem Vicious circle of declining abilities to manage ecosystem assets Work place, occupational exposures

Table 4.1 (continued)

Health Issues	Relevant Ecosystem Services	Health Effects, Health Outcomes from Ecosystem Services	
		Benefits if services are maintained or enhanced	Consequences of disruption to the services
8. *Determinants of health:* Places for lifestyles, personal behaviours, and community engagement	Recreational hunting and fishing Water sports Nature study pursuits Educational values Understanding ecosystem behaviour Cultural heritage Contemporary cultural significance, including for arts and creative inspiration, and including existence	Maintenance of recreational opportunities in ecosystems (benefits derived from physical exercise in quality green spaces, blue spaces, blue-green spaces) Educational opportunities; better understandings of ecosystems (improved ability to respond to life threatening events; improved likelihood that ecosystems will be managed using appropriate and/or traditional means) Health benefits associated with opportunities to be creative and productive	Loss of recreational opportunities (decline in stress relief and physical fitness) Loss of educational opportunities (decline in ability to respond to life-threatening events; or to manage the land and water using appropriate and/or traditional means) Mental health issues associated with an alienation from or lack of easy or safe access to culturally significant elements of ecosystems
9. *Determinants of health:* Access to sites that provide products of medicinal and/or traditional value	Other products and resources, including genetic material (provisioning; medicinal products) Cultural heritage Spiritual and religious values Valuable knowledge systems and importance for research	Provision and development of pharmaceuticals associated with medicinal plants Access to and use of indigenous/traditional/herbal treatments and associated practices and language Maintenance of indigenous/traditional practices associated with specific products and ceremonies	Cultural diversity and biological diversity are reciprocally developed and maintained. The links between culture, language and biodiversity reflect a wider connection between nature and a community's sense of place. Loss of practices, language, and its concomitant cultural loss, may be a source of considerable demoralization and anguish

BOX 4.1 CROSS-CUTTING THEMES FOR ECOSYSTEM SERVICES AND SETTINGS FOR HEALTH – THE EXAMPLE OF A RICE PADDY

In many parts of the world, families draw much of their livelihood from a rice paddy. With its combined dynamics of land, water, and human interactions a rice paddy provides a setting that exemplifies the socio-ecological context for health and the influences of ecosystem services. Members of the family may have individual circumstances that might make them more susceptible to ill health (or wellness), such as a genetic factor, advanced age, or pregnancy. Their health might also be determined by lifestyle factors like their diet, their alcohol consumption, or the amount of sleep they get each night. Social and community influences might determine when in the day, or during the week, in what type of weather, they may be working in the fields, and the length of time they can work for. This might then expose them differentially to living and working conditions such as pesticide applications, mosquitoes (and potentially arboviral diseases), or machinery accidents. The likelihood of exposure will change according to socio-economic conditions (the need to cultivate to sustain a community's livelihood), cultural conditions (observance of religious festivals, planting and harvesting customs), and environmental conditions (such as climate change, migratory birds, etc.), and knowledge systems for all of these, and the ability of the family to earn its living will also be influenced by government constraints and market conditions. Drawing on the systems-based understanding of the rice paddy as a socio-ecological setting for health, it becomes clear that any project or proposal that changes the ecosystem services in this system, has the potential to create impacts on health through a range of interrelated pathways. Understanding these relationships and pathways provides a more detailed understanding of potential health impacts in the short, medium and long term.

Source: Adapted from Horwitz et al. (2012).

health where these ES are disrupted (Table 4.1). The nine major classes of health issues do not stand as mutually exclusive of one another, and some significantly cut across others, such as access to sites to derive products of medicinal and/or traditional value. Depending on one's cultural perspective this could be considered a core requirement equivalent to food and water; could demonstrate a response to, or prevention of, hazardous exposures; and could also be linked to the dynamics of livelihoods, lifestyles and identities described above. Another example is given in Box 4.1.

Population health data, including the incidence and prevalence where diseases and exposures are concerned, and the geographical nature of ill health, may be available to provide statistical appraisal for some of the predicted health outcomes in Table 4.1. Similarly, lifestyle, livelihood and socio-economic data may be available through population census statistics, sufficient to enable an assessment of health inequalities and where interventions through ES might be targeted.

These linkages serve to emphasize two key features of particular relevance to the scoping phase of HIA. First, they provide an effective framing of the environment in terms of communicating with and influencing stakeholders and decision-makers (Baker et al., 2013). In particular, this framing makes explicit a range of social-ecological dynamics such as, for example, connections between ES and the social determinants of health. A reasonable case can be made that stakeholders and the public are well placed to engage with this alternative description, and that the language of core requirements, health risks, livelihoods and lifestyles (see Table 4.1) may resonate as an appropriate and accurate description of complex pathways between ecosystems and health.

Second, they allow for a more complete, holistic and integrated consideration of a socio-ecological system (per Baker et al., 2013), and 'the interplay of behavioural, biological, physical and social interactions that affect, sustain or are modified by living organisms including humans' (Michener et al., 2001, p. 1018). However, this advantage would be countered if the list of ES were used to account for each of them separately, as if they were individual items; doing so would mask interactions between them, and reduce our perception of the social ecological system and how it functions (ibid.).

The framework presented in Table 4.1 proposes a series of steps to orienting an HIA scoping phase, including documenting the health issues that might arise, assigning ES to each, and seeking to maintain and enhance those services to minimize consequences and maximize benefits. It should be noted, however, that an impact assessment done on these grounds alone would be dangerously naive and simplistic. An understanding of ecosystems as settings for health (see Figure 4.1) and the ES associated with this (see Table 4.1), both point to another more potent analysis that needs to be undertaken to achieve an integrated and systemic understanding of health impacts. Being non-linear, inter-related, and non-mutually exclusive, there will invariably be situations where the subject (project, development, proposed policy) of the HIA will produce some beneficial health outcomes and some negative health consequences, and where some outcomes and ES will be valued more highly by different stakeholders. These differentials inevitably lead to the need to consider trade-offs as an integral component of any impact assessment, not the least of which is an HIA.

4.4 ECOSYSTEM SERVICES VALUATION AND TRADE-OFFS IN HIA

Using ES to scope an HIA may be of particular utility where conflicts between environmental, economic, social and health outcomes are apparent, since arguments can be based on a common currency of services whose explicit purpose is to provide for human well-being, without the mediating arguments of financial outcomes. Ideally, HIA will reveal maximum health benefits for maximum numbers of people while ensuring that ES are neither degraded nor disrupted. However, the situation may not be that simple, as was pointed out by Corvalan et al. (2005) and Horwitz and Finlayson (2011): positive health outcomes can accrue from enhanced ES as expected; however, they might accrue from degraded ES as well. And likewise, negative health impacts are not the only product of degraded ES – poor health outcomes can also occur when some ES are enhanced.

So the principal trade-off of interest for HIA is when we seek to enhance (the benefits of) one or more other ES and achieve a positive health outcome, while recognizing that a degrading or disrupting of one or more ES will result, and this may produce poorer health outcomes for others, either immediately or at some time in the future. A proposal to drain significant areas of wetland for malaria control could, for example, be motivated by lowering of the burden of this disease, as well as the possibility of more arable land being made available, enhancing provisioning services and the health benefits of improved livelihoods and local economies, but the benefits of draining the wetland are at the expense of the loss of supporting and regulating services of the wetland (nutrient cycling, carbon storage, flood mitigation and so on). Ultimately, these lost or degraded ES might have higher-order (unintended and unpredictable, distal, longer-term and even geographically displaced) impacts on the determinants of health in the medium and long term.

Like others (see Campbell et al., 2010) we acknowledge that recognizing trade-offs is no simple matter because a trade-off perceived by one stakeholder may not be recognized by another. So undertaking a process by which the trade-offs and their consequences can be fairly negotiated becomes a central concern. Representation of marginalized stakeholders, increased transparency of information, and engaging with the core pursuits of other sectors, will be key components of such a process (Horwitz et al., 2015; Mahboubi et al., 2015).

Attention to trade-offs is not new to the broader assessment literature. Bond et al. (2012) argue that sustainability assessment must adopt formal mechanisms for managing unavoidable trade-offs in an open, participative and accountable manner. Retief et al. (2013) state that one of the difficulties of making trade-off decisions resides around commodities, currencies and non-commodities where decisions concerning trade-offs between non-commodities are more difficult and take longer than those considering commodities and currencies, and that policy-makers in particular shy away from making trade-off decisions about non-commodities. In contrast, the case can be made that where moralities are involved, decision-making around trade-offs becomes easier; since health outcomes are normative, their moral basis has some clarity. Accordingly, we propose that one of the conspicuous advantages in the use of ES in HIA is that they render complex ecosystem, society, economy and human health interactions accessible to health economics and moral determination. This adds weight to the argument by Retief et al. (2013) that ES have the capacity to transcend different value systems.

Gibson (2006) has put forward trade-off decision rules designed to ensure that sustainability assessment processes deal better with and account for any sustainability trade-offs. In order to enhance the utility of ES as a framing tool for HIA, these rules could be modified so that those designing and conducting Health Impact Assessments can be guided by a health-oriented variation of the following:

- *Net gains:* Any acceptable trade-off must deliver net sustainability gains (over the long term); this means that health consequences should be net positive where diminished or enhanced ES or a bundle of services are concerned.
- *Burden of argument:* The proponent of the trade-off must be required to provide justification.
- *Avoidance of significant adverse effects:* No trade-off involving significant adverse effect is acceptable unless all alternatives are worse.
- *Protection of the future:* No displacement of significant adverse impact from present to future can be justified unless all alternatives are worse.
- *Explicit justification:* All trade-offs must be explicitly justified (including a context-specific account of priorities and sustainability decision criteria).
- *Open process:* Stakeholders must be involved in trade-off making through open and effective participatory processes.

HIA would justify their arguments using burden of disease measures, like disability-adjusted life years (DALYs), and quality-of-life assessment (e.g., WHOQOL – a method to determine an individual's quality of life in the context of their culture and value systems – WHO, 2012) to establish where net positive gains might lie.

Related issues raise concerns that using ES as a framing and scoping tool for HIA could, if held open to market-based mechanisms alone, block interventions that seek to address health equities. This dilemma arises due to the fact that when ES are commodified they can become the basis for new socio-economic hierarchies characterized by the repositioning of existing social actors, the emergence of others and very likely the reproduction of power relations in access to wealth and environmental resources. In this scenario, it is hard to see how commodification of ES will address structural poverty conditions, or other inequities generated by resource distribution, especially where these upstream socio-economic conditions can be implicated in health outcomes. An example is presented in the next section of this chapter.

4.5 HIA AND RESOURCE DEVELOPMENT: COAL SEAM/FRACKING

Resource development is highly relevant for an exploration of ES and impact assessment in general. This is because resources provide the foundational natural capital from which ES are derived and, as discussed above, where some ES are enhanced (in this instance as resource extraction), there can be both positive and negative health outcomes. Here Parkes (2016) points out the importance of combining 'an awareness of the cumulative impacts of resource development with a consideration of the determinants of health from a variety of perspectives – including mental health, Aboriginal health, and social and environmental justice concerns' (p. 118).

Coal seam gas production is a topical example of resource development where an HIA could be applied, and as such a useful case study. Coal seam gas is generated through a process of the engineered injection of fluids to hydraulically fracture coal beds for release of methane gas. Impacts associated with the process can occur at any of the phases of coal seam gas exploration, assessment, production, transportation and delivery. To date, particular attention has tended to focus on those impacts that relate to human health, the environment, loss of other land and water amenities, and water resources themselves.

A risk assessment approach to such a scenario might be something like the following (as summarized from O'Kane, 2014, p. iv for coal seam gas production from rural areas in eastern Australia):

The likelihood and consequences (severity) of the risks will depend on local conditions (e.g. geology, hydrogeology, location, environment, etc.). Potential risks to the broader environment are those resulting in impact to surface water, groundwater, soil, or air quality, produced by one or more of the following: drilling, well integrity and fracture stimulation; seam depressurization; spills and leaks; and produced water and solids. Potential biophysical risks to human health occur through four exposure pathways, routes from release of a contaminant(s) to a person(s): through water, soil and air, and indirectly in food. Human health risks and effects are traditionally gauged by dosage (so environmental dilution needs to be known). Exposure pathways can be understood through the modelling of water and air movement, or ecological webs, which requires knowledge of the local environment and the potential contaminants.

An HIA framed by ES would see this traditional risk assessment approach as constrained. A more fulsome treatment might pursue the four areas of considerations made by the Chief Medical Officer of Health for New Brunswick, Canada, albeit for shale gas development (OCMOH, 2012):

- protection of health and community well-being related to changes in the social environment (e.g., recommendations for optimizing equitable distribution of risks and rewards and identifying a role for Public Health in community planning);
- protection of health related to changes in both the social and physical environments (e.g., recommendations for developing a requirement to submit a Health Impact Assessment (HIA) as part of the standard project registration process);
- protection of health related to changes in the physical environment (e.g., recommendations for monitoring networks for ambient air and water quality); and
- protection of future generations (e.g., a plan to anticipate and mitigate 'boomtown effects' and through strategic Health Impact Assessments).

These types of considerations lend themselves to an HIA framed by ES. Using the same example as risk assessed by O'Kane (2014), namely where coal seam gas production is proposed for a rural area in eastern Australia where crop and livestock agriculture is currently the dominant economic activity, a range of ES can be hypothesized to be either maintained/enhanced, or disrupted/diminished; for all nine health issue classes, examples of health outcomes from these ES changes could be outlined. Table 4.2 elaborates on such a hypothetical treatment.

The hypothetical framing shown in Table 4.2 has highlighted positive health and well-being outcomes arising from local economic development (including contributions to agriculture, livelihoods and lifestyles), while suggesting that poor health outcomes are possible due to exposures to toxicants, psychosocial consequences of changes to place, cultural alienation from place and potential exposure to physical hazards. In an HIA, these factors would need to be explored more fully using economic modelling, epidemiological methods and risk assessment processes along with qualitative approaches that can help create a more fulsome appreciation of the

Table 4.2 Application of ES framing for health issues as a way of scoping health impacts, using the hypothetical situation of coal seam gas production on farmland in Eastern Australia

Health Issues	Ecosystem Services Enhanced or Maintained by Coal Seam Gas Production (Hypothetical)		Ecosystem Services Degraded or Eroded by Coal Seam Gas Production (Hypothetical)	
	Relevant ecosystem services	Potential health outcomes via:	Relevant ecosystem services	Potential health outcomes via:
1. *Core requirements:* Contributions to hydration, safe water and air	Indirect – provision of energy to harvest/ produce water, and treat wastewater	Decreased exposures to health risks; associated livelihoods and lifestyles that benefit all locally and distally, and for as long as the energy continues to be provided	Drinking water for humans/livestock Groundwater replenishment Water purification/waste treatment or dilution Air purification	Decreased local availability of potable water Increased likelihood of local exposures to contaminants Associated livelihoods and lifestyles affected locally: psychosocial health consequences
2. *Core requirements:* Contributions to nutrition and microbiomes	Water for irrigated agriculture	Where water produced is of quality or quantity suitable to irrigate, local health benefits accrue in terms of production and income; improved financial security and infrastructure for producers and local communities	Water for irrigated agriculture Food for humans Food for livestock Soil, sediment and nutrient retention	Where water produced is salty or contaminated it can't be used to irrigate, quality soils not useable, fewer local benefits in terms of production, income and livelihoods; psychosocial health consequences

Table 4.2 (continued)

Health Issues	Ecosystem Services Enhanced or Maintained by Coal Seam Gas Production (Hypothetical)		Ecosystem Services Degraded or Eroded by Coal Seam Gas Production (Hypothetical)	
	Relevant ecosystem services	Potential health outcomes via:	Relevant ecosystem services	Potential health outcomes via:
3. *Health risk:* (Sites of) Exposure to infectious diseases	Biological control agents for pests/vectors/pathogens Soil, sediment and nutrient retention	Maintenance of existing regulatory mechanisms, enabling control of pests/vectors/pathogens	Other hydrological services: hydrological maintenance of biogeochemical processes	Increased surface water for vector breeding
4. *Health risk:* (Sites of) Exposure to pollution or toxicants	Water purification/waste treatment or dilution Soil, sediment and nutrient retention Air purification	Decreased local and distant exposures as for (1) above Maintenance of existing regulatory mechanisms	Air or water purification/ waste treatment or dilution Other hydrological services: hydrological maintenance of biogeochemical processes	Increased local exposures as for (1) above
5. *Health risk:* Settings for mental health and psychosocial well-being	Importance for education, knowledge systems, and research 'sense of place' values	Exploration and development of coal seam gas leads to improved local understandings of surface and groundwater hydrology Changed sense of place values	Aesthetic and 'sense of place' values Contemporary cultural significance, including for arts, creative inspiration, and existence values Spiritual and religious values	Changed sense of place values Further alienation from land in terms of taboos and losses of cultural awareness, losses of local connections, belonging and knowledge

6. *Health risk*: (Sites of) Exposure to physical hazards	Local climate regulation/ buffering of change	Restoration of vegetation on land surface (undisturbed by below ground operation) buffers physical hazards	Climate regulation Flood control, flood storage Soil, sediment and nutrient retention	Contributes to global climate change; changed regional climate, more severe weather patterns Vulnerability to excess water, possible human exposures, losses of soil and sediment
7. *Determinants of health:* Places where people derive their livelihood	Provisioning services maintained Cultural services maintained	As for (2) above – benefits from local economic activities	Provisioning services disrupted Cultural services disrupted	As for (1) above As for (5) above
8. *Determinants of health:* Lifestyles (and personal behaviours)	Water sports Educational values	Possibility of physical exercise benefits from additional (discharged) surface waters	Cultural heritage Contemporary cultural significance, including for arts and creative inspiration, and including existence Recreational hunting and fishing Nature study pursuits	As for (5) above
9. *Determinants of health:* Availability of medication and traditional values	Other products and resources, including genetic material (provisioning; medicinal products) Valuable knowledge systems and importance for research	Possibility of petrochemical by-products for pharmaceutical benefit As for (5) above, improved local knowledge of surface and groundwater hydrology	Cultural heritage Spiritual and religious values Important (often alternative) knowledge systems	As for (5) above

combined impacts on all classes of health issues – including the (social and ecological) determinants of health. The importance of considering positive and negative dynamics across short, medium and long time-frames has also been highlighted in the context of considering cumulative determinants of health impacts (Kinnear et al., 2013; Parkes, 2016). Attention to cumulative impacts also reinforces the relevance of a setting-based approach that recognizes the interacting impacts on ES across time and space across landscapes and regions.

4.6 CHALLENGES AND CONCLUDING REMARKS

The case study presented above shows the utility of framing ES, mainly as a heuristic device or a prompting tool, to ensure that scoping embraces a more systemic understanding of impacts on health, and recognizes the interactions implied by understanding both ES and a socio-ecological and settings-based approach to health. In order to complete Table 4.2, a knowledge of the local setting – the biophysical, social and cultural characteristics, as well as at least a rudimentary understanding of the nature of the proposal – is required. The framing shows a potential to accentuate key features, particularly some health outcomes that might be more important, and/or the ES that might be made more or less accessible.

There are some things that such framing cannot do, however. In completing the table – or framing and scoping our understanding of health impacts in this way, we are forced to make judgements about the spatial and temporal scales that are important. This is particularly important for resource development, where the livelihood benefits (from refining, transport, market etc.) might accrue well away from the local zone of production. Temporal changes might also be distorted, with some impacts being short or long term, and these might interact with local and distant effects.

Several questions remain unanswered from this consideration of ES in a healthy settings context for HIA: most significantly perhaps, are ES relevant to *all* plans, programmes or projects or *all* institutional contexts for HIA? And, as thinking matures about ecosystems services and their relationships with health and well-being, could the ideas presented here provide the basis for a scoping framework that would be universal for a standalone HIA? The application will be particularly relevant to contexts wherever resource development is proposed. Beyond this it could have relevance to any context where natural capital and biodiversity, however remotely or indirectly, underlie human endeavours.

Harris-Roxas et al. (2012) said that if HIA were to become more routine in the already complex set of planning and assessment processes of the government and the private sectors, the field would need to demonstrate how conducting an HIA is likely to add new information and improve the outcomes of the decision to which it is oriented. This chapter has sought to demonstrate that by using ES to be explicit about the ecological and economic conditions that underpin health consequences, and by making such an assessment accessible via the scoping phase of HIA, there is a greater potential to not only understand the scope of interrelated health impacts, but also to link health to other sectors for better integrated decision-making.

This framing carries the added benefit of recognizing the ecosystem as setting for human health, which *demands* a socio-ecological approach since it puts people (socio) into the ecosystem (ecological). In addition, a socio-ecological approach to understanding ecosystems as settings for health can be usefully informed by understanding of social ecological systems (*sensu* Folke et al., 2010), and that the (ecological economics) idea of ES embeds the social and ecological, since it involves imposing a human construct and value system on ecological realities. It also makes explicit where trade-offs might be made, and where assumptions around these for human health and well-being might need to be questioned.

In this context, one advantage revealed by this analysis is using ES as a point of reference could help broaden the scope of all impact assessments, including environmental and sustainability ones, towards a consideration of health impacts. The move towards sustainability assessment may represent an opportunity for HIA in that it suggests a more integrated, holistic approach to assessment (Bond et al., 2012), in which health is a key consideration, though as noted earlier there will remain some tensions with integration of HIA with other forms of impact assessment, including the risks that health considerations may be overwhelmed or subsumed by competing priorities (Harris-Roxas et al., 2012). A combined attention of ecosystems services and the settings for health offers a starting point to offsetting these challenges and encouraging a more integrative approach to scoping Health Impact Assessment.

NOTE

1. This chapter purposefully refers to three closely related terms that are used in congruence with their original use in the literature: a socio-ecological approach to health proposed by the Ottawa Charter (*sensu* WHO, 1986); the dynamics and development of complex social-ecological systems (SES) are associated with the theoretical framework of resilience thinking (*sensu* Folke et al., 2010); and Baker et al.'s (2013) reference to the (complexity of) the socio-ecological system also echoes an increasing emphasis on Coupled Human and Natural Systems (CHANS) to advance sustainability (*sensu* Liu et al., 2007).

REFERENCES

Albrecht, G., G.-M. Sartore and L. Connor et al. (2007), 'Solastalgia: the distress caused by environmental change', *Australian Psychiatry*, **15**(S1), S95–S98.

Baker, J., W.R. Sheate and P. Phillips et al. (2013), 'Ecosystem services in environmental assessment – help or hindrance?', *Environmental Impact Assessment Review*, **40**(1), 3–13.

Bangkok Charter (2006), 'The Bangkok Charter for Health Promotion in a Globalized World', *Health Promotion International*, **21**(S1), 10–14.

Bond, A., A. Morrison-Saunders and J. Pope (2012), 'Sustainability assessment: the state of the art', *Impact Assessment and Project Appraisal*, **30**(1), 53–62.

Briggs, D.J. (2008), 'A framework for integrated environmental health impact assessment of systemic risks', *Environmental Health*, **7**, 61.

Campbell, B.M., J.A. Sayer and B. Walker (2010), 'Navigating trade-offs: working for conservation and development outcomes', *Ecology and Society*, **15**(2), 16.

Charron, D. (2012), *Ecohealth Research in Practice: Innovative Applications of an Ecosystem Approach to Health*, New York: Springer.

Cook, A., J. Watson and P. van Buynder et al. (2008), '10th anniversary review: natural disasters and their long-term impacts on the health of communities', *Journal of Environmental Monitoring*, **10**, 167–75.

Corvalan, C., S. Hales and A. McMichael (2005), *Ecosystems and Human Well-being: Health Synthesis*, Geneva: World Health Organization.

Cunningham, A. (2015), 'Wetlands and people's wellbeing: basic needs, food security and medicinal properties', in C.M. Finlayson, P. Horwitz and P. Weinstein (eds), *Wetlands and Human Health*, Amsterdam: Springer.

Dewulf, J., L. Benini and L. Mancini et al. (2015), 'Rethinking the area of protection "natural resources" in life cycle assessment', *Environmental Science & Technology*, **49**(9), 5310–17.

Dooris, M. (2006), 'Healthy settings: challenges to generating evidence of effectiveness', *Health Promotion International*, **21**(1), 55–65.

Folke, C., S.R. Carpenter and B. Walker et al. (2010), 'Resilience thinking: integrating resilience, adaptability and transformability', *Ecology and Society*, **15**(4), 20.

Gibson, R.B. (2006), 'Sustainability assessment: basic components of a practical approach', *Impact Assessment and Project Appraisal*, **24**(3), 170–82.

Harris-Roxas, B. and E. Harris (2013), 'The impact and effectiveness of health impact assessment: a conceptual framework', *Environmental Impact Assessment Review*, **42**, 51–9.

Harris-Roxas, B., F. Viliani and A. Bond et al. (2012), 'Health impact assessment: the state of the art', *Impact Assessment and Project Appraisal*, **30**(1), 43–52.

Horwitz, P. and C.M. Finlayson (2011), 'Wetlands as settings: ecosystem services and health impact assessment for wetland and water resource management', *BioScience*, **61**(9), 678–88.

Horwitz, P. and A. Roiko (2015), 'Wetlands as sites of exposure to pollution and toxicants', in C.M. Finlayson, P. Horwitz and P. Weinstein (eds), *Wetlands and Human Health*, Amsterdam: Springer.

Horwitz, P., C.M. Finlayson and R. Kumar (2015), 'Interventions required to enhance human well-being by addressing the erosion of ecosystem services in wetlands', in C.M. Finlayson, P. Horwitz and P. Weinstein (eds), *Wetlands and Human Health*, Amsterdam: Springer.

Horwitz, P., M. Finlayson and P. Weinstein (2012), *Healthy Wetlands, Healthy People: A Review of Wetlands and Human Health Interactions*, Ramsar Technical Report No. 6, Gland and Geneva, Switzerland: Secretariat of the Ramsar Convention on Wetlands and The World Health Organization.

Kinnear, S., Z. Kabir and J. Mann et al. (2013), 'The need to measure and manage the cumulative impacts of resource development on public health: an Australian perspective', in A.J. Rodriguez-Morales (ed.), *Current Topics in Public Health*, Rijeka, Croatia: InTech Publishers.

Lee, J., N. Röbbel and C. Dora (2013), 'Cross-country analysis of the institutionalization of health impact assessment', *Social Determinants of Health Discussion Paper Series No. 8 (Policy & Practice)*, accessed 24 February 2015 at http://www.who.int/social_determinants/publications/9789241505437/en/.

Liu, J., T. Dietz and S.R. Carpenter et al. (2007), 'Coupled human and natural systems', *AMBIO: A Journal of the Human Environment*, **36**(8), 639–49.

Mahboubi, P., M.W. Parkes and H.M. Chan (2015), 'Challenges and opportunities of integrating human health into the environmental assessment process: the Canadian experience contextualised to international efforts', *Journal of Environmental Assessment Policy and Management*, **17**(04), 1550034.

Michener, W.K., T.J. Baerwald and P. Firth et al. (2001), 'Defining and unravelling biocomplexity', *BioScience*, **51**(12), 1018–23.

Millennium Ecosystem Assessment (MA) (2005), *Ecosystems and Human Well-being: Synthesis*, Washington, DC: Island Press.

Office of the Chief Medical Health Officer (OCMOH) (2012), *Chief Medical Officer of Health's Recommendations Concerning Shale Gas Development in New Brunswick*, Fredericton, NB, accessed 14 May 2015 at http://sustainabilityresearch.wp.rpi.edu/files/2013/04/NewBrunswickFrackingReportSept2012.pdf.

O'Kane, M. (2014), *Independent Review of Coal Seam Gas Activities in NSW. Managing Environmental and Human Health Risks from CSG Activities*, Sydney: New South Wales Government, accessed 16 august 2015 at http://www.chiefscientist.nsw.gov.au/reports/coal-seam-gas-review.

Parkes, M.W. (2016), 'Cumulative determinants of health impacts in rural, remote and resource-dependent communities', in M.P. Gillingham, G.R. Halseth and C.J. Johnson and M.W. Parkes (eds), *The Integration Imperative – Cumulative Environmental, Community and Health Impacts of Multiple Natural Resource Developments*, Cham, Switzerland: Springer.

Parkes, M. and P. Horwitz (2009), 'Water ecology and health ecosystems as "settings" for health and sustainability', *Health Promotion International*, **24**(1), 94–102.

Poland, B. and M. Dooris (2010), 'A green and healthy future: the settings approach to building health, equity and sustainability', *Critical Public Health*, **20**(3), 281–98.

Retief, R., A. Morrison-Saunders and D. Geneletti et al. (2013), 'Exploring the psychology of trade-off decision-making in environmental impact assessment', *Impact Assessment and Project Appraisal*, **31**(1), 13–23.

Rook, G.A. (2013), 'Regulation of the immune system by biodiversity from the natural environment: An ecosystem service essential to health', *Proceedings of the National Academy of Sciences*, **110**(46), 18360–18367.

Utzinger, J., K. Wyss and D.D. Moto et al. (2005), 'Assessing health impacts of the Chad–Cameroon petroleum development and pipeline project: challenges and a way forward', *Environmental Impact Assessment Review*, **25**(1), 63–93.

Walker, B., C.S. Holling and S.R. Carpenter et al. (2004), 'Resilience, adaptability and transformability in social-ecological systems', *Ecology and Society*, **9**(2), 5.

Winkler, M.S., M.J. Divall and G.R. Krieger et al. (2010), 'Assessing health impacts in complex eco-epidemiological settings in the humid tropics: advancing tools and methods', *Environmental Impact Assessment Review*, **30**(1), 52–61.

Winkler, M.S., G.R. Krieger and M.J. Divall et al. (2012), 'Health impact assessment of industrial development projects: a spatio-temporal visualization', *Geospatial Health*, **6**(2), 299–301.

WHO (1986), *Ottawa Charter for Health Promotion*, Geneva: World Health Organization.

WHO (2012), *The World Health Organization Quality of Life (WHOQOL)*, accessed 12 January 2016 at http://www.who.int/mental_health/publications/whoqol/en/.

5. Matching an ecosystem services approach with Social Impact Assessment*

Leena Karrasch

5.1 INTRODUCTION

During the last century, humans increasingly used ecosystems to fulfill their demands. Degradation of ecosystems either through overexploitation of resources or by urban development still leads to a substantial and sometimes irreversible loss in biodiversity. At the same time, the awareness of the inherent value of nature and the infinite state of the natural resources is raised. Ecologists in particular debated the relationships between the stability of ecosystems and their complexity. A first step towards understanding ecological economics was the concept of the 'safe minimum standard' (Ciriacy-Wantrup, 1952) outlining the importance of conserving biodiversity for human life. Some decades later, the work of Costanza et al. (1997) and Daily (1997) introduced ecosystem goods and services into the scientific community. Based on their initial work, the United Nations Millennium Ecosystem Assessment Board published a report in 2003 called *Ecosystems and Human Well-Being. A Framework for Assessment* (MA, 2003). This was a major step towards anchoring the consequences of ecosystem change for human well-being and addresses the ecosystem services (ES) approach to scientists and practitioners.

ES are by definition the benefits people obtain from ecosystems and play a key role in describing the connection between nature and human civilization. The ES approach can help to analyze complex human–environment systems, to structure human activities and to develop impact assessments with regard to changing land use. Following the MA (2003), ES are classified into four categories: (1) provisioning services, (2) regulating services, (3) cultural services, and (4) supporting services.

Nevertheless, the ES approach focuses more on the supply side of ecosystems, rather than on the demand side of communities. Additionally, its scientific character hampers the communication process with laypersons and its incorporation into spatial planning activities and adaptive strategy development (Fürst et al., 2013; Ruckelshaus et al., 2013; Karrasch et al., 2014). Further changes in the environment will take place and these interventions will affect nature and humans. Therefore, an approach is required that identifies the social consequences of such interventions as well.

The Social Impact Assessment provides a framework that includes social consequences of an intervention and moves towards a more sustainable and equitable environment. The elements of the Social Impact Assessment contribute to a comprehensive view of possible effects and aim to maximize the benefits of interventions. The Interorganizational Committee on Guidelines and Principles for Social Impact Assessment (1995, pp. 11–12) defined social impacts as 'consequences to human populations of any public or private action that alter the ways in which people live' and the Social Impact Assessment as a concept to 'assess or estimate, in advance, the social consequences that are likely to follow from specific policy actions'. The concept evolved from the National Environmental Policy Act of the USA in 1969. Moving from a regulatory context towards stakeholder engagement focusing on the needs, interests and aspirations of communities, the Social Impact Assessment is currently well recognized in social science practice (Vanclay and Esteves, 2011).

Climate and anthropogenically introduced alterations will influence land management in the European North Sea coastal regions, requiring a realignment of adaptive and ecosystem-based strategies to manage human activities. The low-lying coastal areas in particular are confronted with natural and anthropogenic change processes. The IPCC *Fifth Assessment Report* relates a high confidence that risks in these areas as well as pressures on ecosystems will increase (IPCC, 2014).

Although the ES approach is increasingly used in environmental sciences and management to stress the ecological benefits that nature provides for people (Fisher et al., 2009; Abson et al., 2014), the interaction with practitioners demonstrated that they think and act in terms of socio-economic considerations (Karrasch et al., 2014). Therefore, direct inclusion of ES in local decision-making to promote an ecosystem-based management remains difficult (Müller and Burkhard, 2012; Smythe and Thompson, 2015). Observations from collaboration with stakeholders demonstrated that local decision-making cannot be based on the ES approach directly (Karrasch et al., 2014). There is a need for enhanced knowledge and integrated understanding on the complex interrelationships within specific socio-ecological systems in order to make the concept of ES relevant and applicable to practical cases (Potschin and Haines-Young, 2013).

Necessary change processes will lead to interventions in current land use and may lead to beneficial and adverse impacts on both ecosystems and human well-being. Since social demands and ecological supplies are not independent variables, it is necessary to raise the awareness of stakeholders and scientists towards the linkages of the natural and social spheres. Therefore, the development of new, interdisciplinary and adaptive land use strategies requires a balanced approach, taking the interaction of socio-economic and ecological concerns into account.

This chapter presents a conceptual framework to guide a regional adaptive governance process including the ES approach and Social Impact Assessment. The presented framework was developed and tested with 14 regional stakeholders (Karrasch et al., 2014). These stakeholders were decision-makers and they had the ability to take actions and decisions influencing the development of the case study area. Additionally, they were experts in their specific sectors (water management, nature conservation, agriculture, policy, and tourism) due to their individual local and sector-specific knowledge, and contributed to the development of concrete and action-oriented suggestions on local and regional levels towards sustainable land use management.

In this framework, the Social Impact Assessment was used as a bridging concept to make the ES approach more accessible by overcoming the communication gap between scientists and local actors (Section 5.2). The framework's application to a participatory case study demonstrates how the framework promotes the incorporation of ES into participatory spatial planning activities and strategy development processes (Section 5.3). This chapter concludes with the statement that the operationalization of the ES approach and Social Impact Assessment clarifies the strong linkages between social demands and ecological supplies (Section 5.4). Therefore, the Social Impact Assessment is a tangible concept as a bridge to access the ES approach, making it usable for adaptive, sustainable and ecosystem-based land use management.

5.2 A CONCEPTUAL FRAMEWORK – FORCING ECOSYSTEM-BASED PARTICIPATORY ADAPTIVE STRATEGY DEVELOPMENT BY MATCHING ECOSYSTEM SERVICES AND SOCIAL IMPACTS

The aim of the conceptual framework is to promote human well-being and sustainable development in participatory regional land use management and adaptive strategy development processes. The work with a diverse group of stakeholders is a challenge. Every stakeholder has a different background and a different knowledge base. Depending on his or her expertise, knowledge creation leading to ecosystem-based land use management is necessary. Therefore, a participatory planning process is needed that integrates the concepts of the ES approach and social impact analysis as a basis for the development of regional adaptive land use management strategies. Participation refers to possibilities set up by scientists, here termed knowledge broker, for stakeholders to express their knowledge, experiences, preferences and needs, and to achieve better development outcomes for the affected region (Salter et al., 2010). The knowledge broker is acting at the interface between science and practice. The task is to moderate and guide the participatory process, and to link scientific and stakeholder knowledge. The objective is to integrate all interested and affected parties (e.g., Pielke, 2007).

In order to reach the goal of sustainable and adaptive development, societal and environmental concerns have to be balanced. Recently, the international debate on integrating ES in impact assessments, especially Strategic Environmental Assessments (e.g., Geneletti, 2011, 2013; Partidário and Gomes, 2013; UNEP, 2014) and Environmental Impact Assessments (e.g., Karjalainen et al., 2013), gained more attention.

Even though there is an increasing awareness about including ES in management processes to support spatial planning and development practices and policies, there is still a lack of emphasizing the social demands of communities. There is a need for enhanced knowledge and integrated understanding of the complex interrelationships within specific social-ecological systems in order to make the concept of ES relevant and applicable to practical cases.

In this framework, the ES approach is used as a strategy to integrate natural supplies – the ES – into decision-making processes. Since ES provide products, regulate natural processes, enrich cultural activities, and support processes within ecosystems, they are regarded as the core of environmental issues in strategy development and spatial planning activities. On the process level, the ES approach is designed to advise decision-makers and to raise their awareness concerning the linkages between changes in ecosystems and human well-being. It helps to understand trade-offs concerning the environment, enhances effective response options on a governance level, and integrates land, water and resource management (MA, 2003).

In addition to ES, human well-being depends on the demands, perceptions, interests, aspirations and needs of society (Burdge, 2003; Vanclay and Esteves, 2011). While the ES approach focuses on people's dependency on the ES, the Social Impact Assessment identifies, analyzes, and evaluates the social impacts of projects on people and focuses the process of managing social issues. It provides a qualitative evaluation of social impacts as a tool for decision-makers and planning processes and triggers adaptation measures (Burdge, 2003). The Social Impact Assessment is an adaptive management process aiming for the development of a knowledge base to determine possible projects' impacts on a community, increasing understanding of how an intervention will affect a community (capacity building), provisioning a qualitative evaluation of social impacts as a tool for decision-makers, and the proposal of alternatives to guide adaptive governance and management processes to maximize the positive outcomes of interventions (Vanclay and Esteves, 2011).

The ES approach and Social Impact Assessment are essential strategies to promote sustainable development. The conceptual framework, based on a participatory process, clarifies the relationships between the community's demand side (referred to as constitutive and indicative social impacts) and the ecological supply side (ES) to guide adaptive governance processes. We propose a classification of ES along functional lines according to the Millennium Ecosystem Assessment Report (MA, 2003) in provisioning, regulating, cultural and supporting services. Based on Vanclay (2002), categories of social impacts have been developed according to personal, development, economic, climate change, landscape, and ecological impacts.

Figure 5.1 shows the aggregation of ES, indicative social impacts and constitutive social impacts, embedded in spatial planning activities and strategy development. In general, the framework suggests that constitutive impacts should be obligatory for the strategy development of spatial planning activities. These impacts exert influence on more than one category of ES and are relevant for the scope of the whole planning process, while direct indicative social impacts influence and depend on specific ES.

In this framework, 'constitutive personal impacts' describe the individual context of the stakeholders and influence the way decisions are taken. Key elements are the attachment to place, obligations to ancestors and traditions, the possibility to be actively involved in decision-making processes, the positive outcome of actions that could be taken, and the general trust in policy and political decision-makers.

The 'constitutive development impacts' focus on the relevance of impacts due to new developments. Crucial are the expectations of the stakeholders concerning the actual need for adaptation measures, the positioning of possible adaptation measures, avoidance of conflicts in different land uses and compliance with property rights and values as well as the need for a long-time view in decision-making.

All these impacts together deliver a frame for adaptive management processes taking the stakeholder's relationships, concerns and prospects into account. Nevertheless, the indicative impacts describe the stakeholder's demands on a more detailed level. A categorization into economic, climate change, cultural, and ecological impacts is possible (based on

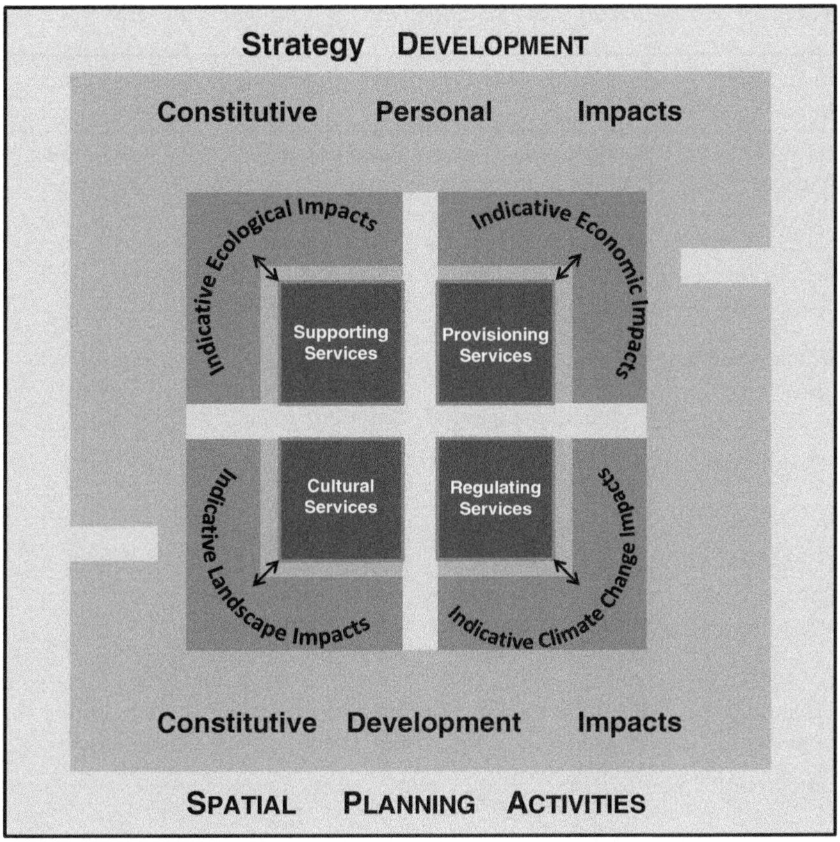

Source: Modified after Karrasch et al. (2014).

Figure 5.1 Conceptual framework to include the ES approach into spatial planning activities and strategy development

Vanclay, 2002). This categorization is necessary to combine the different aspects of stakeholder demands with the four different categories of ES (Table 5.1). A categorization, in general, helps to structure the management process and to focus on specific elements (Karrasch et al., 2014). The direct, indicative social impacts influence and depend on one category of ES (Table 5.1). We have to keep in mind that social impacts describe feelings, aspirations and demands of the community, whereas the ES are the natural resources. For example, indicative economic impacts – like the prosperity of the region – relate to social impacts in terms of economic activities and the material well-being of the community. They influence

Table 5.1 Relationships of exemplary indicative impacts and ES

Indicative Impacts	Social Demands	Ecological Supply	Ecosystem Services
Economic impacts	Security of income Economic prosperity of the region Maintenance of agricultural land Security of drinking supply	Food production Forage production Green energy production Fresh water for agriculture and livestock	Provisioning services
Climate change impacts	Combat negative influences of climate change Perception of safety	Prevention of saltwater intrusion Wave attenuation Hazard regulation by water retention Reduction of greenhouse gases	Regulating services
Landscape impacts	Esthetic quality, beauty of landscape Presence of tourists Cultural-historical values	Community identification and connection Recreation, tourism	Cultural services
Environmental impacts	Beauty of nature Organic management, sustainable use of resources	Biodiversity Soil fertility	Supporting services

Source: Karrasch et al. (2014).

and depend on provisioning ES – the products people obtain from ecosystems – like food production. Provisioning ES form the basis of the regional economy and describe the supply side (see also Burkhard et al., 2012). The supply is related to the demand, which is determined by the indicative economic impacts. This example in terms of supply and demand is representative for the other ES categories.

Table 5.1 gives examples of how the demands of society and stakeholders are matched with ES, and thus the supply nature provides for the people. As described in Vanclay (2002), indicative economic impacts refer to the material welfare of the community and its inhabitants. These are directly interlinked with provisioning services, which include the products people obtain from ecosystems (MA, 2003). The indicative economic impacts emphasize the importance of strongly supply-directed needs and therefore provisioning services.

Climate change impacts are described by the stakeholders' need to combat negative influences of climate change and to ensure the perception of safety. These impacts are matched with regulating ES, assigned to the regulation of ecosystem processes. An increase in such services, like wave attenuation provided, for example, by salt marshes, or hazard regulation by water retention or polder areas, might also lead to a positive increase in climate change impacts.

The quality of the living environment, including cultural-historical values, esthetic appeal, the beauty of landscape, as well as the presence of tourists, is dependent on more non-material benefits of the ecosystems (ibid.). Such cultural ES affect individual behavior with regard to community identification and opportunities for recreation and cultural development are connected with the indicative landscape impacts focusing on the community's quality of daily living and supporting cultural identity.

Furthermore, an increase in the beauty of nature, organic agricultural management, and the sustainable use of resources is desirable. These aspects are in the opinion of the stakeholders determined by a huge variety of different plant and animal species and the fertility of soils. These supporting services, also responsible for ecosystem resilience, are connected with the environmental impacts described by the stakeholders' feelings and aspirations concerning environmental issues.

The integration of social impacts and ES contributes to operationalizing the ES approach and including it in adaptive strategy development and spatial planning activities. The framework assists in understanding the stakeholders' demands, interests and needs. It summarizes the relationships between people and the environment by focusing on the interlinkages between societal demands and environmental supply. The framework structures and organizes such complex interactions in order to guide an adaptive governance approach for sustainable development and enhances the communication between researchers and practitioners.

5.3 APPLICATION TO A CASE STUDY – STAKEHOLDER PARTICIPATION IN INTEGRATIVE PLANNING AND ASSESSMENT

This section presents and discusses how the framework was applied to a case study. Subsection 5.3.1 introduces the case study region and possible land management scenarios. Subsections 5.3.2 and 5.3.3 describe the participatory process and the methodological process necessary to develop the framework. Subsections 5.3.4 and 5.3.5 present the practical application of the framework, give examples of how ES are matched with social impacts, and delineate practical implications of the framework.

5.3.1 Case Study Region

The case study area (Figure 5.2), namely the community of Krummhörn, is located at the Ems-Dollart estuary (north-western German North Sea coastal region). Similar to the surrounding areas, vast parts are below sea level and vulnerable to the impacts of climate change and socio-economic change processes. The community has to face these two major drivers of adaptive land use management and there is still a need for the development of alternative strategies to cope with the future challenges.

The first driver, climate change, is characterized by primary effects like changes in temperature (Beniston et al., 2007), sea level and storm events (Gaslikova et al., 2013), and precipitation (Jacob et al., 2008). The low-lying coastal area has to deal with increasing inland water levels leading to more frequent flood events, increasing pumping costs, or the salinization of groundwater, to name some possible impacts of climate change. In addition to these points, socio-economic change processes relating to secondary effects relate to regional spatial planning activities, like changes in population structure (decline), economic growth (increasing vulnerability), and land use (scarcity of space and resources).

Contemporary coastal land management concentrates almost exclusively on the protection against floods by dikes and groundwater regulation via drainage. The rising sea level and changes in hydrological cycles may restrict the efficiency of these measures in the future, including the effectiveness of current land and water management (Ahlhorn, 2009; Bormann et al., 2012). For this reason, adaptive strategies for future land management have to be developed.

The joint research project COMTESS (Sustainable Coastal Land Management: Trade-offs in Ecosystem Services) aimed at developing action and decision-oriented suggestions on local and regional levels to promote sustainable management of vulnerable coastal landscapes in response to expected climate and land use change. Therefore, different land management scenarios have been developed to design different narratives of possible future developments providing a discussion basis and framework with which to face future challenges (Figure 5.3). These pathways were based on the present land management of the community of Krummhörn. Based on the scenarios, it is possible to define consequences for planning and actions in regional land use management.

The 'trend' scenario focused on business as usual. The region of

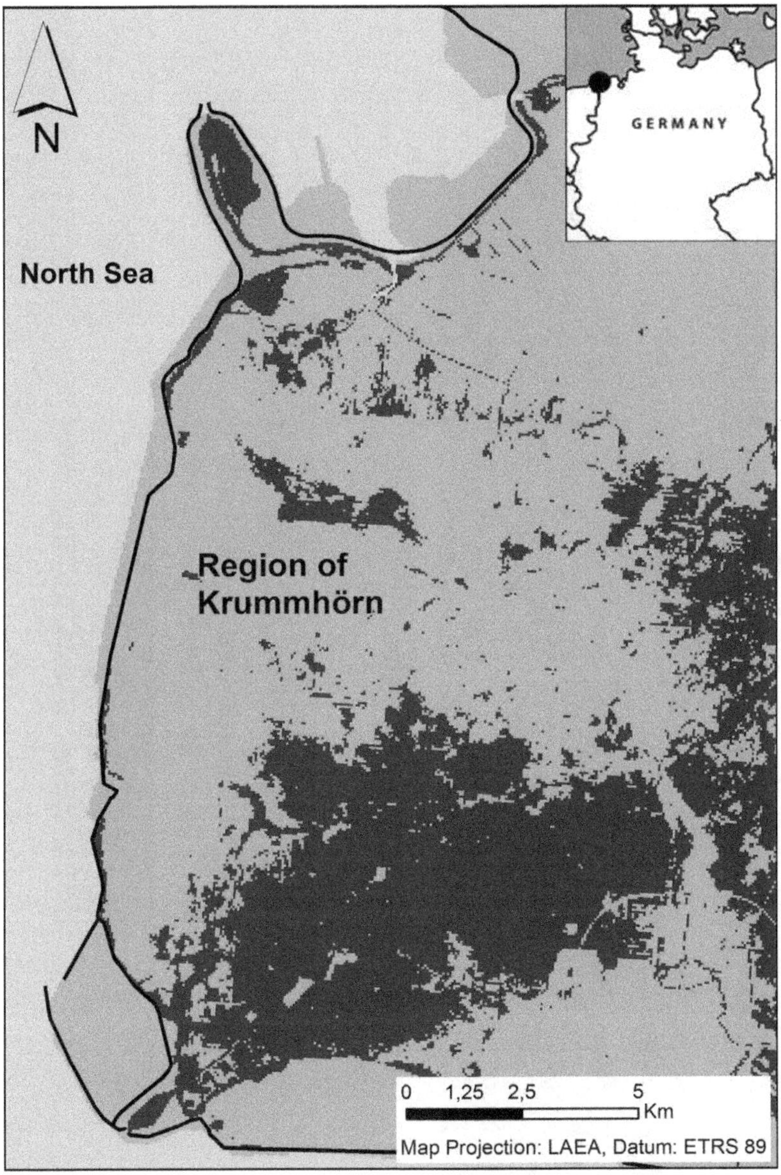

Note: Vast areas of the community are below sea level (dark gray) and the whole region is embanked by a dike line (black line) to prevent flooding.

Figure 5.2 The low-lying North Sea coastal region of Krummhörn is located in the north-western part of Germany

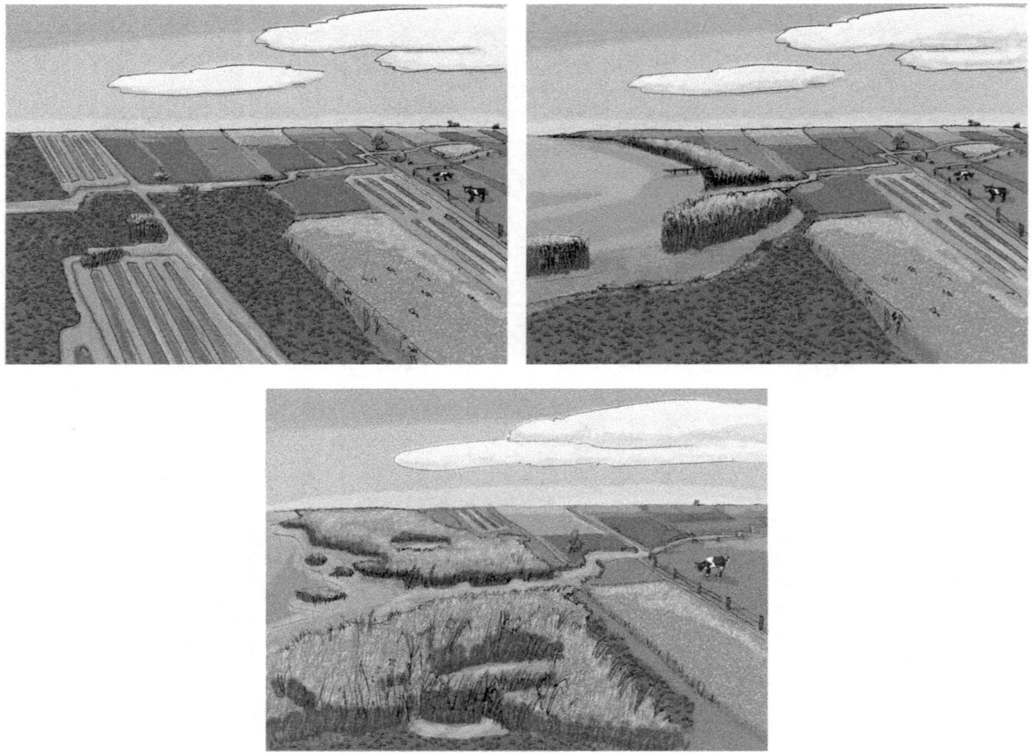

Figure 5.3 *Visualization of the three land management scenarios. Left: trend scenario, right: water management scenario, bottom: carbon sequestration scenario.*

Krummhörn was primarily used for dairy and arable farming. This primary use was complemented by nature conservation and tourism. All activities were expected to continue in the future. Flood events could substantially disrupt the contemporary land use in the future and lead to higher drainage costs, which will probably reduce the ecological and economic viability of current land management practices.

The two alternative scenarios, called 'water management' and 'carbon sequestration', offered a pool of ideas for how to cope with and adapt to the challenges of climate change and socio-economic change processes. The water management scenario focused on the establishment of polder areas. Open waters and extensive reed stands will be created. This scenario primarily aimed to restrict floods to the polder area in the case of increasing precipitation and limited dike overtopping or breaches, using the retained water in dry periods for irrigation measures and the reeds for generating green energy. Additionally, subsurface saltwater intrusion in the area could be prevented.

The carbon sequestration scenario was similar to the former one, but the rewetted polder areas will be extensively covered by reed fens instead of open water. The aim was to yield active peat formation to store atmospheric CO_2. This corresponded to carbon-optimized land management. Additionally, former natural vegetation could be restored.

5.3.2 Participatory Process

In order to develop adaptive strategies and spatial planning activities taking socio-economic and ecological impacts into account, the application of adaptive management (deliberation) and governance (stakeholder engagement) processes was advisable (see Nyberg, 2009). One cornerstone of the research project COMTESS was the active involvement of local stakeholders (here referred to as sector representatives). All affected sectors, namely water management, nature conservation, agriculture, tourism and policy, were included and were represented by 14 local and regional decision-makers (see Karrasch et al., 2014). Stakeholder engagement (see Tompkins et al., 2008) was an important tool for implementing sustainable development and formalizing the link between cross-sectoral interests towards consensus-based decision-making.

Such a governance process included continued development based on sequences of consecutive steps and interrelated results. The sector representatives were actively involved in developing future visions and aims. It was important to involve them from an early stage. Therefore, the moderator, referred to here as a knowledge broker, initiated discussions intended to come up with sound, comprehensible, and desirable results based on the sector representatives' knowledge and experiences.

The inclusion of local knowledge, experiences, and perceptions supported the acceptance and legitimation of new planning activities and adaptation strategies. Since participatory processes are mostly voluntary activities, motivation was fundamental. Therefore, it was important that the knowledge broker ensured that the group of the sector representatives was heterogeneous and balanced, that all interests were taken into account, and that decision-making processes were transparent.

Furthermore, practical applications have shown that the sector representatives preferred, for instance, visualizations to define actual situations. Pictures are adequate tools to explain land management opportunities in combination with concise and coherent sentences connected to emotions (Trumbo, 2000). Therefore, scenarios were used (see Section 5.3.1) as a visual communication tool. Additionally, an ongoing information process and feedback was very important to avoid frustration and misunderstandings during the whole process. The participatory process enabled the sector representatives to engage with possible problems and opportunities in spatial planning activities and adaptive strategy development. The increase in responsibility and trust of the sector representatives resulted in a corresponding increase in positive engagement and commitment. Another advantage was that conflicting goals and trade-offs were recognized at an early stage and cooperation between the different interests was promoted. Due to participatory governance processes, a shift from more technical (scientific) solutions towards more flexible results and solutions focusing on the needs, knowledge, and visions of the different sector representatives was possible. This promoted innovative ideas and partnerships among them. Examples of such rethinking of technical flood protection measures (like strengthening the pumping stations vs polder areas for water retention) were provided by the land management scenarios (Section 5.3.1).

The knowledge broker used the proposed three land management scenarios as a basis and starting point for the sector representatives to investigate their opinions, preferences, interests and needs. The aim was to introduce the land management scenarios in order to examine the effects of established and innovative measures to promote possible spatial planning activities and adaptive strategy development.

5.3.3 Methodological Process

This chapter places its emphasis on three management elements according to Social Impact Assessment (Taylor et al., 2003). The first element is the scoping process, which deals with the identification of key issues, possible impacts, and stakeholder analysis. Second, profiling gives an overview and analyzes the social context and trends. The third element is the formulation of alternatives by examining and comparing potentials for change.

The methodological process behind the proposed conceptual framework started with scientific-based input and moved towards intensive stakeholder engagement (Figure 5.4). The framework distinguishes two main actors: (1) the knowledge broker (i.e., scientists, responsible for the science-based input) and (2) experts (representatives of relevant sectors of the society, stakeholder-based input).

The first step, called setting, included (1) the assessment of investigation units and background data, (2) categorization of ES and social impacts based on theoretical investigations and (3) a stakeholder analysis. This step of the framework was 'science driven'. The input of the sector representatives within this step was rather low. In this process, the data gathering concerning the case study region by the knowledge broker was important. Reviewing background data concerning the selected case study area and defining the scope of possible spatial planning activities in order to develop adaptation strategies was a main task of the knowledge broker. In addition, to prepare the participative stakeholder process and to structure the results of the interviews, categorizing expected ES and social impacts of possible interventions was required. It was important to work with open categories, because the applied participatory approach in the following steps might lead to adjustments, and addition or removal of ES or social impacts depending on the local knowledge of the sector representatives. Furthermore, a stakeholder analysis needed to be conducted. For the whole process, it was important to include all relevant actors of the study region to gain a holistic overview of all the different interests.

The next step was the profiling. Here, semi-structured interviews delivered the data for the following data selection and analysis. The questionnaire addressed several topics, including the personal background of the interviewee, his or her understanding of sustainable development and the evaluation of the different scenarios (Section 5.3.1). It was the task of the sector representatives to select and define region-specific ES. The knowledge broker provided examples of ES. Additionally, data concerning sector-specific aspirations, desires and interests were gathered. The profiling had a high stakeholder-based input, but it was the task of the knowledge broker to analyze the interview data. The theoretically developed categories of social impacts and ES served as substructure and guidelines for the data analysis process. The ES and social impacts considered as relevant by the sector representatives were used as keywords to quantify the results of the interviews and to become indicators in the next step.

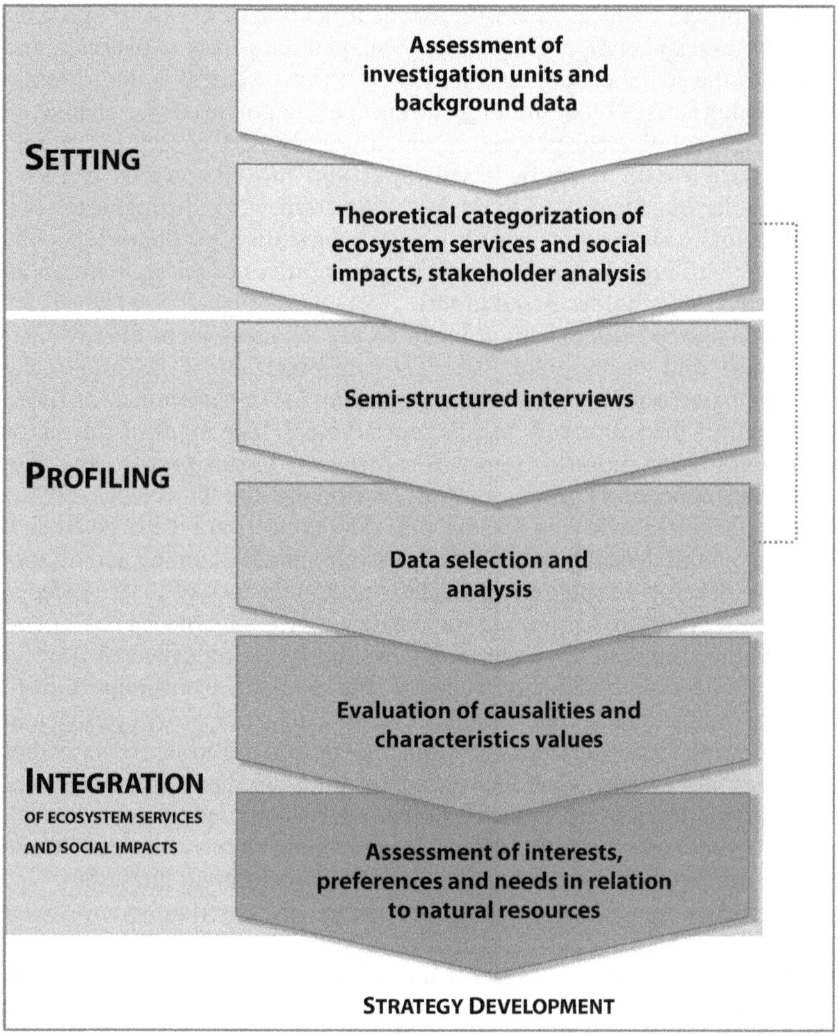

Source: Karrasch et al. (2014).

Figure 5.4 Methodological process scheme: the stakeholder engagement increased step by step

This third step was described by the integration of ES and social impacts. By doing an evaluation of causalities and characteristic values described by the sector representatives (semi-structured interviews), it was possible to make an assessment of their interests, preferences and needs in relation to natural resources and, therefore, ES. In this step, the stakeholder input was very high and the knowledge broker's task was to structure the results.

The following section provides an insight into the practical application of this method and gives many examples of the linkages between ES and social impacts.

5.3.4 Practical Application – Matching Ecosystem Services with Social Impacts

The investigation showed that sector representatives can deal more successfully with socio-economic terms than with the concept of ES. One important reason is that the sector representatives are not trained to think in complex terms of ES and the approach is not part of their everyday activities. The use of social impacts leads to a better understanding of the ES approach as foundation for planning processes (Karrasch et al., 2014).

Therefore, the application of the Social Impact Assessment provided some advantages. For instance (the enhancement of) positive and (the reduction of) negative impacts concerning spatial planning activities and strategy development were analyzed and assessed. The adaptive management process focused on stakeholder engagement, including local knowledge, experiences, preferences and needs. In particular, the stakeholder-driven definition of social impacts and ES raised awareness concerning possible interventions or impacts of spatial planning activities and strategy development.

The defined criteria describing each social impact (which are used as indicators to analyze the data) are regionally specific and were used as examples of how social impacts are defined and perceived. Furthermore, the investigation showed that ES, which were not necessarily understood or applied by every sector representative, were sometimes part of social impacts (see examples below). This promoted the interlinking of social impacts and ES and it confirmed that they are inherently connected.

The criteria for social impacts were dependent on regional characteristics. In the case of a vulnerable low-lying rural coastal region, the main focus was on maintaining the economic activities and related typical landscape features within the community.

Constitutive social impacts – foundation for strategy development

Regarding the 'constitutive personal impacts', the investigation showed that 11 of the 14 sector representatives have spent almost their entire lives in the community of Krummhörn. Therefore, their identification with the community and the attachment to place was high. The sector representatives were rooted in East Frisia and showed a strong sense of place and regional belonging (e.g., Paasi, 2003). In addition, the 'obligation to ancestors' was emphasized. History and traditions played a significant role in the cultural and societal community life of the area. Centuries ago, and up to the present day, the East Frisian inhabitants reclaimed land from the sea and formed today's geographical extension. The 'fight against the water', especially storm surges, confronted the people with many challenges (Lamb, 1991). For this reason, the sector representatives feared that research projects might reconsider existing flood protection measures in favor of a retreat approach, giving land back to the sea. Another indicator of tradition was the East Frisian dialect, which is regularly spoken by locals. Active participation in decision-making was seen as obligatory for further spatial planning activities and strategy development and highly important to all but one of the sector representatives. Positive outcomes of potential change processes were required from all participants, even though some sector representatives had little trust in politicians because they feared the potential negative effects of policy-making.

The category 'constitutive development impacts' relates to expected effects of changing land use as well as the perceptions and aspirations of the sector representatives concerning new strategies and new developments in general. All sector representatives saw the need for adaptation measures concerning climate change and they had positive aspirations towards them. In general, there was uncertainty about the impacts on populated areas and impacts on agricultural areas caused by climate change. Two sector representatives shared the opinion that populated areas would suffer from climate change (policy and nature conservation sector). One sector representative saw negative effects for agricultural areas (nature conservation sector), whereas another one perceived that the effects of climate change were not relevant and might even be positive for the region (agricultural sector).

The positioning of 'new land use strategies and measures' played another overarching role. It was important for the sector representatives that all interests and needs were considered so that they could play an active role in decision-making, and so that the processes were transparent and new measures were not established frivolously. Today, there are already competing land use activities and the sector representatives expected that this would increase in the future because land is a finite resource. Resettlement and expropriation caused by new developments were sensitive topics, and in the opinion of the sector representatives, planning activities and strategy development should be in compliance with property rights and values. In addition, every sector representative expressed the fear of losing land and perspectives, and they considered this an important issue. In their opinion, these social effects of an intervention were negative. It was important for everyone to engage in long-term planning while thinking about the development of new, adaptive strategies as opposed to short-term solutions that could be counter-productive in the long term.

Taking the definition of ES into account (benefits people obtain from ecosystems and a basis for human well-being), it becomes clear that constitutive social impacts influence and depend upon provisioning, regulating, cultural and supporting services. Going one step further, the indicative social impacts showed more direct linkages with specific ES.

Indicative social impacts – the link to ecosystem services

The category 'indicative economic impacts' related to social impacts in terms of economic activities, material well-being and the welfare of the community. The economic prosperity and resilience of the region combined with regional marketing as well as the security of (future) income were very important aspects and regarded as positive by most of the sector representatives. These indicative impacts directly related to agricultural production. Without using arable land and grassland for food and forage production, the economic impacts on the society would be huge, as farming is one of the main income sources in the region. The maintenance of agricultural land and grassland was important, especially for the agricultural sector, since it enhances economic prosperity, landscape value and the traditional land use of the region. Food, forage and green energy production are important ES and the sector representatives highly emphasized the security of food and drinking water supply. Only the nature conservation sector did not emphasize the necessity of food production. The others feared that food production would not be sufficient in the future. The argument referred to most often was the intended increase in the (world) population and the correspondingly increasing demand. Due to the fertile soil conditions, the Krummhörn region is ideal for the production of food. Almost all sector representatives emphasized the competing demand between the production of biogas and food.

The increasing cultivation of maize for the generation of bio-energy was regarded as problematic. Green energy production using crops, especially maize, was a good example of a gap between its relevance and the perceived social impacts. The sector representatives stressed that green energy production was of high relevance and expected to increase in the future. Even though wheat and other crops were usually cultivated in monocultures, not all sector representatives approved of the development of maize monocultures. Although maize provides an important energy source, it took up arable cropland. Furthermore, the cultivation of maize was associated with the intensive use of fertilizers and pesticides. The quality of the soil, biodiversity, and water management were diminished and damaged.

Sector representatives from the agricultural sector emphasized that the cultivation of maize was of low future relevance for the case study region. The reason is that the region of Krummhörn has heavy, wet soils, making harvesting (depending on the weather conditions) almost impossible. The sector representatives thought critically about the establishment of more biogas plants. The input was out of all proportion to the output and the biogas plants were seen as a 'promoter' of climate change (nature conservation sector). They mentioned that politicians set the wrong goals. In their opinion, biogas plants would not be profitable without subsidies. The political situation led to a distortion of competition between the different forms of cultivation (agricultural sector). The cultivation of reeds for bioenergy or industrial use was seen as problematic (proposed in the alternative land management scenarios, Section 5.3.1). The sector representatives outlined that new special-purpose machines and a new infrastructure would be necessary for the harvesting. This related to the investment of money. The feasibility of using reeds for green energy generation was regarded as low. The sector representatives did not believe that the energy content provided by reeds was sufficient for biogas plants. Additionally, the nature conservation sector concluded that the use of reeds would only be possible with a guarantee that birds would be protected.

The sector representatives who saw the relevance for the freshwater retention for irrigation measures and drinking water supply argued that it might only be useful in the distant future. Because in some German areas water is already a limiting factor, the retention of freshwater might be a forward-looking instrument. Some sector representatives from the nature conservation sector emphasized that stored freshwater used for irrigation would be more sustainable than the use of groundwater. In addition, the retained water could be used for drinking water abstraction. In the region of Krummhörn, the marshes retain a lot of water and have the capacity for more water retention (water management sector). During dry periods, the water level has to be regulated or remains consistent due to water inputs from treatment plants or natural springs, for example. In case of an increasing groundwater level, the main water users have to stop their water abstraction and the sewage needs to be adjusted (water management sector). Some sector representatives from water management, agriculture and policy thought that drier periods would have positive effects at least with regard to agricultural land use. Those sector representatives from the nature conservation and tourism sectors who saw no relevance for freshwater retention agreed in that they did not anticipate increasing droughts or overwhelming costs in the construction of an irrigation system.

The 'indicative climate change impacts' contained two main aspects highly relevant for all sector representatives. First, to combat negative influences of climate change, for example increasing sea and inland water levels, dike overtopping, or droughts. Due to changes in current land use (proposed by the alternative land management scenarios, Section 5.3.1), possible negative impacts of climate change might be reduced, and at the same time the feeling of safety could be strengthened. Second, the perception of safety was described as the people's feeling of hazard exposure and personal safety. Changes in land use, both technical and non-technical, might lead to a decreasing risk perception. Regarding ES, especially hazard regulation by water retention, wave attenuation and prevention of saltwater intrusion was relevant but not widely recognized by the sector representatives. The development of water retention areas for inland floods, especially important during extreme weather events, and saltwater intrusion due to dike overtopping, was only seen as relevant by the water management sector. In the case of changing land use conditions, the sector representatives doubted that the development of a polder area to store water would have negative influences on current land uses, especially agriculture. They feared that their living conditions would suffer as a result. The existence of vegetation structures like salt marshes supported wave attenuation and promoted coastal protection. The benefit of this service, even though it is directly related to the social impacts of climate change, was only recognized by the agricultural sector. Only the prevention of saltwater intrusion using freshwater polders to decrease the groundwater level and increase the hydrostatical pressure was recognized by the sector representatives from water management, agriculture and policy.

The sector representatives in general felt strongly that it is necessary to distinguish between short-term and long-term expectations regarding the perception of safety and need for adaptation measures. They did not expect a great danger within the next 50 years. The water management sector in particular emphasized that if current coastal protection measures were continued, safety would be ensured. Alternatively, strengthening green dikes and increasing soft protection measures, concrete, asphalt layers or sheet pile walls were regarded as possible 'hard' adaptation measures in case of increasing hazard exposure (except from the nature conservation sector). Only four sector representatives thought that polder areas would be useful to relieving flooding (water management, agricultural and policy sector). The sector representatives did not think that rising costs would have a negative effect on the willingness to pay for flood protection measures. They argued that the people prefer safety and therefore they are willing to pay for it. If they refused, they would have to leave the region.

Regarding 'indicative landscape impacts', the esthetic quality and beauty of the landscape was precious. The residents of the area identified themselves with the landscape. New developments changing visual impressions would lead to changes in well-being and sense of place. A visible change in the landscape might lead to an 'alienation of the culture' (nature conservation sector). The esthetic quality of the landscape was of crucial importance to the maintenance of a livable environment. The sector representatives from the agricultural, political and tourism sectors emphasized that East Frisia is a human-made landscape and not 'natural'. Important elements of the landscape were the diversity of different land uses, the pastoral, traditional, open and idyllic character, as well as the terp[1] villages (cultural-historical value) representing the typical East Frisian landscape.

The maintenance of the esthetic value was of high importance, not only for community identification and connection but also for recreation and tourism. In the opinion of the tourism sector, the farmers contributed notably to the maintenance of the landscape. Changes within the landscape have to be planned carefully. Water is an important element of the landscape. The increase of water areas in the landscape (alternative land management scenarios, Section 5.3.1) was regarded with skepticism. On the one hand, water was an important element for tourism. The establishment of water areas offered new opportunities like more residences, fishing, water sports, and camping and might enhance the value of the landscape as well. On the other hand, new water areas might negatively influence the current landscape by reducing agricultural areas (policy sector). The presence of tourists and recreation opportunities depends on the landscape and the community's unique ecosystems. The tourism sector described tourism as 'white industry'; it is an important branch of the economy and maintained employment as well as income rates. For the community, tourism is expected to increase in the coming years. Additionally, increasing leisure residences, cultural entertainment, and activities such as biking, boat trips and health tourism were favored by the tourism sector. Nevertheless, in the opinion of some sector representatives, especially nature conservation, the presence of tourists might disrupt daily life and influence the landscape and environment negatively. Due to changing land use, the attractiveness of the landscape might decrease.

In general, it was emphasized that the consequences of a uniform landscape would be negative. All sector representatives preferred mosaic structures with a diversity of land use elements, without extreme forms of one land use. In particular, the tourism sector worked with clichés and typical regional images like grazing cows, water, and coastal areas, as well as the pastoral, traditional, structured, and idyllic character of the community of Krummhörn. So far, sustainable tourism (e.g., Butler, 1999) was only a small part of the community's tourism strategy and not economically valuable (tourism sector). The sector representatives would like to see an expansion of sustainable tourism and hoped that the 'quality increases and not the intensity' (nature conservation sector). Sector representatives of all sectors, except for the nature conservation sector, emphasized the need for inclusive approaches to nature conservation that are open to tourism. Closed areas would have negative influences on tourism. Nature conservation and tourism have to be connected by intelligent measures, like guided tours or visual aids explaining the environment.

'Indicative ecological impacts' were described by beauty of nature and organic management and sustainable use of resources. In the sector representatives' opinion, the beauty of nature was highly connected with biodiversity. In their definition, biodiversity was analogous to a high species diversity that improved the visible value of the environment. The sector representatives from water management, nature protection, and agriculture gave statements: new developments should not endanger the habitats of plants and animals. Newly developed areas might lead to an increasing biodiversity and new species could immigrate. All sector representatives were in favor of modes of production that can exist in harmony with nature, accompanied by the inclusion of nature conservation areas in farming activities, and sustainable land use. They also thought of promoting regionally produced products.

5.3.5 Practical Implications

Using the examples of how ES are matched with social impacts (see also Table 5.1), local spatial planning activities and strategy development could be developed in a more ecosystem-based way. An additional outcome of the study is that the sector representatives' viewpoints were not as different as expected. They were representing different sectors of society and had similar expectations and needs in a wide range of areas. A possible explanation for this result is that traditional conflicts and a lack of communication between different sectors, for instance agriculture and nature conservation, had led to prejudices between the sectors. A participatory process (as performed in this investigation) contributed to a better mutual understanding and exchange between the different sectors. Additionally, it led to the building of new partnerships and alliances.

A discrepancy between perceived and actual perceptions on the part of the sector representatives towards each other was observed. A practical example is that one person proposed the establishment of a biosphere reserve in the community. The concept of biosphere reserve is based on strengthening sustainable land use development including communities and sciences. It is an approach that is intended to include nature conservation and socio-economic development as well in the management of human and nature activities, treating them as interacting entities. The main characteristics are the roles of development (linking environment with actions), conservation (ecosystems), logistics (international networks), and on a landscape level, the establishment of core areas (high protection standard), buffer zones (conservation interests), and transition areas (sustainable land management) (Batisse, 1990).

Sector representatives from the other sectors rejected this suggestion categorically. They assumed, for instance, that the establishment of a biosphere reserve would restrict development or new development in the region and only the interests of nature conservation would play a role. During the group meetings, however, the person representing the nature conservation sector had the opportunity to explain the concept behind the term 'biosphere reserve'. The result was a greater understanding and acceptance of this concept on the part of the other sector representatives, thus facilitating continuing inter-sectoral communication and exchange.

Additionally, the stakeholder dialogue enhanced awareness raising with regard to the need for adaptive strategies that led to sustainable development and human well-being, as well as the understanding of what impacts human interventions may have on natural systems. The process of investigating alternatives named by the sector representatives provided options for future change. The sectoral representatives gave statements about their expectations and aspirations regarding the future, complementing the science-driven part of the investigation.

Furthermore, the ES approach has been anchored in the practical planning process by clarifying the demand of the sector representatives tailored to the (natural) supply of resources. Regarding current land use management, the demand for resources is increasing, while the possibilities of supply are decreasing. In order to overcome this problem, a consensus between different forms of land use will have to be found, and approaches for sustainable land use will have to be developed. Scenarios showing possible development possibilities were an adequate tool (see Section 5.3.1). Scenario planning supported decision-making with regard to uncertainty and integrated new strategies in order to justify possible consequences (see Peterson et al., 2003). Furthermore, the use of the social categories and impacts made the scientific language in terms of ES more easily understandable to laypersons and substantially improved the communication process. The sector representatives needed to feel that their opinions contributed to the research process and that they were actively involved.

The practical application of the framework showed that indicative social impacts not only provided direct relationships with ES, but they may also be interlinked with more than one category of ES and evolve an indirect character as well. Therefore, it is advisable to handle social impacts in a more flexible way. One example is the stakeholders' interest and need to combat negative influences of climate change. This social impact was strongly connected with the water levels in the region and directly influenced and was influenced by regulating services. Therefore, interventions (e.g., the establishment of a polder area) to combat the negative influences of climate change indirectly affect land use, especially agricultural production, and linked provisioning ES. Another example is provided by the indicative environmental impacts, which were directly related to supporting services. Changes in land use might increase or decrease biodiversity and soil fertility, leading indirectly to changes in tourism (willingness to observe endangered species) and therefore cultural services.

5.4 CONCLUSION

The framework provides a platform from which to facilitate the incorporation of the ES concept into spatial planning activities and adaptive strategy development, necessary to meet the requirements of ecosystem-based land use management. It contributes successfully to the concept of sustainable development by considering environmental and socio-economic impacts in an integrated way. The interdisciplinary and transdisciplinary framework is helpful to addressing and incorporating the mutual interdependencies of nature and society. It promotes a holistic understanding of change processes, enhancing the interactions between science and society by engaging local and regional stakeholders.

Following a path from social impacts towards ES enables stakeholders to understand the concept of ES. The translation of social demands into an ES approach helps to develop an appropriate communication strategy and raises awareness concerning the importance of ecosystems for human well-being. Stakeholders thinking within social units is common because they identify themselves with social impacts caused by consequences of interventions. The application of the social impacts assessment within the sphere of planning creates possibilities to integrate the ES approach into sustainable planning processes. It provides a way to tailor the scientific language to the different stakeholders and, thereby, improve the embedding of ecosystem knowledge in communication processes.

Since social demands and environmental supplies are interdependent variables, the framework describes a feedback loop as well. From the understanding of ES, there is also the opportunity to understand ES in relation to the landscape and their influence on social aspects by increasing or decreasing a certain service due to adaptive land use management. This improves the stakeholders' understanding of the consequences of strategy development and spatial planning activities. In order to apply the framework in other regions, there is a need for critical reflection and adaptation to the specific regional contexts, creating opportunities for transformations and scientifically sound best practice.

It became clear that it is crucial to always take stakeholders' wishes and needs into account when developing new strategies because their willingness to adopt or implement these innovations increases according to their level of participation. Additionally, the framework creates collaboration processes between researchers and practitioners, supporting joint fact-finding, co-creation of knowledge, and bridge building for the mutual understanding of the issues and challenges to the development strategies for sustainable land use management.[2]

The practical application proved that such a framework could be used to assist planners and researchers in establishing a more integrated and participatory development of adaptation strategies considering the complex interactions between nature and society. The operationalization of the ES approach and Social Impact Assessment demonstrated that social demands and provision of ES are closely interlinked. Social impacts can be used as a 'map' to anchor the ES approach in spatial planning activities and strategy development processes to support regional sustainable development.

NOTES

* This project was part of the collaborative research project COMTESS (Sustainable Coastal Land Management: Trade-offs in Ecosystem Services). The author would like to express her gratitude for the support she has received through Contract No. 01LL0911 of the German Federal Ministry of Education and Research, part of the funding measure 'Sustainable Land Management'. The development of the framework was only possible with the active participation of the local and regional stakeholders. I am thankful for the time and effort they spent during the research process. I would also like to thank Udo Schotten (http://www.kunst.uscho.de/) for drawing the land management scenarios (Figure 5.3), and Christian Aden for helping to design Figure 5.2.

1. An artificial dwelling hill or mound built high for safety from floods in periods of rising sea levels. The original word for terp is warft or wurt, meaning village in old Frisian.
2. Two years after this study (but with ongoing working processes with the same group of decision-makers, in total 4 years) the results of the whole participatory process have been implemented in official and formal decision-making. The draft version of the spatial regional plan of the county of Aurich, to which the Krummhörn community belongs, emphasizes that a major task of the regional planning is to secure the necessary spatial requirements to deal with increasing sea and inland water levels. Therefore, in a low-lying part of the Krummhörn, priority areas for water retention have been designated to ensure a sustainable and adaptive land use management.

REFERENCES

Abson, D.J., H. von Wehrden and S. Baumgärtner et al. (2014), 'Ecosystem services as a boundary object for sustainability', *Ecological Economics*, **103**, 29–37.

Ahlhorn, F. (2009), *Long-term perspective in coastal zone development: multifunctional coastal protection zones*, Berlin: Springer-Verlag.

Batisse, M. (1990), 'Development and implementation of the Biosphere Reserve concept and its applicability to coastal regions', *Environmental Conservation*, **17**(2), 111–16.

Beniston, M., D.B. Stephenson and O.B. Christensen et al. (2007), 'Future extreme events in European climate: an exploration of regional climate model projections', *Climatic Change*, **81**(Supplement 1), 71–95.

Bormann, H., F. Ahlhorn and T. Klenke (2012), 'Adaptation of water management to regional climate change in a coastal region – hydrological change vs. community perception and strategies', *Journal of Hydrology*, **454–455**, 64–75.

Burdge, R.J. (2003), 'The practice of social impact assessment – background', *Impact Assessment Project Appraisal*, **21**(2), 84–8.

Burkhard, B., F. Kroll and S. Nedkov et al. (2012), 'Mapping ecosystem service supply, demand and budgets', *Ecological Indicators*, **21**(1), 17–29.

Butler, R.W. (1999), 'Sustainable tourism: a state-of-the-art review', *Tourism Geographies: An International Journal of Tourism Space, Place and Environment*, **1**(1), 7–25.

Ciriacy-Wantrup, S.V. (1952), *Resource Conservation: Economics and Policies*, Los Angeles, CA: University of California Press.

Costanza, R., R. d'Arge and R.S. de Groot et al. (1997), 'The value of the world's ecosystem services and natural capital', *Nature*, **387**(6230), 253–60.

Daily, G.C. (ed.) (1997), *Nature's Services: Societal Dependence on Natural Ecosystems*, Washington, DC: Island Press.

Fisher, B., R.K. Turner and P. Morling (2009), 'Defining and classifying ecosystem services for decision making', *Ecological Economics*, **68**(3), 643–53.

Fürst, C., I. Grabemann and N. Groll (2013), 'Integrated land use and regional resource management – a cross-disciplinary dialogue on future perspectives for a sustainable development of regional resources', *Journal of Environmental Management*, **127**(Supplement), S1–S5.

Gaslikova, L., I. Grabemann and N. Groll (2013), 'Changes in North Sea storm surge conditions for four transient future climate realizations', *Natural Hazards*, **66**(3), 1501–18.

Geneletti, D. (2011), 'Reasons and options for integrating ecosystem services in strategic environmental assessment of spatial planning', *International Journal of Biodiversity Science, Ecosystem Services & Management*, **7**(3), 143–9.

Geneletti, D. (2013), 'Assessing the impact of alternative land-use zoning policies on future ecosystem services', *Environmental Impact Assessment Review*, **40**, 25–35.

Interorganizational Committee on Guidelines and Principles for Social Impact Assessment (ed.) (1995), 'Guidelines and principles for social impact assessment', *Environmental Impact Assessment Review*, **15**(1), 1–43.

IPCC (2014), *Climate Change 2014: Impacts, Adaptation, and Vulnerability. Part A: Global and Sectoral Aspects. Contribution of Working Group II to the Fifth Assessment Report of the Intergovernmental Panel on Climate Change*, edited by C.B. Field, V.R. Barros and D.J. Dokken et al., Cambridge, UK and New York: Cambridge University Press.

Jacob, D., H. Göttel and S. Kotlarski et al. (2008), 'Klimaauswirkungen und Anpassung in Deutschland – Phase 1: Erstellung regionaler Klimaszenarien für Deutschland' [Impacts and adaptation in Germany – phase 1: preparation of regional climate scenarios for Germany], *Umweltforschungsamt des Bundesministerium für Umwelt, Naturschutz und Reaktorsicherheit*, Forschungsbericht 204 41 138 UBA-FB 000969.

Karjalainen, T.P., M. Marttunen and S. Sarkki et al. (2013), 'Integrating ecosystem services into environmental impact assessment: an analytic–deliberative approach', *Environmental Impact Assessment Review*, **40**, 54–64.

Karrasch, L., T. Klenke and J. Woltjer (2014), 'Linking the ecosystem services approach to social preferences and needs in integrated coastal land use management – a planning approach', *Land Use Policy*, **38**, 522–32.

Lamb, H.H. (1991), *Historic Storms of the North Sea, British Isles and Northwest Europe*, New York: Cambridge University Press.

Millennium Ecosystem Assessment (MA) (ed.) (2003), *Ecosystems and Human Well-Being. A Framework for Assessment*, Washington, DC: Island Press.

Müller, F. and B. Burkhard (2012), 'The indicator side of ecosystem services', *Ecosystem Services*, **1**(1), 26–30.

Nyberg, J.B. (2009), 'Adaptive management: strategies for coping with change and uncertainty', in K.S. Bawa and R. Seidler (eds), *Dimensions of Sustainable Development Vol. II: Adaptive Management: Strategies for Coping with Change and Uncertainty*, accessed 15 January 2016 at http://www.eolss.net/Sample-Chapters/C13/E1-46B-14-00.pdf.

Paasi, A. (2003), 'Region and place: regional identity in question', *Progress in Human Geography*, **27**(4), 475–85.

Partidário, M.R. and R.C. Gomes (2013), 'Ecosystem services inclusive strategic environmental assessment', *Environmental Impact Assessment Review*, **40**, 36–46.

Peterson, G.D, G.S. Cumming and S.R. Carpenter (2003), 'Scenario planning: A tool for conservation in an uncertain world', *Conservation Biology*, **17**(2), 358–66.

Pielke, R.A. Jr. (2007), *The Honest Broker: Making Sense of Science in Policy and Politics*, Cambridge, UK: Cambridge University Press.

Potschin, M. and R. Haines-Young (2013), 'Landscapes, sustainability and the place-based analysis of ecosystem services', *Landscape Ecology*, **28**(6), 1053–65.

Ruckelshaus, M., E. McKenzie and H. Tallis (2013), 'Notes from the field: lessons learned from using ecosystem service approaches to inform real-world decisions', *Ecological Economics*, **115**(C), 11–21.

Salter, J., J. Robinson and A. Wiek (2010), 'Participatory methods of integrated assessment – a review', *WIREs Climate Change*, **1**(5), 697–717.

Smythe, T.C. and R. Thompson (2015), 'Conceptualizing coastal ecosystem-based management: a mental models approach', *Society and Natural Resources*, **28**(1), 38–56.

Taylor, N., C. Goodrich and G. Fitzgerald et al. (2003), 'Undertaking longitudinal research', in H.A. Becker and F. Vanclay (eds), *The International Handbook of Social Impact Assessment – Conceptual and Methodological Advances*, Cheltenham, UK and Northampton, MA, USA: Edward Elgar Publishing Limited, pp. 13–25.

Tompkins, E.L., R. Few and K. Brown (2008), 'Scenario-based stakeholder engagement: incorporating stakeholders' preferences into coastal planning for climate change', *Journal of Environmental Management*, **88**(4), 1580–92.

Trumbo, J. (2000), 'Seeing science: research opportunities in the visual communication of science', *Science Communication*, **21**(4), 379–91.

UNEP (2014), *Integrating Ecosystem Services in Strategic Environmental Assessment: A Guide for Practitioners*, a ProEcoServ report by D. Geneletti, accessed 16 January 2016 at http://www.ing.unitn.it/~genelab/documents/GuidelineESintoSEA.pdf.

Vanclay, F. (2002), 'Conceptualising social impacts', *Environmental Impact Assessment Review*, **22**(3), 183–211.

Vanclay, F. and A.M. Esteves (2011), 'Current issues and trends in social impact assessment', in F. Vanclay and A.M. Esteves (eds), *New Directions in Social Impact Assessment: Conceptual and Methodological Advances*, Cheltenham, UK and Northampton, MA, USA: Edward Elgar Publishing, pp. 3–19.

6. Economic evaluation of the impacts of transportation infrastructures on ecosystem services*

Léa Tardieu

6.1 INTRODUCTION

Managing land use/cover to fulfill multiple objectives of society is becoming a major challenge to policy-makers. Optimal land management requires a combined consideration of the value of multiple objectives (Polasky et al., 2008). However, some land components, such as the majority of ecosystem services (ES), are not directly traded in markets and lack directly apparent signals of value. As a consequence, the full variety of the impacts of land-use planning choices is difficult to determine. Land-use decisions intended to maximize only observable objectives are likely to promote a decline in ES provision and consequently in human well-being (Nelson et al., 2009; Daily et al., 2011; Polasky et al., 2011). The problem is standard in environmental economics, leading to the non-market valuation techniques that enable the evaluation of non-marketed goods and services.

However, the use of ecosystem service concepts to support real-life decision-making processes is still limited (Laurans et al., 2013). Studies assessing ES remain restricted to the illustration of the importance of preserving ecosystems, assuming that it may have an indirect influence on decision-making (Von Haaren and Albert, 2011; Kumar et al., 2013). According to Haines-Young and Potschin (2010), the main reason that ES are rarely addressed for more practical decision-making is that the currently available models (ecological or economic) are becoming increasingly complex, with little consideration for whether they might be applied to practical cases. The exact opposite reason is also used in the literature to explain this issue: the spatial and temporal dynamics of the services provision and their value still have to be better accounted for to yield more robust results regarding real spatial trade-offs (Nelson et al., 2013). The final reason put forward is that ES valuation is sometimes perceived as not suited to certain types of ES that possess particular characteristics and therefore might be considered not valuable.

The inclusion of ES in spatial planning is, nevertheless, increasingly viewed as a means to create a more rational organization of land uses, to balance demands for development with the need to protect the environment (De Groot et al., 2010; Geneletti, 2011; Polasky et al., 2011). Assessing the impacts of policies or development projects on a wide range of ES can be integrated into more cost-effective policy implementation, in which synergies might emerge and yield win-win situations (De Groot et al., 2010; Geneletti, 2013). The ES assessment is thus intended to explain trade-offs that will enable decision-makers to monitor multiple objectives. However, the inclusion of ES might be expected to influence biodiversity and level of ES, provided that the usefulness of the concept is demonstrated and made operational for practical decision-making.

Here I strengthen the case for the consideration of ES in implementation choices related to transport infrastructure projects. The application of ES mapping in the case of transport planning creates new opportunities to monitor and manage services threatened by conversion. The consideration of ES can be useful at different stages of transportation projects:

- during preliminary studies, to design first avoidance measures;
- during the detailed Environmental Impact Assessment, to compare implementation options, precise avoidance measures and to better locate mitigating measures that can be further tested by cost-effectiveness analysis (Naidoo and Ricketts, 2006);
- during the cost–benefit analysis, to provide indicative evaluations that may help improve guidance of decision-makers and the quality of information for the general public.

Consideration of ES also leads to new challenges beyond the ones in general mapping exercises. New challenges are mainly related to the significant length and linear shape of the infrastructure and to the funnel characteristic of the project process. To be useful to transport planners, particular attention also needs to be paid to the application of methods in the real legal framework, from which they cannot deviate. In this chapter, I try to give an overview of the issues and studies needed to map and manage ES in the particular case of transportation projects.

The remainder of the chapter is organized as follows. Section 6.2 presents a summary of the legal framework of transport infrastructure projects in France. This presentation is made to illustrate how the legal context influences the most appropriate way of incorporating ES into transport project practices. Section 6.3 describes the major challenges that arise during the assessment of ES threatened by this type of project. Section 6.4 presents a real case study (a high-speed rail project in France). Finally, Section 6.5 discusses the results and the possibility of a systematic integration of ES loss in transport project evaluation and concludes on different research avenues related to this topic.

6.2 LEGAL FRAMEWORK FOR TRANSPORTATION INFRASTRUCTURE PROJECTS

In France, transport infrastructure projects are framed according to two different trade-off tools: Environmental Impact Assessment (EIA) and cost–benefit analysis (CBA). At the end of the process, the document prepared for public inquiry is subject to public consultation. It includes the EIA study, the environmental authority notice and the CBA. The public inquiry aims at collecting public opinion and at providing a notice regarding the pursuit of the project (favorable, favorable with conditions or unfavorable). If the notice is favorable, it is designated as a 'project with public utility', and the final implementation option is validated. Details on the content of the EIA and CBA follow.

6.2.1 Environmental Impact Assessment (EIA)

The EIA aims at studying the overall consequences of a project to identify the least harmful implementation option, that is, which minimizes the environmental impacts, and to design environmental measures to implement to the final selected option. In France, several laws frame the process. Since the passing of the 1976 Nature Protection Law (Article 2 of Law No. 76-629 on 10 July 1976), any infrastructure project has to take into account legal environmental measures articulated in three successive measures: avoidance, mitigation and compensatory measures.

Implementation options have to be designed, first of all, by avoiding as far as possible regulated protected areas. Avoidance criteria are diverse and country specific but focus primarily on zones containing endangered species, protected areas (e.g., Natura 2000 sites), biological reserves, and zones containing key ecological connectivity. Further, projects try to avoid particularly sensitive areas in relation to different themes: water resources, natural environment, human environment (urban areas, agricultural areas); landscape and cultural heritage; and finally the impact of noise disturbance and pollution. Sensitivities are represented through qualitative indicators graded from low to strong sensitivities.

Further, project designers have to mitigate the residual impacts by adapting the project characteristics to correct the identified damages. Specifically mitigating measures are wildlife crossing structures or fauna passages to reduce fragmentation, noise barriers to reduce noise disturbance, and fences to reduce fauna mortality. If habitat loss cannot be mitigated, residual impacts have to be compensated for by promoting favorable environmental actions (e.g., species habitat creation and ecosystem restoration) that must be at least equivalent to the loss incurred.

Avoidance measures are particularly used during the stages of selection of implementation options (the screening, scoping and reporting stages). Mitigation measures are designed at the final route option (the monitoring and consultation stage and more precisely specified after the public inquiry). Both measures are developed mainly using geographical information systems (GIS) and by mapping the key components (and sensitivities) of the landscape (João and Fonseca, 1996; Antunes et al., 2001). As the project proceeds, the accuracy of the potential impact identification increases. In practice, the object of the analysis moves from 10–15-km-wide passage corridors during preliminary studies to 1–5-km spindles, and finally to reference routes throughout the selected spindle that are 100 m wide. Hence, the spatial scale of outputs varies from 1:100000 to 1:50000 during the first stages of the project process and down to 1:5000 for the last stages.

Before the consultation stage, an EIA report must be provided, synthesizing all the environmental studies realized from the beginning of the process. The report must provide justifications for all the choices made (choice of development facilities, selection of the option, and all project-related measures). The results are presented through detailed maps describing the different environmental sensitivities. A general mapping of the sensitivities, in which the aggregation of impacts is based on expert judgment, is then presented, and the route option choices are partly justified on this basis. The environmental study is then submitted to the environmental authority. The EIA report and the environmental authority notice are then included in the general document required for public inquiry.

6.2.2 Cost–Benefit Analysis (CBA)

The CBA compares the monetary values associated with the positive and negative effects of the project. It enables decision-makers to determine if a project is acceptable from an economic and social point of view (the opportunity of the project) and to what extent (the profitability of the project). For the different effects to be co-measurable, they are valued in a common unit: money.[1]

Markets generate the relative value of all traded goods and services as relative prices. Prices are therefore very useful in comparing different effects, but they also give some indications of the relative scarcity of goods and services. However, in some cases, market prices are bad indicators of the marginal social cost or benefit, as is the case in the absence of a market (Hanley and Spash, 1993; Pearce et al., 2006). To be accurate the analysis must consider the (non-marketed) external effects of transport projects such as environmental externalities. The main stages described below are common to all CBAs; however, the French legal framework for project evaluations requires the methodological options described.

Scenario and baseline condition for the project evaluation

To monetize the impact of the infrastructure project, a scenario must be defined with regard to the macro-economic framework of the project. Specifically, it must provide the evolution trajectories of the national parameters common to all sectors over the project life span period (demographic and macroeconomic indicators). Transport projects are then compared to a business-as-usual scenario whereby the project is not being implemented, still accounting for other existing or planned transport infrastructure. The effect of the project is assessed as the difference between both scenarios.

Impacts accounted for in the current CBA

For a high-speed rail project, the different regulatory costs and benefits have to be accounted for with respect to different categories of stakeholders: the rail transportation actors, actors from other modes, the users of the modes, the government, and finally the third parties subject to external effects. The negative impacts (costs) are mainly investment costs, investments in rolling stock, and operating expenses. The benefits resulting from the infrastructure are mainly the revenues generated by the use of the infrastructure and the time saved by travelers.

In France, the first formal integration of environmental effects in CBA was made following the Boiteux (2001) report, which produced a methodology to assess noise pollution and local and global pollution due to the functioning infrastructure. The change towards railroad infrastructure is considered a benefit due to lower road and air traffic and cost in the case of road infrastructure. The valuation of these effects has become mandatory in France since the 'Instruction Framework on Economic Valuation Methods for Major Transport Infrastructure Projects' (Robien, 2005).

Infrastructure lifespan and public time discount rate

Infrastructure projects in France are usually assessed over a 50-year period for railroad projects and over a 100-year period for road projects. Over the project life span, flows of costs and benefits are discounted. Indeed, the CBA should be carried out in real terms (correcting for inflation-related distortions); that is, the analyst should convert nominal values into real (relative) prices using prices indexes. Public sector projects in France use the regulatory public time discount rate as defined in the Lebègue (2005) report. The rate is the function of three terms: (1) the rate of pure time preference, (2) an elasticity of marginal utility of consumption associated with (3) an income growth to represent a wealth effect. The Instruction Framework prescribes a 4 percent discount rate regarding development projects in the first 30 years, then a 3.5 percent rate between 30 and 50 years, and finally a 3 percent rate above 50 years.

Non-marketed effects evolution

The legal framework also plans the evolutions of the relative prices of environmental effects to be used in the projects. The noise effect should evolve at the same rate as the gross domestic product (GDP) per capita. The carbon value progresses as described in the Quinet (2009) report: at 5.8 percent before the year 2030 to reach the value of the time discount rate after the year 2030, following the Hotelling rule. For local pollution, the evolution depends on (1) pollutant emissions, assumed to decrease due to the anticipated technical progress, −5.5 percent for light-weight vehicles and −6.5 percent for heavier vehicles, and (2) the health impact measured by the 'value of life', increasing with GDP per capita.

CBA acceptance criteria

For a given project, various criteria can be used. The first one is the net present value (NPV) of the project, defined as the difference between benefits and costs aggregated over the project life span. The aggregation of flows generated at different dates is captured using the discount rate defined above. The higher the NPV, the more the situation in which the project is implemented can be evaluated as preferable over the baseline condition.[2] The criterion for project acceptance is a non-negative NPV: in this case, the project is deemed to result in an improvement in social welfare. However, it does not ensure by any means that there is no other more profitable project. Two supplementary indicators can be analyzed: the internal rate of return and the cost–benefit ratio (refer to Hanley and Spash, 1993 for more details).

Sensitivity analysis is finally required to check the relative sensitivity of computations with respect to data inputs. Such an analysis reveals the degree of robustness of the results with respect to the uncertainty related to the different assumptions made during the process.

6.3 CHALLENGES IN INTEGRATING ECOSYSTEM SERVICES IN TRANSPORTATION INFRASTRUCTURE PLANNING

6.3.1 Infrastructure Impacts on ES

Linear infrastructures can cause serious environmental impacts, which in turn influence ES supply. In Europe, infrastructure development is identified as one of the major drivers of habitat loss and fragmentation (Hicks and Peymen, 2002). Impacts are faced at various stages of infrastructure projects: construction, use, and induced impacts from areas converted *a posteriori* for related developments.

The direct impacts are related to the destruction of natural or semi-natural ecosystems by the infrastructure through soil tillage, sealing and conversion into metaled/sealed surfaces (Trocmé et al., 2002; Geneletti, 2006). In addition, the conversion affects the physical environment at the edges of the infrastructure axis due to the need to clear, level, fill and cut natural material. Construction works change soil patterns, landscape relief, surface, groundwater flows and microclimate, altering ecological functions and habitat composition, which in turn affect ES provision.

Changes in hydrology particularly affect wetlands and riparian habitats; for example, changes caused by embankments (Findlay and Bourdages, 2000) and cuttings, which may drain aquifers. This increases the risk of soil erosion and extensive earth-slides that have the potential to pollute watercourses with sediments (Forman and Deblinger, 2000). Surface water canalization into ditches can also significantly change water run-off and debris flows and thereby modify natural disturbance regimes in riparian networks (Jones et al., 2000). Clearance changes local climate conditions by increasing light intensity, reducing air humidity, and creating a greater daily variation in air temperature. These changes are naturally stronger when a road crosses forested habitats. All these micro-environmental changes are known to affect plant growth and performance (Tardieu et al., 2014).

The area covered by roads and railways is, however, not a reliable measure of the loss of natural habitat and ecological functions disturbance. It is now well recognized, and proven by empirical studies, that the disturbance of surrounding wildlife, vegetation, hydrology and landscape spreads much wider than the area that is physically occupied (Van der Zande et al., 1980; Reijnen and Foppen, 1994; Angold, 1997; Forman and Alexander, 1998; Meunier et al., 1999; Jones et al., 2000; Trombulak and Frissell, 2000). This contributes far more to the overall loss and degradation of habitat than the road/railway body itself (for a review, see Trocmé et al., 2002). These extended effects are generally called 'indirect impacts' provoked by habitat fragmentation and other disturbances in terms of noise and visual nuisance, which act to reduce the suitability of adjacent areas for wildlife (Hicks and Peymen, 2002). Beyond the reduction of the size of crossed habitats, infrastructure affects the structure and functioning of the landscape as a whole by acting as a barrier to the movement and migration of animal and plant species. This may result in isolation and threaten the survival of vulnerable species.

Hence, at the beginning of the process, the loss of habitat is the driving force threatening ecological functions and ES provision in the landscape. Toward the end of the process, isolation effects become stronger (ibid.).

6.3.2 Particular Challenges Faced in the Infrastructure Case

The incorporation of the spatial context in economic valuation brings difficulties. The spatial character of the valuation requires additional analysis at each step compared to a traditional valuation. To assess an ES quality or quantity change, the tasks are performed in an overlapping way: the definition of the biophysical structures, process flows in baseline conditions to assess the change after the project implementation in a spatially explicit way; the definition of ecological/biophysical models to assess changes in services provision in quantity or quality; the identification of potential beneficiaries and their demand according to the spatial context and socioeconomic characteristics; and the determination of the accurate value to assess the service in accordance with the ecosystem spatial context. The process therefore requires a substantial amount of data at every step.

The application of this framework to infrastructure construction leads to specific issues. Moreover, some steps are added such as the impact character of the project, the profile of the temporal changes of values or a better way to adapt the output to the decision-making process. The main issues are identified below.

Challenges related to the impact type of the infrastructure and temporal trend

Infrastructures cross varied multifunctional landscapes and different ecosystems providing more than one service Indeed, the length of transport infrastructures is significant, especially when it comes to highways or high-speed railways connecting large cities or countries. The mapping exercise is thus even more complicated. Assessing the loss for an entire implementation option requires multiple ecological data (on climate, hydrology, topography, land use and cover), socioeconomic data, and population repartition data to produce layers that effectively identify conflicts and synergies across the landscape.

The spatial extent of impacts on the flow of ES must be defined It can be presumed that the multiple effects on ES have different spatial extents. The impacts on ES flows (the ES loss incurred) can be classified in the same order as impacts; that is, in terms of direct ES loss through the conversion and additional indirect ES loss through impacts on landscape connectivity (see Tardieu et al., 2015).

The extent of direct losses will differ according to the infrastructure type and has to be defined according to the project stage in which the assessment is made.[3] Indirect effects associated with modifications to landscape connectivity may impact ES to a different extent (Mitchell et al., 2013). The destruction of an ecosystem within a zone may cause a decline in ES provision of a landscape element exterior to the zone. ES loss will be a function of impacts on the ecological interdependencies among the different elements of a landscape. Moreover, some species and ecosystems may need a minimum surface to complete their ecological functions (Muradian, 2001; Groffman et al., 2006). Hence, when part of a habitat or ecological network is lost due to development, species responses and ecosystem function may exhibit a non-linear response to land conversion due to threshold behavior. Finally, when the ES supply depends on the quality of the ecosystem at a particular point, the loss can be related to whether the infrastructure is more or less distant to the point of interest (e.g., recreational forest impacted by visual or noise disturbance). In this case, the perturbation of a particular unique landscape element may have a large effect relative to the disturbance surface area (Van der Zee, 1990).

Disregarding these effects is likely to result in an underestimation of the ES loss. The determination of the type of flows that are impacted and the extent to which they are impacted are then critical in the assessment.

Changes in ecosystem flows can differ in the degree to which they are impacted First, a total loss can be assumed for all directly impacted ES flows over the infrastructure lifespan (Braat and Ten Brink, 2008). The soil below the infrastructure is converted for a long period, and functions underlying all services are threatened. For the indirect losses suffered around the infrastructure axis, the ecosystem can recover its service flows during the infrastructure lifespan thanks to its resilience. The characterization of the impact should then take this possibility into account. Beyond ecosystem resilience, impacts on natural areas are reduced through environmental measures. Some of the ES impacted may then not be totally lost over the lifespan period in the vicinity of the infrastructure. Finally, the land cover is likely to change during the infrastructure lifespan due to other land-use change or to climatic change. This can be accounted for by building different scenarios[4] and by assessing the loss according to these different scenarios, including the business-as-usual scenario.

Finally, when the ES flows are considered to be lost, questions of how to discount values over time arise On the one hand, the costs and benefits associated with these long-term consequences tend to see their weight in decision-making decreased when time discounting is applied. On the other hand, there is uncertainty about the future consequences of the losses and the behavior and expectations of future generations for some natural assets. This raises the problem of choosing a discount rate that can be applied to environmental goods and services and choosing the evolution of their relative values over time.

Challenges related to the evaluation process of infrastructure projects

Implementation options are defined progressively The precision of the land-use change scenario depends on when the assessment is achieved during the project process. As explained in Section 6.2.1, the object of the impact analysis moves from wide passage corridors to 100-m-wide corridors. The infrastructure land-take is detailed during the process, covering land from 100 m around the axis to the precise land-take, including quarries and land arrangements. Finally, measures to reduce environmental impacts are defined and located late in the process, enabling little proper analysis of the residual impact in contrast to when the assessment is made in the preliminary stages of the project. This has repercussions on the definition of biophysical structures and process changes and then on the overall assessment of the loss.

The analysis of ES loss should be exhaustive As opposed to other spatial assessments focusing on one particular ecosystem delivering multiple services or a particular service delivered by multiple ecosystems, the assessment needs, as far as possible, to be exhaustive in order to assess multiple services of multiple ecosystems. This is due to the need to compare multiple implementation options for the infrastructure.

The analysis of ES loss should be discriminative Even if the spatial assessment needs to be as exhaustive as possible, implementation options are likely to be geographically close, and the analysis thus needs to be sufficiently precise at each step of the cascade framework to enable discrimination between options. This can be achieved by using precise land-use/cover typologies and by defining spatial conditions of service presence according to ecosystems types and their locations when ecological models are not used.[5] However, the more precise the typology is, the more difficult it is to find accurate data. This can be done via an accurate ES supply mapping technique, and finally, by taking into account the beneficiaries and their spatial context.

Scientific reliability and reproducibility should be balanced Accounting for ES in a spatial manner is complex, costly and time consuming, as we have observed particularly with regard to the case of infrastructure projects. Balancing scientific reliability (list of services covered, precise methodologies for mapping, sensitivity analysis) against replicability of methods remains a challenge. Using techniques such as benefit transfer can be an option. However, given the large amount of data required, the use of elaborated transfer techniques may be difficult (as meta-analytic transfers). Value transfer or transfer based on function seems to be the easiest approach to replicability. One possibility could be to base value transfer expressed in biophysical units on elaborated assessments of biophysical changes.

6.4 CASE STUDY

Here, the usefulness of ES loss assessment will be presented through a real case study: a high-speed rail project in France.[6] The analysis of ES integration is made at different key stages of the project given the data available at these stages. The integration is made on the basis of the results obtained in Tardieu et al. (2015), assessing multiple ES with different impact extents (direct and indirect). Specifically, the integration is made first for the stage of preliminary environmental studies within different passage corridors. Then, the inclusion of ES is done during the detailed EIA, when the different implementation options of the project are compared (realized here in a zone of the project having three optional routes). The usefulness of consideration of ES to monitor a mitigation measure is also tested. Finally, the inclusion of ES is applied to the CBA process, assessing the effects of the final option chosen for the project.

6.4.1 Ecosystem Service Indicators for the Preliminary Environmental Study

Decision-makers usually require easy-to-use decision support tools that can be quickly incorporated into project processes (Ruckelshaus et al., 2015).

It is therefore necessary to develop ways to create new indicators that consist of combinations of ES indicators and combinations of indicators traditionally used in transport planning. Such indicators would constitute a practical instrument that would provide a quick scan of the initial features of an area via a characterization of related ES supply and demand. One simple way to design this first tool is to map the ordinary ES 'hotspots' in terms of presence in the area considered. An advantage of this approach is that it can be automatically incorporated into the existing environmental vulnerability maps to illustrate baseline conditions.

ES presence, derived from land-use/cover data (CORINE Biotope typology), according to the analysis of presence provided in Tardieu et al. (2015), can be mapped to illustrate areas that are likely to exhibit a particular loss if the project is implemented. To do so, a score of 1 is applied when the service is supplied unconditionally (e.g., carbon storage). Where the ES is provided only under spatial conditions by an ecosystem type, the score of 1 is applied only if the spatial condition is fulfilled (e.g., water flow regulation is only performed by wetlands related to a water system; or recreation services are demanded only if the recreational site is situated at proximity to a city) and 0 otherwise.

Indicators are aggregated over all services assessed and mapped to illustrate baseline conditions in the selected corridor (Figure 6.1). The indicator map represents the presence of ES according to the ecosystem types and the spatial conditions. The map indicates that a maximum of ten services are potentially supplied in the area. This mapping exercise may help in the design of first avoidance measures.

6.4.2 Ecosystem Service Loss in the Detailed Environmental Impact Assessment

The comparison of implementation options can be made by comparing ES loss induced by each route option for the infrastructure, highlighting the one that minimizes the total loss. To perform this, the characterization of the spatial extent of impacts on ES has to be achieved: direct impacts and additional indirect impacts on landscape connectivity threatening the deliverance of some ES at a larger scale. Impact characterization, biophysical data or ecological functions, and economic values per ecosystem, service and route options are described in Tardieu et al. (2015) in terms of annual flows. An example of the methodology used to assess the loss of the air quality regulation service (by pollutant removal by vegetation) is provided in Box 6.1.

To consider long-term ES losses per route option, an assumption on ES value evolution can be made, as is the case for other external effects in

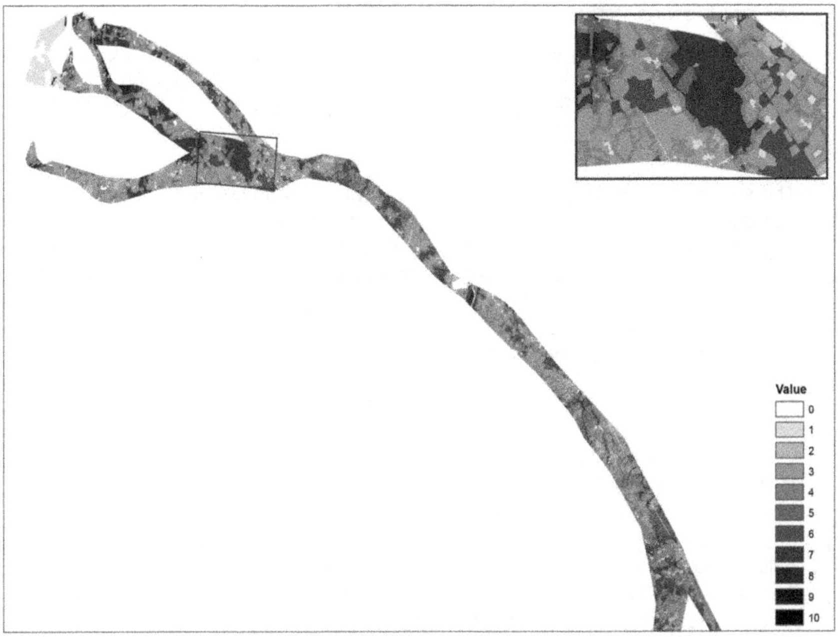

Figure 6.1 *Indicator map for preliminary environmental studies: ES presence*

the project. Environmental economics has a long tradition of considering different evolutions for manufactured goods and for environmental goods (Krutilla and Fisher, 1975). As opposed to manufactured goods benefits, environmental benefits are likely to grow over time. Hanley and Spash (1993) give at least three reasons to explain this growth of environmental relative prices. First, because of increasing relative scarcity, natural areas decline and have a limited substitutability, the remaining areas therefore become more valuable, and the willingness to pay (WTP) for their preservation increases due to the law of diminishing marginal utility. Second, because information on the role of ecosystems for human well-being is constantly increasing, people are becoming better informed and are more likely to have higher WTP for natural resources. Third, an increased material prosperity may augment WTP for scarce non-market goods.

In the literature, one approach that has been adopted is to assume a positive income elasticity of WTP for environmental goods that is a measure of how willingness to pay is affected by changes in income. Some recommendations on the specification of income elasticity of WTP have been made. For instance, Pearce et al. (2006) provide empirical estimates

BOX 6.1 EXAMPLE OF THE METHODOLOGY USED IN THE CASE OF THE AIR QUALITY REGULATION SERVICE SUPPLIED BY URBAN AND PERI-URBAN FORESTS

The service is considered as potentially present for urban and peri-urban forests, as they contribute to air purification by filtering or eliminating a number of pollutants and particles (McPherson, 1997). We relied on the Lovett (1994) model based on deposition velocity and local pollutant concentration to determine the amount of pollutant removed by peri-urban forests in the study area. The pollutant removal fluxes F_i of pollutant *i* is equal to the product between the dry disposition velocity (V_{di}) and the hourly concentration of pollutant in air (C_i), as follows:

$$F_i = Vd_i * C_i$$

The principal pollutants are: CO (carbon monoxide), nitrogen dioxide (NO_2), O_3 (ozone), particulate matter less than 10 μm (PM10) and sulfur dioxide (SO_2). We relied on the French local AASQA ('Associations Agréées pour la Surveillance de la Qualité de l'Air': monitoring air quality associations) for the yearly average pollutant concentrations in the air in the study area. To monetize the service we transferred the damage cost value for a marginal pollution change per ton of pollutant in order to value the cost in terms of impacts on health, crops and materials (Watkiss et al., 2006 for the UK Department for Environment, Food & Rural Affairs [Defra]). Values are situated at 1.1 euros/kg for NO_2, 55 euros/kg for PM10 and 1.9 euros/kg for SO_2. The impact is considered as direct; that is, contained in the area of urban and peri-urban forest removed for the infrastructure.

Source: Tardieu et al. (2015).

and suggest using a value of elasticity for environmental goods that is lower than 1 and in the interval 0.3–0.7. This is supported by Kriström and Riera (1996), Hökby and Söderqvist (2003) and Jacobsen and Hanley (2009). WTP for environmental goods then grows less than proportionally with respect to income.

The annual values of ES loss per route option (expressed in real terms) then evolve as a function of, first, an increase in relative prices (represented by the income elasticities of WTP and the real income rate per capita), and, second, the public discount rate. For the evolution of relative prices, and for the example, we first assume that every type of ES has a relative price evolving over time, even for ecosystems with no conservation status. In the case where services are assumed to be technically substitutable, the evolution of the relative prices is interpreted as the loss of an option value for which an income elasticity of 0.3 is applied. For other services, we assume that the income elasticity of WTP is higher than that of the

Table 6.1 ES long-term and annual loss per route option (in thousand euros 2010)

Loss in Thousands Euros for the Year 2010		Route 1	Route 2	Route 3
Provision	55 years	85.1	*81.9*	87.4
	Annual	3.1	*2.9*	3.1
Regulation	55 years	6493.2	6955.7	*4514.6*
	Annual	232.4	249.1	*161.0*
Cultural	55 years	1370.8	*1215.8*	1891.1
	Annual	46.4	*41.2*	64.1
Total	55 years	7949.1	8253.4	*6493.2*
	Annual	281.9	293.2	*228.2*
Total per km	55 years	393.5	412.4	*290.6*
	Annual	13.9	14.6	*10.2*

Note: Figures in italics highlight the least impacting route.

first case, and we use an elasticity of 0.5. This applies to local climate, pollination and recreation services. A low elasticity of substitution is then presumed, which implies a larger increase in the relative price because the well-being provided is not easily compensated for by increases in consumption levels of other goods. Values then evolve annually according to the factor $(1 + e_k.\mu)^t$, as suggested by Pearce et al. (2006), where e_k denotes the income elasticity of WTP for the ES k considered, μ the real income rate per capita, and t the year of consideration.

The long-term ES losses per route option and per ES (15 ES) are presented in Table 6.1. This zone is interesting in terms of choice because it exhibits contrasting environmental constraints and because one of the options is slightly longer (2 supplementary km for route 3). It is thus interesting to see how the ES approach will respond to these characteristics. The total monetary loss for the different route options yields to the same trade-off among options as the annual values. Trade-offs regarding ES loss are just strengthened. The results indicate a large predominance of regulation services, with a loss ranging from 4 to 6 million euros for 55 years. Route 3 is the least impacting route in terms of aggregated values over 55 years. The major increase between annual and long-term values is observed for recreation services without changing the ranking among options. The repartition of the loss per service is presented in Figure 6.2. We underline the predominance of the global climate service, which is highly valued in comparison to other services. Here, the air quality regulation is also important given the presence of peri-urban forests in the area.

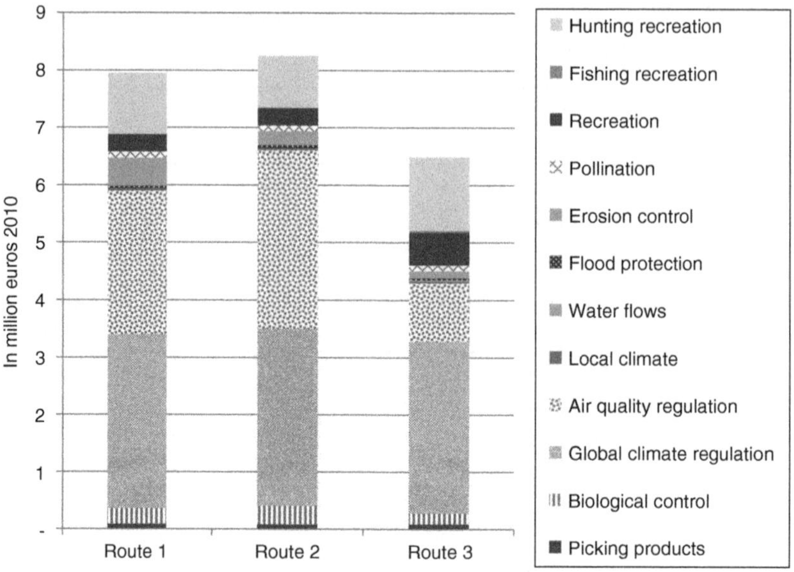

Figure 6.2 Repartition of loss per ES type

By comparing the results presented in Table 6.1 to the real EIA conducted in the area, several issues regarding coherence and incoherence can be raised. With respect to the global consistency, the EIA- and the ES-based approach come to the same conclusion; that is, route 3 is the least impacting route. Both approaches also indicate an important impact on natural areas in route 2. However, some incoherence with regard to specific points can also be noted. First, a low impact on natural areas is detected for route 1, while the ES-based approach has an important impact on ES (and particularly on regulating services). Concerning the impacts on recreation, a strong impact is observed for route 2, while a low impact is identified for route 1. When we assess the recreation service, the impact between the two routes is not as clearly discriminating. A quite low impact on agriculture is identified for route 2; however, the ES approach has a strong impact on pollination, local climate regulation, erosion control and biological control for this option. Finally, impacts on landscape connectivity for wild fauna are identified as low for route 1 and strong for route 2. More broadly, The ES approach indicates that the overall impact on ES related to landscape connectivity (water flow and flood regulation, pollination, hunting, fishing, and general outdoor recreation) may also be important in route 1.

Hence, the ES-based approach supports some arguments by giving

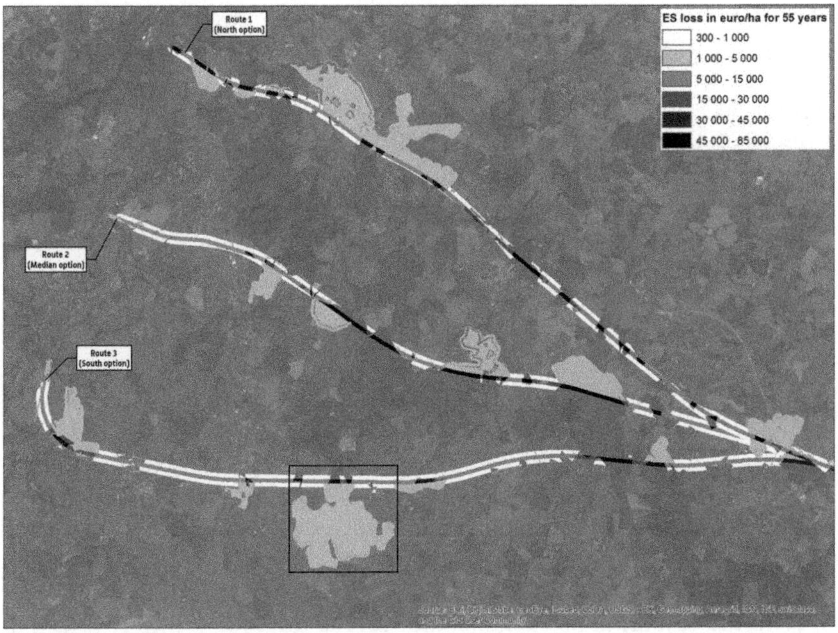

Figure 6.3 *ES loss over 55 years in the studied zone*

quantitative information (such as the identification of route 3 as the least impacting route) and challenges others. It puts environmental issues into perspective within a larger territory, which involves more stakeholders and is less confined to the project area. In terms of additional information to the general public, stakeholders and decision-makers, it seems important to communicate information on the differences between options.

Figure 6.3 provides a map of the direct and indirect ES loss (over a 55-year period) in the zone. This type of map may be provided to help locate the best options for mitigating measures. This gives the opportunity to compare the cost of one mitigating measure with the cost of ES loss. The measure analyzed here is a wildlife passageway for large game fauna to implement in the highlighted zone of Figure 6.3. To do so, the difference between the loss before and after the construction of the wildlife passageway is computed. The optimal position of the corridor (in terms of habitats and migration) in the highlighted zone is determined with Optiflux software. The service is then valued (as in Tardieu et al., 2015) as a loss of huntable territory. The difference in hunting recreation service delivery, between the situation without infrastructure and with infrastructure, is estimated at approximately 17000 euros (2010) for one year and approximately 507 000 euros (2010) for 55 years. The cost of the wildlife passageway is estimated at 650 000 euros. The construction of the passageway is not economically justified over a 55-year period if we consider only this service; however, the conclusion is reversed if benefits are evaluated from $t = 0$ to infinity (1.16 million euros). It appears to be cost efficient after the 81st year. It can be noted that the only service considered (and valued) here relating to wild game fauna is the hunting recreation service; however, other services can be considered such as habitat services or genepool protection.

6.4.3 Ecosystem Service Loss Related to the Final Option for CBA

To conclude this case study, the NPV test is applied to the final route option chosen for the infrastructure. The CBA is computed for 55 years (five years of construction and 50 of operation), assuming that the infrastructure will be operating by year 2020 (an assumption made for the real project evaluation). The loss of ES and the construction costs start from the first year of construction, whereas benefits and operating costs start from the moment at which the infrastructure is in operation.

With the assumptions explained above, the total loss of ES is valued at 44 million euros for the year 2010. Regarding the project overall NPV, the project was already not profitable by having a negative NPV. Accounting for ES loss results in the project being even less profitable, increasing the project negative NPV by 19 percent. However, the loss of ES represents a small part in the total CBA. Still, it is interesting to observe the weight of ES loss in the third party CBA (incurring externalities) as presented in Figure 6.4. Accounting for ES loss diminishes the third-party balance by 42 percent, which is not negligible. It can also be observed that the ES loss is in the same order or greater than the other externalities accounted for in the current CBA.

Beyond the simple aggregation of the economic loss associated to ES loss (which can be highly criticized), the assessment allows pair-wise comparisons between the different external effects assessed for the infrastructure. For instance, the benefit in terms of air quality induced by the high-speed rail construction (due to smaller road traffic and then a smaller emission of local pollutant) can be balanced against the cost incurred by threatening urban and peri-urban forests that regulate air quality. Here the benefit is totally cancelled by the costs. For the case of global climate regulation, the benefit of the high-speed rail construction (also due to reduced road traffic and then a lower emission of carbon dioxide) is reduced by either two-thirds or totally by the sealing and destruction of several ecosystems. This depends on whether we value the loss of carbon

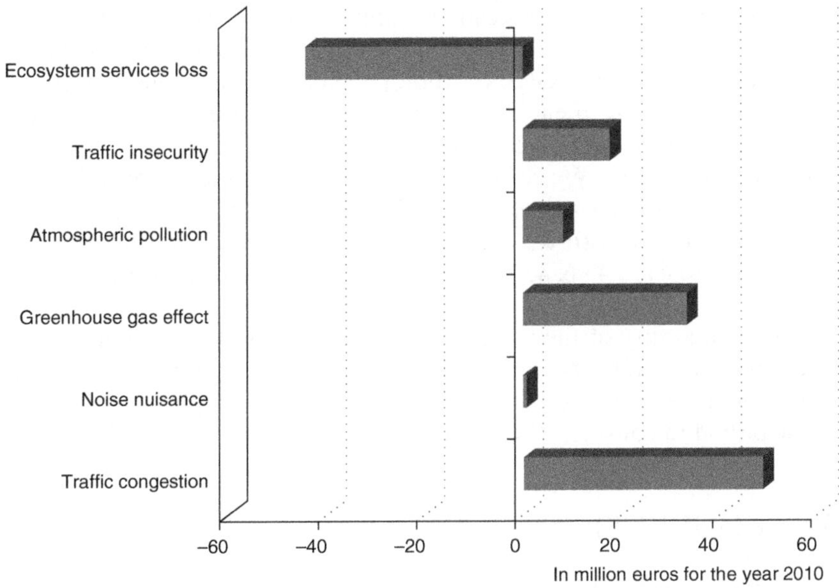

Figure 6.4 CBA for third parties including ES loss (55 years)

stocks induced by the construction in the first years of the infrastructure or if we value the loss of a storage service performed by different ecosystems during the entire infrastructure lifespan (see Tardieu et al., 2013).

A simple sensitivity analysis of the total ES loss estimated follows according to different changes in assumptions previously made and their influence on the total NPV of ES loss.[7] The NPV of ES loss is re-computed with different values of the following:

- *Project lifespan:* Development benefits will not last forever because the infrastructure schemes have a fixed life span. To be applicable to the infrastructure CBA, we calculate the loss of ES over the 55-year period; however, the loss is likely to be incurred for a period far longer than this. When the ES loss is evaluated over an infinite horizon, it results in a loss of 69.2 million euros.
- *Discount rate:* Using a lower discount rate (3 percent) in computation, decreasing at the same pace as the public discount rate, results in an increase in the ES loss from 44 million euros to 46.4 million euros. Using a 4 percent constant discount rate results in a loss of 38.9 million euros.
- *Land-use/cover typology:* When the ES loss is estimated with a less precise scale typology, it results in an underestimation of the loss

(41 million euros). This can be mainly explained by the fact that the CORINE Land Cover (CLC) typology (1:25 000) overlooks spatial details that the CORINE Biotope typology can detect (1:5000). More precisely, the CLC typology cannot capture the same ES presence as the CORINE Biotope typology.

- *Relative prices:* We consider different relative prices, depending on four levels of income elasticity of WTP applied to all services (0, 0.3, 0.5 and 0.7); with these elasticities, the values of the ES losses are equal to 41.5, 43.6, 45.2 and 46.9 million euros, respectively.

For the majority of the assumptions made, the NPV variation is quite negligible. However, accounting for the loss in a very long-term period changes the assessment of the loss significantly. This emphasizes the issue of the period to consider when accounting for ES loss.

6.5 CONCLUSIONS

6.5.1 The Added Value of Ecosystem Service Consideration in the Actual Framework

This contribution supports the point that ecosystems and their capacity to produce ES flows benefiting society, may constitute an important criterion in transportation planning. Generally, the ES consideration does enable us to identify impacts that are not already characterized in the actual framework. Some ES can be partially accounted for through existing environmental measures. However, given that measures are not assessed with an ES approach, the restoration of the service is not an objective; impacts on ES provision are likely to be maintained. Further, the ES approach accounts for multi-scaled impacts because ES are supplied and demanded at local, regional, national and global scales. Finally, it allows for a better identification and improves the specification of direct and indirect impacts because it offers a more functional approach between the natural environment and human well-being. This could contribute to widening the scope of both the EIA and CBA processes. More specifically, consideration of ES can guide, as can be observed in the case study, spatial planners at different stages of the EIA process of transport infrastructure projects.

First, a quick scan of key ES can be provided for preliminary environmental studies through maps of the baseline conditions of the area considered. This can be made through indicators that may provide information for preliminary avoidance measures. Further, it may constitute a new discrimination criterion and a new environmentally based rationale during the comparison of implementation options. This may result in a more holistic mitigation strategy focusing on ecosystem functions, helping to ensure the sustainability of project-related outcomes. This can result in a stronger framework for avoidance and mitigation strategy and in a better understanding of management impacts.

The ES valuation would, moreover, allow for a more accurate comparison of environmental measures to be implemented according to their relative costs and benefits, which in turn would better inform the choice of the appropriate measures. Then, the environmental studies and the economic valuation eventually become complementary: the environmental studies suggest effective environmental measures; and the economic valuation identifies the most cost-effective one. Finally, it can answer public concerns related to transport projects. Indeed, at least in France, a majority of public debate reports highlight that the chief public concerns relate primarily to issues of landscape planning, environmental management, and health in particular for the case of large-scale projects. The analysis puts forward additional information, enlightening not only decision-makers but also the general public at every stage of the process.

In the case of CBA, no impact of the infrastructure construction on ES is taken into account in the actual process. However, the ES approach provides a way to understand the most important interactions between society and the environment, identifying the issues that matter for decision-making. Expressing the ES loss in monetary terms makes it possible to account for the economic losses induced by the final option of the project and contrasts them (in the CBA computation) with other external effects of the project. Although valuation techniques face various challenges addressing uncertainty, irreversibility and resilience, the estimates derived can thus be used to provide policy-makers with an order of magnitude of the loss incurred and thus of information on the importance of ES on a regular basis (Kumar et al., 2013).

6.5.2 Limitations

The frontiers of ES scarcity are difficult to define, and this represents a limitation to economic valuation. Many issues are raised and relate to either the supply side or the demand side of ES (Baumgärtner et al., 2006). On the supply side, ecological systems are governed by different scale effects, discontinuities, thresholds, minimum viable population sizes, limited resilience, and irreversibility for some of them. On the demand side, the distinction is based on whether the satisfaction of needs is considered 'necessary for life' or not. Thus, when scarcity concerns a limited or non-substitutable means for the satisfaction of basic needs and cannot be obtained by additional production, one may speak of absolute scarcity, and economic valuation cannot be applied. Notions of scarcity are, moreover, strongly dependent on time and spatial scales. Technical progress may result in new technologies over time, which then yields production of substitutes for previously absolutely scarce goods, which would in turn become scarce only in a relative sense. In the same way, a good that is absolutely scarce on a given spatial scale may be relatively scarce on a larger spatial scale (or conversely). These issues are related to the discussion on weak and strong sustainability initiated by Dasgupta and Mäler (2000).

Moreover, when practicing economic valuation, it inevitably forms part of a logical process of CBA. Criticism of general CBA needs therefore to be incisive. Beyond the imprecision it holds, the use of CBA tends to give more weight to the preferences of the richest agents because they have a generally higher willingness to pay (as income elasticity for normal goods makes WTP increase with income). Social choice theory considers the possibility of attributing unequal weights to the gains and losses of different groups within the society (Fleurbaey et al., 2013). This consideration might be important for ES analysis. The calibration of weights needs further study to take accurate account of individual preferences and adjust social welfare function according to inequality aversion. So far, such weighting applied to practical CBA is unusual in OECD countries (Hanley and Spash, 1993).

Finally, with regard to ES valuation, there is always the risk of misguiding policy-makers because valuation is an uncertain process. Moreover, interest groups may want to use the estimates based on prices to justify their own interests. Therefore, valuing the flows and stocks of nature needs to be scientifically credible with a clear objective. However, other effects are valued with non-market valuation methods and are integrated in the project process even if they are imprecise. This is the case for valuation of time saving, which plays a major role in project evaluation.

6.5.3 To take Services into Account in Project Evaluation or Not?

Estimates of ES impacts depend on the assumptions made. In the absence of a market validation, methods necessarily lack robustness. With regard to moving from principles to practice, we may feel that the data and empirical or theoretical studies are insufficient to make a reliable assessment. Waiting might seem more reasonable, but meanwhile, the nuisance inflicted on ES is still excluded from the project assessment and is counted as zero in the computations. It can therefore be possible to base such assessment on the rationale that economies and human well-being critically depend on biodiversity and ecosystems and that the trend of erosion is still too great. I believe that it is now essential to translate current knowledge into actions that will influence development planning processes and particularly those applying to transportation planning. Moreover, we address here ecosystems that are not protected by any conservation status and are thus not directly considered by environmental measures. This contribution is thought to improve existing instruments for projects assessments to consider these types of ecosystems.

6.5.4 Research Avenues

In terms of environmental measures, research could be conducted to determine the residual impact on services delivery remaining after implementing these measures. This would require that mitigation and compensation measures are identified, located, and their costs determined at the stage of study, thereby clarifying the assessment of the net loss caused by the infrastructure implementation.

Regarding impacts, one could expand the impact type considered to take into account the cumulative impacts with neighboring development projects. The adaptation could be made quite simply provided that the information is available (location, commissioning year, etc.). Taking into account the impacts induced by the infrastructure seems more difficult to achieve because the information may not be available at the time of the study. The consideration of multiple scenarios for a more precise indication of the impacts incurred during the infrastructure lifespan would also be a plus. Finally, the positive impacts of an infrastructure could be considered in analyses, as, for example, in the case of roadsides colonized as a new habitat or serving as corridor transition for species habitat or migration function.

For the consideration of services, a more extensive assessment would obviously be desirable. This can be achieved by provoking thought on how one can include services such as habitat maintenance or ecosystem dis-services (such as pollen, which causes allergies in urban areas).

More generally, further research could be conducted to examine how to improve the link between the spatial assessment of the services supply, demand and values. This research would allow for a better characterization of the relationship between a change in ES supply and the associated change in welfare. A possible option could involve the combination of mapping techniques for services supply, with techniques enabling the differentiation in willingness to pay per service unit provided by multiple ecosystems in type, size, or location. The development of functions to scale up assessments into large areas would also be required for this type of project.

NOTES

* This work was financed by Egis Structures et Environnement. The author would like to thank Bernard Chevassus-au-Louis and Serge Garcia for their constructive comments on a preliminary version of the manuscript. The author gratefully acknowledges Davide Geneletti for the invitation to write this chapter and for his useful comments and suggestions that led to significant improvement of the manuscript.

1. As argued by Hanley and Spash (1993), this is merely a device of convenience rather than an implicit statement that only money is important.
2. This criterion is based on the Kaldor-Hicks principle.
3. For instance, with buffer zones in the first stages, generally larger for highways than for high-speed rails, and precise project land-take in the last stages.
4. Different models to build scenarios are already available – for example, ALARM, ECOCHANGE. Scenarios can also be adapted to integrate other infrastructures planned.
5. Ecological models and toolkits, for example, InVEST (Kareiva et al., 2011), ARIES (Bagstad et al., 2013) and TESSA (Peh et al., 2013), can calculate the presence given the biophysical data inputs. When no models are used to determine the presence, spatial conditions of presence can be specified by spatial assumptions, for example, regulation of water flows is delivered by a wetland only if it is related to a river system.
6. I am not allowed to disclose more information on the case study due to a contractual commitment with Egis Structures et Environnement.
7. Sensitivity analysis for route selection is already given in Tardieu et al. (2015).

REFERENCES

Angold, P.G. (1997), 'The impact of road upon adjacent heathland vegetation: effects on plant species composition', *Journal of Applied Ecology*, **34**, 409–17.

Antunes, P., R. Santos and L.S. Jordão (2001), 'The application of geographical information systems to determine environmental impact significance', *Environmental Impact Assessment Review*, **21**(6), 511–35.

Bagstad, K.J., G.W. Johnson and B. Voigt et al. (2013), 'Spatial dynamics of ecosystem service flows: a comprehensive approach to quantifying actual services', *Ecosystem Services*, **4**, 117–25.

Baumgärtner, S., C. Becker and M. Faber et al. (2006), 'Relative and absolute scarcity of nature. Assessing the roles of economics and ecology for biodiversity conservation', *Ecological Economics*, **59**(4), 487–98.

Boiteux, M. (2001), *Transport: choix des investissements et coûts des nuisances* [Transport: Investment Choices and Nuisance Costs], Paris: La Documentation Française.

Braat, L. and P. ten Brink (2008), *The Cost of Policy Inaction: The Case of Not Meeting the 2010 Biodiversity Target*, Brussels/Wageningen: DG Environment European Commission.

Daily, G.C., P. Kareiva and S. Polasky et al. (2011), 'Mainstreaming natural capital into decisions', in P. Kareiva, H. Tallis and T.H. Ricketts et al. (eds), *Natural Capital: Theory and Practice of Mapping Ecosystem Services*, Oxford, UK: Oxford University Press, pp. 3–14.

Dasgupta, P. and K.G. Mäler (2000), 'Net national product, wealth and social well-being', *Environment and Development Economics*, **5**(1), 69–93.

De Groot, R.S., R. Alkemade and L. Braat et al. (2010), 'Challenges in integrating the concept of ecosystem services and values in landscape planning, management and decision making', *Ecological Complexity*, **7**(3), 260–72.

Findlay, C.S. and J. Bourdages (2000), 'Response time of wetland biodiversity to road construction on adjacent lands', *Conservation Biology*, **14**(86–94).

Fleurbaey, M., S. Luchini and C. Muller et al. (2013), 'Equivalent income and fair evaluation of health care', *Health Economics*, **22**(6), 711–29.
Forman, R.T.T. and L.E. Alexander (1998), 'Roads and their major ecological effects', *Annual Review of Ecology and Systematics*, **29**, 207–31.
Forman, R.T.T. and R.D. Deblinger (2000), 'The ecological road-effect zone of a Massachusetts (USA) suburban highway', *Conservation Biology*, **14**(1), 36–46.
Geneletti, D. (2006), 'Some common shortcomings in the treatment of impacts of linear infrastructures on natural habitat', *Environmental Impact Assessment Review*, **26**(3), 257–67.
Geneletti, D. (2011), 'Reasons and options for integrating ecosystem services in strategic environmental assessment of spatial planning', *International Journal of Biodiversity Science, Ecosystem Services & Management*, **7**(3), 143–9.
Geneletti, D. (2013), 'Assessing the impact of alternative land-use zoning policies on future ecosystem services', *Environmental Impact Assessment Review*, **40**, 25–35.
Groffman, P.M., J.S. Baron and T. Blett et al. (2006), 'Ecological thresholds: the key to successful environmental management or an important concept with no practical application?', *Ecosystems*, **9**(1), 1–13.
Haines-Young, R. and M. Potschin (2010), 'The links between biodiversity, ecosystem services and human well-being', in D. Raffaelli and C. Frid (eds), *Ecosystem Ecology: A New Synthesis*, Cambridge, UK: Cambridge University Press, pp. 110–39.
Hanley, N. and C.L. Spash (1993), *Cost–Benefit Analysis and the Environment*, Aldershot, UK and Brookfield, VT, USA: Edward Elgar Publishing.
Hicks, C. and J. Peymen (2002), 'Habitat fragmentation due to existing transportation infrastructure', in M. Trocmé, S. Cahill and H. de Vries et al. (eds), *COST 341 – Habitat Fragmentation due to Transportation Infrastructure: The European Review*, Luxembourg: Office for Official Publications of the European Communities, pp. 73–113.
Hökby, S. and T. Söderqvist (2003), 'Elasticities of demand and willingness to pay for environmental services in Sweden', *Environmental and Resource Economics*, **26**(3), 361–83.
Jacobsen, J.B. and N. Hanley (2009), 'Are there income effects on global willingness to pay for biodiversity conservation', *Environmental and Resource Economics*, **43**(2), 137–60.
João, E. and A. Fonseca (1996), 'The role of GIS in improving environmental assessment effectiveness: theory vs. practice', *Impact Assessment*, **14**(4), 371–87.
Jones, J.A, F.J. Swanson and B.C. Wemple et al. (2000), 'Effects of roads on hydrology, geomorphology, and disturbance patches in stream networks', *Conservation Biology*, **14**(1), 76–85.
Kareiva, P., H. Tallis and T.H. Ricketts et al. (2011), *Natural Capital: Theory and Practice of Mapping Ecosystem Services*, Oxford, UK: Oxford University Press.
Kriström, B. and P. Riera (1996), 'Is the income elasticity of environmental improvements less than one?', *Environmental and Resource Economics*, **7**(1), 45–55.
Krutilla, J.V. and A.C. Fisher (1975), *The Economics of Natural Environments*, Baltimore, MD: Johns Hopkins Press.
Kumar, P., S.E. Esen and M. Yashiro (2013), 'Linking ecosystem services to strategic environmental assessment in development policies', *Environmental Impact Assessment Review*, **40**, 75–81.
Laurans, Y., A. Rankovic and R. Billé et al. (2013), 'Use of ecosystem services economic valuation for decision making: questioning a literature blindspot', *Journal of Environmental Management*, **119**, 208–19.
Lebègue, D., L. Baumstark and P. Hirtzman (2005), *Révision du taux d'actualisation des investissements publics* [Revision of the Discount Rate for Public Investments], Paris: La Documentation Française.
Lovett, G.M. (1994), 'Atmospheric deposition of nutrients and pollutants in North America: an ecological perspective', *Ecological Applications*, **4**(4), 629–50.
McPherson, E.G., D. Nowak and G. Heisler et al. (1997), 'Quantifying urban forest structure, function, and value: the Chicago urban forest project', *Urban Ecosystems*, **1**(1), 49–61.

Meunier, F.D., C. Verheyden and P. Jouventin (1999), 'Bird communities of highway verges: influence of adjacent habitat and roadside management', *Acta Oecologica – International Journal of Ecology*, **20**(1), 1–13.

Mitchell, M.G.E., E.M. Bennett and A. Gonzalez (2013), 'Linking landscape connectivity and ecosystem service provision: current knowledge and research gaps', *Ecosystems*, **16**(5), 894–908.

Muradian, R. (2001), 'Ecological thresholds: a survey', *Ecological Economics*, **38**(1), 7–24.

Naidoo, R. and T.H. Ricketts (2006), 'Mapping the economic costs and benefits of conservation', *PLoS Biology*, **4**(11), e360.

Nelson, E., N. Bhagabati and D. Ennaanay et al. (2013), 'Modeling terrestrial ecosystem services', in S.A. Levin (ed.), *Encyclopedia of Biodiversity*, 2nd edition, Waltham, MA: Academic Press, pp. 347–61.

Nelson, E., G. Mendoza and J. Regetz et al. (2009), 'Modeling multiple ecosystem services, biodiversity conservation, commodity production, and trade-offs at landscape scales', *Frontiers in Ecology and the Environment*, **7**(1), 4–11.

Pearce, D.W., G. Atkinson and S. Mourato (2006), *Cost Benefits Analysis and the Environment*, Paris: OECD.

Peh, K.S.H, A. Balmford and R.B. Bradbury et al. (2013), 'TESSA: a toolkit for rapid assessment of ecosystem services at sites of biodiversity conservation importance', *Ecosystem Services*, **5**, 51–57.

Polasky, S., E. Nelson and J. Camm et al. (2008), 'Where to put things? Spatial land management to sustain biodiversity and economic returns', *Biological Conservation*, **141**(6), 1505–24.

Polasky, S., E. Nelson and D. Pennington et al. (2011), 'The impact of land-use change on ecosystem services, biodiversity and returns to landowners: a case study in the State of Minnesota', *Environmental and Resource Economics*, **48**(2), 219–42.

Quinet, A., L. Baumstark, J. Célestin-Urbain et al. (2009), *La valeur tutélaire du carbone* [The Shadow Price of Carbon], Paris: La Documentation Française.

Reijnen, R. and R. Foppen (1994), 'The effect of car traffic on breeding bird populations in woodland. Evidence of reduced habitat quality for willow warblers (*phylloscopus trochilus*) breeding close to a highway', *Journal of Applied Ecology*, **31**, 85–94.

Robien, G. (2005), *Instruction Cadre relative aux méthodes d'évaluation économique des grands projets d'infrastructures de transport* [Instruction Framework on Economic Valuation Methods for Major Transport Infrastructure Projects], Paris: Ministère de l'Equipement, des Transports, du Logement, du Tourisme et de la Mer.

Ruckelshaus, M., E. McKenzie and H. Tallis et al. (2015), 'Notes from the field: lessons learned from using ecosystem service approaches to inform real-world decisions', *Ecological Economics*, **115**, 11–21.

Tardieu, F., B. Parent and C.F. Caldeira et al. (2014), 'Genetic and physiological controls of growth under water deficit', *Plant Physiology*, **164**(4), 1628–35.

Tardieu, L., S. Roussel and J.M. Salles (2013), 'Assessing and mapping global climate regulation service loss induced by terrestrial transport infrastructure construction', *Ecosystem Services*, **4**, 73–81.

Tardieu, L., S. Roussel and J.D. Thompson et al. (2015), 'Combining direct and indirect impacts to assess ecosystem service loss due to infrastructure construction', *Journal of Environmental Management*, **152**, 145–57.

Trocmé, M., S. Cahill and H. de Vries et al. (2002), *COST 341 – Habitat Fragmentation Due to Transportation Infrastructure: The European Review*, Luxembourg: Office for Official Publications of the European Communities.

Trombulak, S.C. and C.A. Frissell (2000), 'Review of ecological effects of roads on terrestrial and aquatic communities', *Conservation Biology*, **14**(1), 18–30.

Van der Zande, A.N., W.J. ter Keurs and W.J. van der Weijden (1980), 'The impact of roads on the densities of four bird species in an open field habitat – evidence of a long distance effect', *Biological Conservation*, **18**(4), 299–321.

Van der Zee, D. (1990), 'The complex relationship between landscape and recreation', *Landscape Ecology*, **4**, 225–36.
Von Haaren, C. and C. Albert (2011), 'Integrating ecosystem services and environmental planning: limitations and synergies', *International Journal of Biodiversity Science, Ecosystem Services & Management*, **7**(3), 150–67.
Watkiss, P., M. Holland and F. Hurley et al. (2006), *Damage Costs for Air Pollution*, report for Defra, London: AEA Technology.

7. Addressing biodiversity and ecosystem services in Life Cycle Assessment*

Assumpció Antón, Danielle Maia de Souza, Félix Teillard and Llorenç Milà i Canals

7.1 INTRODUCTION

7.1.1 Brief Introduction to Life Cycle Assessment

Life Cycle Assessment (LCA) is a method to quantitatively assess the environmental impacts of goods and processes from 'cradle to grave'. LCA models are underpinned by cause–effect relationships in different environmental compartments, which enable us to understand the environmental consequences of human actions (Hellweg and Milà i Canals, 2014). According to the ISO standard 14040 (ISO, 2006a), an LCA study is divided into four phases. The first phase is the description of the goal and scope of the study, which includes determination of the objectives and assumptions of the study and sets the system boundaries (system processes to be included in the analysis). The second phase, life cycle inventory (LCI) analysis, consists of the compilation of inputs (resources, energy) and outputs (emissions, waste) for each of the processes included within the defined system boundaries. In the third phase, life cycle impact assessment (LCIA), inventory flows (inputs and outputs) are assigned to the different impacts and characterized to a unit common to all contributions within the impact category (e.g., kg CO_2 equivalents for greenhouse gases contributing to the impact category 'global warming'). Finally, in the fourth stage, interpretation, the results from the previous phases are evaluated in relation to the goal and scope in order to draw conclusions and recommendations (ISO, 2006a). Focusing on the third stage and according to ISO 14044 (ISO, 2006b) the indicator of an impact category can be chosen anywhere along the impact pathway, which links inventory data to damage on 'Areas of Protection' (AoP): human health, natural environment and resources. The relationship between inventory data and the magnitude of impacts on the AoP in LCIA are expressed in terms of characterization factors (CF) in relation to a reference substance or flow (e.g., the value of global warming CF for methane is 34 kg CO_2 equivalent per kg of CH_4 with a time horizon of 100 years; Pennington

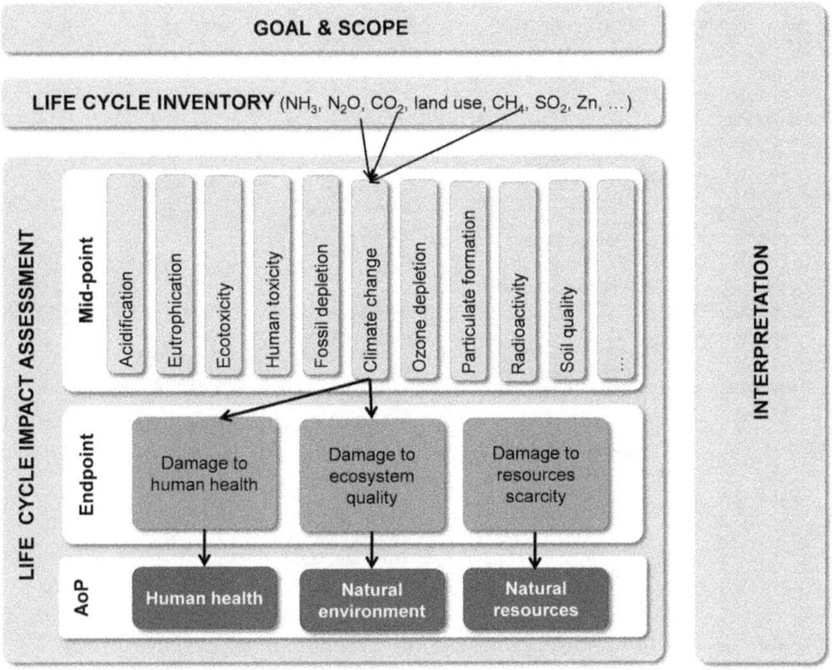

Note: Arrows are shown as an example of pathway; in this case for climate change mid-point impact, from greenhouse gases to damage to human health and natural environment.

Figure 7.1 Life Cycle Assessment scheme including the different stages: goal and scope definition; inventory; impact assessment modelling from mid-points, endpoints, to Areas of Protection and interpretation

et al., 2004; IPCC, 2013). Characterization at mid-point level models the impact using an indicator located somewhere along (but before the end of) the mechanism. Traditional characterization methods are examples of mid-point modelling (e.g., climate change, eutrophication, toxicity). Characterization at the endpoint level requires modelling all the way to the damage on the entities described by the AoP (EU-JRC, 2010) (Figure 7.1).

Within the LCA framework, biodiversity and ecosystem services (ES) loss resulting from environmental interventions (e.g., emission of acidifying substances, transformation or occupation of land) has been suggested as suitable indicators to assess endpoint damages related to natural environment AoP (Figure 7.2). Both biodiversity and ES are complex and multi-faceted concepts, with high specificity on a local scale. Thus,

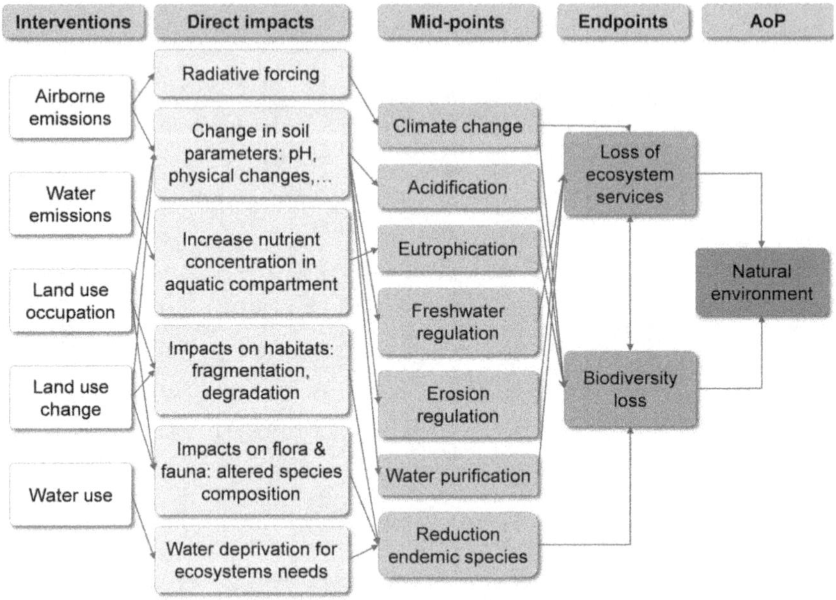

Figure 7.2 Major impact pathway flows leading from environmental interventions converted to mid-point and endpoint impact categories to natural environment AoP

it is very difficult to measure and aggregate impacts occurring across the product's life cycle, on a global scale, using a single metric, as is required in LCA (Milà i Canals and de Baan, 2015).

7.1.2 Impact Mechanisms Affecting Biodiversity and Ecosystem Services Included in LCA

There are many impact pathways leading to effects on biodiversity and ES. In LCA, many of these have been quantified, particularly for biodiversity. Impacts linked to chemical pollution have been included for a long time in LCA, with potentially disappeared fraction (PDF) of species used and even recommended (EU-JRC, 2011) as an indicator. Pollution impacts due to acidification and eutrophication have been modelled estimating the potentially not occurring fraction (PNOF) of species, referring to the difference between 1 and the ratio of species present at determined conditions (e.g., determined pH) to the highest possible number of species in optimum conditions. Stressor–response relationships representing the influence of pH and total phosphorus (P) concentration on PNOF (for acidification and eutrophication) are provided by Azevedo, van Zelm and Hendriks (2013) and Azevedo, van Zelm and Elshout (2013), respectively.

The consensus model agreed in LCA for toxicity impacts, USEtox (Rosenbaum et al., 2008) provides ecosystem CF for aquatic ecotoxicity. The calculation of these CF is based on the potentially affected fraction of species (PAF) that are expected to be (potentially) affected per unit mass of a chemical emitted.

Climate change is another major driver of biodiversity loss. Within LCA, De Schryver et al. (2009), based on the work of Thomas et al. (2004), approximated the effects on terrestrial ecosystems per unit greenhouse gas emission expressed as PDF of species, including plants, butterflies, birds and mammals.

Water use can also affect ecosystems as a result of changes in surface and groundwater withdrawals and flow regimes, and loss of freshwater quality. Freshwater deprivation, which is associated with water scarcity and water quality, eventually leads to a change in species diversity in aquatic and terrestrial ecosystems (Kounina et al., 2013). Milà i Canals et al. (2009) suggested the use of the freshwater ecosystem impact (FEI), an indicator based on environmental water requirements (EWR), as defined by Smakhtin et al. (2004) for all world river basins. Pfister et al. (2009) modelled the impacts of freshwater use to the endpoint level of natural ecosystems, using net primary production (NPP) as a proxy for ecosystem quality. Pfister et al.'s (2009) method was regionalized to provide spatially differentiated impact factors at the watershed and country level. Hanafiah et al. (2011) derived global CF per watersheds, relating PDF of fish species richness to river discharge. Van Zelm et al. (2011) estimated CF based on the effect of groundwater extraction on PNOF of terrestrial species for the Netherlands. Verones et al. (2013) derived site-specific CF for impacts of water consumption for wetlands, taking into account an effect factor that includes species richness, vulnerability and habitat loss risk for each wetland and five different taxa: waterbirds, non-residential birds, amphibians, reptiles and water-dependent mammals.

Nowadays there is not yet a common unit to express the different endpoint indicators in the natural environment AoP. For the different environmental impacts we have seen indicators could be expressed in either PAF, PDF or PNOF. In the PROSUITE project framework (Dong et al., 2013) it has been suggested that an equivalence between PAF, PDF and PNOF (i.e., PAF = PDF = PNOF) should be assumed.

Land use and land use change have been considered the most important drivers of biodiversity loss, followed by climate change (MA, 2005). Major efforts within the LCA framework have therefore been devoted to developing methods to characterize land use impacts on biodiversity as well as on ES. These are explained in the next section.

7.2 INDICATORS TO ADDRESS BIODIVERSITY LOSS IN LCA

Research on biodiversity indicators in LCA has been ongoing for more than 15 years, mainly focused on the assessment of land use impacts, with a few reviews on the topic (Lindeijer, 2000; Curran et al., 2011; Koellner et al., 2013b; Souza et al., 2015). However, no consensus has yet been reached on the use of a specific method (Milà i Canals and de Baan, 2015), and this keeps limiting the inclusion of this important impact pathway in LCA studies, hampering its applicability as a decision-making tool.

Indicators used to measure biodiversity can be described in terms of three levels (genes, species, ecosystems) and three attributes (composition, function, structure). Composition refers to the quantity and variety of biodiversity elements, such as species present or habitat type. Function describes the ecological and evolutionary process acting among elements (e.g., predator–prey relationship); and structure, the physical organization of elements, such as spatial configuration (Noss, 1990; Curran et al., 2011). Most of the current methods in LCA are based on indicators that reflect changes in compositional aspects of biodiversity, namely species richness (EU-JRC, 2011). Figure 7.3 shows the main CF proposed in the LCA framework arranged according to their different attributes, because it is difficult to establish the exact boundaries between the different indicators we have included compositional and indicators as attributes for species and differentiate them from functional and structural ones valid for species and ecosystems indicators.

7.2.1 Species Level

Compositional indicators

Relative species richness, expressed as m^2/yr of PDF of species, has been largely used as CF to express damage at endpoint. Historically, PDF has been interpreted as the fraction of species that has a high probability of no occurrence in a region, due to unfavourable conditions (Goedkoop and Spriensma, 2000).

Species richness is strongly scale dependent and increases in a non-linear way when larger areas are analysed. A strong pattern in ecology is the species–area relationship (SAR), used to predict the loss of species from an area; that is, the number of species found in a region is a positive function

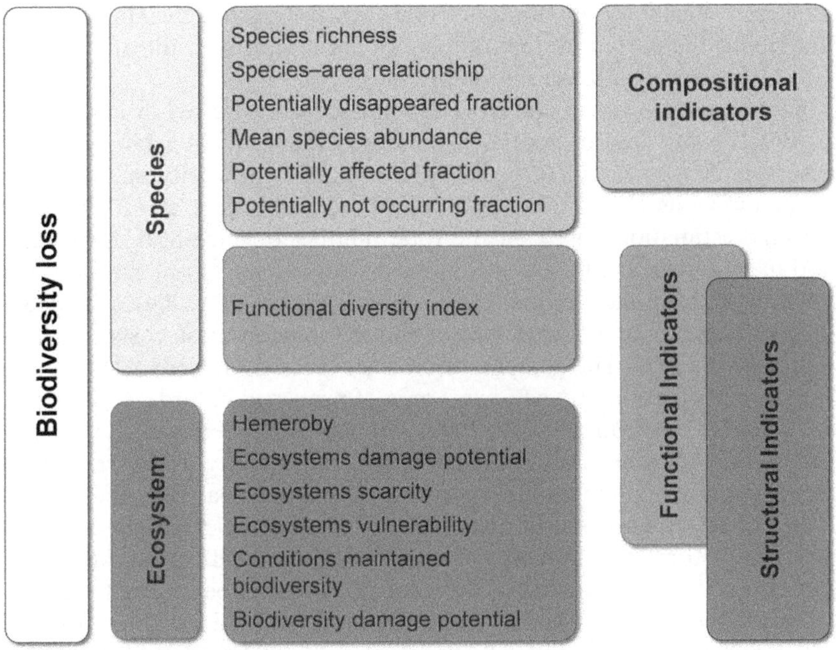

Figure 7.3 Biodiversity loss indicators suggested in the LCA framework according to species and ecosystems attributes

of the area of the region. SAR models are commonly used to predict species extinction caused by habitat loss, by expressing the number of new species, S, of an area, A, as a function of the number of original species, S_0, occurring in the original habitat area, A_0 (Koh et al., 2010).

De Baan, Alkemade and Koellner (2013) compared four different biodiversity indicators to the species richness (derived from SAR). These were (1) Fisher's α (sampling corrected alpha diversity), (2) Shannon's entropy (diversity measure), (3) mean species abundance of original species, and (4) Sørensen's S_s (dissimilarity measure). The authors recognized the importance of developing CF that include information on habitat diversity, especially indicators 3 and 4. The lack of data led them to conclude that in the meantime relative species richness could be a suitable indicator as far as regional differences in absolute species richness could be included.

As described in the previous section, PAF has been used as biodiversity indicator for ecotoxicological impacts. It is defined as the percentage of species within a community or taxonomic group that are expected to be exposed above a level that affects them. The PAF is based on the principles of species sensitivity distribution (SSD), which is a statistical distribution describing the variation among a set of species, as a result of the toxic effect of a certain substance or mixture.

Some methods try to convert endpoint score, expressed in PAF, PDF or PNOF, into loss of species in final ecosystem quality damage using a species density factor (Goedkoop et al., 2009) or attributing them an economic value (Vogtländer et al., 2004). Goedkoop et al. (2009) made a rough estimation based on the total number of terrestrial, freshwater and marine registered species combined with the terrestrial area and the volume of fresh and marine waters. Vogtländer et al. (2004) provided a factor based on the prevention costs and the compensation costs of degradation of 'nature'. The compensation costs (eco-costs) were calculated as the costs of creation of a new extra area of protected nature.

GLOBIO3 (Alkemade et al., 2009), a model developed to assess human-induced changes in biodiversity, describes biodiversity as the remaining mean species abundance (MSA) of original species, relative to their abundance in pristine areas. The latter are assumed to be areas that have not yet been disturbed by human activities for a prolonged period. MSA represents the average response of the total set of species belonging to an ecosystem. Alkemade et al. (2009) emphasized that MSA does not completely cover the complex biodiversity concept, and complementary indicators should be included, when used in extensive biodiversity assessments.

Functional indicators

Functional diversity (FD) is a component of biodiversity that involves the characteristics (functional attributes or traits) of organisms that have effects on ecosystem processes or services and that determine their responses to their abiotic and biotic environment (Petchey and Gaston, 2006; Harrington et al., 2010; Aubin et al., 2013). These features drive the interaction among species and between them and their environment (Mouillot et al., 2007; Lavorel et al., 2011) and help to understand the relation between species richness and ecosystem functions (Cadotte et al., 2011; Clark et al., 2012). Functional traits are the measurable characteristics or features of organisms that reflect their response to variations in the environment (response traits) or their effects (effect traits) and reveal their effects on different ecosystem processes (Harrington et al., 2010).

FD has been suggested as a promising complement to taxonomic measures (De Bello et al., 2010; Cadotte et al., 2011) or even as a more consistent indicator, in order to assess biodiversity loss (Díaz and Cabido, 2001; Petchey and Gaston, 2006; Flynn et al., 2009). Compared to species richness (SR), FD measures can reflect and explain specific responses to changes in the environment, even when small changes in richness occur.

Although there is a link between function at species and ecosystem level, the use of FD to predict ecosystem functioning (Norman et al., 2003) is not an easy task and needs an understanding of communities and ecosystems and the choice of methods to quantify it (Petchey and Gaston, 2006).

Within the LCA framework, Souza et al. (2013) proposed CF for biodiversity loss using the FD Index from Petchey and Gaston (2006) for three taxonomic groups: mammals, birds and plants. Data on different studies and information compiled by Gibson et al. (2011) and Flynn et al. (2009) were used, while data on functional traits were gathered in national databases from different countries.

To date, there is as yet no consensus on the method to quantify FD (Mouchet et al., 2010). This consists of a first challenge in using FD indices in LCA. Second, although these indices allow a better detection of community response to changes, they are data hungry (e.g., data on species richness, functional traits, abundance for different land use types).

7.2.2 Ecosystem Level

Although impacts such as eutrophication or land use change cause deep structural modifications to habitats, very few attempts have been done in LCA that aim to include such structural pathway (Curran et al., 2011). At the species level, structure relates to the spatial configuration of habitats (e.g., fragmentation, patchy arrangement). At the ecosystem level, structure is closely related to function and LCA indicators have been proposed at the interface of these two attributes.

Brentrup et al. (2002) suggested the hemeroby concept as a measure of the 'naturalness' of an area. The authors assigned CF for the 'naturalness degradation potentials' (NDP) to different levels of hemeroby for different land use types.

Michelsen (2008) proposed a new method to assess biodiversity indirectly by means of three factors: ecosystem scarcity (ESc), ecosystem vulnerability (EV) and conditions for maintained biodiversity (CMB). The first was an adaptation of the method developed by Weidema and Lindeijer (2001), who expressed ESc as the inverse value of the potential area of a biome. Michelsen (2008) adapted the concept to the ecoregion level and normalized it in relation to the potential area of the most widespread ecoregion. EV was introduced as an indicator to give information about the present total area pressure to an ecosystem type and relate the existing area of an ecosystem to its potential area. However, it is difficult to find data on EV for all existing ecoregions. Coelho et al. (2014) performed a sensitivity test of different approaches to calculating EV; this showed that present values for EV at ecoregion level are probably too crude to give a realistic picture of the onsite situation in specific cases. CMB is an index that provides the relative importance of the sum of different key factors (e.g., amount of decaying wood, areas set aside, introduction of alien tree species) for an ecosystem.

Koellner and Scholz (2008) developed CF for Europe, linking numerous classes of land use/cover and intensities to biological diversity, expressed as an ecological damage potential (EDP) indicator. Mueller et al. (2014) calculated a biodiversity damage potential (BDP) of land use as the sum of land use occupation and transformation impacts. They adapted the relative species richness method proposed by De Baan, Alkemade and Koellner (2013), applying a biodiversity weighting factor based on absolute species richness, irreplaceability and vulnerability. Souza et al. (2015) discuss some of the limitations and drawbacks in using each of these approaches.

7.3 INDICATORS TO ADDRESS ECOSYSTEM SERVICES IN LCA

ES are the benefits that humans obtain from ecosystems processes and conditions (MA, 2005). They include provisioning, regulating, supporting and cultural services.

Accounting for ES in decision-making tools such as LCA is a challenge, as not all the services are easily quantified (Zhang et al., 2010; Bakshi and Small, 2011). Moreover as a holistic representation of the provided services is not feasible, aggregation of raw data and decisions on specific indicators is advised. Ideally, ES should be represented by a multi-indicator, but the aggregation of multiple services represents a challenge. In practice, LCA methods have attempted to address one or a few of such ES (see Figure 7.4). Indeed, impacts on species habitat directly affect biological diversity, as well as species diversity may influence ES provision (Wall and Nielsen, 2012). The evidence of this close relationship is reflected in the quantification of some existing indicators, which often encompass not only biodiversity attributes, but also different ES.

Zhang et al. (2010) carried out a critical review on LCA methods accounting for ES and identified important gaps and potential for research. We explain in more detail the inclusion of different services in LCA in each of the following sections. Currently, the scope of LCA does not encompass the inclusion of non-material environmental aspects, such as cultural ES, and, therefore, we do not discuss them in detail.

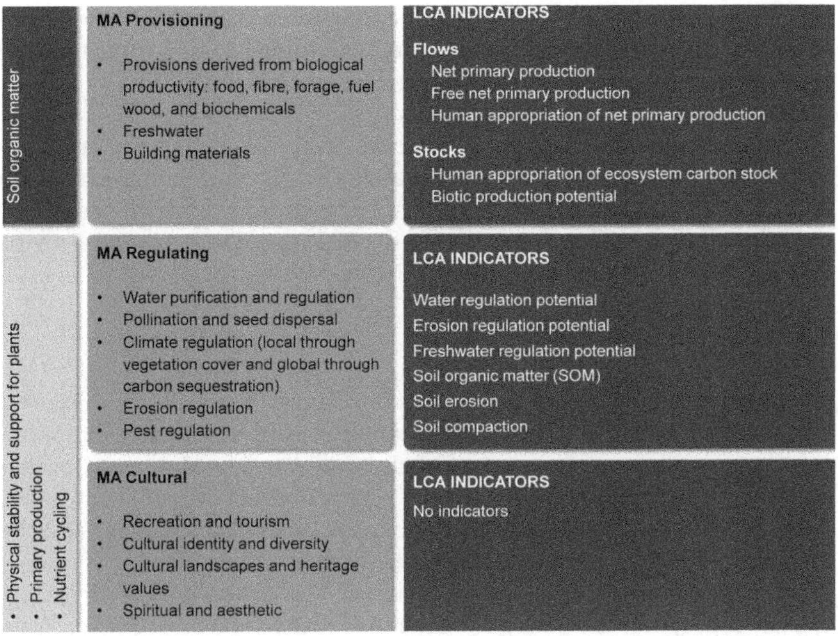

Figure 7.4 Relationship between Millennium Ecosystem Assessment ES and LCA indicators

7.3.1 Provisioning Services

According to the Millennium Ecosystem Assessment, provisioning services refer to the delivery of ecosystem products: food, fibre, fuel, genetic resources, ornamental resources and freshwater (Figure 7.4). Therefore, many authors have resorted to indicators related to biomass in order to quantify impacts on provisioning services. One of the earliest of such indicators in LCA is net primary production (NPP), and the amount of NPP returned to nature after human consumption of NPP: free NPP (fNPP) (Cowell and Clift, 1997; Lindeijer, 2000; Weidema and Lindeijer, 2001). Haberl et al. (2007) suggested human appropriation of NPP (HANPP) as a measure of impact from humans on ecosystems (in a way, opposed to fNPP), and provided a world map with estimated values of HANPP. These indicators related to NPP are measures of the flow of biomass production (measured in change of biomass produced per unit of area per year), which are affected by short-term fluctuations in weather conditions or nutrient inputs, among others (Figure 7.4). Other authors have suggested that stock-based indicators (measured in change of biomass per unit of area) are more suited to indicating the long-term capacity of ecosystems to deliver provisioning services; for example Brandão (2012) puts forward human appropriation of ecosystem carbon stock (HAPECS) as a longer-term indication of provisioning services. Brandão and Milà i Canals (2013) suggested that the impacts of land occupation and transformation should be estimated on its long-term ability to produce biomass, referred as biotic production potential (BPP). These authors recommended soil organic carbon (SOC) as an indicator for BPP, providing references to the link between SOC and productivity (Figure 7.4).

7.3.2 Regulating Services

Regulating services are the benefits obtained from the regulation of ecosystem processes, such as carbon sequestration of greenhouse gases and water purification. These services are often not duly considered in the decision-making context as their value is usually poorly understood and their change over time is usually slower in comparison with provisioning services. However, they are valuable to the conservation and maintenance of ecosystem functions (e.g., biodiversity regulation). The regulation of soil quality is a key aspect of maintaining soil functions such as soil biological activity, nutrient cycling, gases and water regulation. Soil quality, defined as the 'capacity of a living soil to function, within natural or managed ecosystem boundaries, to sustain plant and animal productivity, maintain or enhance water and air quality, and promote plant and animal health' (Karlen et al., 1997 in Doran, 2002, p. 4), is recognized as essential for the development of human activities such as food production (Garrigues et al., 2013).

The inclusion of different regulating services within the LCA framework was first proposed by Cowell and Clift (1997), who emphasized the need for further methodological development in LCA in order to include soil quality and nutrient depletion as abiotic resource depletion. Cowell and Clift (2000) proposed three impact categories to consider soil quantity and quality in LCA: soil erosion, change in organic matter and compaction.

Milà i Canals, Romanyà and Cowell (2007) proposed SOM as a single indicator for soil quality in LCA, in order to enable the assessment over the whole life cycle. Garrigues et al. (2012) suggested that a minimum data set (MDS) of soil quality indicators were needed in order to best assess impacts on soil quality, arguing that a more aggregated indicator simplifies the characterization of impacts too much.

Other soil ecological indicators have also been proposed. Núñez et al. (2013) developed a global method to characterize occupation impacts related to life cycle of products and processes on erosion regulation, at endpoint level. The effect of soil erosion on ecosystem quality is expressed using a growth-based value potential net primary production depletion (NPPD) related to the SOC lost through the soil eroded.

Saad et al. (2013) developed a characterization method, with spatially differentiated factors, based on three ES: erosion regulation potential, freshwater regulation potential and water purification potential. Nine input parameters, related to soil properties and landscape and climatic conditions, were used in the method.

New proposals and developments are still facing problems with the current land use framework applied in LCA and the level of detail demanded in modelling (Garrigues et al., 2012, 2013).

7.3.3 Supporting Services

The ecosystem also provides supporting services needed for the production of other ES. Some of these services include habitat provisioning, photosynthesis, maintenance of genetic biodiversity, water cycling and soil formation (MA, 2005). Changes in these services usually have indirect impacts on or provide benefits to human well-being.

As mentioned previously, Milà i Canals, Romanyà and Cowell (2007) proposed a methodology using soil organic matter (SOM) as a single indicator for life support functions (LSF) in LCA, since it is believed to play important roles in biotic production, climate regulation and maintenance of substance cycles. This indicator has the clear advantage of simplifying the modelling enough to make it operational, and in fact it is the only indicator for soil quality/ES so far implemented in LCA software and recommended in policy documents (EU-JRC, 2011).

Although SOM was pointed out as the best stand-alone indicator for soil quality for this model, some controversy exists with regard to the indicators to be used. Some important aspects of soil ecological quality have not been included in this method, such as soil erosion, soil compaction, toxic substances, acidification and salinization, depletion of nutrients and groundwater (Milà i Canals, Bauer and Depestele et al., 2007; Mattila et al., 2011).

7.4 CHALLENGES TO INCLUDING IMPACT ASSESSMENT ON BIODIVERSITY AND ECOSYSTEM SERVICES IN LCA

7.4.1 Data Availability

Data availability and feasibility of collecting new data are often limiting factors when conducting an LCA. This is particularly true regarding biodiversity and ES assessment, as data collection is time and resource consuming and demands a high level of expertise. Several sources can be useful for collecting data to developing and applying CF. The global biomes and ecoregions dataset (Terrestrial Ecoregions of the World – Olson et al., 2001), for example, shows the potential natural vegetation (PNV) reference state and includes biogeographical differentiation of the impacts on biodiversity.

Remote sensing can also be used to obtain data on biodiversity. An extensively studied measure is the Normalized Difference Vegetation Index (NDVI), derived from the red/near-red reflectance ratio. NDVI is a measure of 'greenness', which can be interpreted as a vegetation density measure (Weiss et al., 2001) and has a strong correlation with vegetation productivity (Pettorelli et al., 2005). As a proxy for vegetation productivity, it provides a description of one functional dimension of biodiversity. For other dimensions, such as composition (e.g., species richness), data availability is also the highest for vascular plants, which explains why these taxa have been used in many land use assessments (Schmidt, 2008). However, data also exist on the distribution range of other taxa (e.g., through the WWF Wildfinder database). On a European scale, the Pan-European Common Bird Monitoring Scheme coordinates national monitoring schemes and generates national and continental population trend indices for bird species (Gregory et al., 2009). The IUCN publishes – and constantly updates – a global Red List of species, classifying them according to their risk of extinction (Hambler and Canney, 2013). Finally, meta-analysis of the scientific literature can be used to obtain large-scale biodiversity data. This method has been used to compute global land use CF, mean species abundance (Alkemade et al., 2009) and population trends, for example, the Living Planet Index (Butchart et al., 2004).

Data availability is also limiting at the level of linking specific processes in the life cycle to the locations/regions where they occur (see, e.g., Milà i Canals et al., 2013), thus impeding the accuracy of biodiversity and ES assessment in LCA.

A further scientific and policy agreement could prioritize the need for data collection and an international database.

7.4.2 Reference State

It is commonly accepted that impact assessment in LCA requires comparing the environmental quality linked to the studied system to a reference situation or state (Milà i Canals, Bauer and Depestele et al., 2007). Selecting a suitable reference state to assess the difference in quality of the current land use and land use changes is a challenge because it has a strong influence on the results of the assessment and their interpretation. Three main alternatives for this reference state have been proposed: the natural or quasi-natural vegetation, historic land use and current land quality states.

The PNV has been used in several recent studies to describe the mature state of vegetation in the absence of human interventions (Chiarucci et al., 2010). One limitation of the PNV is that there may be multiple potential equilibria for the final state (Souza et al., 2015). After land recovery, the species richness and composition in the recovered land could differ with the original natural land cover (Curran et al., 2014). Several factors could influence this difference in species composition, such as the time of occupation and land use history (Souza et al., 2015). Aspects such as the presence of species, intraspecific genetic variability and landscape structure in the surrounding region will determine the potential for recolonization. Using the PNV as a reference may be more adapted to support decisions considering long-term effects of land use policies than shorter-term effects of land use change (Milà i Canals et al., 2013).

Alternative options for the reference are a historic state or the current mix of land uses, as proposed for Europe by Koellner and Scholz (2008). De Schryver et al. (2010) and De Baan, Alkemade and Koellner (2013) proposed a combination of both a current and a semi-natural reference state. De Baan et al. (ibid.) defined a reference situation as the current natural mix of natural land covers (forest, wetlands, shrubland, grassland, bare area, snow and ice, lakes and rivers). Choosing these options would result in higher impact for land use changes occurring in areas with remaining natural land covers. It should be more adapted to supporting decisions addressing short-term land use change or aiming to preserve traditional land uses that differ from the PNV, such as agricultural landscapes, or extensively managed forests. Land cover data exist for recent years at continental (e.g., Europe, CORINE Land Cover for years 2000, 2006 and 2012) to global scale (e.g., Global Land Cover for year 2000, Global Map for year 2003), which could be used to describe a proxy for the current state.

In any case we recommend that the goal and scope sections of an LCA should clearly state which reference state has been considered.

7.4.3 Spatial Differentiation

For global impact categories like climate change and stratospheric ozone depletion the resulting impacts are independent of the location where the emission occurs (Finnveden et al., 2009). For other impacts, such as biodiversity loss or ecosystem damage, neglecting specific details of the emitting source or receiving environment at regional or local scales may lead to more uncertain results.

The need for spatial differentiation is very clear for land use impacts; that is, the results of transforming 1 ha of area in the tropics are different from those in a desert area. Land use can alter the species richness in an occupied area or even in the surrounding region. The local impact describes the change in species richness on an occupied area, compared to the species richness on the baseline land. The regional impact describes the marginal species change in the surrounding area (De Schryver et al., 2010).

Local and regional impacts of land use are linked to the ecological concepts of α, β and γ species diversity: α diversity is the species diversity of a local community or habitat, β diversity is the diversity (dissimilarity) between communities, and γ is the diversity of the whole region or landscape (Whittaker, 1960). LCA frameworks usually do not consider β and γ diversity. Geyer, Stoms and Lindner (2010) proposed an interesting step towards accounting for regional impacts by coupling LCA and geographical information systems (GIS) to model land use scenarios, although they only considered α diversity. Focusing on α diversity only enables us to describe regional impacts through the relative species richness of the different land uses and their relative areas in the region. This approach does not allow the consideration of the most important aspect of regional diversity, which is the dissimilarity of the communities of species within the different land uses composing the region. For this reason, future approaches should also include β and γ diversity.

7.4.4 Biodiversity Coverage

As we have seen in the previous section, among the three levels of biodiversity (genes, species and ecosystem) the genetic level has not been addressed yet in existing LCA methodologies, which are mainly focused on species-level indicators. More recently, methodologies have also been developed on the ecosystem level.

At the species level, the vast majority of biodiversity indicators used in LCA are based on species richness or abundance and therefore these provide an incomplete vision of this dimension. First, only few taxonomic groups are often used to build indicators. Most indicators use vascular plants as proxy for other groups, mainly due to the higher data availability. Other taxa were also addressed in a few studies, such as molluscs and moss (Koellner, 2003; Koellner and Scholz, 2008); mammals, birds, amphibians and insects (De Schryver et al., 2009); multiple arthropod and vertebrate taxa (Alkemade et al., 2009; De Baan, Mutel and Curran, 2013). The focus on vascular plants also relies on the assumption that their diversity is reasonably well correlated with other terrestrial taxa (Weidema and Lindeijer, 2001; Michelsen, 2008). Addressing other biodiversity dimensions could be a way to overcome this limitation. Souza et al. (2013) showed that CF for functional diversity of species were significantly different from those for species richness and that their use in LCA could be useful to establishing links with key ecosystem functions. Second, species richness and abundance indicators do not reflect the conservation value of species, neglecting aspects such as endemism, rarity, threatened status or cultural importance. Conversely, some species may have a lower conservation value because they are non-native, generalist or more common, and even a negative conservation value, as invasive species. There have been initial efforts to take into account these differences in conservation ecology, for instance by weighting species richness for IUCN Threatened Species (Michelsen, 2008; Souza, 2010) or by differentiating between endemic and non-endemic species loss (Souza, 2010; De Baan, Mutel and Curran, 2013).

At the ecosystem level, LCA methodologies focus mainly on the functional aspects of biodiversity, as explained in Section 7.2.1. The structure dimension is, however, rarely taken into account in LCA, although it has a pivotal importance at the landscape level. In certain landscapes, agricultural ones in particular, habitat is not continuous, but spatially structured into patches or a mosaic. The landscape pattern determines the structure at the species level and also influences the function of the ecosystems (Tscharntke et al., 2005). By focusing on simple land use elementary flows, that is, land area transformed or occupied, the current framework (Koellner et al., 2013b) fails to account for spatial structure. Therefore the need to better account for spatial structure, and its underlying ecological processes, is an important challenge for LCA methodologies. Only Alkemade et al. (2009) computed CF for fragmentation by roads and other infrastructures. LCA could also use the concept of countryside SARs that take into account multiple habitats and connectivity (Pereira and Daily, 2006).

In summary, there is a need to develop LCA methodologies to broaden the biodiversity coverage to species other than plants and to the functional and structural dimensions of biodiversity.

7.4.5 Geographic Coverage

Most regional LCA methodologies have been developed for Europe (e.g., Koellner, 2003; Koellner and Scholz, 2008; Goedkoop et al., 2009) although a few studies covered other regions, such as South-East Asia (Schmidt, 2008) or the Americas (Souza et al., 2013). Several recent methodologies cover the global scale and emphasize the importance of uncertainty assessment of the results (Azevedo, Henderson and van Zelm, 2013; Azevedo, van Zelm and Hendriks, 2013; Brandão and Milà i Canals, 2013; De Baan, Alkemade and Koellner, 2013; De Baan, Mutel and Curran, 2013; Saad et al., 2013; Verones et al., 2013).

Another important aspect related to geographic coverage is biogeographical differentiation; that is, the fact that the same type of land use may have different impact on biodiversity at different locations of the globe (Koellner et al., 2013b). This aspect is particularly important for methodologies adopting a wide geographical coverage. For instance, the mean species abundance (MSA) of Alkemade et al. (2009) is defined for a global scale but without any biogeographical differentiation. It means that the biodiversity MSA value of undisturbed forest, or the biodiversity loss when these forests are converted to pasture is the same in Europe and Latin America despite wide knowledge of greater levels of biodiversity in the latter (Gaston, 2000). De Baan, Alkemade and Koellner (2013) developed global land use CF accounting for regional differences among nine different biomes. Koellner et al. (2013a) proposed five different levels of biogeographical differentiation: first distinguishing between terrestrial, freshwater, and marine biomes and further specifying climatic regions, specific biomes, ecoregions, and finally indicating the exact geo-referenced information of land use.

Koellner et al. (2013a) also proposed several levels for a harmonized land use/cover classification. Four nested levels describe land cover/use classes with varying levels of intensiveness, ranging from coarse ($n = 11$) to refined ($n = 74$) classes. For certain land uses, such as those related to agriculture, there may be significant variations in the management practices. For instance, intensive vs extensive crops or irrigation are practices that have an important impact on biodiversity (Matson et al., 1997; Malmqvist and Rundle, 2002). Land use intensity is also a key variable influencing biodiversity in grassland (Vickery et al., 2001), cropland (Stoate et al., 2001) and used forests (Putz et al., 2001). For these land uses, it is very likely that difference in ecosystem quality could be lower between two distinct land use classes than within the same land use class with important variation in intensity and production practices.

In terms of geographic coverage it would be also desirable to agree the level of detail that LCA is most suited to cover, as a compromise between the detailed level, often desired for a more meaningful impact assessment, and the level of information available in life cycle inventories (see, e.g., Milà i Canals et al., 2013).

7.4.6 Non-linear Damage Relationships

In current LCA methodologies a linear scaling of impact is assumed. Yet, non-linear responses of species or ecosystem to disturbance and habitat loss in particular are common. Non-linear relationships in impacts on biodiversity and ES mean that converting 1 m^2 of forest does not necessarily yield 1000 times smaller impacts than converting 1000 m^2. Milà i Canals and De Baan (2015) provide further discussion on the topic of scale.

A simple pattern of non-linear response is when habitat fragmentation worsens the effect of habitat loss. Increased fragmentation leads to smaller and more isolated patches of the original habitat, for which the island biogeography theory predicts reduced number of species (Levins, 1969). Reduced biomass (Laurance, 1997) and species loss (review in Turner, 1996) in fragmented habitats have been widely evidenced. In practice, many types of land transformation are accompanied by habitat fragmentation. For instance, most of the pasture expansion into tropical forest occurs in a diffuse manner, causing fragmentation of the original forest habitat (Wassenaar et al., 2007). In this case, the linear damage relationships assumed by current LCA methodologies would systematically underestimate the impacts on biodiversity. Chaplin-Kramer et al. (2015) have recently explored the effect of considering boundary effects from land use change in the LCA results.

Due to the availability of information in this area, the non-linear damage relationships would hence be a promising aspect to be assessed. However, it would be important to define the limits of acceptability between possible simplification and complexity required.

7.5 CONCLUSIONS

ES and biodiversity loss have been proposed as suitable endpoint indicators to quantify damage in the AoP of natural environments. Due to the different nature (scale, location, impact pathway, etc.) of damages, both are complex issues, difficult to measure with simple indicators. In addition, the need for a large dataset of species and land data for different biogeographical areas around the world, as well as high requirements of data accuracy for local to regional conditions, represent important challenges for the scientific community.

To date most of the current methods in LCA are based on indicators that reflect changes in compositional aspects of biodiversity – mainly changes in species richness. However, there is a common agreement that complementary indicators should be included. Research on functional diversity indicators may be particularly promising given their link with ES. For ES, even though indicator sets have been explored to account for their multiple dimensions, only single simple indicators such as SOM have actually been made operational in LCA studies. The extensive ongoing research during the last few years has led to substantial advances in the consideration of biodiversity and ES in LCA results, but further consensus is still required for them to become commonly included in LCA studies.

NOTE

* The views expressed in this chapter are those of the authors and do not necessarily reflect those of their institutions.

REFERENCES

Alkemade, R., M. van Oorschot and L. Miles et al. (2009), 'GLOBIO3: a framework to investigate options for reducing global terrestrial biodiversity loss', *Ecosystems*, **12**(3), 374–90.

Aubin, I., L. Venier and J. Pearce et al. (2013), 'Can a trait-based multi-taxa approach improve our assessment of forest management impact on biodiversity?', *Biodiversity and Conservation*, **22**(12), 2957–75.

Azevedo, L.A., A.D. Henderson and R. van Zelm et al. (2013), 'Assessing the importance of spatial variability versus model choices in life cycle impact assessment: the case of freshwater eutrophication in Europe', *Environment Science & Technology*, **47**(23), 13565–70.

Azevedo, L.A., R. van Zelm and P.M.F. Elshout et al. (2013), 'Species richness – phosphorus relationships for lakes and streams worldwide', *Global Ecology and Biogeography*, **22**(12), 1304–14.

Azevedo, L.B., R. van Zelm and A.J. Hendriks et al. (2013), 'Global assessment of the effects of terrestrial acidification on plant species richness', *Environmental Pollution*, **174**, 10–15.

Bakshi, B. and M.J. Small (2011), 'Incorporating ecosystem services into life cycle assessment', *Journal of Industrial Ecology*, **15**(4), 477–8.

Brandão, M. (2012), 'Food, feed, fuel, timber or carbon sink? Towards sustainable land use – a consequential life-cycle approach', PhD thesis, Guildford, UK: Centre for Environmental Strategy; Division of Civil, Chemical and Environmental Engineering; Faculty of Engineering and Physical Sciences, University of Surrey.

Brandão, M. and L. Milà i Canals (2013), 'Global characterisation factors to assess land use impacts on biotic production', *The International Journal of Life Cycle Assessment*, **18**(6), 1243–52.

Brentrup, F., J. Kfisters and J. Lammel et al. (2002), 'Life cycle impact assessment of land use based on the hemeroby concept', *The International Journal of Life Cycle Assessment*, **7**(6), 339–48.

Butchart, S.H.M., A.J. Stattersfield and L.A. Bennun et al. (2004), 'Measuring global trends in the status of biodiversity: Red List indices for birds', *PLoS Biology*, **2**(12), e383.

Cadotte, M.W., K. Carscadden and N. Mirotchnick (2011), 'Beyond species: functional diversity and the maintenance of ecological processes and services', *Journal of Applied Ecology*, **48**(5), 1079–87.

Chaplin-Kramer, B., R. Sharp and L. Mandle et al. (2015), 'Understanding how spatial patterns of agricultural expansion impact biodiversity and carbon storage at a landscape level', *PNAS*, **112**(24), 7402–7.

Chiarucci, A., M.B. Araújo and G. Decocq et al. (2010), 'The concept of potential natural vegetation: an epitaph?', *Journal of Vegetation Science*, **21**(6), 1172–8.

Clark, C.M., D.F.B. Flynn and B.J. Butterfield et al. (2012), 'Testing the link between functional diversity and ecosystem functioning in a Minnesota grassland experiment', *PLoS ONE*, **7**(12), e52821.

Coelho, C.R.V. and O. Michelsen (2014), 'Land use impacts on biodiversity from kiwifruit production in New Zealand assessed with global and national datasets', *The International Journal of Life Cycle Assessment*, **19**(2), 285–96.

Cowell, S.J. and R. Clift (1997), 'Impact assessment for LCAs involving agricultural production', *The International Journal of Life Cycle Assessment*, **2**(2), 99–103.

Cowell, S.J. and R. Clift (2000), 'A methodology for assessing soil quantity and quality in life cycle assessment', *Journal of Cleaner Production*, **8**(4), 321–31.

Curran, M., L. de Baan and A.M. de Schryver et al. (2011), 'Toward meaningful end points of biodiversity in life cycle assessment', *Environmental Science & Technology*, **45**(1), 70–79.

Curran, M., S. Hellweg and J. Beck (2014), 'Is there any empirical support for biodiversity offset policy?', *Ecological Applications*, **24**(4), 617–32.

De Baan, L., R. Alkemade and T. Koellner (2013), 'Land use impacts on biodiversity in LCA: a global approach', *The International Journal of Life Cycle Assessment*, **18**(6), 1216–30.

De Baan, L., C.L. Mutel and M. Curran et al. (2013), 'Land use in life cycle assessment: global characterization factors based on regional and global potential species extinction', *Environmental Science & Technology*, **47**(16), 9281–90.

De Bello, F., S. Lavorel and S. Díaz et al. (2010), 'Towards an assessment of multiple ecosystem processes and services via functional traits', *Biodiversity and Conservation*, **19**(10), 287–93.

De Schryver, A.M., K.W. Brakkee and M.J. Goedkoop et al. (2009), 'Characterization factors for global warming in life cycle assessment based on damages to humans and ecosystems', *Environmental Science & Technology*, **43**(6), 1689–95.

De Schryver, A.M., M.J. Goedkoop and R.S.E.W. Leuven et al. (2010), 'Uncertainties in the application of the species area relationship for characterisation factors of land occupation in life cycle assessment', *The International Journal of Life Cycle Assessment*, **15**(7), 682–91.

Díaz, S. and M. Cabido (2001), 'Vive la différence: plant functional diversity matters to ecosystem processes', *Trends in Ecology & Evolution*, **16**(11), 646–55.

Dong, Y., A. Laurent and M.Z. Hauschild (2013), *Recommended Assessment Framework, Characterisation Models and Factors for Environmental Impacts and Resource Use*, report prepared within the EC 7th Framework Project: PROSUITE: Development and application of a standardized methodology for the PROspective SUstaInability assessment of TEchnologies, accessed 17 January 2016 at http://www.prosuite.org/c/document_library/get_file?uuid=ed02587a-4787-4bc2-904b-cb878fac9616&groupId=12772.

Doran, J.W. (2002), 'Soil health and global sustainability: translating science into practice', *Agriculture, Ecosystems & Environment*, **88**(2), 119–27.

EU-JRC (2010), *International Reference Life Cycle Data System (ILCD) Handbook. General Guide for Life Cycle Assessment – Detailed Guidance*, Luxembourg: European Commission – Joint Research Centre – Institute for Environment and Sustainability.

EU-JRC (2011), *International Reference Life Cycle Data System (ILCD) Handbook – Recommendations for Life Cycle Impact Assessment in the European Context*, Luxembourg: European Commission – Joint Research Centre – Institute for Environment and Sustainability, Publications Office of the European Union.

Finnveden, G., M. Hauschild and T. Ekvall et al. (2009), 'Recent developments in life cycle assessment', *Journal of Environmental Management*, **91**(1), 1–21.

Flynn, D.F.B., M. Gogol-Prokurat and T. Nogeire et al. (2009), 'Loss of functional diversity under land use intensification across multiple taxa', *Ecology Letters*, **12**(1), 22–33.
Garrigues, E., M.S. Corson and D.A. Angers et al. (2012), 'Soil quality in life cycle assessment: towards development of an indicator', *Ecological Indicators*, **18**(1), 434–42.
Garrigues, E., M.S. Corson and D.A. Angers et al. (2013), 'Development of a soil compaction indicator in life cycle assessment', *The International Journal of Life Cycle Assessment*, **18**(7), 1316–24.
Gaston, K.J. (2000), 'Global patterns in biodiversity', *Nature*, **405**(6783), 220–27.
Geyer, R., J. Lindner and D. Stoms et al. (2010), 'Coupling GIS and LCA for biodiversity assessments of land use. Part 2: impact assessment', *The International Journal of Life Cycle Assessment*, **15**(7), 692–703.
Geyer, R., D.M. Stoms and J.P. Lindner et al. (2010), 'Coupling GIS and LCA for biodiversity assessments of land use. Part 1: inventory modeling', *The International Journal of Life Cycle Assessment*, **15**(5), 454–67.
Gibson, L., T.M. Lee and L.P. Koh et al. (2011), 'Primary forests are irreplaceable for sustaining tropical biodiversity', *Nature*, **478**(7369), 378–83.
Goedkoop, M. and R. Spriensma (2000), *The Eco-Indicator 99. A Damage Oriented Method for Life Cycle Impact Assessment*, Amersfoort, the Netherlands: PRé Consultants B.V.
Goedkoop, M., R. Heijungs and M. Huijbregts et al. (2009), *ReCiPe 2008 A Life Cycle Impact Assessment Method which Comprises Harmonised Category Indicators at the Midpoint and the Endpoint Level*: Amersfoort/Leiden/Nijmegen/Bilthoven: PRé Consultants/CML, University of Leiden/Radboud University/RIVM.
Gregory, R.D., S.G. Willis and F. Jiguet et al. (2009), 'An indicator of the impact of climatic change on European bird populations', *PLoS ONE*, **4**(3), e4678.
Haberl, H., K.H. Erb and F. Krausmann et al. (2007), 'Quantifying and mapping the human appropriation of net primary production in earth's terrestrial ecosystems', *Proceedings of the National Academy of Sciences*, **104**(31), 12942–7.
Hambler, C. and S.M. Canney (2013), *Conservation*, Cambridge, UK: Cambridge University Press.
Hanafiah, M., M. Xenopoulos and S. Pfister et al. (2011), 'Characterization factors for water consumption and greenhouse gas emissions based on freshwater fish species extinction', *Environmental Science & Technology*, **45**(12), 5272–8.
Harrington, R., C. Anton and T. Dawson et al. (2010), 'Ecosystem services and biodiversity conservation: concepts and a glossary', *Biodiversity and Conservation*, **19**(10), 2773–90.
Hellweg, S. and L. Milà i Canals (2014), 'Emerging approaches, challenges and opportunities in life cycle assessment', *Science*, **344**(6188), 1109–13.
IPCC (2013), *Climate Change 2013: The Physical Science Basis* (Intergovernmental Panel on Climate Change. Working Group I), Cambridge, UK: Cambridge University Press.
ISO (2006a), *ISO-14040. Environmental Management – Life Cycle Assessment – Principles and Framework*, Geneva: International Organization for Standardization.
ISO (2006b), *ISO-14044. Environmental Management – Life Cycle Assessment – Requirements and Guidelines*, Geneva: International Organization for Standardization.
Karlen, D.L., M.J. Mausbach and J.W. Doran et al. (1997), 'Soil quality: a concept, definition, and framework for evaluation', *Soil Science Society of America*, **61**, 4–10.
Koellner, T. (2003), *Land Use in Product Life Cycles and Ecosystem Quality* (Europäische Hochschulschriften/European University Studies/Publications Universitaires Européennes series, Vol. 3012), Brussels: Peter Lang.
Koellner, T. and R.W. Scholz (2008), 'Assessment of land use impacts on the natural environment part 2: generic characterization factors for local species diversity in Central Europe', *The International Journal of Life Cycle Assessment*, **13**(1), 32–48.
Koellner, T., L. de Baan and T. Beck et al. (2013a), 'Principles for life cycle inventories of land use on a global scale', *The International Journal of Life Cycle Assessment*, **18**(6), 1203–15.
Koellner, T., L. de Baan and T. Beck et al. (2013b), 'UNEP-SETAC guideline on global land use impact assessment on biodiversity and ecosystem services in LCA', *The International Journal of Life Cycle Assessment*, **18**(6), 1188–202.

Koh, L.P., T.M. Lee and N.S. Sodhi et al. (2010), 'An overhaul of the species – area approach for predicting biodiversity loss: incorporating matrix and edge effects', *Journal of Applied Ecology*, **47**(5), 1063–70.
Kounina, A., M. Margni and J. Bayart et al. (2013), 'Review of methods addressing freshwater use in life cycle inventory and impact assessment', *The International Journal of Life Cycle Assessment*, **18**(3), 707–21.
Laurance, W.F. (1997), 'Biomass collapse in Amazonian forest fragments', *Science*, **278**(5340), 1117–18.
Lavorel, S., K. Grigulis and P. Lamarque et al. (2011), 'Using plant functional traits to understand the landscape distribution of multiple ecosystem services', *Journal of Ecology*, **99**(1), 135–47.
Levins, R. (1969), 'Some demographic and genetic consequences of environmental heterogeneity for biological control', *Bulletin of the ESA*, **15**(3), 237–40.
Lindeijer, E. (2000), 'Biodiversity and life support impacts of land use in LCA', *Journal of Cleaner Production*, **8**(4), 313–19.
Malmqvist, B. and S. Rundle (2002), 'Threats to the running water ecosystems of the world', *Environmental Conservation*, **29**(2), 134–53.
Matson, P.A., W.J. Parton and A.G. Power et al. (1997), 'Agricultural intensification and ecosystem properties', *Science*, **277**(5325), 504–9.
Mattila, T., T. Helin and R. Antikainen et al. (2011), *Land Use in Life Cycle Assessment*, Helsinki: The Finnish Environment Insitute.
Michelsen, O. (2008), 'Assessment of land use impact on biodiversity', *The International Journal of Life Cycle Assessment*, **13**(1), 22–31.
Milà i Canals, L. and L. de Baan (2015), 'Land use', in M. Hauschild and M. Huijbregts (eds), *Life Cycle Impact Assessment, LCA Compendium – The Complete World of Life Cycle Assessment*, Dordrecht: Springer, pp. 197–222.
Milà i Canals, L., C. Bauer and J. Depestele et al. (2007), 'Key elements in a framework for land use impact assessment within LCA', *The International Journal of Life Cycle Assessment*, **12**(1), 5–15.
Milà i Canals, L., J. Chenoweth and A. Chapagain et al. (2009), 'Assessing freshwater use impacts in LCA: part I – inventory modelling and characterisation factors for the main impact pathways', *The International Journal of Life Cycle Assessment*, **14**(1), 28–42.
Milà i Canals, L., G. Rigarlsford and S. Sim (2013), 'Land use impact assessment of margarine', *The International Journal of Life Cycle Assessment*, **18**(6), 1265–77.
Milà i Canals, L., J. Romanyà and S.J. Cowell (2007), 'Method for assessing impacts on life support functions (LSF) related to the use of "fertile land" in life cycle assessment (LCA)', *Journal of Cleaner Production*, **15**(15), 1426–40.
Millennium Ecosystem Assessment (MA) (2005), *Ecosystems and Human Well-being: Synthesis*, Washington, DC: Island Press.
Mouchet, M.A., S. Villéger and N.W.H. Mason et al. (2010), 'Functional diversity measures: an overview of their redundancy and their ability to discriminate community assembly rules', *Functional Ecology*, **24**(4), 867–76.
Mouillot, D., N.H. Mason and J.B. Wilson (2007), 'Is the abundance of species determined by their functional traits? A new method with a test using plant communities', *Oecologia*, **152**(4), 729–37.
Mueller, C., L. de Baan and T. Koellner (2014), 'Comparing direct land use impacts on biodiversity of conventional and organic milk – based on a Swedish case study', *The International Journal of Life Cycle Assessment*, **19**(1), 52–68.
Norman, M.W.H., K. MacGillivray and J.B. Seel et al. (2003), 'An index of functional diversity', *Journal of Vegetation Science*, **14**(4), 571–8.
Noss, R.F. (1990), 'Indicators for monitoring biodiversity: a hierarchical approach', *Conservation Biology*, **4**(4), 355–64.
Núñez, M., A. Antón and P. Muñoz et al. (2013), 'Inclusion of soil erosion impacts in life cycle assessment on a global scale: application to energy crops in Spain', *The International Journal of Life Cycle Assessment*, **18**(4), 755–67.

Olson, D.M., E. Dinerstein and E.D. Wikramanayake et al. (2001), 'Terrestrial ecoregions of the world: a new map of life on earth: a new global map of terrestrial ecoregions provides an innovative tool for conserving biodiversity', *BioScience*, **51**(11), 933–8.

Pennington, D., G. Finnveden and E. Lindeijer et al. (2004), 'Life cycle assessment part 2: current impact assessment practice', *Environment International*, **30**(5), 721–39.

Pereira, H.M. and G.C. Daily (2006), 'Modeling biodiversity dynamics in countryside landscapes', *Ecology*, **87**(8), 1877–85.

Petchey, O.L. and K.J. Gaston (2006), 'Functional diversity: back to basics and looking forward', *Ecology Letters*, **9**(6), 741–58.

Pettorelli, N., J.O. Vik and A. Mysterud et al. (2005), 'Using the satellite-derived NDVI to assess ecological responses to environmental change', *Trends in Ecology & Evolution*, **20**(9), 503–10.

Pfister, S., A. Koehler and S. Hellweg (2009), 'Assessing the environmental impacts of freshwater consumption in LCA', *Environmental Science & Technology*, **43**(11), 4098–104.

Putz, F., G. Blate and K. Redford (2001), 'Tropical forest management and conservation of biodiversity: an overview', *Conservation Biology*, **15**(1), 7–20.

Rosenbaum, R.K., T.M. Bachmann and L. Swirsky Gold et al. (2008), 'USEtox – the UNEP-SETAC toxicity model: recommended characterisation factors for human toxicity and freshwater ecotoxicity in life cycle impact assessment', *The International Journal of Life Cycle Assessment*, **13**, 532–46.

Saad, R., T. Koellner and M. Margni (2013), 'Land use impacts on freshwater regulation, erosion regulation, and water purification: a spatial approach for a global scale level', *The International Journal of Life Cycle Assessment*, **18**(6), 1253–64.

Schmidt, J.H. (2008), 'Development of LCIA characterisation factors for land use impacts on biodiversity', *Journal of Cleaner Production*, **16**(18), 1929–42.

Smakhtin, V., C. Revenga and P. Döll (2004), *Taking into Account Environmental Water Requirements in Global-scale Water Resources Assessments. Comprehensive Assessment Report 2*, Colombo, Sri Lanka: IWMI.

Souza, D.M. (2010), 'Proposta de um modelo de caracterização de impactos do uso da terra, segundo indicadores de biodiversidade, em AICV: cálculo de fatores de caracterização para ecorregiões brasileiras' [Proposal of a model to characterize impacts of land use on biodiversity in LCIA: calculation of characterization factors for Brazilian ecoregions], PhD, Programa de Pós-Graduação em Engenharia Ambiental, Florianópolis, Brazil: Universidade Federal de Santa Catarina.

Souza, D.M., D.F.B. Flynn and F. DeClerck et al. (2013), 'Land use impacts on biodiversity in LCA: proposal of characterization factors based on functional diversity', *The International Journal of Life Assessment*, **18**(6), 1231–42.

Souza, D.M., R.F.M. Teixeira and O.P. Ostermann (2015), 'Assessing biodiversity loss due to land use with life cycle assessment: are we there yet?', *Global Change Biology*, **21**(1), 32–47.

Stoate, C., N.D. Boatman and R.J. Borralho et al. (2001), 'Ecological impacts of arable intensification in Europe', *Journal of Environmental Management*, **63**(4), 337–65.

Thomas, C.D., A. Cameron and R.E. Green et al. (2004), 'Extinction risk from climate change', *Nature*, **427**(6970), 145–8.

Tscharntke, T., A.M. Klein and A. Kruess et al. (2005), 'Landscape perspectives on agricultural intensification and biodiversity – ecosystem service management', *Ecology Letters*, **8**(8), 857–74.

Turner, I. (1996), 'Species loss in fragments of tropical rain forest: a review of the evidence', *Journal of Applied Ecology*, **33**(2), 200–209.

Verones, F., D. Saner and S. Pfister et al. (2013), 'Effects of consumptive water use on biodiversity in wetlands of international importance', *Environmental Science & Technology*, **47**(21), 12248–57.

Vickery, J.A., J.R.B. Tallowin and R.E. Feber et al. (2001), 'The management of lowland neutral grasslands in Britain: effects of agricultural practices on birds and their food resources', *Journal of Applied Ecology*, **38**(3), 647–64.

Vogtländer, G., E. Lindeijer and J.P.M. Witte et al. (2004), 'Characterizing the change of land-use based on flora: application for EIA and LCA', *Journal of Cleaner Production*, **12**(1), 47–57.

Wall, D.H. and U.N. Nielsen (2012), 'Biodiversity and ecosystem services: is it the same below ground?', *Nature Education Knowledge*, **3**(12), 8.

Wassenaar, T., P. Gerber and P.H. Verburg et al. (2007), 'Projecting land use changes in the neotropics: the geography of pasture expansion into forest', *Global Environmental Change*, **17**(1), 86–104.

Weidema, B.P. and E. Lindeijer (2001), *Physical Impacts of Land Use in Product Life Cycle Assessment*, final report of the EURENVIRON LCAGAPS sub-project on land use, Lyngby, Denmark.

Weiss, E., S.E. Marsh and E.S. Pfirman (2001), 'Application of NOAA-AVHRR NDVI time-series data to assess changes in Saudi Arabia's rangelands', *International Journal of Remote Sensing*, **22**(6), 1005–27.

Whittaker, R.H. (1960), 'Vegetation of the Siskiyou Mountains, Oregon and California', *Ecological Monographs*, **30**(3), 279–338.

Zelm, R. van, A.M. Schipper and M. Rombouts et al. (2011), 'Implementing groundwater extraction in life cycle impact assessment: characterization factors based on plant species richness for the Netherlands', *Environment Science and Technology*, **45**(2), 629–35.

Zhang, Y., S. Singh and B.R. Bakshi (2010), 'Accounting for ecosystem services in life cycle assessment, part I: a critical review', *Environmental Science & Technology*, **44**(7), 2232–42.

PART II

APPLICATIONS IN DIFFERENT SECTORS

8. Impacts of urban development on biodiversity and ecosystem services

Berit Balfors, Juan Azcárate, Ulla Mörtberg, Mårten Karlson and Sara Odelius Gordon

8.1 INTRODUCTION

Urbanization has increased rapidly worldwide and this development is expected to continue. In the beginning of the twentieth century, around 10 per cent of the human population lived in urban areas, while today it is estimated that more than half of the world's population lives in cities (Grimm et al., 2008), an estimate that is expected to increase to nearly 70 per cent in 2050 if current urbanization trends continue (Seto et al., 2013). Hence, sustainable development challenges will be increasingly concentrated in cities, which call for policies that can lead the way towards sustainable societies (UN, 2014).

While urban areas only constitute around 3 per cent of the global terrestrial area (McGranahan et al., 2007) they contain the majority of the human population and are centres of human action. These actions lead to emissions and consumption of natural resources that generate global impacts, for instance already a decade ago urban activities were estimated to account for 78 per cent of carbon emissions, 60 per cent of residential water use, and 76 per cent of wood used for industrial purposes (Brown, 2001). Thus, to meet the demands of urban populations, cities rely on external commodities and shape land-use decisions and environmental change, on not only local but also regional and global scales.

The urban domination of resource flows and the general localization pattern of cities have significant consequences. A majority of the world's cities are situated along coastlines and major river systems, in areas with highly productive ecosystems and abundant natural resources that are also strategic from a transportation point of view (Luck, 2007). At the global level and also at regional and municipal levels, there is a general orientation of urbanizing areas towards highly productive ecosystems, in proximity to rivers and coasts, reliable water sources and high-fertility soils – ecosystems that support high levels of biodiversity and ecosystem services (ES) (Falkenmark and Chapman, 1989; Ricketts and Imhoff, 2003) (Figure 8.1). Hence, urbanization contributes to the loss of

Source: Photograph: Ulla Mörtberg.

Figure 8.1 *Railway bridge in urban green area in Stockholm*

biodiversity and due to the continuing urbanization trend the impact on biodiversity is likely to continue.

The spatial correlation between urban development and highly productive ecosystems with high biodiversity, together with the complex response to urbanization, provide opportunities for planning as an enabler for sustainable urban development. In view of this, approaches to planning and impact assessment are necessary, that take the scale of ecological processes into account across administrative borders. For this purpose, a landscape approach can be useful that addresses scales of, for example, watersheds, animal movements and human perceptions and actions. A landscape approach includes natural, semi-natural and artificial multi-functional ecosystems in urban environments, often called 'green areas' (Tzoulas et al., 2007), which support ecosystem functions that can be studied in a landscape perspective. Tools have thus been developed to integrate biodiversity and ES in planning and impact assessment of urban development options, promoting a sustainable urban development.

This chapter addresses impacts of urbanization on biodiversity and the capacity of urban green areas to provide ES. A brief description of the role of biodiversity and ES in urbanizing areas provides a framework for a landscape approach in biodiversity assessments and for the practical examples from the Stockholm region presented. Related to the assessment, tools for predicting and assessing biodiversity impacts at a landscape level are discussed as well as planning and management of urban green areas. The chapter concludes with lessons learned and key recommendations for best practice.

8.2 IMPACTS OF URBANIZATION ON BIODIVERSITY

Urbanization is a dominant agent of loss, fragmentation and degradation of habitats with significant impacts on biodiversity (Marzluff and Ewing, 2001; Ricketts and Imhoff, 2003, Müller et al., 2013). In central urban areas, habitat is continuously lost and fragmented due to urban compaction, while in peri-urban areas, urban sprawl not only converts land into new built-up areas and infrastructure, but also changes land use in the surroundings. As rural land becomes peri-urban, it is increasingly supporting the city with industries, waste dumps, building materials and water supply as well as recreation areas. These changes on local to regional scales can have both positive and negative impacts on biodiversity, with rich opportunities for planning to influence the direction of change (Forman, 2008).

The overall loss of natural and semi-natural habitats and shrinkage of habitat patches is a main driver of biodiversity decline, since habitat size is related to the probability of extinction of local populations. In addition, fragmentation increases the distance between habitat patches, which together with barrier effects of urban land cover decreases habitat connectivity in the landscape, with lower possibilities for colonization as a consequence. Habitat size and connectivity together are main predictors of species abundance, diversity and genetic exchange, and influence the probability of species persistence in the landscape (Hanski and Ovaskainen, 2003; Holderegger and Di Giulio, 2010).

With fragmentation, the proportion of edges between habitat and non-habitat will increase at the expense of core habitat. Conditions near edge habitat will differ from that of core habitat, in that edge habitat, for example, has a different microclimate with, in the case of forest habitat edges, higher temperature and wind speed, and disturbances from the surroundings will be higher. Certain edge effects, such as microclimate, may impact in a range of tenths of metres, while others can be more long-ranging. Consequently, habitat near edges will change and can provide a high diversity on local level, while species tied to core habitat will be negatively affected and the result on a regional level may still be detrimental (Di Giulio et al., 2009). Furthermore, habitat that becomes incorporated into urban settlements or is affected by urban sprawl will change, even if it is protected, as a result of urban disturbances and pressures (McDonald et al., 2009). These can consist of, for instance, air and water pollution, noise, vibration and artificial light disturbances, changes in microclimate and hydrology, as well as impacts from urban predators and recreation pressure (Forman, 2008; Mörtberg, 2009; Lehvävirta et al., 2014).

A consequence of urbanization is an expanding network of transportation infrastructure that reduces habitat connectivity as roads and railways constitute movement barriers, cause habitat degradation through noise and pollution, and also cause road mortality. Simultaneously, with construction of transport infrastructure new habitat is created that can have positive impacts on some species. However, several studies have found an overall declining diversity and abundance of birds and mammals with increasing proximity to transport infrastructure (Bissonette and Rosa, 2009; Eigenbrod et al., 2009; Avon et al., 2010; Benítez-López et al., 2010). Impacts of transport infrastructure on biodiversity are further elaborated in Forman et al. (2003); Fahrig and Rytwinski (2009); Holderegger and Di Giulio (2010); Karlson et al. (2014).

Another effect of urbanization is the increase in number and extent of non-native invasive species (e.g., McDonald and Urban, 2006). Moreover, there is a suite of skilled generalist species that are present in most cities around the world, which can be called 'urban adaptors'. Meanwhile, urbanization often leads to the loss of 'sensitive' species dependent on larger areas of more natural habitat. The net result is sometimes termed 'biotic homogenization' (McKinney, 2006).

The general pattern of urbanization impacts on biodiversity is complicated, and differs depending on scale, properties of biomes, habitat and species as well as with site history (Müller et al., 2013). However, the overall trend is considered to be a global decline of biodiversity and a homogenization due to urbanization (Hobbs et al., 2006; Müller et al., 2013). The changing density of species in cities can best be explained by anthropogenic features such as land cover and city age, rather than by non-anthropogenic factors such as geography, climate and topography. Despite this, cities still retain a high diversity of native species (Aronson et al., 2014), thus providing opportunities for planning and management that would be significant for biodiversity on all levels, regional and global (Balfors et al., 2005).

8.3 A LANDSCAPE APPROACH TO BIODIVERSITY ASSESSMENT

Planning decisions, which lead to changes in the composition and pattern of the landscape, impact on ecological processes and thus biodiversity and ES. The impacts may occur on-site, just affecting the area of development, but often they occur also off-site, since ecological processes like species persistence and dispersal often work on large scales. Therefore, a site-based approach is not sufficient for the consideration of biodiversity and ES in impact assessment. Instead, it is necessary to consider the quality, quantity, and spatial cohesion of ecosystems and the persistence requirements of species and communities in the entire landscape.

A landscape approach calls for methods for assessing the impacts of human actions on biodiversity at a landscape level. Such methods should allow an analysis of cumulative impacts of many single planning decisions, which may lead to substantial alterations including the loss, isolation, and disturbance of natural habitats. Many of the processes involved have a temporal and spatial dimension and are possible to quantify, analyse, and visualize with geographical information systems (GIS) combined with spatial ecological models. This allows for localization and quantification of predicted effects of urbanization on biodiversity components over landscape and regional scales. Also, the long-term effects of planning scenarios can be addressed.

8.3.1 Spatial ecological models

There are a multitude of GIS-based spatial ecological models that are designed to address ecological processes and biodiversity in relation to land use change, such as urbanization and related infrastructure, which can be used for impact assessment (e.g., Gontier et al., 2006; Mörtberg et al., 2007; Zetterberg et al., 2010; Segan et al., 2011; González del Campo, 2012). Such spatial models address ecosystem processes on different levels, focusing on, for example, (1) ecosystems distribution, fragmentation, rarity, and so on (Geneletti, 2004; Cushman et al., 2008), (2) species and community distribution, diversity, abundance, rarity, habitat suitability and connectivity and so on (Mörtberg et al., 2007; Phillips and Miroslav, 2008; Gontier et al., 2010; Bacaro et al., 2012), or (3) both (Gontier et al., 2006; Alkemade et al., 2009; Di Minin and Moilanen, 2012).

Data to support such assessments are increasingly available through national and international sources involving classified remote sensing datasets on urban development as well as other land cover and land use (e.g., Hedblom and Mörtberg, 2011). However, such ecological knowledge may not be readily available, for instance concerning detailed habitat types that would require new classifications from remote sensing data. More detailed information on habitat types will facilitate the assessment of the relative amount of habitat, which can provide a better understanding of the rarity of habitat in an area (e.g., Geneletti, 2004).

Habitat amount and connectivity can be assessed using landscape metrics, such as mean habitat patch size, number of patches and edge/area ratios (McGarigal et al., 2012), from which indices useful for impact assessment can be selected (Cushman et al., 2008). Karlson and Mörtberg (2015), for instance, used such indices to find vulnerable habitat types in relation to impacts of transport infrastructure.

Addressing biodiversity components such as species and communities and their spatial distribution would potentially relate to both the spatial and non-spatial properties of a project or plan. In order to find spatial distribution data for a baseline situation, field inventories can be conducted, which offers several advantages in terms of data accuracy. However, if this is not feasible, the spatial distribution of a biodiversity component, as well as anticipated changes due to urban development alternatives, can be predicted in different ways.

When occurrence data on a biodiversity component are available, their spatial distribution can be predicted through their correlation with independent variables, describing their habitat in the baseline situation and for scenarios. These could be based on habitat quality, quantity and connectivity (Mörtberg et al., 2007; Drielsma and Ferrier, 2009; Gontier et al., 2010; Bacaro et al., 2012), while urban disturbances and pressures can also be included in such models (e.g., Mörtberg, 2009). If empirical data on the distribution of a biodiversity component are not available, a predicted spatial distribution could be estimated through expert knowledge, based on resource requirements and movement capacity (Gontier et al., 2006).

One type of spatial ecological model that has potential to be used as decision support is based on graph theory, the so-called ecological network model (Saura and Pascual-Hortal, 2007; Saura and Rubio, 2010; Saura et al., 2011). In such models questions concerning suitable and available habitat can be addressed, taking both habitat amount and connectivity into account simultaneously. Ecological network models have been developed and applied to decision support problems in urban settings (e.g., Zetterberg et al., 2010; Mörtberg et al., 2013; Bergsten and Zetterberg, 2013) and could be very suitable for impact assessment.

An approach that could be advantageous in a planning context is to model the predicted spatial distribution for prioritized groups of species characterized by their ecological profiles (Vos et al., 2001; Angelstam et al., 2004; Mörtberg et al., 2012). An ecological profile is a representation of several biodiversity components that share important attributes, such as resource requirements and dispersal capacity. This could, for instance, be a functional group of species that all depend on coniferous forest of a certain age and composition and that move similar distances within the landscape. This approach can help both setting priorities and to overcome data scarcity, and have been applied in several case studies addressing urban development (Mörtberg et al., 2007) and transport infrastructure (Karlson and Mörtberg, 2015), and can be linked to an environmental assessment process.

8.3.2 Integrating Multiple Criteria in Decision Support Tools

For integrated assessment of urbanization scenarios, multiple and sometimes conflicting criteria need to be handled together and evaluated in decision-making. For this purpose, multi-criteria analysis (MCA) methods have been widely used for handling complex issues (e.g., Segan et al., 2011; Coutinho-Rodrigues et al., 2011; González del Campo, 2012). Compared with single criteria approaches, MCA is designed to take complex factors into consideration and to perform an integrated analysis. A multitude of tools have been developed to carry out MCA analysis both in spatial (e.g., Geneletti, 2010; Orsi and Geneletti, 2010; Bagli et al., 2011) and non-spatial (e.g., González et al., 2013) approaches.

An elaboration of these approaches is to translate all indicators into one 'currency', as in cost–benefit analysis. Economic valuation of biodiversity and ES has been seen as a promising approach to improving sustainable management and decision-making (Helm and Hepburn, 2012; Laurans et al., 2013; Gomez-Baggethun and Barton, 2013).

8.4 URBAN GREEN AREAS AND THEIR ECOSYSTEM SERVICES

Urban green areas include natural reserves, parks, green infrastructure, green hearts, greenbelts, green wedges and greenways, which have been used in city planning to connect urban areas with their suburbs and rural areas, and enhance recreation and other ES (Sandström, 2002; Colding, 2011). Urban green areas can also include more informal areas such as brown sites (Kabisch and Haase, 2013), productive or abandoned agricultural land (Li et al., 2005; La Greca et al., 2011; Yokohari and Bolthouse, 2011), and artificially created green areas (Jim, 2004; Tian et al., 2012).

Urban green areas are increasingly being considered an integral part of city development and a key issue for city sustainability (Pickett et al., 2011). They are also more often seen as vital city attractiveness factors (Douglas and Ravetz, 2011) and as an issue to address, especially in spatially expanding cities and in cities experiencing population growth (James et al., 2009). Moreover, due to changes in lifestyles, regarding expectations of a higher quality of life, including good health, unpolluted environments and access to nearby recreation possibilities, there are claims and increasing societal support for more green areas in and around cities (Sandström, 2002; De Ridder et al., 2004; Li et al., 2005).

The urban ES provided by urban green areas deliver an ample range of benefits for cities and their inhabitants. In essence, all these urban ES and their interconnections are important for and benefit human well-being and life quality in cities (Niemelä et al., 2010; Colding, 2011; Beatley and Newman, 2013; Marcus et al., 2013). In particular, the availability of and access to regulating and cultural ES can be seen as important (Gómez-Baggethun et al., 2013). Regulating ES, such as air quality, climate and water regulation, are fundamental to managing and providing safe, healthy and good living environments in urban areas (McDonald and Marcotullio, 2011). Cultural ES, which are ecosystems' contributions to the non-material benefits that arise from human–ecosystem interactions (Chan et al., 2012), are basic for the quality of life in cities and are increasingly being sought by urban dwellers (De Groot et al., 2005; Tzoulas et al., 2007). Cultural ES are usually categorized as recreational, educational, cultural and historical, artistic and spiritual (Chiesura and De Groot, 2003). These categories, together with other categories of urban ES, are, however, context dependent, as are their provided benefits (Chan et al., 2012; Gómez-Baggethun et al., 2013).

In general, cities can benefit from urban green areas and their ES in terms of economics, health and quality of life, and ecology (Jansson, 2014). Cities can benefit economically from urban ES by increasing city attractiveness, property prices, consumption, employment, food production, as well as cost savings in health care, disaster prevention and infrastructure and service maintenance. The availability of and accessibility to urban ES can define city characters and also have several positive economic effects (Dunn and Henegham, 2011). For instance, green cities can attract technology and knowledge creation (James et al., 2009), provide status (Pauleit and Breuste, 2011), and lead to higher property prices (Ståhle, 2008), in some cases by between 5 and 20 per cent depending on the green area and the neighbourhood (SCBD, 2012).

In cities, the regulating ES provided by green areas deliver various types of ecological benefits – for example, air filtration, microclimate regulation, noise reduction, water purification and supply, storm water regulation and flood control, sewage water treatment, soil erosion control, and biodiversity maintenance (Douglas and Ravetz, 2011). For example, urban and peri-urban green areas have the capacity to maintain water supply security in cities (Beatley and Newman, 2013), providing city inhabitants with fresh water for drinking and for many other activities, and influencing the amount and quality of the supplied water (McDonald et al., 2013). Moreover, vegetated areas in cities can contribute to climate change mitigation by acting as carbon storage and by minimizing the urban heat island effect through evapotranspiration and solar energy absorption (SCBD, 2012).

Cultural ES contribute to health and improve well-being and quality of life for people in cities. Urban nature serves a regenerative function, lowering stress and providing relaxation from daily routines, which is extremely important especially in urbanizing contexts (Chiesura and De Groot, 2003). Contact with urban green areas has also been seen as helping to restore attention capacity and emotion regulation, countering depression and anxiety (Tzoulas and Greening, 2011). People that live in proximity to urban green areas tend to spend more time outdoors and are more physically active, taking part in walking, biking and sporting activities, and viewing and appreciating nature (Beatley and Newman, 2013).

Contact with urban nature is also important for urban dwellers in order to develop and strengthen social relationships and community bonds (Tzoulas et al., 2007). For instance, weaving community gardens into the urban fabric is important to promoting community cohesion and rebuilding productive urban landscapes (Yokohari and Bolthouse, 2011). Gardening can also increase longevity of the elderly and contribute to cognitive development in children, to whom green areas are vital for learning (Tzoulas and Greening, 2011).

8.4.1 Planning and Management of Urban Green Areas

One key message from *Cities and Biodiversity Outlook* (SCBD, 2012) is that ES must be integrated into urban policy and planning. This integration requires an in-depth understanding of how ES are manifested in urban contexts and how they are affected by urban development. In cities, the tasks of structuring and designing urban green areas should focus on maximizing the multiple services and benefits that these areas can provide (Douglas and Ravetz, 2011). Urban green areas could, for instance, be planned and managed to simultaneously motivate human physical activity and promote biodiversity.

However, green areas in city contexts are typically fragmented and disconnected, conflicting with the urban green area planning and management goal of maximizing multiple ES and benefits (James et al., 2009). There is also a lack of knowledge around the quality of different ecosystems from both a biodiversity and an ES perspective, such as how to manage them to increase their performance. Other challenges for urban green area planning and management are: how to identify and balance ES and benefit trade-offs (Grimm et al., 2008); how to assess and manage urban green areas' perceived ecosystem disservices such as increased fear, feelings of insecurity, and spread of diseases associated with urban green areas (Tzoulas et al., 2007; Lyytimäki and Sipilä, 2009); and how to develop urban green areas that are shaped by human actions and are subject to socio-demographic and environmental change (Chiesura and De Groot, 2003; James et al., 2009; Niemelä et al., 2010).

An understanding of how the biophysical and human processes in urban settings interact with each other across multiple spatial and temporal scales can contribute to address the above-mentioned challenges (Pickett et al., 2011; SCBD, 2012). Moreover, gaining a wider understanding of how society values urban ecosystem functions and services can affect how pressures or impacts on urban green areas are viewed, and what actions can be designed to manage these pressures (Niemelä et al., 2010). For instance, including information on ES in city development assessments and decision-making can be used to identify both where human action has adversely impacted a service and where intervention might give rise to additional benefits (Azcárate, 2015). In urban planning, consideration of green areas at an early stage could lead to a move from planning at the project level to strategies at the urban landscape level (Pauleit et al., 2011), and to address conflicts between urban development and green areas, which is an important part of including and maximizing biodiversity and ES in urban development (Hostetler et al., 2011).

8.5 APPLICATIONS IN THE STOCKHOLM REGION

The Stockholm County consists of 678 500 hectares of land and water, extending 180 km from north to south, and representing around 2 per cent of Sweden's total land area. The county includes 26 municipalities and embraces varying landscapes such as large forested areas, open agricultural areas, and built up areas that respectively account for 50, 20 and 14 per cent of the land area (Statistics Sweden, 2012). Within the Regional Development Plan for the county, green areas in the form of ten 'green wedges' were identified that stretch from the region's rural–urban fringe to the centre of Stockholm City, the capital of Sweden (ORPUT [Office of Regional Planning and Urban Transportation], 2010). The green wedges are remnants of formerly connected natural and semi-natural habitats having high ecological, recreational and cultural significance (Elmqvist et al., 2004; Colding, 2013) that are built up by core areas, with high biodiversity and recreational values, and connecting dispersal links. Apart from having green wedges at a close distance, urban residents in the Stockholm region have access to other types of green areas, such as nature reserves, city parks and natural shorelines, as well as a large number of golf courses, private gardens and allotment gardens that also provide opportunities for biodiversity and ES in the urban region (Barthel et al., 2005; Colding el al., 2006). From a European perspective, the Stockholm region and its urban areas can be considered to provide a high coverage of green areas (Elmqvist et al., 2004; Colding 2013). However, these green areas are affected by continuous and rapid urbanization (Bolund and Hunhammar, 1999; Mörtberg, 2009; Borgström, 2011; Colding, 2013).

The Stockholm region is a fast-growing and dynamic region that attracts companies and new residents. It has been estimated that the migration into Stockholm County will be as much as 250 000–400 000 more households by the year 2030 (ORPUT, 2010), which is expected to significantly increase the region's total population of approximately 2 million people (Statistics Sweden, 2012). With an increase in population, the demand for and pressure on urban resources, from housing to public transport and recreational possibilities, will increase and additional capacity will be required to support the current pace of urbanization in the region. To handle these increases, and to avoid urban sprawl, which can, for instance, affect the green wedges, it is planned that certain urban areas or urban nodes in the Stockholm region will be densified and strengthened with additional transport infrastructure (ORPUT, 2010).

In these urban nodes, densification activities aim at increasing housing possibilities and a variety of urban functionalities, but there is a risk that these activities might cause undesired effects on the green areas located within the affected areas. Moreover, since the Stockholm region is increasingly connected to the much larger region of Mälardalen, involving several counties, fast trains and other transport modes, this could mean that valuable green areas in the region, within and outside of its green wedges could become increasingly fragmented and decrease in size, which can negatively affect biodiversity and ES on a regional scale.

In the context of the Stockholm region three case studies were selected to illustrate different landscape approaches to integrating urbanization impacts on biodiversity and ES in urban planning (Figure 8.2). The first case study addresses the formulation of regional landscape strategies using a landscape approach in a study area covering the larger part of six municipalities situated north of Stockholm City. The second case study concerns the construction of a new railway connecting the important urban development nodes of Arninge and Arlanda. This case study involves development of methods using spatial ecological and geological models to investigate important ecological characteristics that may be affected by the planned construction. Spatial ecological and geological models are used to test alternative corridors for the railway. The third case study analysed the potential provision of cultural ES of various green areas as well as the pressures that these green areas could face with the implementation of city densification in Metropolitan Stockholm. In this way the impacts of

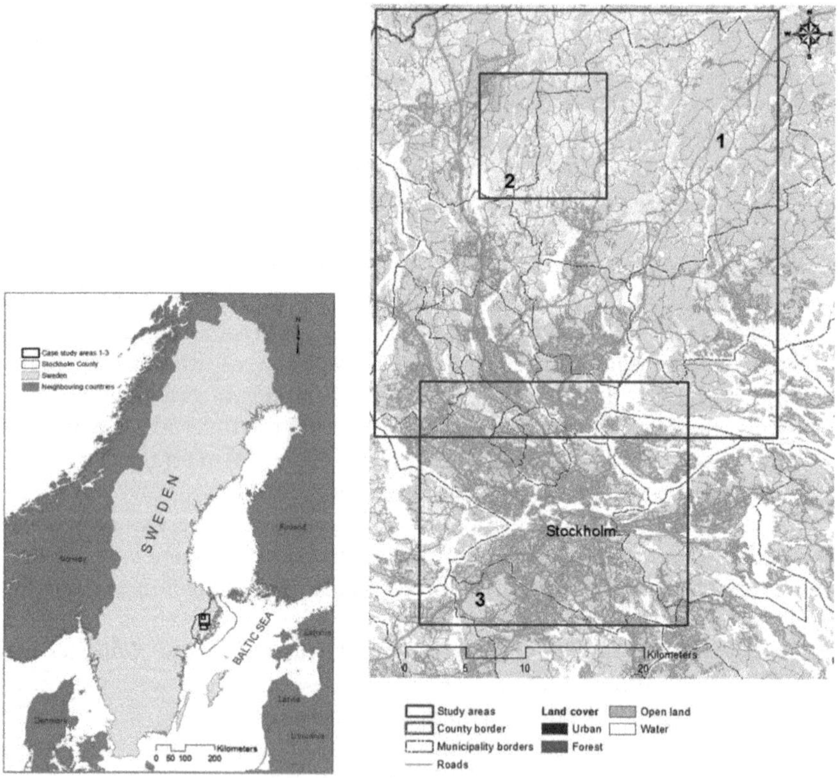

(a) (b)

Sources: Spatial data ©Lantmäteriet [I2014/00591].

Figure 8.2 Location of Stockholm County and the three case study areas

single planning projects can be linked to the overall status and needs on the metropolitan level.

8.5.1 Regional Landscape Strategies

The conservation of biodiversity is the main concern of one of the Swedish national environmental quality objectives called 'A Rich Diversity of Plant and Animal Life' (Government of Sweden, 2004). Within this objective, a target is to develop regional landscape strategies for the consideration of biodiversity on a landscape scale. For this purpose, the Swedish government commissioned seven county councils to develop methods in different pilot projects. One of these pilot projects was carried out in a rapidly urbanizing area north of Stockholm City (Figure 8.2). The study area embraced six municipalities in suburban and peri-urban settings with a mosaic of built-up areas, transport infrastructure, forest, agricultural land and lakes, with high values from biodiversity and ES perspectives. The aim of the project was to develop methods for creating regional landscape strategies for this area, with the target to create a joint policy document for the landscape. The strategies should provide an umbrella for a wide array of planning and conservation measures, for example within biodiversity conservation, conservation of historical landscapes, the programme for the countryside, regional and municipal planning and urban development (Stockholm County Administrative Board, 2007).

The regional landscape strategies were formulated using a landscape approach, integrating biodiversity objectives together with cultural history and recreation values. For the objectives, operational targets and indicators were formulated, the latter in the form of ecological profiles, historical landscapes and recreation values and availability. Spatial ecological and landscape models were applied to study the baseline situation for the selected indicators. An integrated assessment of the baseline situation compared to the targets was performed and was followed by a formulation of strategies for the landscape to reach those targets.

The project approach allowed a broad participation of stakeholders through several meetings and workshops involving different competences from the Stockholm County Administrative Board, together with the Stockholm County Council, regional representatives of the Swedish Forest Agency, the Swedish National Road Administration, the Swedish National Heritage Board, planners and ecologists from six municipalities, the Federation of Swedish Farmers and several NGOs. The stakeholders were engaged in the formulation of suitable planning objectives and targets, provision of local ecological knowledge and final integration of results into a spatial visualization of strategies.

The targets related to biodiversity that were selected in the participative process were the conservation of three main habitat types and the potential for sensitive species tied to these nature types to persist in the landscape. The habitat types consisted of coniferous forest with old pine trees, deciduous forest with old oaks and natural grasslands. The ecological profiles were groups of species tied to these habitat types with low to intermediate dispersal or movement capacity.

The results of the spatial ecological modelling were habitat networks for each ecological profile representing resource and/or dispersal limited species of targeted habitat types in the region. These were integrated with information on remnants of historical landscapes and recreation values and availability, which resulted in a final map with core areas and connecting links for these values. In this way, synergetic effects of several different values could be outlined. Examples of this were the outlined core areas with historical landscapes and natural grasslands, a large forest area with high recreation values, shorelines and castles with assemblages of old oak trees and green areas in suburban areas with high recreation values and availability. The latter correlated to a large extent with the green wedges in the Regional Development Plan (ORPUT, 2010). However, the core areas differed in certain locations between the regional landscape strategies and the green wedges, which would be expected since the scope of the current study was substantially broader and a result of a broad stakeholder engagement.

The strategies formulated in the resulting policy document (Stockholm County Administrative Board, 2007) concerned the following: (1) Strengthening the status of the landscape in urban planning: for example, important landscape structures and the peri-urban landscape functions should have a clearer role in urban, regional and infrastructure planning. A holistic and integrated view over administrative borders should be developed and harmonized databases on landscape values set up. Disturbing activities such as industries and waste dumps need to be planned from a regional perspective. (2) Conserving and developing good habitat for people and for biodiversity: there is a need to develop baseline data on important cultural and ecological structures in the landscape, concerning biodiversity, recreation and health. Likewise, there is a need to plan for measures to mitigate barrier effects with regard to biodiversity and recreation. Further, there is a need to plan forestry for multiple goals including economic, biodiversity, recreation, cultural and aesthetic values. This concerns, for example, both coniferous forest and the oak stands. Also natural grasslands need long time management plans, as well as nature close to residential areas with high recreation values, which is of key importance and needs plans that take into account people's movements in their urban or suburban landscape.

The broad participative process benefited a more unified view on the landscape values and enhanced the cooperation of the different sectors. Further, it was important for gathering local ecological knowledge and knowledge on recreation values and barriers in the landscape. The process also enhanced strategic thinking among responsible stakeholders, development of visions and identification of synergies concerning landscape values. Finally, the regional landscape strategies have the potential to be a valuable tool for planning, since through the presentation of core areas and connecting links over administrative boundaries, the municipalities' areas and values can be seen in a broader perspective, that is, in a landscape and regional perspective.

8.5.2 Arninge–Arlanda Railway

Within the Regional Development Plan for Stockholm County (ORPUT, 2010), it was proposed that a selection of urban nodes, which are already important centres of urban activities, were to be further developed. Two such urban nodes situated north of Stockholm City are Arninge and Arlanda Airport, both of which are important commercial centres, while the latter is the largest international airport in Sweden. In an effort to meet future demands on public transport between these urban nodes, a feasibility study for new public transportation links, either a railway connection or a new bus line between the two nodes was initiated by Stockholm Public Transport, directed by Stockholm County Council. Within the study, three alternative railway corridors were proposed (Håkansson et al., 2012) (Figure 8.3). Since all the three alternatives for the proposed connection crossed an area with high biodiversity, cultural and recreational values, trade-offs could be anticipated between economic development, the climate-smart solution of supplying public transport, and biodiversity objectives.

Compared with other planning efforts concerning biodiversity and ES, all three railway corridors crossed straight through a broad dispersal link outlined in the green wedges (ORPUT, 2010). In addition, the northernmost corridor crossed an area of national interest for cultural history. Furthermore, the same corridor passed through one of the core areas formulated by the regional landscape strategies described in Case Study 1. At an early planning stage, the plans with the three proposed railway corridors were subjected to a public exhibition, which generated high interest from the public and NGOs. Following the exhibition, the northern corridor, which was fragmenting the area of national interest for cultural history, was removed, and the remaining corridors adjusted to further avoid this area (Håkansson et al., 2012).

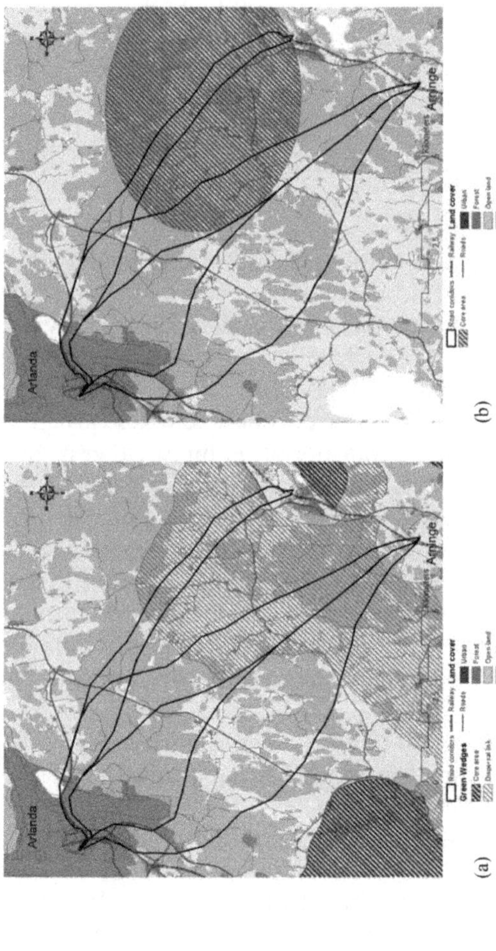
Arlanda
Arninge
Road corridors
Railway
Roads
Green Wedges
Core area
Dispersal link
Land cover
Urban
Forest
Open land
(a)
(b)

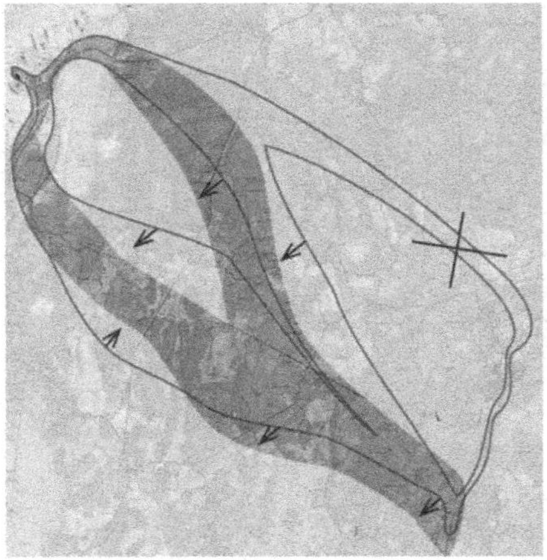

(c)

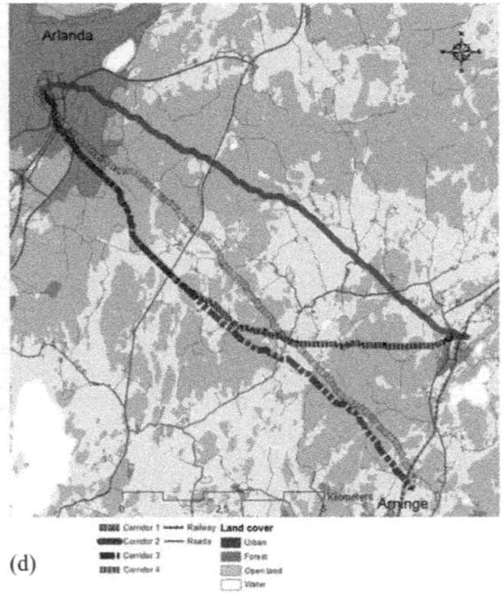

(d)

Sources: (a) ORPUT (2010); (c) Håkansson et al. (2012). Spatial data ©Lantmäteriet [I2014/00591].

Figure 8.3 *(a) The three proposed railway corridors between Arninge and Arlanda crossed through a dispersal link of the green wedges. (b) The northern corridor crossed a core area of the regional landscape strategies (Case Study 1), where an area of national interest for cultural values is also situated. The middle corridor was impacting on the area. (c) During the feasibility study, after the public hearing, the northern corridor was removed and the middle corridor adjusted to avoid the sensitive area. (d) Using spatial multi-criteria analysis (SMCA) and least cost path for designing road corridors. Weighting the geological criteria high results in Corridors 1 and 2, weighting the ecological criteria high results in Corridors 3 and 4, and the northern area would be avoided.*

A similar planning problem was addressed in another research study on developing novel methods for planning and designing road corridors, aiming to integrate disparate sustainability issues such as resource efficiency, biodiversity and travel time, already at the design stage of planning transport infrastructure corridors (Karlson, 2015). Resource efficiency was addressed by integrating spatial models of geological suitability and mass balance into the design, to find the corridor with least resource load. Biodiversity was addressed by integrating spatial ecological models of habitat networks for prioritized groups of species; that is, ecological profiles related to coniferous forest, southern broadleaved forest, natural grasslands and wetlands. The habitat networks consisted of habitat patches and dispersal links, and the road corridor with least impact on these was found. An integration step between resource efficiency and biodiversity was performed using MCA and by applying different weights that gave different results. When resource efficiency was given a high rank, together with travel time, a northern alternative was chosen. By contrast, when biodiversity was given a high rank, together with travel time, the northern parts were avoided; see Figure 8.3a.

Thus, effects of habitat loss and fragmentation on the habitat networks of prioritized groups of species can be quantified and used in the design of transport corridors (ibid.). More detailed information on the overall performance of the habitat networks and the impacts of removing certain patches and links can also be integrated (e.g., Zetterberg et al., 2010). However, a limitation of the use of habitat networks in corridor design is that thresholds for when significant impacts on biodiversity components occur are still largely missing. Another related problem is that since all proposed railway corridors will cross outlined dispersal links, either in the current method development study, as outlined in the green wedges or in the regional landscape strategies (Case Study 1), threshold values for habitat networks would be needed in order to address the overall impacts on species persistence in the studied landscapes. From landscape and regional perspectives, all the proposed alternatives may be unsustainable and the optional transportation mode in the form of a new bus line on existing and already planned roads may be a better alternative. Mitigation measures to avoid fragmentation, informed by the habitat network models, could also be effective. To sum up, the methodology has great potential to integrate biodiversity considerations in designing planning alternatives and to localize critical areas for mitigation measures, supporting sustainable urban development.

8.5.3 Densification of the Stockholm Metropolitan Region

One of the urban nodes where urban development is particularly emphasized in the Regional Development Plan (ORPUT, 2010) embraces the

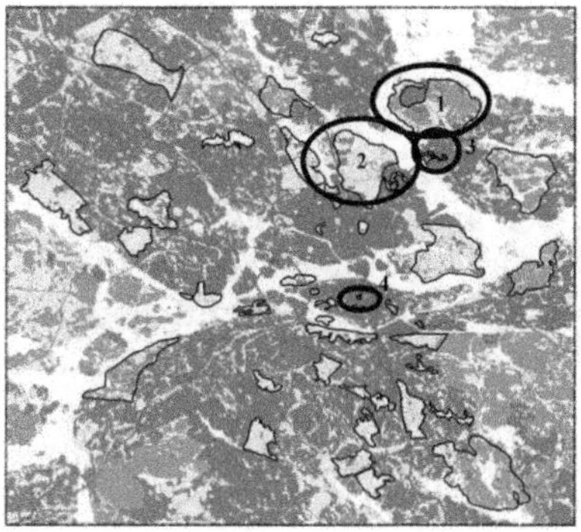

(a)

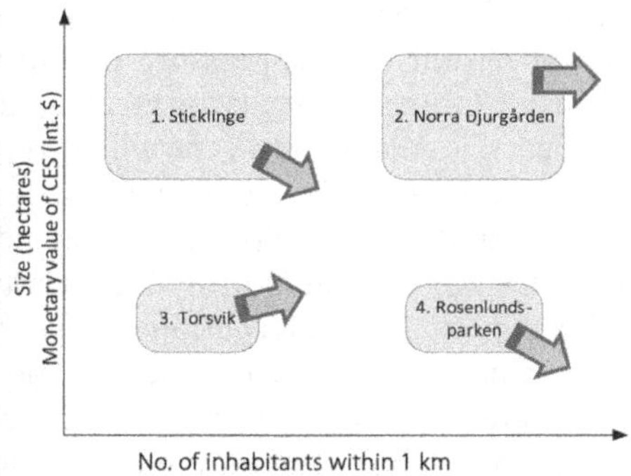

(b)

Source: Spatial data ©Lantmäteriet [I2014/00591].

Figure 8.4 *(a) The analysed green areas within Metropolitan Stockholm. (b) From the analysis of the potential provision of cultural ES for each green area, in hectares and in monetary terms, combined with the number of inhabitants living within 1 km distance of the green areas, four case studies were selected for further analysis of their planning situation.*

central parts of the Stockholm region. In this node, Stockholm City and surrounding municipalities have initiated different urban development projects in areas close to the inner city of Stockholm to meet the urgent need for new housing in the region. These projects result in a densification of this central node, which is in line with the intentions of the Regional Development Plan. Moreover, densification provides preconditions for sustainable mobility as expressed in 'The Walkable City – Stockholm City Plan', which was adopted in 2010. Due to these developments, which primarily take place within existing residential areas and on brown sites close to the inner city, several green areas will be affected by ecosystem loss and fragmentation, and/or by increasing numbers of citizens living in adjacent areas. This means that ES provided by the green areas will be affected and that the pressure on the green areas will increase. The provision of cultural ES is not well integrated in the planning process for individual development projects. Hence, incremental planning and decision-making does not provide information for an overall assessment of the impacts of ongoing densification projects on cultural ES at local and regional levels.

In order to relate single planning projects to a landscape perspective, the potential cultural ES provision of each green area needs to be quantified. In a research project, a set of 44 green areas in the central and suburban parts of Metropolitan Stockholm have been studied, for which the potential provision of and pressure on cultural ES was estimated (Azcárate, 2015). While the capacity of green areas to provide cultural ES is related to their size, the habitat type will also matter. Furthermore, there may be a limit to how many people may use a green area before the pressure will start degrading the cultural ES. The potential economic value of the cultural ES was estimated using De Groot et al.'s (2012) global ES valuation scheme with 2007 international dollars per hectare and year as the unit. The values were applied to the size and habitat type composition for each green area. In addition, the potential pressure from users of cultural ES, in the form of the human population living within 1 km from each green area, was also estimated. Based on the combined results and ongoing city compaction plans, planning cases related to four of the green areas were selected for extended analysis. The selection represented four different types of urban green areas: two with low cultural ES values and two with high cultural ES values, two with a high number of inhabitants within 1 km and two with a low number of inhabitants within 1 km (Figure 8.4).

The green areas varied in size from 2.3 to 633 hectares, and the number of citizens within 1 km varied from over 10 000 potential users per hectare to less than 100 users per hectare. The cultural ES monetary value per hectare and habitat type was the same in any location, but the total value differed with size and habitat type composition. The values of cultural ES that were provided by the green areas varied between 557 (50–3035) Int$/year and 669 700 (108 144–2 929 100) Int$/year (Azcárate, 2015). The results of the monetary evaluation are highly uncertain, taking into account the uncertainties within the database reported by De Groot et al. (2012). However, these estimates give an indication and facilitate a discussion on the actual benefits provided by cultural ES in an urban landscape, in terms of city attractiveness, recreational and health benefits, and other benefits that can provide an input into urban planning.

A comparison of the provision of cultural ES by the four cases is presented in Figure 8.4b, showing that the two large green areas provide high total cultural ES values, while they differ strongly in pressure. Likewise, the two small areas provide low total cultural ES values, while they also differ widely in pressure. Based on a review of planning documents and interviews with urban planners, the planned increases of housing, urban population and other urban activities were summarized for each of these cases, as well as related changes of green area size and management. All four cases would imply an increase in the number of potential visitors, while the change in size of the green areas differed widely. In two cases the green area would shrink (Sticklinge and Rosenlundsparken), while in one case the green area would be indifferent (Norra Djurgårdsstaden) and in another case (Torsvik) increase, due to the development of a new park on a planned ecoduct across a major road.

The provision-pressure analysis illustrates the challenge of compacting cities, to provide both room for an increasing number of citizens as well as for cultural ES, on limited space. This challenge raises questions on whether increasing habitat quality can compensate for shrinkage of green areas, how much pressure is possible before cultural ES starts degrading, and how to plan for sufficient levels and benefit-sharing of cultural ES.

The landscape approach to identifying potential provision and pressure concerning cultural ES can serve to initiate an active engagement of concerned actors in strategic dialogues on complex issues surrounding green area and ES planning in compacting city contexts. However, to further such strategic dialogues, and to include green areas and ES issues early on and throughout the entire urban planning processes, the use of an ES-based Strategic Environmental Assessment (SEA) approach could be useful.

Several ES-based SEA guidelines (Slootweg and Van Beukering, 2008; OECD, 2010; Slootweg et al., 2010; UNEP, 2014) as well as various ES-based SEA methodological frameworks (Geneletti, 2011; Helming et al., 2013; Partidário and Gomes, 2013) have been developed. Combining these guidelines with the cultural ES provision-pressure analyses could frame the planning of city densification projects in the Metropolitan Stockholm area, and contribute to identifying key measures concerning green area design, size, connectivity, composition, quality, uses and trade-offs. Moreover, the combination of the guidelines and the cultural ES analyses would enable the formulation of strategies for including measures to secure the provision of cultural ES at local and regional scales of urban planning.

8.6 LESSONS LEARNED AND KEY RECOMMENDATIONS FOR BEST PRACTICE

A consistent assessment of potential impacts can strengthen biodiversity considerations in the planning process and contribute to the preservation of biodiversity in the long term. This is also in line with international policy ambitions, which require protection of green areas so as to maintain biodiversity in urban regions. In addition, a more informed urban design can enhance new development projects that can save or increase ES performance.

To support the impact assessment of human actions on biodiversity at a landscape level there is a need for methods that analyse cumulative impacts – for example, GIS-based methods such as spatial ecological models. In the case studies a number of advantages of spatial ecological models were identified, that is (1) to model biodiversity components on a landscape level, (2) to study quantity, quality and connectivity of related habitat types in the landscape, (3) to put single sites into a landscape perspective, (4) to evaluate planning alternatives and to visualize and communicate the results. A challenge with these GIS-based methods is the strong visual capability that can be misleading if the inherent uncertainties are not properly communicated. Further, it is difficult to predict how viable populations of different species are developing over a time. However, it is possible to analyse where in the landscape the support for different functional groups of species would be best. With this knowledge, it is possible to prioritize and provide decision support for planning measures on protection, restoration and management of biodiversity components and to analyse and evaluate planning alternatives.

In the case for ongoing city densification development projects in Metropolitan Stockholm, a GIS-based method was applied to facilitate an ES provision-pressure analysis. The analysis provided a comprehensive understanding of how urban green areas can or cannot cope with increasing city densification. Moreover, specific issues such as recommended green area sizes, tolerable wear and tear levels, appropriate ecological compensation measures, required habitat compositions, and key ecosystems services can be identified based on ES provision-pressure analyses. In addition, information on ES provision-pressure relations can be used to strengthen evaluation of development scenarios in Strategic Environmental Assessments and support urban planning processes in Metropolitan Stockholm in formulating strategies for urban densification.

In the case study of the Arninge–Arlanda railway, the analysis of the impacts of fragmentation on biodiversity supported the comparison of alternative rail infrastructure corridors. Enhanced knowledge on how transport infrastructure would affect habitat networks and including this knowledge in environmental assessment processes can then improve the planning of infrastructure projects, strengthen biodiversity considerations in ongoing urbanization processes and ensure the urban green areas' ability to provide ES. This is in accordance with the goals of EU requirements on environmental assessments, as well as with the Swedish environmental quality objective of maintaining and providing access to natural environments near and within urban areas that are valuable for outdoor recreation, cultural heritage and biodiversity.

Besides the assessment of impacts on green areas, which improves the knowledge base in urban planning, there is a need for dialogue on biodiversity and ES in urban development processes and increased collaboration between stakeholders at different planning levels to further coordinated actions for enhanced biodiversity and ES in urban planning. Experiences from the case study on regional landscape strategies showed that a broad participation process fostered the development of strategies, visions and synergies concerning landscape values. Moreover, the collection of local ecological knowledge benefited from an active stakeholder involvement. Also, the dialogue that resulted in a broader view on the landscape values – a landscape perspective – facilitated the identification of core areas and connecting links over administrative borders, like municipal borders.

The case studies from Stockholm showed that a landscape approach provides opportunities to address different scales of ecological processes and hence strengthen important landscape structures and functions in urban, regional and infrastructure planning. Furthermore, a landscape approach can facilitate multi-scalar studies by providing a link between different planning actions at the local, regional and national levels, which facilitates a long-term sustainable management of biodiversity and ES in urban development.

REFERENCES

Alkemade, R., M. van Oorschot and L. Miles et al. (2009), 'GLOBIO3: a framework to investigate options for reducing global terrestrial biodiversity loss', *Ecosystems*, **12**(3), 374–90.

Angelstam, P., T. Edman and M. Dönz-Breuss et al. (2004), 'Land management data and terrestrial vertebrates as indicators of forest biodiversity at the landscape scale', *Ecological Bulletins*, **51**, 333–49.

Aronson, M.F.J., F.A. La Sorte and C.H. Nilon et al. (2014), 'A global analysis of the impacts of urbanization on bird and plant diversity reveals key anthropogenic drivers', *Proceedings of the Royal Society B*, **281**(1780), 20133330.

Avon, C., L. Bergès and Y. Dumas et al. (2010), 'Does the effect of forest roads extend a few meters or more into the adjacent forest? A study on understory plant diversity in managed oak stands', *Forest Ecology and Management*, **259**(8), 1546–55.

Azcarate, J. (2015), 'Beyond impacts: contextualizing strategic environmental assessment to foster the inclusion of multiple values in strategic planning', TRITA-LWRPHD-2015:02, doctoral thesis, Stockholm: KTH Royal Institute of Technology.

Bacaro, G., D. Rocchini and A. Ghisla et al. (2012), 'The spatial domain matters: spatially constrained species rarefaction in a free and open source environment', *Ecological Complexity*, **12**, 63–9.

Bagli, S., D. Geneletti and F. Orsi (2011), 'Routing of power lines through least-cost path analysis and multicriteria evaluation to minimise environmental impacts', *Environmental Impact Assessment Review*, **31**(3), 234–9.

Balfors, B., U. Mörtberg and M. Gontier (2005), 'Impacts of region-wide urban development on biodiversity in strategic environmental assessment', *Journal of Environmental Assessment Policy and Management*, **7**(2), 229–46.

Barthel, S., J. Colding and T. Elmqvist et al. (2005), 'History and local management of a biodiversity-rich, urban cultural landscape', *Ecology and Society*, **10**(2), 10.

Beatley, T. and P. Newman (2013), 'Biophilic cities are sustainable, resilient cities', *Sustainability*, **5**(8), 3328–45.

Benítez-López, A., R. Alkemade and P.A. Verweij (2010), 'The impacts of roads and other infrastructure on mammal and bird populations: a meta-analysis', *Biological Conservation*, **143**(6), 1307–16.

Bergsten, A. and A. Zetterberg (2013), 'To model the landscape as a network: a practitioner's perspective', *Landscape and Urban Planning*, **119**, 35–43.

Bissonette, J.A. and S.A. Rosa (2009), 'Road zone effects in small-mammal communities', *Ecology and Society*, **14**(1), 27.

Bolund, P. and S. Hunhammar (1999), 'Ecosystem services in urban areas', *Ecological Economics*, **29**(2), 293–301.

Borgström, S. (2011), 'Urban shades of green', dissertation, Stockholm: Department of Systems Ecology, Stockholm University.

Brown, L.R. (2001), *Eco-Economy: Building an Economy for the Earth*, New York: Norton.

Chan, K.M.A., T. Statterfield and J. Goldstein (2012), 'Rethinking ecosystem services to better address and navigate cultural values', *Ecological Economics*, **74**, 8–18.

Chiesura, A. and R. de Groot (2003), 'Critical natural capital: a socio-cultural perspective', *Ecological Economics*, **44**, 219–31.

Colding, J. (2011), 'The role of ecosystem services in contemporary urban planning', in J. Niemelä, J.H. Breuste and G. Guntenspergen et al. (eds), *Urban Ecology, Patterns, Processes, and Applications*, Oxford, UK: Oxford University Press, pp. 228–37.

Colding, J. (2013), 'Local assessment of Stockholm: revisiting the Stockholm urban assessment', in T. Elmqvist, M. Fragkias and J. Goodness et al. (eds), *Urbanization, Biodiversity and Ecosystem Services: Challenges and Opportunities*, London: Springer, pp. 313–35.

Colding, J., J. Lundberg and C. Folke (2006), 'Incorporating green-area user groups in urban ecosystem management', *Ambio*, **35**(5), 237–44.

Coutinho-Rodrigues, J., A. Simão and C.H. Antunes (2011), 'A GIS-based multicriteria spatial decision support system for planning urban infrastructures', *Decision Support Systems*, **51**(3), 720–26.

Cushman, S.A., K. McGarigal and M.C. Neel (2008), 'Parsimony in landscape metrics: strength, universality, and consistency', *Ecological Indicators*, **8**(5), 691–703.

De Groot, R., L. Brander and S. van der Ploeg et al. (2012), 'Global estimates of the value of ecosystems and their services in monetary units', *Ecosystem Services*, **1**(1), 50–61.

De Groot, R., P.S. Ramakrishnan and A.V.D. Berg et al. (2005), 'Cultural and amenity services', in Millennium Ecosystem Assessment Board (ed.), *Ecosystems and Human Well-being: Current Status and Trends*, Washington, DC: Island Press, pp. 455–76.

De Ridder, K., V. Adamec and A. Bañuelos et al. (2004), 'An integrated methodology to assess the benefits of urban green space', *Science of the Total Environment*, **334–335**, 489–97.

Di Giulio, M., R. Holderegger and S. Tobias (2009), 'Effects of habitat and landscape fragmentation on humans and biodiversity in densely populated landscapes', *Journal of Environmental Management*, **90**(10), 2959–68.

Di Minin, E. and A. Moilanen (2012), 'Empirical evidence for reduced protection levels across biodiversity features from target-based conservation planning', *Biological Conservation*, **153**, 187–91.

Douglas, I. and J. Ravetz (2011), 'Urban ecology – the bigger picture', in J. Niemelä, J.H. Breuste and G. Guntenspergen et al. (eds), *Urban Ecology, Patterns, Processes, and Applications*, Oxford, UK: Oxford University Press, pp. 246–62.

Drielsma, M. and S. Ferrier (2009), 'Rapid evaluation of metapopulation persistence in highly variegated landscapes', *Biological Conservation*, **142**(3), 529–40.

Dunn, C.P. and L. Henegham (2011), 'Composition and diversity of urban vegetation', in J. Niemelä, J.H. Brueste and G. Guntenspergen et al. (eds), *Urban Ecology, Patterns, Processes, and Applications*, Oxford, UK: Oxford University Press, pp. 103–15.

Eigenbrod, F., S.J. Hecnar and L. Fahrig (2009), 'Quantifying the road-effect zone: threshold effects of a motorway on Anuran populations in Ontario, Canada', *Ecology and Society*, **14**(1) 24.

Elmqvist, T., J. Colding and S. Barthel et al. (2004), 'The dynamics of social-ecological

systems in urban landscapes. Stockholm and the National Urban Park, Sweden', *Annals of the New York Academy of Science*, **1023**, 308–22.
Fahrig, L. and T. Rytwinski (2009), 'Effects of roads on animal abundance: an empirical review and synthesis', *Ecology and Society*, **14**(1), 1–20.
Falkenmark, M. and T. Chapman (1989), *Comparative Hydrology: An Ecological Approach to Land and Water Resources*, Paris: UNESCO.
Forman, R.T.T. (2008), *Urban Regions: Ecology and Planning Beyond the City*, New York: Cambridge University Press.
Forman, R.T.T., D. Sperling and J.T. Bissonette et al. (2003), *Road Ecology: Science and Solutions*, Washington, DC: Island Press.
Geneletti, D. (2004), 'Using spatial indicators and value functions to assess ecosystem fragmentation caused by linear infrastructures', *International Journal of Applied Earth Observation and Geoinformation*, **5**(1), 1–15.
Geneletti, D. (2010), 'Combining stakeholder analysis and spatial multicriteria evaluation to select and rank inert landfill sites', *Waste Management*, **30**(2), 328–37.
Geneletti, D. (2011), 'Reasons and options for integrating ecosystem services in strategic environmental assessment of spatial planning', *International Journal of Biodiversity Science, Ecosystem Services & Management*, **7**(3), 143–9.
Gómez-Baggethun, E. and D.N. Barton (2013), 'Classifying and valuing ecosystem services for urban planning', *Ecological Economics*, **86**, 235–45.
Gómez-Baggethun E., Å. Gren and D.N. Barton et al. (2013), 'Urban ecosystem services', in T. Elmqvist, M. Fragkias and J. Goodness et al. (eds), *Urbanization, Biodiversity and Ecosystem Services: Challenges and Opportunities*, London: Springer, pp. 175–251.
Gontier, M., B. Balfors and U. Mörtberg (2006), 'Biodiversity in environmental assessment: current practice and tools for prediction', *Environmental Impact Assessment Review*, **26**(3), 268–86.
Gontier, M., U. Mörtberg and B. Balfors (2010), 'Comparing GIS-based habitat models for applications in EIA and SEA', *Environmental Impact Assessment Review*, **30**(1), 8–18.
González del Campo, A. (2012), 'GIS in environmental assessment: a review of current issues and future needs', *Journal of Environmental Assessment Policy and Management*, **14**(1), 1250007.
González, A., A. Donnelly and M. Jones et al. (2013), 'A decision-support system for sustainable urban metabolism in Europe', *Environmental Impact Assessment Review*, **38**, 109–19.
Government of Sweden (2004), *Summary of Government Bill 2004/05:150*, accessed 18 January 2016 at http://www.government.se/contentassets/a575d5a63d7b4e7a853671c3812f116a/environmental-quality-objectivs--a-shared-responsibility-summary-of-government-bill-200405150.
Grimm, N.B., S.H. Faeth and N.E. Golubiewski et al. (2008), 'Global change and the ecology of cities', *Science*, **319**(5864), 756–60.
Håkansson, M., M. Regazzoni and A.I. Lundberg et al. (2012), *Feasibility Study for Railway Link Between Roslagsbanan and Arlanda Airport* (Feasibility study No. SL-2011-05811), Stockholm [in Swedish].
Hanski, I. and O. Ovaskainen (2003), 'Metapopulation theory for fragmented landscapes', *Theoretical Population Biology*, **64**(1), 119–27.
Hedblom, M. and U. Mörtberg (2011), 'Characterizing biodiversity in urban areas using remote sensing', in X. Yang (ed.), *Urban Remote Sensing: Monitoring, Synthesis and Modeling in the Urban Environment*, 1st edition, Hoboken, NJ: John Wiley & Sons, pp. 287–304.
Helm, D. and C. Hepburn (2012), 'The economic analysis of biodiversity: an assessment', *Oxford Review of Economic Policy*, **28**(1), 1–21.
Helming, K., K. Diehl and D. Geneletti et al. (2013), 'Mainstreaming ecosystem services in European policy impact assessment', *Environmental Impact Assessment Review*, **40**, 82–7.
Hobbs, R.J., S. Arico and J. Aronson et al. (2006), 'Ecosystems: theoretical and management aspects of the new ecological world order', *Global Ecology and Biogeography*, **15**(1), 1–7.

Holderegger, R. and M. di Giulio (2010), 'The genetic effects of roads: a review of empirical evidence', *Basic and Applied Ecology*, **11**(6), 522–31.
Hostetler, M., W. Allen and C. Meurk (2011), 'Conserving urban biodiversity? Creating green infrastructure is only the first step', *Landscape and Urban Planning*, **100**(4), 369–71.
James, P., T. Tzoulas and M.D. Adams et al. (2009), 'Towards an integrated understanding of green space in the European built environment', *Urban Forestry and Urban Greening*, **8**(2), 65–75.
Jansson, M. (2014), 'Green space in compact cities: the benefits and values of urban ecosystem services in planning', *Nordic Journal of Architectural Research*, **2**, 139–60.
Jim, C.Y. (2004), 'Green-space preservation and allocation for sustainable greening of compact cities', *Cities*, **21**(4), 311–20.
Kabisch, N. and D. Haase (2013), 'Green spaces of European cities revisited 1990–2006', *Landscape and Urban Planning*, **110**(1), 113–22.
Karlson, M. (2015), 'Road ecology for environmental assessment'. TRITA-LWR PHD-2015:06, doctoral thesis, Stockholm: KTH Royal Institute of Technology.
Karlson, M. and U. Mörtberg (2015), 'A spatial ecological assessment of fragmentation and disturbance effects of the Swedish road network', *Landscape and Urban Planning*, **134**, 53–65.
Karlson, M., U. Mörtberg and B. Balfors (2014), 'Road ecology in environmental impact assessment', *Environmental Impact Assessment Review*, **48**, 10–19.
La Greca, P., D. La Rosa and F. Martinico et al. (2011), 'Agricultural and green infrastructures: the role of non-urbanised areas for eco-sustainable planning in a metropolitan region', *Environmental Pollution*, **159**(8–9), 2193–202.
Landsberg, F., S. Ozment and M. Stickler et al. (2011), *Ecosystem Services Review for Impact Assessment. Introduction and Guide to Scoping*, Washington, DC: World Resources Institute.
Laurans, Y., N. Pascal and T. Binet et al. (2013), 'Economic valuation of ecosystem services from coral reefs in the South Pacific: taking stock of recent experience', *Journal of Environmental Management*, **116**, 135–44.
Lehvävirta, S., F. Vilisics and L. Hamberg et al. (2014), 'Fragmentation and recreational use affect tree regeneration in urban forests', *Urban Forestry & Urban Greening*, **13**(4), 869–77.
Li, F., R. Wang and J. Paulussen et al. (2005), 'Comprehensive concept planning of urban greening based on ecological principles: a case study in Beijing, China', *Landscape and Urban planning*, **72**(4), 325–36.
Luck, G.W. (2007), 'A review of the relationships between human population density and biodiversity', *Biological Reviews*, **82**(4), 607–45.
Lyytimäki, J. and M. Sipilä (2009), 'Hopping on one leg – the challenge of ecosystem disservices for urban green management', *Urban Forestry and Urban Greening*, **8**(4), 309–15.
Marcus, L., B. Balfors and T. Haas (2013), 'Sustainable urban fabric: the development and application of analytical urban design theory', in J. Metzger and A. Rader Olsson (eds), *Sustainable Stockholm: Exploring Urban Sustainability in Europe's Greenest City*, New York: Routledge, pp. 71–101.
Marzluff, J.M. and K. Ewing (2001), 'Restoration of fragmented landscapes for the conservation of birds: a general framework and specific recommendations for urbanizing landscapes', *Restoration Ecology*, **9**(3), 280–92.
McDonald, R. and P. Marcotullio (2011), 'Global effects of urbanization on ecosystem services', in J. Niemelä, J.H. Brueste and G. Guntenspergen et al. (eds), *Urban Ecology, Patterns, Processes, and Applications*, Oxford, UK: Oxford University Press, pp. 193–205.
McDonald, R.I. and D.L. Urban (2006), 'Edge effects on species composition and exotic species abundance in the North Carolina piedmont', *Biological Invasions*, **8**(5), 1049–60.
McDonald, R.I., R.T.T. Forman and P. Kareiva et al. (2009), 'Urban effects, distance, and protected areas in an urbanizing world', *Landscape and Urban Planning*, **93**(1), 63–75.
McDonald, R.I., P.J. Marcotullio and B. Güneralp (2013), 'Urbanization and global trends in biodiversity and ecosystem services', in T. Elmqvist, M. Fragkias and J. Goodness

et al. (eds), *Urbanization, Biodiversity and Ecosystem Services: Challenges and Opportunities*, London: Springer, pp. 31–52.
McGarigal, K., S.A. Cushman and E. Ene (2012), 'FRAGSTATS v4: spatial pattern analysis program for categorical and continuous maps', software program produced by the authors at the University of Massachusetts, Amherst, accessed 17 January 2016 at http://www.umass.edu/landeco/research/fragstats/faq/fragstats_faq.html.
McGranahan, G., D. Balk and B. Anderson (2007), 'The rising tide: assessing the risks of climate change and human settlements in low elevation coastal zones', *Environment & Urbanization*, **19**(1), 17–37.
McKinney, M.L. (2006), 'Correlated non-native species richness of birds, mammals, herptiles and plants: scale effects of area, human population and native plants', *Biological Invasions*, **8**(3), 415–25.
Millennium Ecosystem Assessment (MA) (2005), *Ecosystems and Human Well-being: Synthesis Report*, Washington, DC: Island Press.
Mörtberg, U. (2009), 'Landscape ecological analysis and assessment in an urbanizing environment', in M.J. McDonnell, A. Hahs and J.H. Breuste (eds), *Ecology of Cities and Towns: A Comparative Approach*, Cambridge, UK: Cambridge University Press, pp. 439–55.
Mörtberg, U., B. Balfors and W.C. Knol (2007), 'Landscape ecological assessment: a tool for integrating biodiversity issues in strategic environmental assessment and planning', *Journal of Environmental Management*, **82**(4), 457–70.
Mörtberg, U., J. Haas and A. Zetterberg et al. (2013), 'Urban ecosystems and sustainable urban development – analysing and assessing interacting systems in the Stockholm region', *Urban Ecosystems*, **16**(4), 763–82.
Mörtberg, U., A. Zetterberg and B. Balfors (2012), 'Urban landscapes in transition – lessons from integrating biodiversity and habitat modelling in planning', *Journal of Environmental Assessment Policy and Management*, **14**(1), 1250002.
Müller, N., M. Ignatieva and C.H. Nilon et al. (2013), 'Patterns and trends in urban biodiversity and landscape design', in T. Elmqvist, M. Fragkias and J. Goodness et al. (eds), *Urbanization, Biodiversity and Ecosystem Services: Challenges and Opportunities*, London: Springer, pp. 123–74.
Niemelä, J., S.-R. Saarela and T. Söderman et al. (2010), 'Using the ecosystem services approach for better planning and conservation of urban green spaces: a Finland case study', *Biodiversity and Conservation*, **19**(11), 3225–43.
Office of Regional Planning and Urban Transportation (ORPUT) (2010), *Regional utvecklingsplan för Stockholmsregionen 2010, RUFS 2010* [Regional Development Plan for Stockholm, 2010] [in Swedish], Stockholm, accessed 17 January 2016 at http://www.tmr.sll.se/rufs2010/.
Organisation for Economic Co-operation and Development (OECD) (2010), *Strategic Environmental Assessment and Ecosystem Services*, Paris: OECD.
Orsi, F. and D. Geneletti (2010), 'Identifying priority areas for forest landscape restoration in Chiapas (Mexico): an operational approach combining ecological and socioeconomic criteria', *Landscape and Urban Planning*, **94**(1), 20–30.
Partidário, M.R. and R.C. Gomes (2013), 'Ecosystem services inclusive strategic environmental assessment', *Environmental Impact Assessment Review*, **40**, 36–46.
Pauleit, S. and J.H. Breuste (2011), 'Land-use and surface-cover as urban ecological indicators', in J. Niemelä, G.H. Brueste and G. Guntenspergen et al. (eds), *Urban Ecology, Patterns, Processes, and Applications*, Oxford, UK: Oxford University Press, pp. 19–30.
Pauleit, S., L. Liu and J. Ahern et al. (2011), 'Multifunctional green infrastructure. Planning to promote ecological services in the city', in J. Niemelä, J.H. Breuste and G. Guntenspergen et al. (eds), *Urban Ecology, Patterns, Processes, and Applications*, Oxford, UK: Oxford University Press, pp. 272–85.
Phillips, S.J. and D. Miroslav (2008), 'Modeling of species distributions with Maxent: new extensions and a comprehensive evaluation', *Ecography*, **31**(2), 161–75.
Pickett, S.T.A, M.L. Cadenasso and J.M. Grove et al. (2011), 'Urban ecological systems:

scientific foundations and a decade of progress', *Journal of Environmental Management*, **92**(3), 331–62.
Ricketts, T. and M. Imhoff, (2003), 'Biodiversity, urban areas, and agriculture: locating priority ecoregions for conservation', *Ecology and Society*, **8**(2), 1.
Sandström, U.G. (2002), 'Green infrastructure planning in urban Sweden', *Planning Practice Research*, **17**(4), 373–85.
Saura, S. and L. Pascual-Hortal (2007), 'A new habitat availability index to integrate connectivity in landscape conservation planning: comparison with existing indices and application to a case study', *Landscape and Urban Planning*, **83**(2–3), 91–103.
Saura, S. and L. Rubio (2010), 'A common currency for the different ways in which patches and links can contribute to habitat availability and connectivity in the landscape', *Ecography*, **33**(3), 523–37.
Saura, S., C. Estreguil and C. Mouton et al. (2011), 'Network analysis to assess landscape connectivity trends: application to European forests (1990–2000)', *Ecological Indicators*, **11**(2), 407–16.
Secretariat of the Convention on Biological Diversity (SCBD) (2012), *Cities and Biodiversity Outlook*, Montreal: SCBD.
Segan, D.B., E.T. Game and M.E. Watts et al. (2011), 'An interoperable decision support tool for conservation planning', *Environmental Modelling & Software*, **26**(12) 1434–41.
Seto, K.C., S. Parnell and T. Elmqvist (2013), 'A global outlook on urbanization', in T. Elmqvist, M. Fragkias and J. Goodness et al. (eds), *Urbanization, Biodiversity and Ecosystem Services: Challenges and Opportunities*, London: Springer, pp. 1–12.
Slootweg, R. and P. van Beukering (2008), *Valuation of Ecosystem Services and Strategic Environmental Assessment*, Utrecht: MER.
Slootweg, R., A. Rajvanshi and V.B. Mathur et al. (2010), *Biodiversity in Environmental Assessment. Enhancing Ecosystem Services for Human Well-Being*, Cambridge: Cambridge University Press.
Statistics Sweden (2012) [website – statistics on national, county and municipality levels], accessed 30 June 2014 at www.scb.se.
Stockholm County Administrative Board (2007), 'Det storstadsnära landskapet. Regional landskapsstrategi – en pilotstudie'. *Report No. 2007:34* [in Swedish].
Ståhle, A. (2008), 'Compact sprawl: exploring public open space and contradictions in urban density', dissertation, Stockholm: KTH Royal Institute of Technology.
Tian, Y., C.Y. Jim and Y. Tao (2012), 'Challenges and strategies for greening the compact city of Hong Kong', *Journal of Urban Planning and Development*, **138**(2), 101–9.
Tzoulas, K. and K. Greening (2011), 'Urban ecology and human health', in J. Niemelä, J.H. Breuste and G. Guntenspergen et al. (eds), *Urban Ecology, Patterns, Processes, and Applications*, Oxford, UK: Oxford University Press, pp. 263–71.
Tzoulas, K., K. Korpela and S. Venn et al. (2007), 'Promoting ecosystem and human health in urban areas using green infrastructure: a literature review', *Landscape and Urban Planning*, **81**(3), 167–78.
UN (2014), *World Urbanization Prospects: The 2014 Revision, Highlights* (No. ST/ESA/SER.A/352), New York: United Nations, Department of Economic and Social Affairs, Popular Division.
UNEP (2014), *Integrating Ecosystem Services in Strategic Environmental Assessment: A Guide for Practitioners*, a ProEcoServ report by D. Geneletti, accessed 18 January 2016 at http://www.ing.unitn.it/~genelab/documents/GuidelineESintoSEA.pdf.
Vos, C.C., J. Verboom and P.F.M. Opdam et al. (2001), 'Toward ecologically scaled landscape indices', *The American Naturalist*, **157**(1), 24–41.
Yokohari, M. and J. Bolthouse (2011), 'Planning for the slow lane: the need to restore working greenspace in maturing contexts', *Landscape and Urban Planning*, **100**(4), 421–4.
Zetterberg, A., U. Mörtberg and B. Balfors (2010), 'Making graph theory operational for landscape ecological assessments, planning, and design', *Landscape and Urban Planning*, **95**(4), 181–91.

9. Impacts of agricultural and forest management on biodiversity and ecosystem services*

Christine Fürst, Susanne Frank, Marcos Jiménez, Daniel Alejandro Rozas Vásquez, Katrin Pietzsch and Frank Pietzsch

9.1 INTRODUCTION

Agriculture and forestry are the most important terrestrial land uses relevant for the provision of essential resources, such as food, fodder and timber. Beyond, both play a key role in the water cycle, in local and global climate regulation (Siry et al., 2005; Cooper et al., 2009) and are key elements of the cultural identity particularly in rural environments (Vos and Mekes, 1999). How and with what intensity agricultural and forest management is performed depends upon sustaining soil functioning and the provision and connection of habitats. Particularly in densely settled world regions with high pressures on land, the adaptation of agricultural and forest management practices is often the only opportunity to sustain ecological integrity and satisfy all resource demands (Tscharntke et al., 2012). All in all, a scientifically well-grounded understanding of the role and impacts not only of agriculture and forestry as such, but also of management alternatives within them is requested to support sustainable regional development.

Ecosystem services (ES) (MA, 2005) became a key concept for the accounting of environmental impacts of land use and to communicate the value of natural capitals in EU and international policy processes (Boyd and Banzhaf, 2007). An example is the implementation of the EU Biodiversity Strategy, which necessitates the biophysical mapping and valuation of ES (Maes et al., 2012). The Intergovernmental Platform for Biodiversity and Ecosystem Services, IPBES (Larigauderie and Mooney, 2010), for instance, aims at a regional/subregional and a global assessment of biodiversity and ES and is expected to be of high relevance for future environmental policies. However, an important aspect of the success or failure of implementing ES successfully in planning, decision-making and policy consulting, is how different scales, namely land management at the sectorial (agriculture, forestry) and landscape level, are integrated (De Groot et al., 2010).

So far, land cover and its changes are the most common indicators for assessing the provision of ES (Verburg et al., 2009) and for monitoring biodiversity. Particularly, when aiming at introducing the concept of ES successfully into Strategic Environmental Assessment (SEA) and improving its accounting in Environmental Impact Assessment (EIA), more concrete culturally and economically determined management aspects need to be included to avoid failures in policy, plans, programmes and management consulting (Daily et al., 2009; Geneletti, 2011). Regional varieties in land ownership in agriculture and forestry imply uncertainties considering potential future land cover, land use and land management changes. Therefore, scenario-based assessments of environmental and economic impacts are recommended for deriving best practices and achieving consensus on how to adapt spatial constellations of different land uses and the management therein to achieve sustainability (González-Estrada et al., 2008). In this context, a better integration of spatial planning with SEA would provide valuable opportunities for incorporating the ES approach in multi-level decision processes (Geneletti et al., 2007; Barral and Maceira, 2012; Geneletti, 2013a, 2013b).

In this chapter, we aim to introduce the modelling and impact assessment platform GISCAME and to demonstrate its application in the assessment of agricultural and forest management alternatives in spatial planning contexts. The platform combines land cover/use and land management change scenario analyses with an assessment of their effects on regional targets, which are currently mainly assessed using the ES concept. Based upon participatory and analytical scenario design instruments, highly integrative and complex scenarios combining land cover/use and land management changes can be developed and analysed. The platform provides a visual, qualitative and partially quantitative feedback including ES provision maps, risk maps (water erosion, mass movement) and habitat connection or fragmentation maps. Diagrams provide further information on the overall regional balance in ES bundles and on the temporal variability of provisioning services.

To demonstrate how to work with GISCAME in spatial planning, we selected two application cases in Germany and in Chile where we analysed how agricultural and forestry management practices at regional and local scale can be successfully integrated to improve the consideration of land use and management information in spatial planning and impact assessment (Fürst, Frank and Witt et al., 2013; Koschke et al., 2013; Peña-Cortes et al., 2013; Frank, Fürst and Pietzsch, 2015). In both cases, we simulated a large number of integrative scenarios from which we selected a few to demonstrate how land use and land management adaptations can be combined in scenario analysis and to what extent they help to enhance single services or contribute to a better balance of ES bundles.

Finally, we come up with a suggestion for how to enhance the usability of the ES approach at the interface between spatial planning and impact assessment supported by integrative modelling tools such as GISCAME. To achieve a better consideration of how ES are generated in the interaction between natural capacities and human demands (Burkhard et al., 2010), we recommend establishing interactive feedback processes that reflect the structures of spatial planning and impact assessment and help to guide the planning and assessment process along the cascade from ecosystem process to benefits for human beings (Potschin and Haines-Young, 2011).

9.2 METHODOLOGICAL FRAMEWORK – GISCAME SOFTWARE SUITE

9.2.1 Scientific and Technological Basis

GISCAME is an integrated modelling and assessment platform that has been developed based upon a stakeholder-driven approach since 2007 (Fürst, König and Pietzsch et al., 2010; Fürst, Vacik and Lorz et al., 2010) for supporting spatial planning and regional economic development processes. GISCAME is particularly designed to assess the impact of spatial planning scenarios on the provision of regionally prioritized ES including their interactions with the functional diversity and structure of landscapes and their ecosystems.

Considering the theoretical assessment basis, we refer to CICES (Common International Classification of Ecosystem Services; Haines-Young and Potschin, 2012) and the cascade model developed by Potschin and Haines-Young (2011). We understand 'biodiversity' as the bio(geo) physical structure relevant for the functioning of all essential ES processes that are relevant for the provision of a service. Considering such bio(geo) physical structures, we look at agricultural (Lorenz et al., 2013) and forest management strategies (Fürst et al., 2011, 2012) related to intra and inter annual changes in the abundance of the vegetation cover, its density and its horizontal and vertical structure. Considering agriculture, we also consider different physical soil management strategies such as ploughing, mulching, no-till and others (Koschke et al., 2013; Lorenz et al., 2013). Proximity effects and landscape metrics – as described below – are integrated to adjust the overall results on services provisioning capacities

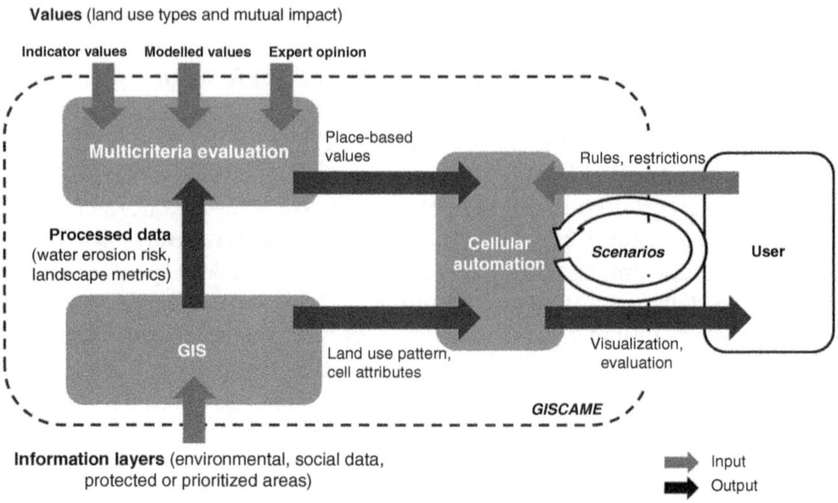

Figure 9.1 *GISCAME concept: key components and their interplay, data flow and interactions between system and user*

at landscape scale, addressing beneficial or disadvantageous effects of spatial interactions and landscape composition and configuration (Fürst, Helming and Lorz et al., 2013).

To enable linking bio-geophysical information with ES provision and participatory or policy-driven scenario building, GISCAME combines geographical information systems (GIS) routines with a cellular automaton (Fürst, König and Pietzsch et al., 2010; Fürst, Volk and Pietzsch et al., 2010) and a hierarchical multicriteria assessment framework (Fürst, Nepveu and Pietzsch et al., 2009; Koschke et al., 2012) that can be iteratively adapted to the specific regional situations (Figure 9.1).

A cellular automaton is a mathematical model for exploring the development of pattern in regular lattices depending on the initial properties of the raster cells, of their neighbours and cell-specific transition probabilities (Ulam, 1952). This modelling approach was adapted for environmental modelling by Tobler (1979) and gained in relevance later on for spatial planning and decision-making through combination with GIS (White and Engelen, 1993, 1994, 1997; White et al., 1997). In GISCAME, this explorative modelling approach is implemented to support the spatial scenario design. Furthermore, in combination with the hierarchical multicriteria assessment approach the cellular automaton model is applied to account for proximity effects: accounting for information on which land use or management classes surround a cell helps to integrate potential boundary effects and spatial interactions that might increase or decrease the capacity of a land use or management type to provide a specific service.

GIS routines are implemented in GISCAME for integrating, combining and merging highly diverse digital information. This includes land cover and land use information including additional attributes that inform on the land management and its intensity and on geophysical parameters such as soil types, topography or climate characteristics. Additionally, we consider as far as possible socio-political characteristics such as land tenure (types) or other place-specific and/or legally framed scenario restrictions. Such restrictions can be, for instance, nature reserves or sites that require a specific land use and management to protect neighbouring areas against risks (avalanches) or disturbances (emissions, noise).

Furthermore, the assessment of the water erosion risk following the RUSLE (Revised Universal Soil Loss Equation; Renard et al., 1991, 1994) and the mass movement risk based upon a sediment routing algorithm are part of the GIS core of the platform.

Completing the above-described assessment of proximity effects relevant for ES, we implemented a set of landscape metrics by which we analysed landscape fragmentation, land cover/use diversity and habitat connectivity. We included the metrics 'patch density', 'Shannon-Wiener Index', 'shape index (of [semi-]natural areas)', 'effective mesh-size of unfragmented areas', 'core area of (semi-)natural areas', and 'connected (semi-)natural areas' (Frank et al., 2011, 2013). In combination with some conventions formulated in the multicriteria assessment framework below, we apply these metrics to refine the assessment of land system capacities to provide ES by including the effects of landscape composition and configuration.

Our multicriteria assessment framework is hierarchically structured, starting from indicators that express the place-specific capability of land cover/use types and management strategies therein to contribute to ES provision. In the multicriteria assessment of land use impacts on ES, we are confronted with the problem that the information basis is highly diverse considering, for example, different indicator systems and units and that it is often incomplete so that some land use classes cannot be assessed. To overcome this problem, we normalize indicator values and project them thereby to a scale from 0 (lowest/no potential) to 100 (highest potential) (Fürst, Volk and Pietzsch et al., 2010, Koschke et al., 2012). The use of such normalized values helps to fill knowledge gaps by expert estimation related to known (measured) land use – ES provision constellations. Furthermore, we obtain a comprehensive calculation basis to scale up the impact of each single land use type to landscape level. Comparably, we transform the landscape metrics values into a qualitative correction.

We introduced the convention that the impact of the land use pattern can reduce or increase the result obtained from upscaling the land use values maximally by 30 value points, while the upper (100) and lower (0) thresholds are maintained (Frank et al., 2011).

With regard to integration of modelled results from forestry and agriculture to account for temporal dynamics, beyond the above-described indicator-based assessment, we also include modelled datasets that enable us to reflect the temporal dynamics of some services, such as the provision of biomass or the storing of carbon in the vegetation layer. We created a template-based interface that allows the specification of what kinds of data (units, temporal resolution, period for which these are available) are integrated so that diverse model outcomes can be integrated. This routine is currently implemented for a variety of forest growth and yield models (Frank, Fürst and Pietzsch, 2015). Interfaces to agricultural models are at work to investigate the regional performance of different cropping systems, fertilization and soil management alternatives. For more integrative scenarios, a harmonization of agricultural and forest modelling outcomes in time and space is still needed.

As most of the agricultural and forest models are only available for single crops or pure stands, we added routines how to combine different species. In the case of the forest models, results from modelling pure stands can be combined with 'mixed stands' by defining the percentage shares of the different tree species. This supports the analysis of scenarios that include stand types that are not yet established to pre-test the effects of different silvicultural strategies. However, we had to accept that this simple routine ignores differences in the growth of mixed stands provoked by interspecific competition so that the uncertainty level in the scenario outcomes is high. To derive information on present stand types and integrate the model outcomes in a spatially explicit way, we base our work on forest inventory data including additional parameters such as stand density, age and yield class.

Subsequently, the modelled growth and yield data are broken down into more decision-relevant information such as the provision of different timber resources (construction, pulp and paper, fuel wood), the carbon sequestration in the stocking volume or the calorific value of the harvestable volume. This transformation is based on statistical information and literature values. The up-scaled results are then displayed per selected time step and scenario, so that opportunities, risks and trade-offs of forest management strategies in combination with land use changes can be assessed over time. This serves, for instance, to answer questions such as, 'To what extent can forest management intensification by shortening rotation periods and/or converting forests stands with fast-growing tree species compensate potential production area losses due to land cover changes (deforestation)?' Vice versa we can analyse how a reduction in management intensity through the prolongation of rotation periods would impact carbon sequestration and the provision of different timber resources from a regional perspective.

9.2.2 Scenario Design for Impact Assessment

A key aspect in impact assessment is the reliability and relevance of the scenarios (Seppelt et al., 2013). Cellular automata are currently mostly applied to explore the change of spatial pattern (e.g., Soares-Filho et al., 2002) based upon rule sets that rely on transition probabilities. Such probabilities can be based upon expert opinion, expectations of stakeholders, or can be taken from time series analyses. The resulting scenarios give evidence on future land use and land management constellations and how these might vary depending on alternative starting situations or modified transition probabilities (Veldkamp et al., 2002). However, this kind of scenario design poses some problems: time series data for land cover/use changes are often not available for long periods or they are exposed to problems of disharmony between land use and land cover classification standards. Consequently, the derived rule sets are not necessarily valid, even more so as they reflect land use and land cover change (LULCC) driver constellations from the past. Also, non-evident drivers such as economic, socio-cultural and political settings cannot easily be correlated with observed changes. They might have various impacts on how land cover/use or land management strategies develop dependent on local or place-specific conditions. Finally, behavioural aspects (agent behaviour) cannot be optimally considered in such scenarios and therefore the communication of and tangibility of this kind of 'automatized' scenario for stakeholders often remains limited (Parker et al., 2003).

Consequently, we added scenario design instruments that support manually drawing land cover/use and land management changes reflecting the expectations of stakeholders as to 'what kind of changes could happen' or curiosity-driven scenarios as to 'what would need to be changed to enhance a specific ES or to improve the balance in ES bundles'. Furthermore, our scenario design instruments enable change in specific management settings such as the soil management in agriculture relevant for soil erosion, or to rebuild and develop road infrastructures relevant for habitat fragmentation risks.

Additionally, we added the opportunity to formulate rule-based scenarios that reflect the impact of multiple and interconnected attributes (Fürst, Frank and Witt et al., 2013). With this kind of scenario instrument, strategic questions can be answered such as 'What would happen if all arable land on fertile soils in slope positions with high or low risk in soil erosion is changed into maize mono-cropping systems?'.

All kinds of derived scenarios – transition probability based, rule based or manually designed – can be used separately or can be combined. All of them include the opportunity to change only the land cover/use and leave management strategies within agricultural and forest classes untouched and vice versa, or to combine both. Additionally, information layers that reflect variable environmental data (e.g., climate data under different climate change scenarios) can be activated, exchanged or excluded. This supports sensitivity analyses of how the capacities of land systems to provide ES depend on different input data and scenarios (Fürst et al., 2014; Schulze et al., 2015).

9.3 CASE STUDY 1 – ASSESSING THE IMPACT OF REGIONAL STRATEGIES FOR ENHANCING BIOMASS PROVISION

In the RegioPower project we assessed strategies for combining management alternatives in forestry and agriculture with afforestation or establishing short rotation coppices on extensively used land for increasing regional biomass provision capacities. Such strategies were analysed along a model region transect from Eastern Finland, Southern Sweden, North-Eastern Germany to Central Slovenia (Frank, Fürst and Japelj et al., 2015). In this chapter we focus on results developed for the model region Upper-Elbe-Valley-Eastern-Ore-Mts, Germany.

An essential aspect of scenario analyses in the Upper-Elbe-Valley-Eastern-Ore-Mts model region was to include land use and land management information with the highest possible spatial and particularly thematic resolution. Due to the need to compare results between the regions along our transect with CORINE Land Cover (CLC) 2012 data as the only universally available reference, we restricted our spatial resolution to 1 ha, though a higher resolution of 25 × 25 m would have been possible. To obtain a sufficiently detailed thematic resolution, we developed, in cooperation with GAF AG (former Euromap GmbH), a regionally specific but still CLC-compatible classification standard (Euromap Land-Cover Classification, EMLC). We combined remote sensing and terrestrial data from forest inventory, biotope and land use mapping and agricultural mapping at field block level (Fürst and Flügel, 2015). Additionally, we included classes for crop rotations (agriculture) and forest stand types (forestry) that represent future management opportunities that are not yet introduced into practice, but are considered to be particularly well adapted to climate change based on field experiments and model-based analyses (Eisenhauer and Sonnemann, 2009; Lorenz et al., 2013) (see Figure 9.2).

In forestry, we also added information on the present stand age and stand density to map spatially explicit information on growth, harvestable assortments and carbon storage. We included information layers on the ownership type, slope, and the legal protection status (e.g., national parks or nature reserves) as potential restrictions for land use or land management change scenarios: these criteria provide evidence of whether management activities such as harvesting will take place and, if yes, how much of the harvestable volume is realized. In small-scale private forests, for instance, harvesting might not even take place. Data on typical harvesting amounts in the different forest ownership types compared to what could be possible based upon the pure natural potential were obtained from expert opinion in the region (Gerold and Thode, 2010). The integration of such social aspects in framing scenarios and analysing their outcomes is essential for regional biomass provision projections to avoid unrealistic numbers being created.

9.3.1 Services and Scenarios for Impact Assessment

To identify key regional development objectives and to derive land use and land management change scenarios to be tested, we conducted interviews with regional experts in a regional focus group. This group included representatives from governmental institutions (forest and agricultural administration), the regional planning authority Upper-Elbe-Valley-Eastern-Ore-Mts., a couple of coordinators of LEADER regions, representatives from agricultural and forest land owner associations, private consultants offering agricultural and forest extension services and NGOs in tourism and nature conservation.

Based on these interviews (and the terminology agreed in this project), we selected the ES 'Wood production', 'Food and fodder production' with regard to our aim to increase the regional biomass provision. The services 'C sequestration', 'Soil erosion reduction' and 'Drought risk regulation' were selected to reflect the regional objective to reduce climate change impacts by enhancing important regulating services. 'Recreation' was selected as most important cultural service. Additionally, we introduced the term 'Ecological integrity', which, according to CICES (Haines-Young and Potschin, 2012), corresponds to essential biophysical processes and structures. In the regionally agreed terminology, the term was used synonymously for 'biodiversity' as no

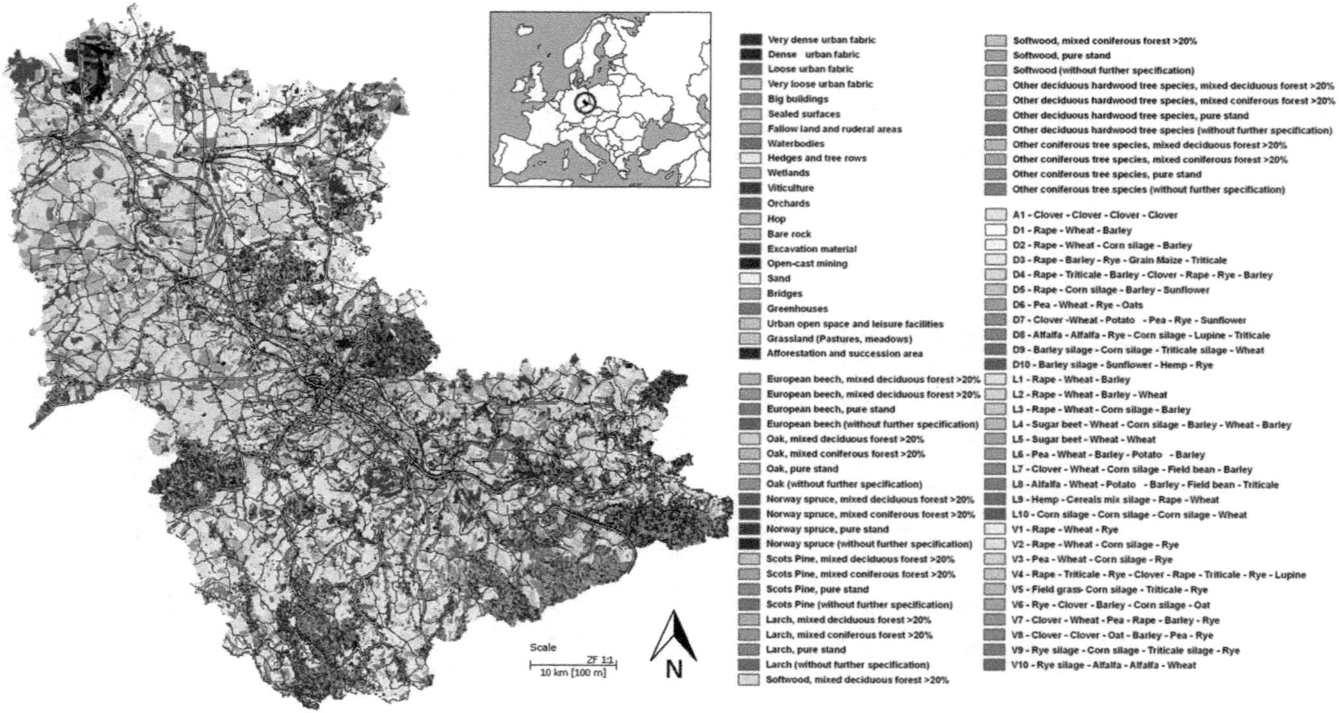

Figure 9.2 *EMLC map of Upper-Elbe-Valley-Eastern-Ore-Mts. and location in Europe*

Table 9.1 *Scenario examples*

Sc2 Optimal Balance of all Ecosystem Services	Sc3 Maximum Increase of Biomass Output
Replacing crop rotations/stand types with negative impact on regulating services	Replacing crop rotations/stand types with low production potential
1. Converting pure Norway spruce stands (8%) into mixed stands (following the silvicultural guidelines; Eisenhauer and Sonnemann, 2009)	1. Converting forest stand types that are expected to be less productive/ more risky under climate change (Norway spruce, 8%; European beech, 4%) with fast-growing tree species (Douglas fir)
2. Replacing homogeneous crop sequences (14%) with sequences with higher crop diversity (e.g., 'Rape-Wheat-Barley' by 'Rape-Triticale-Rye-Clover-Rape-Triticale-Rye-Lupine')	2. Replacing traditional crop rotations including fallow (12%) by a maize-dominated high-performing sequence ('Corn silage-Corn silage-Corn silage-Wheat')

consensus could be reached among the regional focus group members for using the latter term.

Scenarios were consequently outlined with the regional focus groups by formulating strategic requests for changing the shares of different agricultural and forest management types and breaking them down in a spatially explicit way with the GISCAME scenario instruments. Among a variety of scenarios, we selected two most illustrative examples, which combine land use and land management change in agriculture and forestry, for further use in communication support. Table 9.1 explains the two integrated scenarios sc2 and sc3. The reference scenario scl represents the business-as-usual (BAU) scenario, which illustrates expected ES potentials now and in the future, without any land use/land management change.

9.3.2 Scenario Evaluation and Results

The evaluation of the scenarios was made from the following two different perspectives. First, we used the GISCAME multicriteria assessment approach for an overall assessment of the scenario performance in the provision of the above-described ES and in changes of the ecological integrity (Figure 9.3). We applied a set of indicators (Koschke et al., 2012) complemented by expert knowledge from our regional focus group. Indicators were, for instance, the C factor (vegetation and management factor of the

Universal Soil Loss Equation) for soil erosion protection, annual timber yield for wood production, yield in agricultural products for human or livestock alimentation, water retention capacity for drought risk regulation or number of vascular plant species per land use type for ecological integrity. Recreation was purely based on expert opinion (value of a land use type) and was adjusted using landscape metrics to account for the particular character of the study region. Subsequently, we went into more detail to test temporal fluctuations in the two provisioning services regarding an optimal consultation of the land owners (hidden trade-offs of management strategies over time). In this contribution, we focus on the outcomes of the adaptation of the forest management (forest conversion) over a time period of 100 years (Figure 9.3). Here, we applied modelled results complemented by data taken from growth and yield tables for species, where models were not available (Frank, Fürst and Pietzsch et al., 2015).

Our two above-described scenarios (sc2 and sc3) performed differently

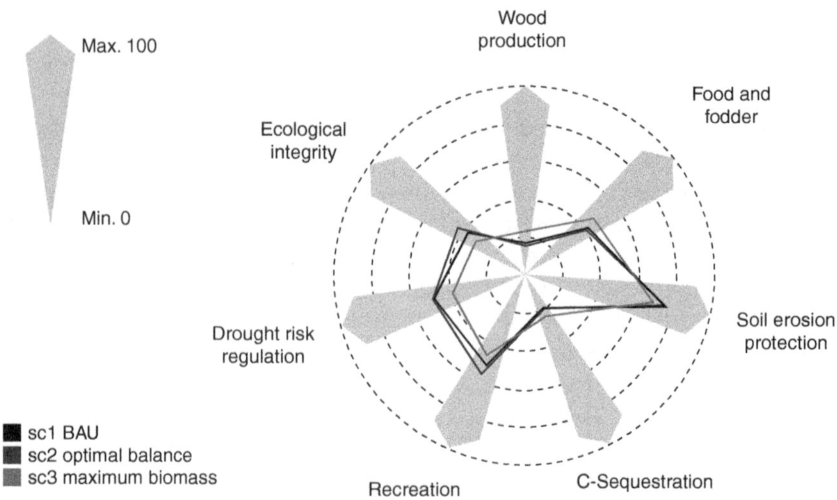

	sc1	sc2	sc3
Wood production	17	16	24
Food and fodder	41	39	47
Soil erosion protection	76	76	70
C-sequestration	20	19	24
Recreation	52	56	46
Drought risk regulation	49	50	40
Ecological integrity	37	43	31

Figure 9.3 Overall performance of the two scenarios compared to BAU

when considering the selected ES bundle and ecological integrity. The intention to maximize the biomass output (sc3) resulted in the expected (but not considerably high) increase in the provision of timber, food and fodder. However, losses in the drought risk regulation capacity through the introduction of higher shares of maize-dominated crop rotations are a trade-off of this particular scenario. Also, the value for recreation was decreased, while carbon sequestration could even slightly benefit from higher timber volumes in Douglas fir stands. In contrast, sc2 with the intention of achieving an overall better balance of all ES only performed slightly better compared to BAU in ecological integrity and recreation.

These results underpin a general problem in scenario-based impact assessment: scenarios that represent potentially realistic management adaptations or land use changes experience real world limits in their impact. Only extreme and therefore not realistic land use changes would result in visible improvements or decreases of ES (Fürst, Frank and Witt et al., 2013).

Therefore, the in-depth analyses of potential temporal fluctuations – here for the timber provision potential – help us to understand potential benefits or risks related to the different scenarios more comprehensively. The bar chart in Figure 9.4 illustrates how yield (Y) and fuel wood production (FW) will perform 20 and 100 years after the simulated change in forest management. Considering the harvestable timber, sc3 turns out to

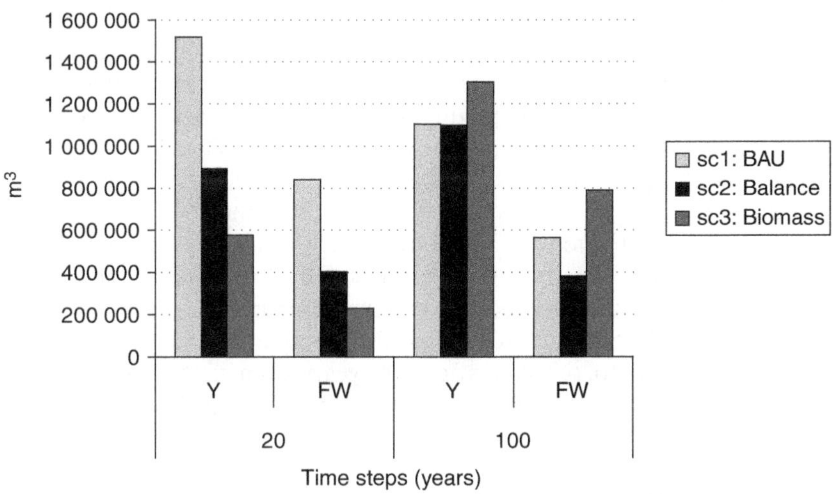

Figure 9.4 Temporal variability of timber provision potentials; harvestable volume (Y) and fuel wood (FW) for the time steps 20 and 100

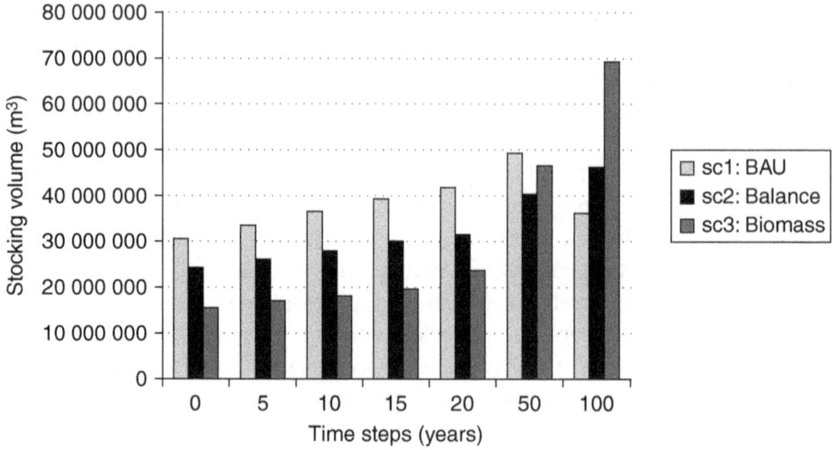

Figure 9.5 Temporal variability of the stocking volume in the three scenarios over 100 years

be most beneficial after 100 years. In particular, fuel wood production will clearly exceed the amounts of sc1 and sc2. The lower performance of sc1 and sc2 in 100 years results particularly from the lack of fast-growing (productive) tree species, and partially (BAU) from the transition of currently middle-aged stands to young stands through the applied silvicultural treatment model.

Figure 9.5 provides more detailed information on the development of the stocking volume in our three scenarios. Forest conversion (i.e., lowering of the stand density and planting of young deciduous trees) particularly impacts the stocking volume in sc2 over five decades. Only after 100 years, will the stocking volume of sc2 exceed that in sc1 due to a higher share of mature stands. In sc1, the average stand age will then be inferior due to harvesting compared to sc2 and sc3, resulting in a lower stocking volume.

In contrast, sc3, which starts with intense forest conversion and replanting, will have an inferior stocking volume compared to sc1 and sc2 over two decades; after 50 and 100 years, the stocking volume in sc3 will be higher compared to sc2 and after 100 years even higher compared to sc1, reflecting the beneficial effect of the conversion to fast-growing tree species.

9.3.3 Regional Recommendations and Conclusions

With the help of the two selected scenarios, we could finally successfully communicate the effects of different land use/management changes taking into account the fact that private forest owners and farmers prioritize timber, food and fodder production. Insights from these exercises provided an informed decision-making basis for updating the regional plan of the model region leading to spatial prioritization of ES and particularly adequate management strategies. Furthermore, we could reveal 'how fast' potential risks or synergies from management changes would occur. This was of particular relevance for communicating successfully with the regional planners: prior to developing a regional plan, they are requested to project future demands in timber, food and fodder and ensure by their plan that these demands can be satisfied either by delineating sufficiently large areas for production or by planning infrastructures that support the import of missing resources. So far, the estimation of how large production areas should be to feed future demands was confronted with high uncertainties, as only statistical numbers and not process-based model projections were included. Consequently, combining mapping and assessment of the ES provision at present with modelling of future fluctuations can be seen as a helpful extension to support the knowledge-based consultation process of spatial planning and decision-making.

9.4 CASE STUDY 2 – MODELLING THE IMPACT OF FOREST AND AGRICULTURAL LAND USE SCENARIOS ON ECOSYSTEM SERVICES PROVISION IN SUPPORT OF STRATEGIC ENVIRONMENTAL ASSESSMENT IN SPATIAL PLANNING – CHILE

In Chile, planning and decision-making is highly centralized. This situation is recognized as one of the major reasons for economic, social, cultural, territorial and ethnic inequalities (OECD, 2009). Spatial planning and its associated decision-making instruments, however, are broken down into four different hierarchical levels defined by the General Law of Urbanism and Constructions (Table 9.2).

Until 2009, environmental aspects were considered in spatial planning through Environmental Impact Assessment (EIA). Addressing only lower tiers in the above-described decision hierarchy, EIA could not ensure an effective safeguard of the environment and ecosystems (Rojas et al., 2013). Consequently, since 2010, Strategic Environmental Assessment (SEA) was made mandatory for the approval of spatial plans as shown in Table 9.2 (except at the national level), allowing the incorporation of sustainability objectives and criteria for the conservation of important nature reserves such as wetlands and forests (Rozas-Vásquez et al., 2014). Furthermore,

Table 9.2 Spatial planning hierarchies and examples of plans and planning instruments

Level of Planning	Relevant Spatial Plans/Planning Instruments
National	General Law for Urbanism and Constructions
	National Urban Development Policy
Regional	Regional Plan for Urban Development
	Regional Plan for Territorial Planning[a]
Intercommunal	Metropolitan Regulating Plan
	Intercommunal Regulating Plan
Communal	Communal Regulating Plan
	Sectional Plan

Note: a. The Regional Plan for Territorial Planning (in Spanish: PROT) is under the responsibility of the Under-Secretary of the Regional and Administrative Development.

Source: Modified from OECD (2014).

the introduction of SEA in spatial planning opened up an opportunity to include ES as a comprehensive framework for impact assessment. However, the selection of suitable indicators for complex interactions between ecosystems and society including trade-off analyses still needs research.

A challenge for Chile is how to organize transdisciplinary participation in SEA; there are serious communication and collaboration gaps among scientists, policy decision-makers and stakeholders. We tried to overcome this problem in the presented case study by introducing the ES concept supported by the application of the GISCAME approach.

To demonstrate how we envisaged enhancing spatial planning processes in Chile, we selected La Araucanía region (Figure 9.6). This region is of outstanding national interest for participatory processes in spatial planning as it has the biggest concentration of Mapuche people, the most important indigenous group (INE, 2007). Furthermore, La Araucanía is characterized by the highest national rate of people living below the breadline (20.7 per cent; OECD, 2009). In terms of land use change, during the last 30 years there has been a considerable expansion in the area of fast-growing exotic species, mainly eucalyptus and pine. This growth has been based on the replacement of native forest, agricultural lands from small landowners and even by drainage of wetlands. A similar trend is more recently seen in the increasing of agriculture land in order to fulfil the demands of Chile's current agro-export model (Geneletti, 2013b; OECD, 2014). These changes in the land use have generated a variety of environmental and social impacts related to biodiversity and habitat connectivity

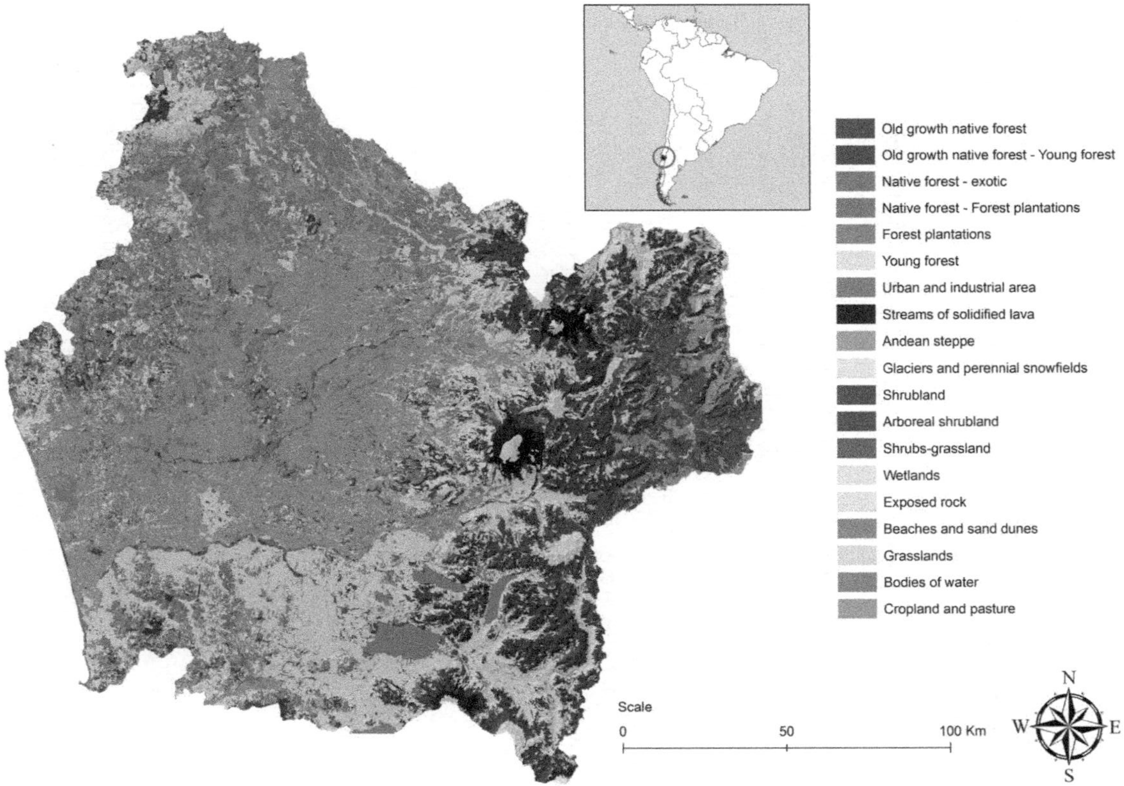

Figure 9.6 *La Araucania region*

loss, water scarcity and soil erosion (Nahuelhual et al., 2012; Miranda et al., 2015).

9.4.1 Ecosystem Services-based Impact Assessment and Scenario Development

Based upon Geneletti (2012, 2014) and the SEA process structure implemented in Chile, we divided our ES-based impact assessment in GISCAME into three steps:

1. Based on interviews with experts, regional scientific-technical reports, and the given legal frame (environmental policies), we selected a bundle of ES that are of relevance for the region.
2. Using interviews with experts and a consensus-building process among scientific experts and regional citizens in a regional focus group, we prioritized key services of particular value for regional decision-making. Based on existing planning regulations (PROT, 2014), literature analysis and prevailing research activities (Koschke et al., 2012), we identified reliable indicators to assess their provision and these were subsequently normalized and included in the GISCAME multicriteria assessment framework (Table 9.3).
3. Together with the experts in our regional focus group, we developed a strategic scenario outline including 'real' land use changes and potential

Table 9.3 Key ES and indicators for La Araucanía region, Chile

ES Group	Ecosystem Service	Indicator
Provisioning	Food	Production of cereals (t/ha)
	Fibre and fuel	Overall biomass production (t/ha)
Regulating	Global climate	Carbon sequestration (% of C in the vegetation cover)
	Flood mitigation	Water retention capacity
	Water erosion control	C (cropping management) and L (topographic) factors of the Revised Universal Soil Loss Equation
Cultural	Landscape aesthetics	Expert opinion (qualitative assessment supported by landscape metrics calculation in GISCAME)
	Recreational value (touristic attractiveness)	Surveys (qualitative assessment)

Note: Also refer to Koschke et al. (2012).

Table 9.4 Scenario examples

Sc2 Optimal Balance of all Ecosystem Services	Sc3a/Sc3b Increase of Biomass Output/ Maximum Increase of Biomass Output
Replacing forest plantations and arable (intensively managed) land 1. Conversion of intensively managed forests (eucalypt and pine plantations, 8%) into near-to-nature forests 2. Conversion of arable land (5%) into pastures	Increasing the share of highly productive land uses *Sc3a: Increase* 1. Converting scrublands, suitable bare soils and native forests (10%) into forest plantations 2. Conversion of pastures (3%) into arable land *Sc3b: Maximum increase* 1. Converting scrublands, suitable bare soils and native forests (20%) into forest plantations 2. Conversion of pastures (6%) into arable land

Note: The numbers in brackets are the percentages of the area undergoing land use change.

management changes in agriculture and forestry. These focused on improving provisioning services or enhancing regulating services including scenarios that tried to balance both ES groups. Trade-offs were assessed for the cultural services. The subsequently formulated scenarios were built on adapted recommendations from the prevailing regional plan (Geneletti, 2012; Peña-Cortés et al., 2013). Table 9.4 provides some examples of integrated scenarios, where sc2 focuses more on management changes in agriculture and forestry, while sc3a and sc3b test the effects of combined land use and land management change. The reference scenario sc1 represents business as usual (BAU).

9.4.2 Scenario Evaluation and Results

Figure 9.7 shows the performance of all scenarios. Compared to BAU (sc1), the overall balance in ES provision could only slightly be improved in sc2, particularly considering the two cultural services and water erosion. However, trade-offs such as losses in the food production were also relatively small. In contrast, particularly in sc3b, which implied rigorous but still realistic land use changes, there was a considerable increase not only in fibre and fuel production, but also – through enhanced carbon sequestration – a higher contribution to global climate change regulation.

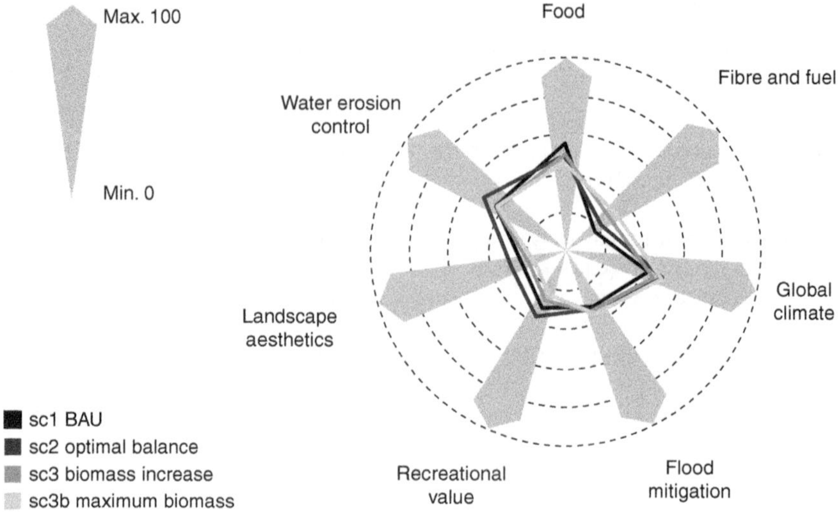

Food	57	52	53	50
Fibre and fuel	20	23	27	34
Global climate	43	49	46	48
Flood mitigation	26	30	30	33
Recreational value	32	39	28	24
Landscape aesthetics	25	30	22	18
Water erosion control	44	49	44	45

Figure 9.7 Overall performance of the two scenarios compared to BAU

Also, flood mitigation could be improved by the relative gain in forest cover, which also compensated for the potentially higher water erosion risks that might have resulted from the conversion of pastures. On the other hand, losses in the two cultural ES and food production were relatively high. In comparison, sc3a only resulted in minor changes in the ES balance, performing slightly better in fibre and fuel provision compared to sc2. Depending on regional priorities, scenarios sc2 and sc3a seemed to be the most appropriate to increase fibre and fuel provision without running the risk of provoking critical losses in the other services.

9.4.3 Regional Recommendations and Conclusions

The presented scenario-based assessment is currently applied to test the impact of different environmental policies at regional level and to provide feedback on how efficient or risky these could be and what options are eligible to realize policy goals in sustainable development. Compared to the Upper-Elbe-Valley-Eastern-Ore-Mts. region, La Araucanía region provides greater opportunities for land use changes. A problematic trend is the rising conversion of near-to-nature forests into forest plantations and the increase of cash crops in agriculture. Through our simulations we could provide recommendations as to what extent the increasing desire for biomass could be satisfied by 'no-net-loss' strategies that include agricultural intensification and establishment of new forest plantations on so far only extensively used agricultural areas. By doing so, the risk of losing near-to-nature forests could be decreased in favour of an increased forest cover that enhances regulating and provisioning services.

9.5 LESSONS LEARNED AND CONCLUSIONS

We intended to demonstrate by two examples in Germany and Chile how to better integrate sectorial management aspects into impact assessment by introducing the ES concept as a general reference. We did so supported by an integrative platform – GISCAME – that provides a structural framework and standardizes how information for impact assessment can be aggregated and implemented in scenario simulation and evaluation. We demonstrated how to make use of a normalized scale to integrate indicator-based assessments and expert opinion to achieve a highly inclusive evaluation of the impact of land cover and land use changes on the provision of ES (e.g., Burkhard et al., 2009). Our application examples highlighted how ES as such, but also underlying indicators, can be implemented and how even temporal fluctuations – here only for timber provision – could be explored to reveal future synergies and risks in matching the demand and the provisioning side of ES.

However, to take full benefit from the ES concept in planning and include agricultural and forest management opportunities better in impact assessment processes, some further adaptations are essential. The chain from agro- and forest ecosystem functions and processes to services needs to be reflected in the methodological concepts of impact assessment to obtain sound and reliable results for supporting spatial planning and environmental decision-making (Leitão and Ahern, 2002). In GISCAME, we account for spatial interactions and landscape composition and configuration using proximity effects and a set of landscape metrics for an adjustment of ES provision capacities of the single land cover/use or land management types according to their spatial constellations. However, this approach needs still to be complemented by an overall integration of larger-scale ecological and hydrological processes and their interactions (Ludwig et al., 2005). This might even necessitate rethinking the regional and thematic scope of plans for appropriate support by modelling and impact assessment tools (Fürst and Flügel, 2015).

The cascade approach suggested in CICES (Potschin and Haines-Young, 2011; Haines-Young and Potschin, 2012) could provide a valuable reference point from which to better connect modelling and impact assessment tools with the ES concept and derive knowledge-based recommendations for spatial planning. We therefore suggest the conceptual framework as displayed in Figure 9.8.

The question of an appropriate regional and thematic scope and the hereon-based projections of land system responses to changes in agricultural and forest management could be based upon combined landscape-scale and ecosystem-scale process models. Their outcomes need to be complemented by indices or metrics that account for boundary phenomena and edge effects at the borders between different ecosystems (Porembski, 2003). Furthermore, upper-scale response units should be considered that are relevant, for instance, for hydrological processes and nutrient cycles, but could also provide help to better integrate agricultural and forest management options in social-ecological system contexts (e.g., Fu et al., 2013). Finally, in this stage of the cascade, functional landscape structures must be addressed that can – beyond agriculture and forestry – also incorporate other land cover/use types fulfilling a comparable or complementary role in connecting habitats and supporting services such as pollination (Balvanera et al., 2005). By combining such different types of metrics and models, the environmental baseline could be delivered for simulating actions and interventions and for subsequently assessing their impact on the provision of ES. For knowledge-based consensus building, information on ES might not be the final step to understanding benefits or threats to human well-being (De Groot et al., 2010). An additionally requested step is therefore the transformation of ES into nationally, regionally or locally applicable value systems (e.g., Cowling et al., 2007). This final step will support participants in spatial planning processes to help understand the impacts and the relevance of LULCC scenarios and the particular role of changes in agricultural and forest management within them for strategic actions and interventions and their relevance for achieving individual or communal targets.

The presented GISCAME platform cannot yet cover all modelling steps along the above-described cascade, therefore it still needs some adaptations to be fully consistent with and support the interaction between agricultural and forest management, spatial planning and impact assessment approaches. Missing elements are still how to model biodiversity at

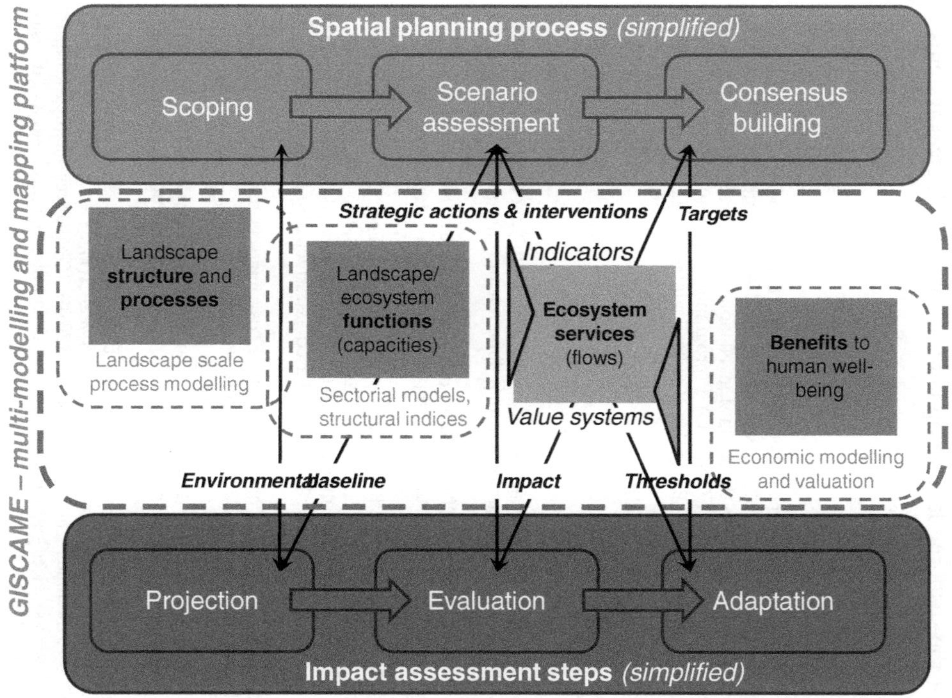

Source: Potschin and Haines-Young (2011); Haines-Young and Potschin (2012).

Figure 9.8 *Linking spatial planning and Strategic Environmental Assessment through a model platform-based assessment of ES following the cascade approach in CICES*

its different scales – from genome to landscape – to address all aspects in landscape and ecosystem functions and their impact on ES provision.

On the other hand, modelling at larger and integrative scales close to policy-making and consulting requires a high level of information aggregation. Scientists and decision-makers will need to cooperate more intensively and develop consensus on what levels of detail and uncertainty are acceptable for the sustainable development of social-ecological systems.

NOTE

* The study cases are taken from the project RegioPower (ERA WoodWisdom/BioEnergy, supported by the German Federal Ministry of Food and Agriculture, 2201991) and from a doctoral study supported by the Comisión Nacional de Investigación Científica y Tecnológica, Chile within the FONDECYT programme. The authors would like to thank Davide Geneletti for initializing this *Handbook* and inviting them to contribute.

REFERENCES

Balvanera, P., C. Kremen and M. Martínez-Ramos (2005), 'Applying community structure analysis to ecosystem function: examples from pollination and carbon storage', *Ecological Applications*, **15**(1), 360–75.

Barral, M.P. and N.O. Maceira (2012), 'Land-use planning based on ecosystem service assessment: a case study in the Southeast Pampas of Argentina', *Agriculture, Ecosystems and Environment*, **154**(7), 34–43.

Boyd, J. and S. Banzhaf (2007), 'What are ecosystem services? The need for standardized environmental accounting units', *Ecological Economics*, **63**(2–3), 616–26.

Burkhard, B., F. Kroll and F. Müller et al. (2009), 'Landscapes' capacities to provide ecosystem services – a concept for land-cover based assessments', *Landscape Online*, **15**, 1–22.

Burkhard, B., I. Petrosillo and R. Costanza (2010), 'Ecosystem services – bridging ecology, economy and social sciences', *Ecological Complexity*, **7**(3), 257–9.

Cooper, T., K. Hart and D. Baldock (2009), *Provision of Public Goods Through Agriculture in the European Union*, London: Institute for European Environmental Policy.

Cowling, R., B. Egoh and A.T. Knight et al. (2007), 'An operational model for mainstreaming ecosystem services for implementation', *PNAS*, **105**(28), 9483–8.

Daily, G., S. Polasky and J. Goldstein et al. (2009), 'Ecosystem services in decision making: time to deliver', *Frontiers in Ecology and the Environment*, **7**(1), 21–8.

De Groot, R.S., R. Alkemade and L. Braat et al. (2010), 'Challenges in integrating the concept of ecosystem services and values in landscape planning, management and decision making', *Ecological Complexity*, **7**(3), 260–72.

Eisenhauer, D.R. and S. Sonnemann (2009), 'Silvicultural strategies under changing environmental conditions – guiding principles, target system and forest development types', *Waldökologie, Landschaftsforschung und Naturschutz*, **8**, 71–88.

Frank, S., C. Fürst and A. Japelj et al. (2015), 'RegioPower – an online platform for interactive tools for participatory land assessment and planning', *GLP News*, No. 11.

Frank, S., C. Fürst and L. Koschke et al. (2011), 'Towards the transfer of the ecosystem service concept to landscape planning using landscape metrics', *Ecological Indicators*, **21**, 30–38.

Frank, S., C. Fürst and L. Koschke et al. (2013), 'Assessment of landscape aesthetics – validation of a landscape metrics-based assessment by visual estimation of the scenic beauty', *Ecological Indicators*, **32**, 222–31.
Frank, S., C. Fürst and F. Pietzsch (2015), 'Cross-sectoral resource management: how forest management alternatives affect the provision of lignocellulosic resources and other ecosystem services', *Forests*, **6**(3), 533–60.
Fu, B., S. Wang and C. Su et al. (2013), 'Linking ecosystem processes and ecosystem services', *Current Opinion in Environmental Sustainability*, **5**(1), 4–10.
Fürst, C. and W. Flügel (2015), 'Assessing the impact of land use changes on providing hydrological ecosystem functions (ESF) and services (ESS)', in L. Chicharo, F. Müller and N. Fohrer et al. (eds), *Ecosystem Services and River Basin Ecohydrology*, London: Springer.
Fürst, C., S. Frank and A. Witt et al. (2013), 'Assessment of the effects of forest land use strategies on the provision of ecosystem services at regional scale', *Journal of Environmental Management*, **127**, 96–116.
Fürst, C., K. Helming and C. Lorz et al. (2013), 'Integrated land use and regional resource management – a cross-disciplinary dialogue on future perspectives for a sustainable development of regional resources', *Journal of Environmental Management*, **127**(Supplement), S1–S5.
Fürst, C., H. König and K. Pietzsch et al. (2010), 'Pimp your landscape – a generic approach for integrating regional stakeholder needs into land use scenario design and sustainable management support', *Ecology and Society*, **15**(3), 34–59.
Fürst, C., C. Lorz and F. Makeschin (2011), 'Integrating land management aspects into an assessment of the impact of land cover changes on ecosystem services', *International Journal of Biodiversity Science and Management*, **7**(3), 168–81.
Fürst, C., G. Nepveu and K. Pietzsch et al. (2009), 'Comment intégrer des considérations multicritères dans la gestion d'un territoire? "Pimp your landscape" – un essai de planification interactive pour satisfaire les besoins des utilisateurs' [How to integrate multicriteria considerations into the management of a territory? 'Pimp your landscape' – an interactive planning attempt to meet the needs of users], *Revue forestière française*, **1**, 21–35.
Fürst, C., P. Opdam and L. Inostroza et al. (2014), 'Evaluating the role of ecosystem services in participatory land use planning: proposing a balanced score card', *Landscape Ecology*, **29**(8), 1435–46.
Fürst, C., K. Pietzsch and S. Frank et al. (2012), 'How to better consider sectoral planning information in regional planning – example afforestation and conversion', *Journal of Environmental Planning and Management*, **55**(7), 1–29.
Fürst, C., H. Vacik and C. Lorz et al. (2010), 'How to support forest management in a world of change? Results of some regional studies', *Environmental Management*, **46**(6), 941–52.
Fürst, C., M. Volk and K. Pietzsch et al. (2010), 'Pimp your landscape! A tool for qualitative evaluation of the effects of regional planning measures on ecosystem services', *Environmental Management*, **46**(6), 953–68.
Geneletti, D. (2011), 'Reasons and options for integrating ecosystem services in strategic environmental assessment of spatial planning', *International Journal of Biodiversity Science, Ecosystem Services & Management*, **7**(3), 143–9.
Geneletti, D. (2012), 'Environmental assessment of spatial plan policies through land use scenarios: a study in a fast-developing town in rural Mozambique', *Environmental Impact Assessment Review*, **32**(1), 1–10.
Geneletti, D. (2013a), 'Ecosystem services in environmental impact assessment and strategic environmental assessment', *Environmental Impact Assessment Review*, **40**, 1–2.
Geneletti, D. (2013b), 'Assessing the impact of alternative land-use zoning policies on future ecosystem services', *Environmental Impact Assessment Review*, **40**, 25–35.
Geneletti, D. (2014), *Integrating Ecosystem Services in Strategic Environmental Assessment: A Guide for Practitioners*, a ProEcoServ report, Nairobi: UNEP.
Geneletti, D., S. Bagli and P. Napolitano et al. (2007), 'Spatial decision support for strategic environmental assessment of land use plans. A case study in Southern Italy', *Environmental Impact Assessment Review*, **27**(5), 408–23.

Gerold, D. and H. Thode (2010), *Clusterinitiative Forst und Holz in Sachsen* [Cluster Analysis Forest & Wood in Saxony], final report, Kesselsdorf: Saxon State Office for the Environment, Agriculture and Geology, p. 159.

González-Estrada, E., L.C. Rodríguez and V.K. Walen et al. (2008), 'Carbon sequestration and farm income in West Africa: identifying best management practices for smallholder agricultural systems', *Ecological Economics*, **67**(3), 492–502.

Haines-Young, R. and M. Potschin (2012), *Common International Classification of Ecosystem Services (CICES): Consultation on Version 4, August–December 2012*, EEA Framework Contract No. EEA/IEA/09/003.

INE (2007), *Región de La Araucanía. Division político, administrativo y censal* [Araucanía Region: Political, Administrative and Census Division], accessed 4 February 2015 at www.inearaucania.cl/archivos%5Cfiles%5Cpdf%5CDivisionPoliticoAdministrativa%5Carauca nia.pdf.

Koschke, L., C. Fürst and S. Frank et al. (2012), 'A multicriteria approach for an integrated land-cover-based assessment of ecosystem services provision for planning support', *Ecological Indicators*, **21**, 54–66.

Koschke, L., C. Fürst and M. Lorenz et al. (2013), 'The integration of crop rotation and tillage practices in the assessment of ecosystem services provision at the regional scale', *Ecological Indicators*, **32**, 157–71.

Larigauderie, A. and A.H. Mooney (2010), 'The Intergovernmental science-policy Platform on Biodiversity and Ecosystem Services: moving a step closer to an IPCC-like mechanism for biodiversity', *Current Opinion in Environmental Sustainability*, **2**(1–2), 9–14.

Leitão, A.B. and J. Ahern (2002), 'Applying landscape ecological concepts and metrics in sustainable landscape planning', *Landscape and Urban Planning*, **59**(2), 65–93.

Lorenz, M., C. Fürst and E. Thiel (2013), 'A methodological approach for deriving regional crop rotations as basis for the assessment of the impact of agricultural strategies using soil erosion as example', *Journal of Environmental Management*, **127**(Supplement), S37–S47.

Ludwig, J.A., B.P. Wilcox and D.D. Breshears et al. (2005), 'Vegetation patches and runoff-erosion as interacting ecohydrological processes in semiarid landscapes', *Ecology*, **86**(2), 288–97.

Maes, J., B. Egoh and L. Willemen et al. (2012), 'Mapping ecosystem services for policy support and decision making in the European Union', *Ecosystem Services*, **1**(1), 31–9.

Millennium Ecosystem Assessment (MA) (2005), *Ecosystems and Human Well-being: Synthesis*, Washington, DC: Island Press.

Miranda, A., A. Altamirano and L. Cayuela et al. (2015), 'Different times, same story: native forest loss and landscape homogenization in three physiographical areas of south-central Chile', *Applied Geography*, **60**, 20–28.

Nahuelhual, L., A. Carmona and A. Lara et al. (2012), 'Land-cover change to forest plantations: proximate causes and implications for the landscape in south-central Chile', *Landscape and Urban Planning*, **107**(1), 12–20.

OECD (2009), *OECD Territorial Reviews: Chile 2009*, Paris: OECD Publishing, accessed 20 January 2016 at http://dx.doi.org/10.1787/9789264060791-en.

OECD (2014), *OECD Rural Policy Reviews: Chile 2014*, Paris: OECD Publishing, accessed 20 January 2016 at http://dx.doi.org/10.1787/9789264222892-en.

Parker, D.C., S.M. Manscon and M.A. Janssen (2003), 'Multi-agent systems for the simulation of land-use and land-cover change: a review', *Annals of the Association of American Geographers*, **93**(2), 314–37.

Peña-Cortes, F., D.A. Rozas-Vásquez and G. Rebodello et al. (2013), 'Territorial planning for coastal zones in Chile: the need for geographical-environmental and natural risk indicators for spatial decision support systems', *International Journal of Geosciences*, **4**(6B), 17–29.

Plan Regional de Ordenamiento Territorial (PROT) (2014), *Escenarios, propuestas y desafíos, región de La Araucanía* [Scenarios, Proposals and Challenges, Araucanía Region], Temuco, Chile: Gobierno Regional de La Araucanía.

Porembski, S. (2003), 'Trends in plant diversity research', *Progress in Botany*, **64**, 506–22.

Potschin, M. and R. Haines-Young (2011), 'Ecosystem services: exploring a geographical perspective', *Progress in Physical Geography*, **35**(5), 575–94.
Renard, K.G., G.R. Foster and G.A. Weesies et al. (1991), 'RUSLE: Revised Universal Soil Loss Equation', *Journal of Soil and Water Conservation*, **46**(1), 30–33.
Renard, K.G., G.R. Foster and G.C. Yoder et al. (1994), 'RUSLE revisited: status, questions, answers and the future', *Journal of Soil and Water Conservation*, **49**(3), 213–20.
Rojas, C., J. Pino and E. Jaque (2013), 'Strategic environmental assessment in Latin America: a methodological proposal for urban planning in the Metropolitan Area of Concepción (Chile)', *Land Use Policy*, **30**(1), 519–27.
Rozas-Vásquez, D., F. Peña-Cortés and D. Geneletti et al. (2014), 'Scenario modelling to support strategic environmental assessment: application to spatial planning of coastal wetlands in La Araucanía region, Chile', *Journal of Environmental Impact Assessment and Policy Management*, **16**(2), 1450014-1–1450014-25.
Schulze, J., R. Martin and A. Finger et al. (2015), 'Design, implementation and test of a serious online game for exploring complex relationships of sustainable land management and human well-being', *Environmental Modelling & Software*, **65**, 58–66.
Seppelt, R., S. Lautenbach and M. Volk (2013), 'Identifying trade-offs between ecosystem services, land use, and biodiversity: a plea for combining scenario analysis and optimization on different spatial scales', *Current Opinion in Environmental Sustainability*, **5**(5), 458–63.
Siry, J.P., F.W. Cubbage and M.R. Ahmed (2005), 'Sustainable forest management: global trends and opportunities', *Forest Policy and Economics*, **7**(4), 551–61.
Soares-Filho, B., G.C. Cerqueira and C.L. Pennachin (2002), 'Dinamica – a stochastic cellular automata model designed to simulate the landscape dynamics in an Amazonian colonization frontier', *Ecological Modelling*, **154**(3), 217–35.
Tobler, W. (1979), 'Cellular geography', in S. Gale and G. Olsson (eds), *Philosophy in Geography*, Dordrecht: Reidel, pp. 379–86.
Tscharntke, T., Y. Clough and T.C. Wanger et al. (2012), 'Global food security, biodiversity conservation and the future of agricultural intensification', *Biological Conservation*, **151**(1), 53–9.
Ulam, S. (1952), 'Random processes and transformations', *Proceedings of the International Congress on Mathematics*, **2**, 264–75.
Veldkamp, T., P.H. Verburg and K. Kok et al. (2002), 'Spatial explicit land use change scenarios for policy purposes: some applications of the CLUE framework', in S.J. Walsh and K.A. Crewes-Meyer (eds), *Linking People, Place, and Policy*, New York: Springer, pp. 317–41.
Verburg, P.H., J. van de Steeg and A. Veldkamp et al. (2009), 'From land cover change to land function dynamics: a major challenge to improve land characterization', *Journal of Environmental Management*, **90**(3), 1327–35.
Vos, W. and H. Meekes (1999), 'Trends in European cultural landscape development: perspectives for a sustainable future', *Landscape and Urban Planning*, **46**(1), 3–14.
White, R. and G. Engelen (1993), 'Cellular automata and fractal urban form: a cellular modelling approach to the evolution of urban landuse patterns', *Environment and Planning A*, **25**(8), 1175–99.
White, R. and G. Engelen (1994), 'Urban systems dynamics and cellular automata: fractal structures between order and chaos', *Chaos, Solitons and Fractals*, **4**(4), 563–83.
White, R. and G. Engelen (1997), 'Cellular automata as the basis of integrated dynamic regional modelling', *Environment and Planning B: Planning and Design*, **24**(2), 235–46.
White, R., G. Engelen and I. Uljee (1997), 'The use of constrained cellular automata for high-resolution modelling of urban land-use dynamics', *Environment and Planning B: Planning and Design*, **24**(3), 323–43.

10. Applications of biodiversity and ecosystem services impact assessment in spatial planning

Leena Kopperoinen, Christian Albert and Pekka Itkonen

10.1 INTRODUCTION

Spatial planning is concerned with spatial impacts of decision-making in sector policies and actions in public and private sectors. It aims at considering and coordinating these impacts in a way that leads to sustainable spatial development (Nadin, 2006, p. 15). As a comparison, land use planning is seen as being more narrowly focused for regulation of land use and development through designation of areas of development and protection, and application of performance criteria (ibid., p. 18). Land use planning has been seen to relate more to setting of rules and regulations by planning authorities, while spatial planning goes further and integrates all the policies and programmes that influence the nature of places and how they function, including those that influence demands on or needs for development but cannot be solved by planning permission alone or in the main. Spatial planning is also seen as more strategic with spatial goals rather than just designating areas either for development or protection as in land use planning. In addition, spatial planning is more interactive, involving mutual learning and working towards the generation of spatial development alternatives in a collaborative political process.

Until now, the consideration of biodiversity and ecosystem services (ES) in spatial planning has been based on environmental objectives and targets as derived from international, European Union, national or subnational government directives and legislation as democratically legitimized aims of land development. Participatory approaches and economic valuation of biodiversity and ES play a growing role in planning and decision-making.

The purpose of this chapter is:

- to present key issues related to including biodiversity and ES in impact assessment of spatial planning;

- to introduce some useful tools for assessing these impacts;
- to showcase three real-life spatial planning processes where impacts have been assessed using various tools; and
- to provide a checklist for implementing impact assessment in practice.

10.2 OVERVIEW OF KEY ISSUES AND TOOLS

Spatial planning is all about reconciliation of various interests and needs. These can have different spatial impacts, which can be synergetic or lead to trade-offs. Impacts can be both positive and negative. The fact is, however, that spatial planning leads to spatial changes, which mean change in ecosystem structure through converting land and water areas for new purposes. In Figure 10.1 we present the impact of spatial planning on biodiversity and ES using the ES cascade model by Potschin and Haines-Young (2011) as a basis from which to concretize the causal impact chain following change in ecosystem structure. Figure 10.1 shows also the different elements that need to be addressed in assessing impacts of spatial planning alternatives.

Biodiversity and ES impact assessment of spatial planning includes all the basic elements of an ordinary impact assessment: ecological, social, cultural and economic impacts from the point of view of biodiversity and ES:

- Ecological impacts are related to:
  - safeguarding biodiversity: species, biotopes, habitats, and genetic diversity;
  - sufficient extent and quality of green infrastructure to host core nature areas, large forest areas and so on;
  - ecological connections;
  - sustaining environmental quality or improving it by sustaining or enhancing the provision of ES.
- Social and cultural impacts include:
  - environmental justice aspects;
  - equity issues;
  - environmental health aspects;
  - taking into account cultural variation in the way nature and its services are perceived.
- Economic impacts mean:
  - costs, benefits and opportunity costs resulting from the spatial plan. However, economic impacts may not be applicable when

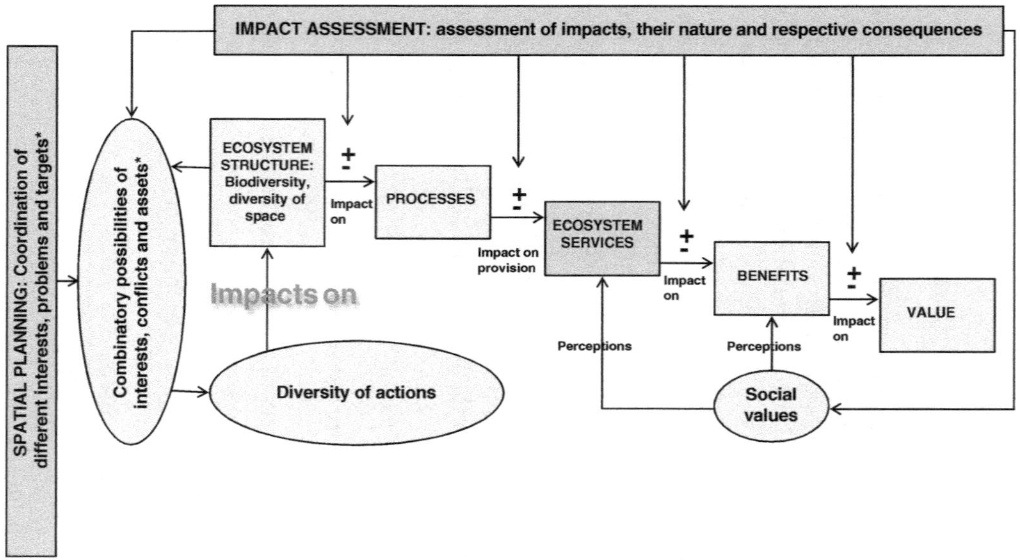

Note: *The figure shows how the interests, problems and targets combined with assets provided by ecosystem structure and biodiversity frame the possible diversity of actions. Decided actions will have a spatial impact on ecosystem structure, which will further cascade down through impacts to ecosystem processes, provision of ecosystem services, availability of benefits and derived values from the benefits. Impact assessment intervenes throughout the cascade.*

Sources:
ES cascade model (ecosytem structure – processes – ES – benefits – value) adopted from Potschin and Haines-Young (2011).
* Inspired by Van Assche et al. (2013).

Figure 10.1 *Impact cascade of spatial planning on biodiversity and ES*

intrinsic values of nature or the well-being of people, even if monetarily calculable, are in question.

Tools for assessing impacts of different spatial planning alternatives on biodiversity and ES are needed to find the most optimal solutions during the planning process. The question of how the concept of ES could best be integrated in spatial planning is an active field of research (Albert et al., 2014). Important research foci concern conceptual frameworks for integrating the concept into existing planning procedures (e.g., Spangenberg et al., 2014; Von Haaren et al., 2014), methods for assessing and valuing ES in the planning context of limited time and resources (e.g., Kopperoinen et al., 2014; Moreno et al., 2014), and critical evaluations of real-world case studies of applying the ES concept in different fields of practice (Hatton MacDonald et al., 2014). Understanding of the best methodologies to assess impacts of spatial planning on biodiversity and ES is constantly evolving, and there is already a great deal of experience in feasible methods (Geneletti, 2011, 2013, 2014; Söderman et al., 2012; Honrado et al., 2013; Partidário and Gomes, 2013). Because spatial planning is about allocation of space to different functions, spatially explicit analysis and impact visualization methods are often found to be the most useful. Therefore, methods based on geographical information systems (GIS) analyses and mapping using spatial datasets are the focus in this chapter. In addition, as collaboration and public participation are seen as fundamental to spatial planning, we also present examples of spatial participatory methods.

10.2.1 GIS Analyses and Mapping

Mapping and assessment of ecosystems and their services are expected to support their maintenance and restoration (European Commission, 2011). The European Commission's Working Group on Mapping and Assessment on Ecosystems and their Services (MAES), set up within the Common Implementation Framework of the Biodiversity 2020 Strategy, has affirmed the utility of mapping (Maes et al., 2013, p. 13):

> Maps are useful for spatially explicit prioritisation and problem identification, especially in relation to synergies and trade-offs among different ecosystem services, and between ecosystem services and biodiversity. Further, maps can be used as a communication tool to initiate discussions with stakeholders, visualizing the locations where valuable ecosystem services are produced or used and explaining the relevance of ecosystem services to the public in their territory. . . However, the mentioned purposes will not be attempted through the sole mapping exercise, but rather through the combination of digital

mapping with the assessment of the supply of ecosystem services related to their demand (including the spatial interactions between them).

The quote above explains the different benefits of mapping also with regard to impact assessment. Spatially explicit methods have a key role when assessing impacts of spatial planning: impacts have to be identified in a spatially explicit way to help understand where and how the impacts take place, whom they concern and if, where and how the spatial plan should be modified. Modelling has a limited role in assessing biodiversity and ES impacts in spatial planning practice. Data availability, planning capacities and the strict timetable of the planning process often do not allow for detailed modelling approaches covering various ES. Therefore, relatively easy approaches are currently in use, often based on standards, look-up tables and so on.

Versatile mapping methods are all worthy ways of assessing impacts of spatial planning on biodiversity and ES. The choice of method(s) depends on the case and the most important place-based questions related to it.

10.2.2 Landscape Structural and Connectivity Analyses of Green Infrastructure

To preserve genetic diversity and to be able to adapt to climate change, it is very important for different species to be able to move from one area to another. The community and urban structure should not become a hindrance to the movement of species. In addition to the intrinsic value of biodiversity, this is also essential to the provision of ES on which our health and well-being depend. The qualities of green infrastructure can be assessed with a variety of methods for the needs of spatial planning. Landscape structural analyses are suitable for evaluating (1) structural connectivity or fragmentation of green infrastructure, (2) permeability of landscape from the point of view of animal movement, (3) potential or functional connectivity of green infrastructure, and (4) diversity of land cover, among others.

These various methods can be applied to carry out temporal and spatial comparison of areas and to prioritize the most important ones. The methods can be used to assess the fragmentation or, on the other hand, connectivity of green infrastructure. In spatial planning, connectivity analysis also helps to examine whether the continuity of green infrastructure enables people to move from one place to another in a natural setting. Pleasant and accessible routes attract people for recreational purposes and for commuting. Through analysis of different spatial plan options with landscape structural methods it is possible to see the impacts that the options have on green infrastructure.

There are dozens of approaches that can be used for analysing impacts of spatial planning on the structural properties and connectivity of a landscape. Among these are the following methods:

- *Morphological spatial pattern analysis (MSPA)* (Soille and Vogt, 2009): Structural connectivity analysis for identification of core areas and varying types of edge and connecting structures of green infrastructure.
- *Landscape permeability analysis*: Assessment of resistance of different land cover types in terms of impeding the movement of certain species. This method uses land cover data and expert knowledge, for example. In order to also take into account the surrounding land cover of each location, a smoothing algorithm, such as CORILIS (Peifer, 2009), may be used.
- *Potential connectivity analysis using graph theoretical approaches* (Calabrese and Fagan, 2004; Pascual-Hortal and Saura, 2006; Laita et al., 2011; Bergsten et al., 2013; Saura et al., 2014): The patches of green infrastructure are regarded as nodes of a network, and the probability of connectedness (links) between each pair of nodes can be derived, for example, using distance decay functions based on data on the dispersal ability of different species. Graph theoretical analyses allow the computational comparison of relative importance of each distinct patch (node) and connection (link) in relation to the overall connectedness of the network.

With these types of analyses it is possible to weigh impacts of, for example, removing or diminishing various green spaces for the overall connectivity of green infrastructure in terms of safeguarding biodiversity as well as for ES supply potential. These analyses are crucial in impact assessment of spatial plans if connectivity is to be maintained or improved by spatial decisions.

10.2.3 Participatory Mapping – Assessing Citizens' Values and Perceptions for Impact Assessment

Sociocultural values are commonly mapped using deliberative or qualitative methods. In sociocultural valuation, the focus is on the importance of, and preferences, needs or demands for nature, expressed by people or groups of people, including plural values (i.e., valuing something for several reasons), through a variety of qualitative and quantitative measures (Chan et al., 2012). When a spatial extent is added to these valuation methodologies, they are also often referred to as demand mapping.

Sociocultural values are difficult – or impossible – to map based on biophysical parameters only. Acknowledging this problem, participatory mapping methodologies and, for example, photo-based methods, have been developed to capture the values (Milcu et al., 2013).

The local experiential knowledge of residents and users of various areas is essential in spatial planning when the target is to ensure an attractive living environment integrating versatile values. Methods by which non-experts or the lay public help to identify spatial information are called public participatory GIS methods (PPGIS) (Brown et al., 2013). Participatory mapping methods comprise Internet-based surveys, interviews, focus group discussions, citizens' juries, community or group process mapping, and modelling from participatory mapping of landscape values (Kyttä and Kahila, 2011; Klain and Chan, 2012; Brown, 2013; Kyttä et al., 2013; Kelemen et al., 2014; García-Nieto et al., 2015). These methods provide systematic identification and measurement of values based on local ecological knowledge and people's experiential values, which are seen as critical to developing place-based solutions to societal problems such as biodiversity loss and to supporting robust and adaptive social-ecological systems and expanding public participation and community consultation (Raymond et al., 2009; Brown, 2013).

Voluntary geographical information (VGI) can be obtained from openly accessible surveys or online services, such as OpenStreetMap. Openly accessible map surveys usually have difficulties in getting a representative sample of respondents. Using whatever type of participatory mapping methods, structured survey design and also gathering background information of the respondents is recommended. Participatory mapping methods can also deploy more traditional paper maps for collecting information. Both online map surveys and paper maps are feasible and their suitability is case specific. Online tools can target larger audiences.[1] On the other hand, some people prefer working with paper maps and they may find it easier to answer more accurately on hard copies than on online maps.

In spatial planning, participatory mapping methods can be used to acquire knowledge about, for example, biodiversity and the demand for ES, especially for cultural ES. Also, information on, for example, barriers of use, perceived shortcomings and development needs of green and blue areas, among other themes, can be gathered. All these aspects are important in assessing impacts of spatial planning on biodiversity and ES.

Using participatory methods is in line with the current paradigm of collaborative planning, which means a shift from planning *for* the people to planning *with* the people. When carried out appropriately, collaborative planning is based on co-production of knowledge and the process ideally builds trust between the planners and citizens as well as other stakeholders. Spatial planning affects people's living environments in many ways and either delivers or denies rights for building and development – socio-cultural values attached to nature and its various benefits are therefore of fundamental importance in impact assessment of spatial planning. By engaging citizens in co-creation of spatial knowledge, locally meaningful places can be identified, general acceptance better achieved and conflicts avoided or mitigated in spatial planning.

10.2.4 Green Factor – Assessing Ecosystem Services at Site Scale

In residential areas, city block or courtyard green infrastructure can be assessed based on the so-called 'green factor' (in some cases, 'green space factor' or 'blue-green factor'), with which the green efficiency and naturalness of the area can be calculated (Becker and Richard, 1990; Kruuse, 2011; Sindell and Thomas, 2011). The green factor can be applied to ensure that a diversity of plants, green roofs, green facades, and permeable surfaces allowing storm water retention are installed. Each of the green or blue elements receives a multiplier – also called a factor – from 0.0 to 1.0. The areas of the elements within a property are multiplied by their corresponding factors and summed. The sum is divided by the total property area to give the overall green space factor. The resulting green factor can be given a specific target score, for example 0.5 in the City of Malmö in Sweden (Kruuse, 2011). In addition, other requirements for green elements can be added. In Malmö, for example, each garden area must have at least ten green measures selected from a list of 35 green points. These points consist of elements such as nesting boxes for every apartment, permeable surfaces, the absence of lawns, and existence of gardens for food cultivation. The objective of putting the green factor into operation in many cities (e.g., Berlin, Seattle, Tokyo, Malmö) has been to increase overall climate change adaptation capacity, storm water retention on-site, microclimate regulation, and attractiveness of city blocks and courtyards (Stenning, 2008; Emanuelsson and Persson, 2014; Järvelä and Tiihonen, 2014).

The green factor can be used setting the targets for spatial planning as well as assessing impacts of suggested spatial plans on a local scale and at block level.

10.3 CASE STUDY 1: ASSESSMENT OF GREEN INFRASTRUCTURE IN THE HELSINKI-UUSIMAA REGION, FINLAND

The Helsinki-Uusimaa region, Finland, consists of 26 municipalities with a total of 1.6 million inhabitants. The 1.1 million inhabitants of the capital region alone (cities of Helsinki, Espoo, Kauniainen and Vantaa) cover 20 per cent of Finland's total population (Statistics Finland, 2014). In addition, the Helsinki-Uusimaa region is among the fastest-growing regions in Europe. Thus, there is a constant urge to densify the urban structure and convert new areas for residential purposes.

In 2012–15 the Uusimaa Regional Council prepared Regional Plan 4 for the Helsinki-Uusimaa Region, which complements the previous regional land use plans. The goal of the plan was to ensure the competitiveness of the region while not exceeding the limits of sustainable development. Regional Plan 4 concentrated on five particular themes: green infrastructure, business and innovation, logistics, wind energy and cultural heritage. The Regional Council cooperated with the Finnish Environment Institute to discover how a regional plan could best ensure the maintenance and enhancement of the provision of ES. With this aim, new methods for assessing and mapping ES provision and demand and impacts of population growth on them were applied and the results were utilized in the planning of the green infrastructure theme of Regional Plan 4. The implemented methods provided a means to assess impacts of planning options in the course of planning process.

As the aim was to provide meaningful information for the process of regional spatial planning, the process involved close interaction between the regional planners and stakeholders. The regional council had gathered wide groups of stakeholders and experts for each of the regional plan's themes. The expert group on regional green infrastructure met eight times during the regional plan development process and the assessment results in each phase were discussed with the expert group in detail. Cooperation with this stakeholder group was indispensable in the co-creation of knowledge.

10.3.1 Mapping Spatial Variation in the Ecosystem Service Provision Potential

The spatial variation in the provision potential of ES was analysed using GreenFrame methodology (Kopperoinen et al., 2014). GreenFrame is an integrated approach to study the variation in the ES provision potential within a study region. The method enables examination of the whole range of ES simultaneously or a selected number of them. The method has been developed to suit the needs of a quickly proceeding spatial planning process. The different phases of the method application are transparent, which makes it possible to track back what is behind the spatial differences in ES provision potential. GreenFrame makes use of a wide variety of spatial data and expert knowledge. Instead of quantifying the actual stocks and flows of ES, the aim is to evaluate areas based on their potential to support the supply of various ES. Spatial data is usually scarce on regulation and maintenance services and intangible services, such as cultural ES. GreenFrame provides a way to infer this information from related thematic data based on assessments from experts and local and regional actors. Qualitative assessments can be complemented with existing quantitative spatial data from the study area. Quantitative data is more often available for tangible provisioning services, such as timber volume. The spatial data and its resolution is chosen based on the scale of spatial planning: in regional planning more general data is applied than in detailed planning where very accurate and fine-resolution data is needed.

Altogether 23 ES groups, derived from the Common International Classification of Ecosystem Services (CICES) (European Environment Agency, 2015), were analysed using a wide variety of spatial data and expert and local actor evaluations (Table 10.1) (see Itkonen et al., 2015 for a more detailed description). Some of the provisioning services could be analysed based on available quantitative data, namely 'surface and groundwater for drinking', 'surface and groundwater for non-drinking purposes', 'materials from plants, algae and animals and genetic materials from all biota' and 'biomass-based energy sources'.

As outputs of these analyses, 23 raster layers of the supply potential of different ES groups were created, as well as composite layers of each of the three ES sections (provisioning services, regulation and maintenance services, cultural ES). Finally, these composite layers were combined into a final synthesis layer, where all three ES sections were included as equally important (Figure 10.2).

10.3.2 Identification of Key Areas of Regional Green Infrastructure

The core of the regional green infrastructure consists of a network of protected areas and other areas with high nature and biodiversity values (European Environment Agency, 2014). Not only do these areas sustain biodiversity, they also provide many other important ES of local, regional and national importance (e.g., water purification). However, there are also large areas providing ES outside the protected areas. These multifunctional areas need to be recognized and taken into account in decision-making and spatial planning. Unfortunately, there is no way to unambiguously determine which areas belong to the key green infrastructure and which areas do not – the examination is always context and scale dependent. In this case, in order to conceive an overall picture of the most important areas of the regional green infrastructure, the following components were identified and selected:

Table 10.1 Potential supply of 23 ES groups

Ecosystem service section	Ecosystem service group
Provisioning	Agricultural and aquacultural products
	Wild plants, animals and their outputs
	Surface and groundwater for drinking
	Surface and groundwater for non-drinking purposes
	Materials from plants, algae and animals and genetic materials from all biota
	Biomass-based energy sources
Regulation and maintenance	Mediation of waste and toxins
	Mediation of smell/noise/visual impacts
	Mass stabilization and control of erosion rates, buffering and attenuation of mass flows
	Hydrological cycle and flood protection
	Mediation of air flows
	Pollination and seed dispersal
	Maintenance of nursery populations and habitats, gene pool protection
	Pest and disease control
	Soil formation and composition
	Maintenance of chemical condition of waters
	Global climate regulation
	Micro- and regional climate regulation
Cultural	Recreational use of nature
	Nature as a site and subject matter for research and education
	Aesthetics and cultural heritage
	Spiritual, sacred, symbolic or emblematic meanings of nature
	Existence and bequest values of nature

Source: The classification is adapted from the Common International Classification of Ecosystem Services (CICES Version 4.3, http://cices.eu).

- The core network of protected and other valuable nature areas was identified.
- The best 20 per cent of the landscape, having the highest provision potential of ten most relevant and important ES in the region, outside the network of protected areas and other valuable nature areas.
- The core nature areas and connecting corridors that were not included in the first two steps were included. These are areas that support the connectivity of the network and provide habitats less

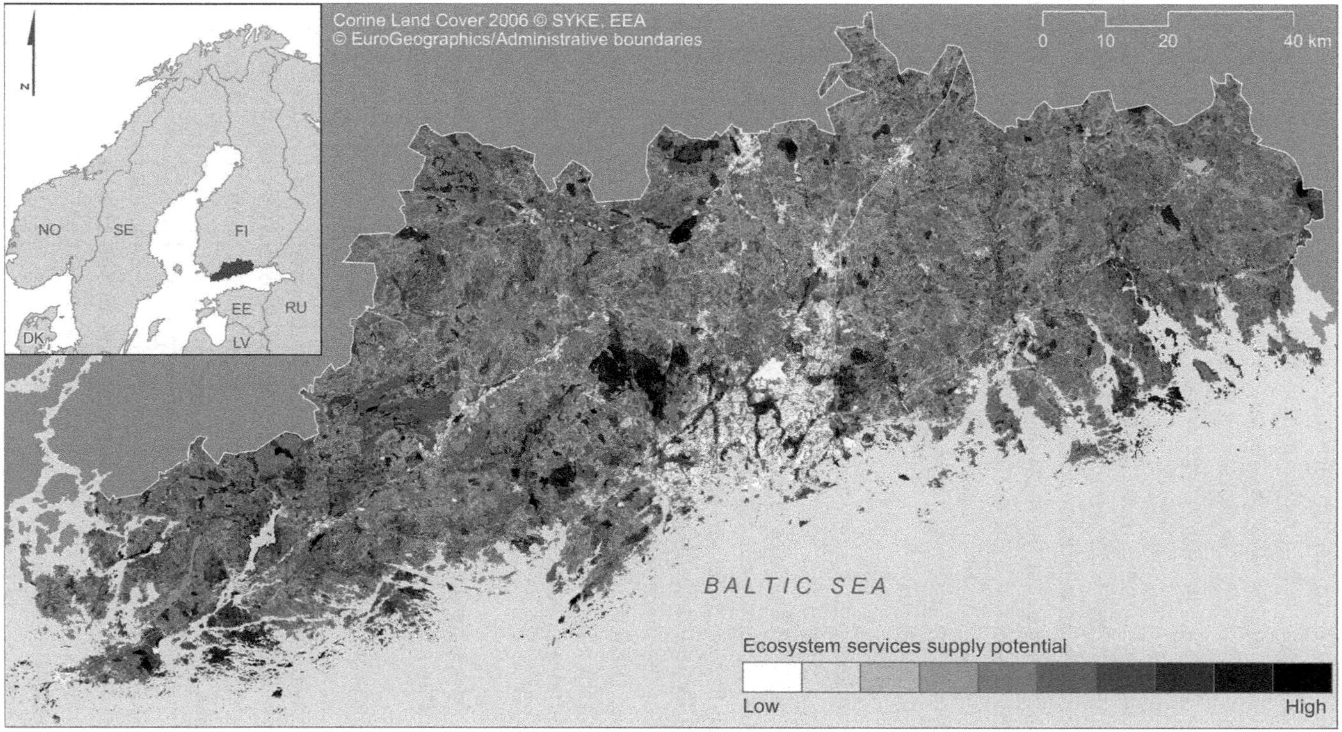

Figure 10.2 *ES provision potential in the Helsinki-Uusimaa region*

susceptible to disturbance and edge effects. Core nature areas and connecting corridors were identified based on land cover classification, using MSPA (Soille and Vogt, 2009). Connectivity beyond the administrative borders of the region was taken into account to prevent arbitrary edge effects from affecting the results of the analysis.

- After combining these three components, separate patches of less than 10 ha were removed, except those belonging to the core network of valuable nature areas. This was done for visual clarity and to focus on the most important targets of regional significance. Small areas should not be neglected in municipal spatial planning, however, and their significance in the municipal green infrastructure should be assessed when preparing local master plans and local detailed plans.

A complementary analysis on the ecological permeability of the landscape was carried out using the Finnish national CORINE Land Cover raster (2006). The impedance, that is, difficulty of movement through different land cover classes, was scored based on expert discussions. The permeability of different patches of land in a certain area is not only affected by the characteristics of the patch itself, but also by the characteristics of the surrounding areas. Hence, the CORILIS spatial smoothing technique (Peifer, 2009) was applied to also take into account land cover of the surrounding areas of each cell in the Land Cover raster. The impact of surrounding pixels decreases as a function of distance within a selected radius.

10.3.3 Mapping the Demand for Ecosystem Services Using Public Participation GIS

After assessing the potential supply of ES, the demand was also assessed. Spatial analyses of the demand within the region was not feasible for all ES, because the service flows from service providing units (e.g., a forest patch) to the actual beneficiaries may have been very complicated and largely determined by other factors than the location of the population. Therefore, the spatial analyses focused solely on the demand for cultural ES. Cultural ES are immaterial and experiential, and their benefits are not in fact restricted to a single location – the beneficiaries can take, for example, the health benefits of recreation, or the experience of admiring scenic beauty with them. However, these immaterial services need to be consumed at a distinct location – recreational activities and impressive scenery require a physical setting.

An online public participation GIS (PPGIS) survey was carried out to examine the region's inhabitants' perceptions on cultural ES. The survey was titled 'The Meanings of Nature for the People of Uusimaa'. It was in operation for eight weeks, and targeted at all residents of Uusimaa, regardless of their background. The respondents were asked to pinpoint targets on a map from the following themes:

- good places or routes for recreation;
- good places to learn from nature;
- very scenic places;
- good vantage points;
- places where history and cultural heritage combine in a way that adds to the value of the place;
- regionally symbolic places;
- places that have unique identity that people attach meanings to and where they can feel attached to their environment (sense of place);
- relaxing/revivifying places;
- places where people can experience holiness;
- places having intrinsic value, to be preserved for future generations.

These themes cover the whole variety of cultural ES. The respondents were allowed to mark places that were not necessarily unspoiled nature, but nature did have to be present in these places and affect their experience. In fact, in addition to the actual demand, these markers also represent the actual supply of cultural ES – the respondents marked locations where they had already consumed these ES. Altogether 5043 point markers were gathered by a total of 555 respondents covering all cultural ES (Figure 10.3).

The sample of the respondents was not fully representative of the demography of the region, but the survey results can be used as a source of supplementary information in collaborative spatial planning. The clusters of point markers constitute hotspots – locations to which many respondents attach several different meanings. A regional plan is a relevant planning instrument for cultural ES, as, for example, many recreation areas, cultural heritage landscapes and aesthetically valuable sites have regional importance.

10.3.4 Mapping the Potential Demand for Ecosystem Services Using Accessibility Analysis

Population pressure and potential demand for local recreation can be assessed, for example, by relating the number of people living in the

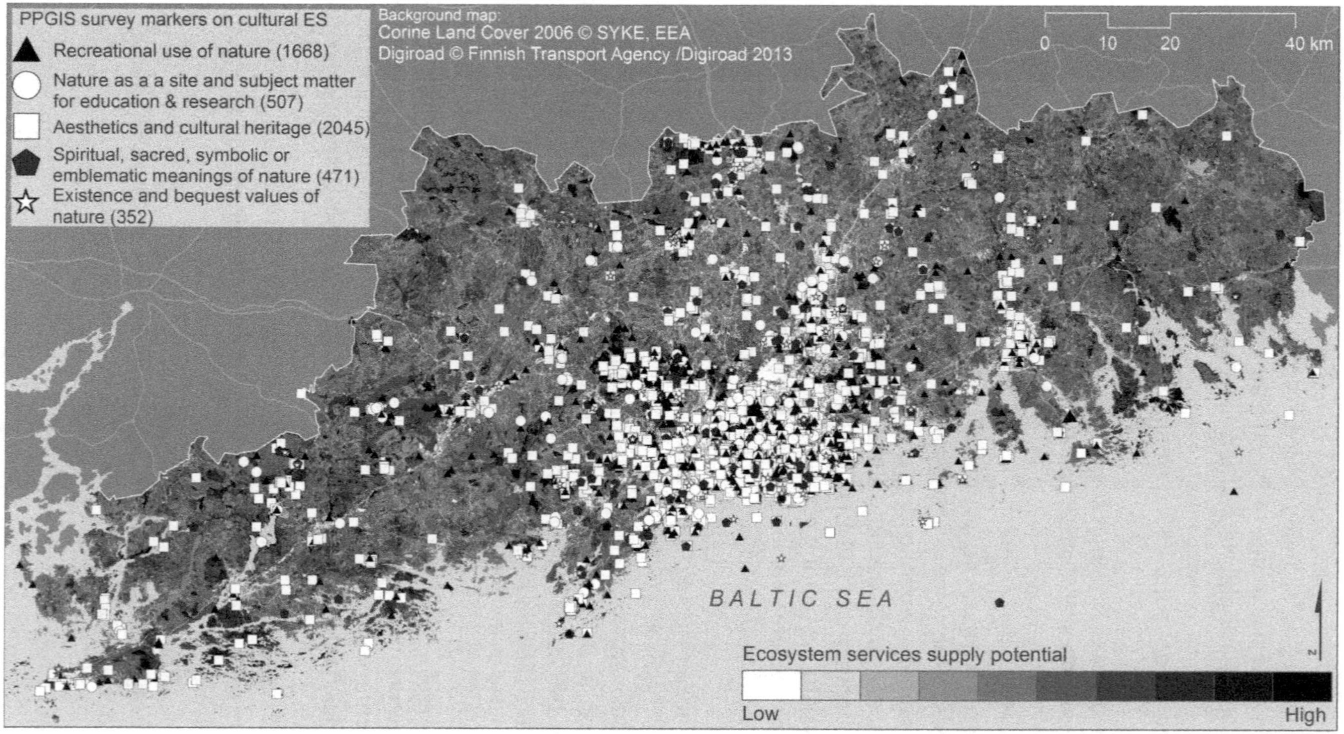

Figure 10.3 *The places marked by the respondents of the PPGIS survey 'The Meanings of Nature for the People of Uusimaa' using GreenFrame analysis on cultural ES provision potential as a background*

immediate vicinity of delineated green infrastructure per area unit. Our approach to analyse the potential demand for cultural ES was to calculate the number of population that can access each location of the region in a given time (10, 20 and 30 minutes) via the transport network. For population data, a 250 m grid was used, containing the population of each grid cell. The estimates of the travel time between pairs of population grid cells were calculated using road network data. The estimates of travel time for each road segment takes into account different road types and the slowing effect caused by traffic in city centres (Figure 10.4).

10.3.5 Relating the Supply of and the Demand for Ecosystem Services

Although different areas may be equally good in terms of accessibility, their attractiveness and ability to supply different ES may vary. Therefore, it is useful to compare the estimates of the potential demand together with the supply potential of cultural ES in each grid cell of the region. This allows inter alia the recognition of locations where high demand for ES meets high supply, or on the other hand, where the green infrastructure providing ES cannot meet the needs of people either in terms of quantity or quality. All these estimates are valuable in the context of impact assessment of spatial planning. Despite densifying urban structure and the trend towards the expansion of urban areas, there are still areas with high ES supply potential in the core of the capital region. These areas are subject to high demand because of their central location and good accessibility, also by public transport.

It must be borne in mind that these accessibility estimates represent the potential demand for the cultural ES based on the population density and accessibility of different areas – not actual visits. Some locations, such as national parks, draw visitors from much larger areas than just their surroundings. When available, data on the admission of national parks and the use of other recreational areas is useful in assessing their actual demand. However, as it is not possible to monitor the use of all areas providing cultural ES, mapping the potential demand is an alternative when assessing the whole region.

10.3.6 Impacts of Expected Population Growth on Green Infrastructure in the Helsinki-Uusimaa Region

Constantly increasing population and land use change are two of the top drivers of change in the Helsinki-Uusimaa region. The Helsinki Metropolitan Area is one of the fastest-growing urban regions in the whole of Europe, which has many indirect impacts on the ecosystems through

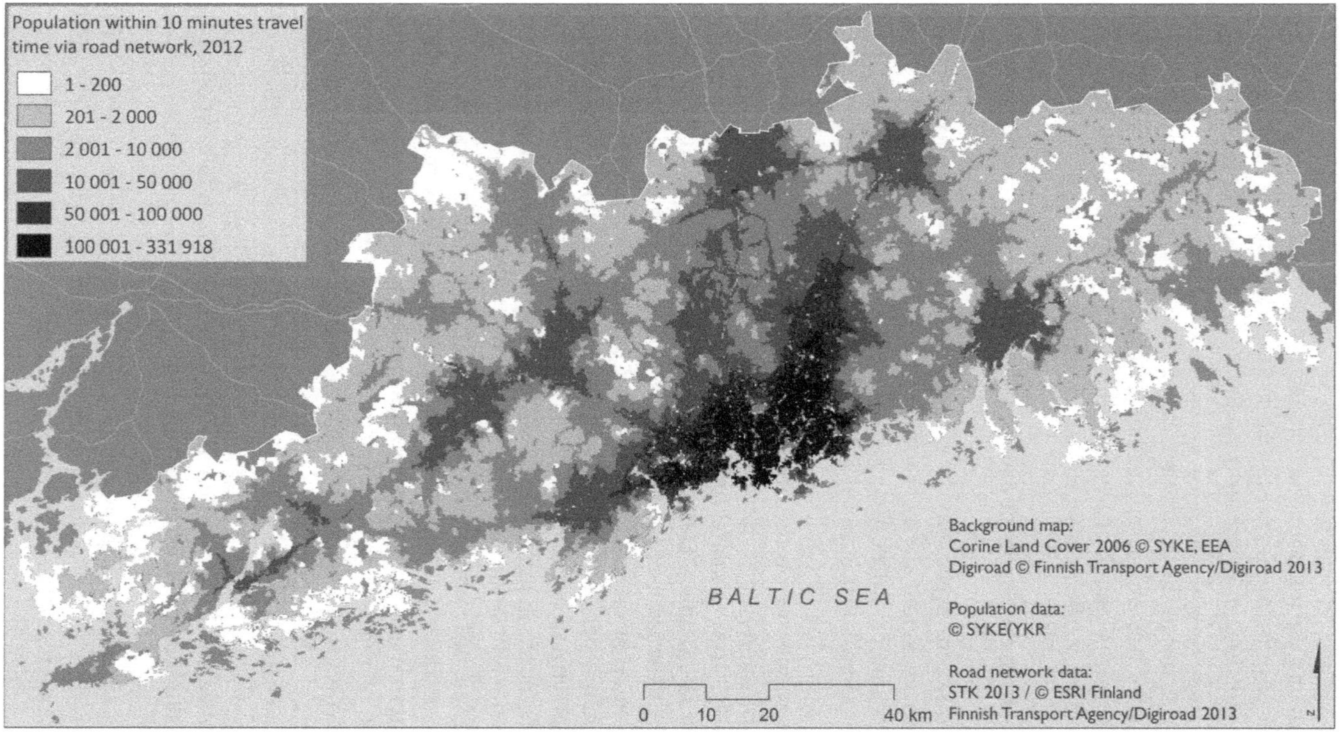

Source: The estimate is based on population data (SYKE/YKR) and road network data (STK 2013/© ESRI Finland, Finnish Transport Agency/ Digiroad 2013).

Figure 10.4 Population within ten-minute travel time via road network in Helsinki-Uusimaa region

spreading built-up areas and growing use of natural resources for building, livelihoods, energy consumption, and recreation. To prevent further urban sprawl and to mitigate climate change by increasing eco-efficiency of cities, densification of urban structure is encouraged. This has a twofold impact on the green infrastructure: ES-providing areas diminish and the number of potential users grows. Paradoxically, prospects, inter alia, for using nature-based solutions for climate change adaptation deteriorate.

To visualize impacts of population growth on green infrastructure and therewith on the ES provision potential, the current situation and the anticipated future change in the population pressure were assessed by calculating the number of the population residing in the immediate surroundings of the key areas of the regional green infrastructure. In addition to the current distribution of population, a scenario of the situation in 2035 was also examined, based on data from the Uusimaa Regional Council that predicts an increase of 300000 new inhabitants from 2006. The possible land use conversions that might change the distribution of the regional green infrastructure were not taken into account because the outline of the new regional land use plan was still in process. It is possible that some valuable green infrastructure areas decrease in size or are even lost due to expansion of urban areas, which would result in even greater demand for the ES provided by the remaining areas of green infrastructure.

10.4 CASE STUDY 2: ASSESSING IMPACTS OF THE LOCAL MASTER PLAN OF THE CITY OF TAMPERE, FINLAND

With more than 220000 inhabitants, Tampere is the third largest city in Finland (Figure 10.5). The larger Tampere city region has more than 360000 inhabitants, which makes it the second largest city region of Finland after the capital region of Helsinki. To prepare the local master plan for the inner city 2040, the City of Tampere wanted to 'assess urban natural diversity' in addition to a more traditional assessment of recreational and green areas (City of Tampere, 2014, p. 13). Thus, an indicator-based ES assessment using GIS was carried out in cooperation between the Finnish Environment Institute SYKE and the City of Tampere (Söderman et al., 2014). In this assessment, GIS-based indicators of ecological sustainable regional land use developed earlier were applied (Söderman and Saarela, 2011; Söderman et al., 2012). The original indicator set consists of more than 80 indicators describing the ecological, social and economic sustainability at city region scale. In this case, the most applicable and relevant indicators of ecological sustainability were chosen based on

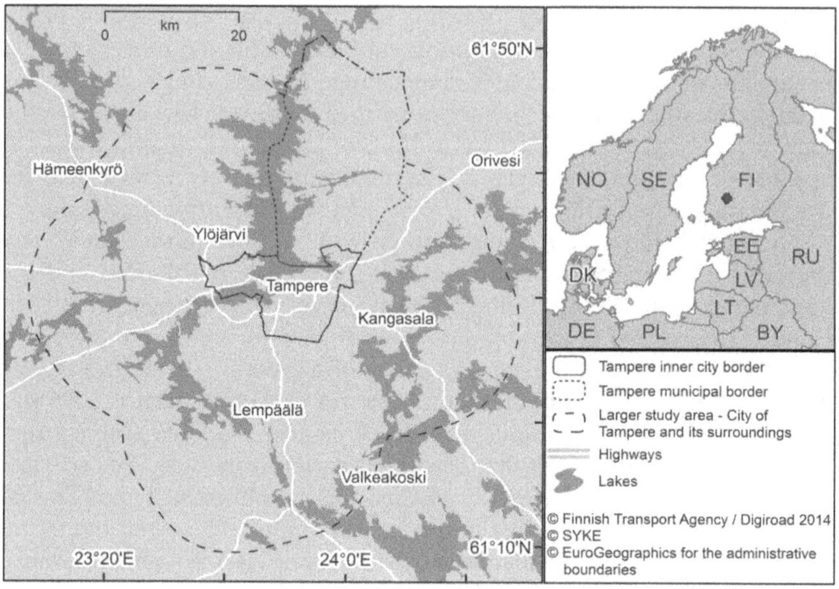

Note: GIS-based indicators were applied in three spatial scales: inner city, municipality, and regional. More than 97 percent of the population of the municipality live in the inner city area.

Figure 10.5 Tampere, Southern Finland

discussions with the City of Tampere. Most indicators were then also applied in more detailed spatial scales of municipality and/or inner city. In addition to computing comparable indicator values, thematic maps were created for all selected indicators. The collection of selected indicators is listed in Table 10.2.

For most of the indicators, two versions were produced: a version not taking into account statutory land use plans, and a version taking into account the zoning in local master and detailed plans. In the latter version, all zones denoted as built areas (although not necessarily yet implemented) were removed from forested areas, core nature areas, and land cover suitable for recreation (on the basis of public right of access). However, this approach might overestimate the adverse effect of future built areas, as some natural areas will most likely remain among the buildings and sealed soil surface within these zones. Using just these indicators, an exhaustive assessment of the supply of and demand for ES cannot be obtained. Instead, they provide comparable information on the impacts of spatial planning on the extent and connectedness of green areas capable of pro-

Table 10.2 Selected indicators of ecological sustainability of land use in assessing the ES and green infrastructure of the City of Tampere

	Spatial Scale of the Analyses		
Indicator	Tampere inner city	Tampere municipality	Tampere and surrounding region
Population density enabling public transportation (min. 20 inhabitants per hectare)		●	
Sparsely built detached housing areas of all populated areas (based on the Finnish Monitoring System of Spatial Structure and Urban Form YKR)		●	●
Traffic research statistics		●	●
The proportion of forests and mires of the total land area		●	●
The proportion of core nature areas of the total forested area (after removing forest edge zone of 250 m in width)		●	●
The MSPA statistics of the structural connectivity of core nature areas and other forested areas		●	●
Forest areas within population centres: the proportion of forest patches of min. 3 ha of total forest area within densely built areas	●	●	
Land cover suitable for recreation: the proportion of patches of min. 1.5 ha	●	●	
Proximity of land cover suitable for recreation: the proportion of population within 300 m of patches of min. 1.5 ha	●	●	
The number of inhabitants in relation to the total area of land cover suitable for recreation (patches of min. 1.5 ha)	●	●	
The proportion of sealed land surface	●	●	
The proportion of ground water formation areas at risk (due to human activity)	●	●	●
The proportion of sealed land surface on ground water formation areas	●	●	●

Note: For each indicator, an indicator value and a thematic map was produced at one or more spatial scales.

Source: Original indicators were derived from Söderman et al. (2012).

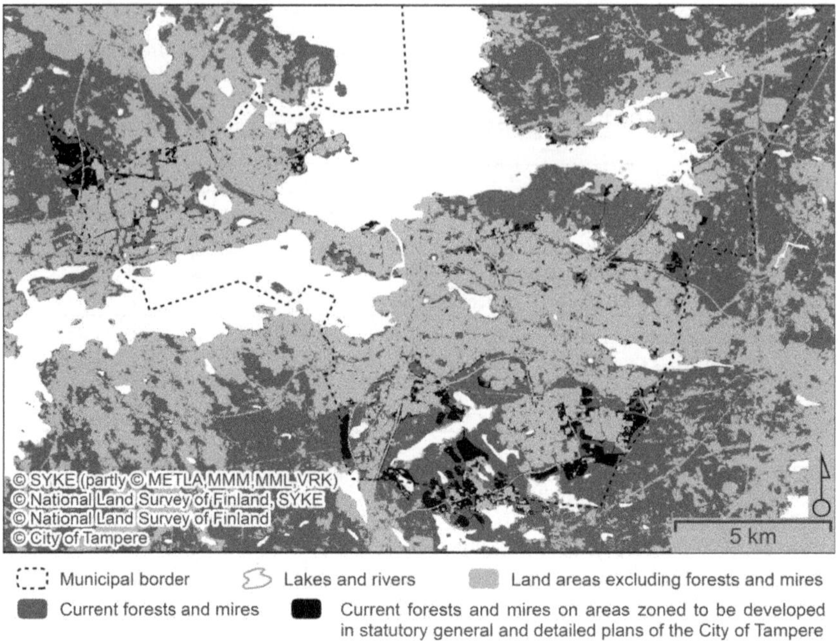

Figure 10.6 Implementation of current land use plans would lead to losses of forest area within the City of Tampere

viding various ES. Many of the indicators are also related to the potential demand for and/or population pressure as a driver of change on ES provided by these areas.

One of the applied indicators was the total area of forests and mires. The forests and mires were extracted from the Finnish National CORINE Land Cover 2006 raster with a pixel size of 25 m. According to the analyses, the current total area of forests and mires within the municipality is 323 km^2, constituting 62 per cent of the total land area. In inner city these areas mainly consist of forest. Complete implementation of statutory general and detailed land use plans would lead to an approximately 6 per cent decrease (19 km^2) in the total forest and mire area (Figure 10.6). Moreover, in the inner city this would mean a 34 per cent decrease in the total forest area of forest patches that are at least 3 ha in size. However, it has to be noted that this approach assumes that all current forests located in areas that are zoned for development will be cleared, which may result in overestimation of forest loss.

The impact of the current land use plans was also assessed on areas suitable for recreation, with a minimum size of 1.5 ha. The extraction from the Finnish CORINE Land Cover raster included sports and leisure facilities, forests, natural grassland, moors and heathland, transitional woodland and shrub, beaches, dunes and sand plains, bare rock, inland marshes and peatbogs. In Finland, all these areas are generally accessible due to public right of access (so-called 'everyman's right'). Currently, the total area of these areas in the inner city is around 57 km^2. Full implementation of the plans would lead to a 25 per cent decrease in areas suitable for recreation (Figure 10.7). Currently, 95 per cent of the population in the inner city live within 300 m from at least one of these areas. The share of the population living within 300 m from these areas after implementing the plans could not be estimated reliably because the locations of new residents were not allocated in sufficient spatial detail in the population projection data.

The areal extent of core nature areas was also assessed. First, the following land cover classes were extracted from the Finnish CORINE Land Cover raster: pastures, forests, natural grasslands, moors and heathland, transitional woodland/shrub, beaches, dunes and sand plains, bare rock, inland marshes, peatbogs and rivers. Then an edge of 250 m was removed from the extraction. The remaining areas within the edge zone were regarded as core nature areas (Figure 10.8). The assessment revealed that the plans would reduce the amount of core nature areas by approximately 6 per cent within the municipality. However, the ratio between edge areas and core areas would stay roughly constant (0.74 edge/0.26 core) – indicating that the impact would not fall solely on contiguous core areas, but on the edge areas as well. A remarkable increase in the edge-to-core ratio would indicate a disproportionately severe fragmentation effect in relation to the areal loss. Also, the analyses on the structural connectivity of the core areas supported this finding. The structural connections between at least two core areas were identified using MSPA (see Soille and Vogt, 2009). The share of core connected network (core areas and their interconnections) of the total land cover extraction remained constant after taking the land use plans into account.

10.5 CASE STUDY 3: CONSIDERATION OF ECOSYSTEM SERVICES IN REGION HANNOVER, GERMANY

'Region Hannover' is a special governance institution in the German local government landscape. In an effort to respond to the challenges of the multiple and diverse relationships between a major city and its hinterland,

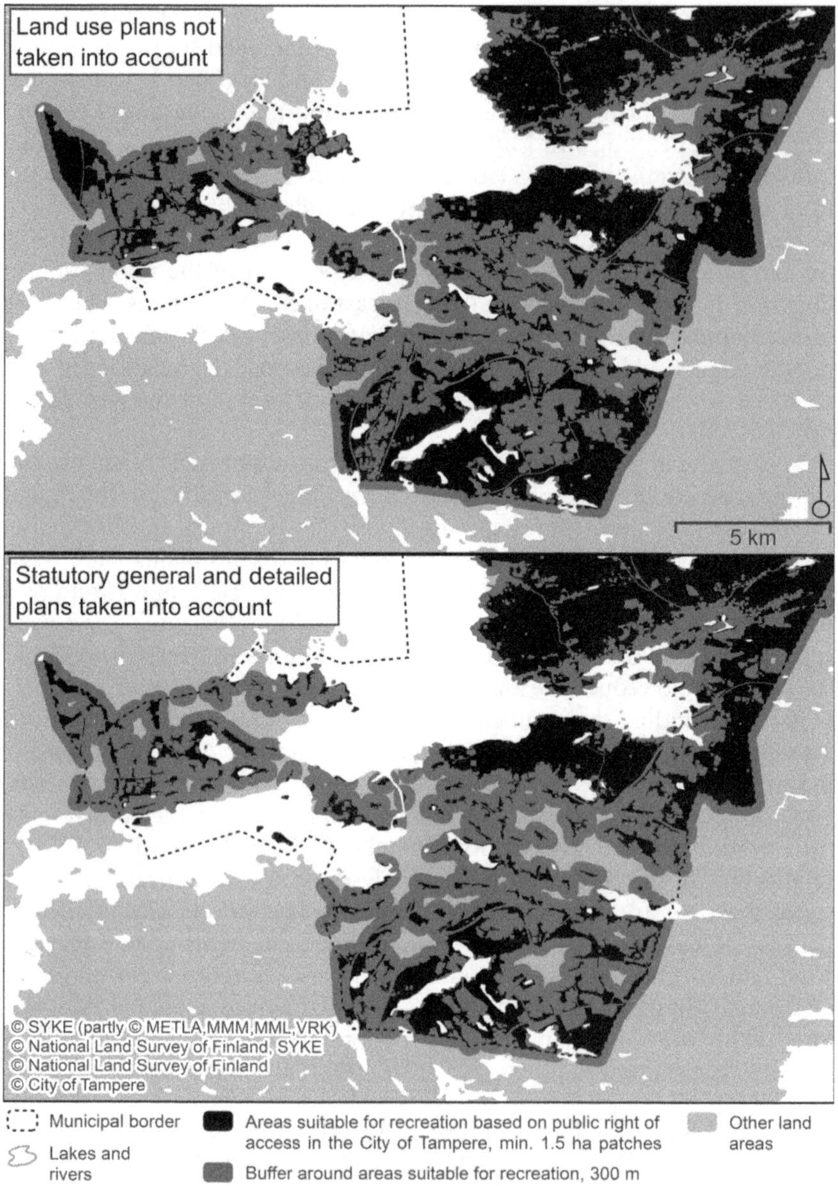

Figure 10.7 *Implementing current plans would lead to a decrease in areas suitable for recreation in the inner city of Tampere*

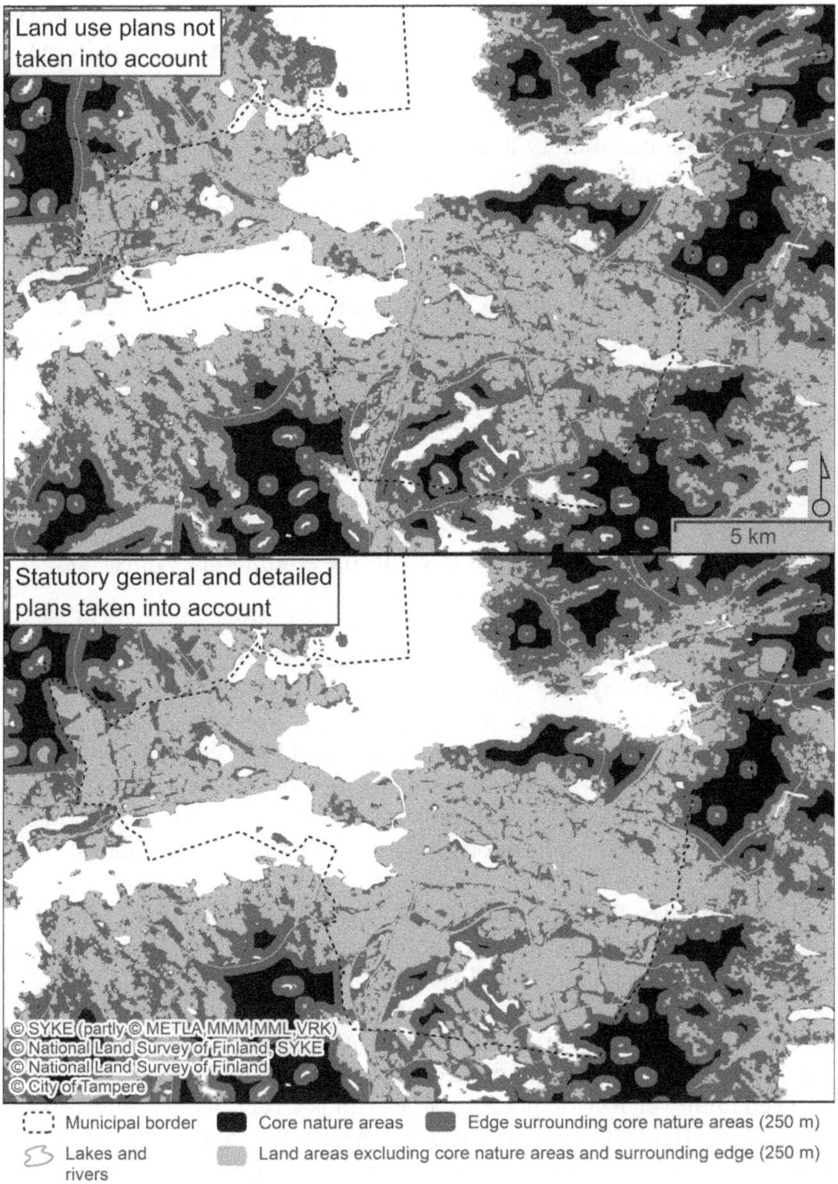

Figure 10.8 *Implementation of current land use plans would result in approximately equal decreases in core nature areas and edge areas surrounding them in the municipality of Tampere*

Region Hannover was founded in 2001 as a new regional government and administration. The region consists of the City of Hannover and 20 towns and local authorities of the former County of Hannover. In this way, Region Hannover now addresses most local government responsibilities that transcend community and city borders. Region Hannover stretches for about 60 km from west to east and about 55 km from north to south. It is home to more than one million inhabitants (Region Hannover, 2009).

Regional planning in Region Hannover focuses on guiding and shaping spatial development towards the stated vision and resolving emerging land use conflicts. A regional vision statement, the 'Leitbild', is developed and regularly updated to provide a policy framework for orientation and action. The strategy for attaining this vision through spatial regulation and planning is described in the regional plan (Regionales Raumordnungsprogramm, RROP), which incorporates all planning and land-use policies of significance for regional development. Among others, the regional plan designates priority areas for specific land uses or interests, including settlement areas, nature, recreation areas, and wind energy zones. Subordinate plans of the city and the communities must take into account and align with the regional plan.

One option for considering biodiversity and ES in regional planning is through protecting priority areas for nature and landscape (Natura 2000 network and nature conservation sites – about 3.5 per cent of the region's area) and areas of particularly high diversity, uniqueness and beauty (landscape conservation areas, about 44 per cent of the region's area) (Region Hannover, 2009). In areas outside these more or less protected zones, the regional planning authority aims to support sustainable spatial development by harmonizing social and economic interests with ecological considerations.

The Hannover Regional Landscape Plan provides important information on biodiversity and ES for the regional planners to consider when designating protected areas and weighing competing interests; that is, for example, when assessing how different spatial planning alternatives would impact on biodiversity and ES. As with the regional plan, the regional landscape plan is regularly updated, the latest version being from 2013.

The Hannover Regional Landscape Plan is structured as follows: (1) a characterization of the planning area, (2) a summary of legal requirements, (3) an in-depth description of the current condition of nature and landscape, (4) a concept of nature conservation targets and (5) suggestions for the implementation of the concept of targets. The plan considers various ES with reference to the above-described concept of landscape functions. A selection of mapped landscape functions is provided in Figure 10.9.

The development of the concept of conservation targets is still a sectoral

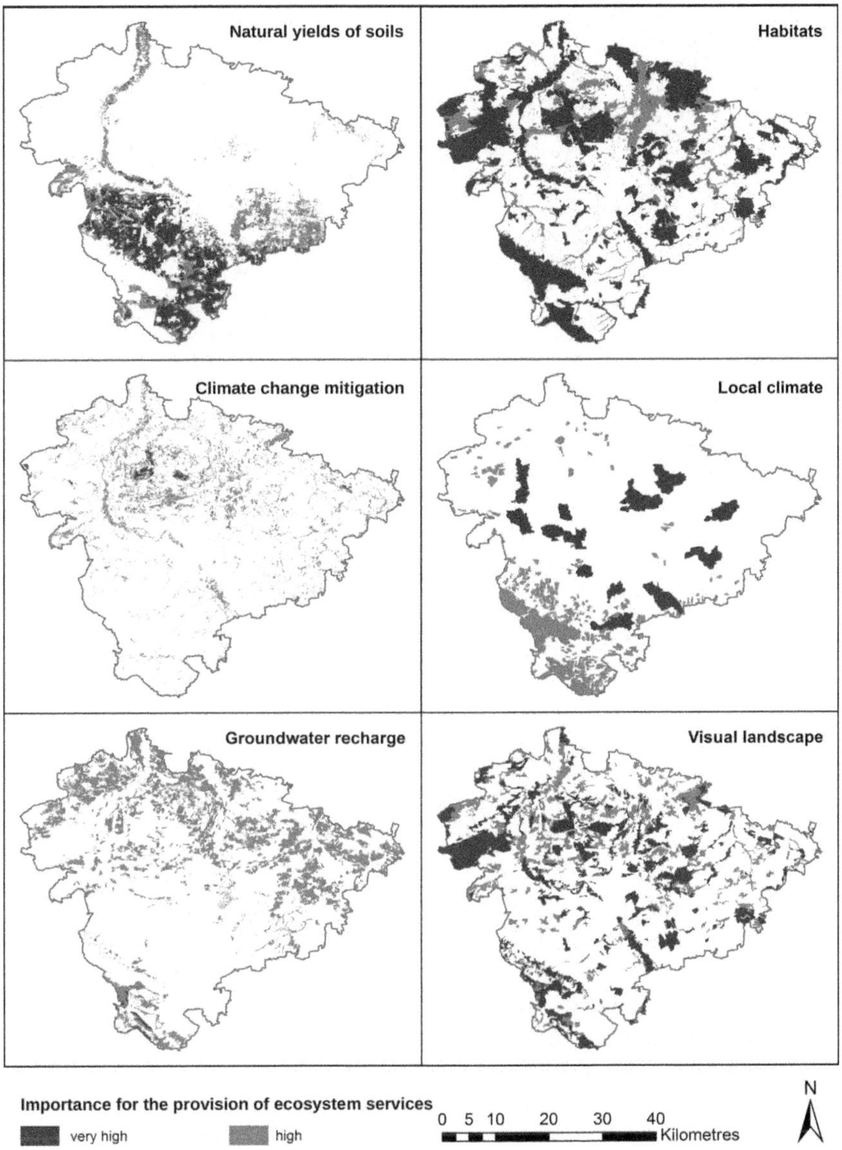

Source: Region Hannover (2013); Albert and Von Haaren (2014).

Figure 10.9 Areas of very high or high importance for the provision of different ES in the region of Hannover

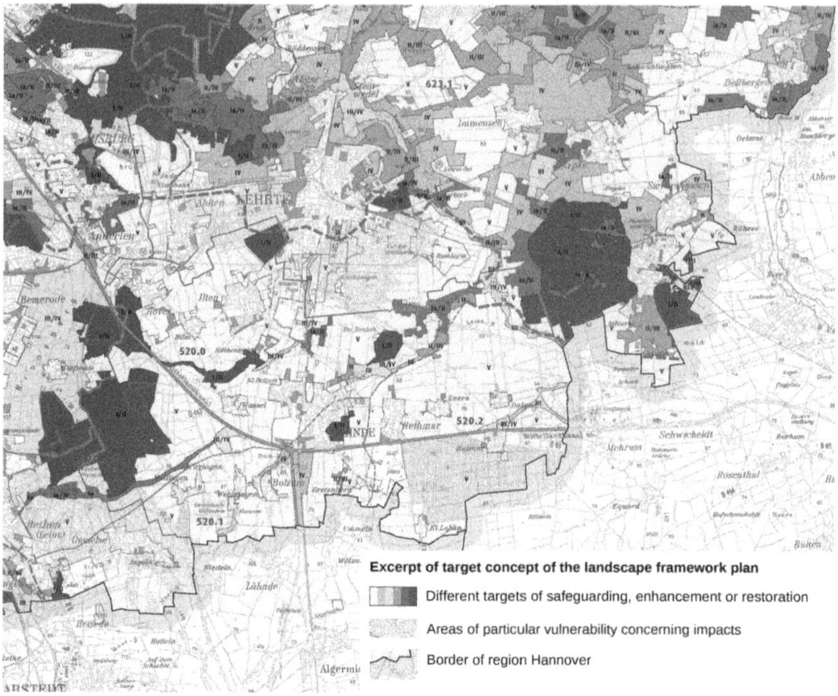

Source: Region Hannover (2013, amended).

Figure 10.10 Excerpt of the concept of nature conservation targets in the landscape framework plan of Region Hannover

activity from the perspective of nature conservation. It describes specific, spatially explicit objectives for different landscape functions, and creates an integrated plan of nature conservation targets including a regional habitat network plan (Figure 10.10).

The case of the Region Hannover illustrates how maps showing the spatial variation in relative importance of providing various ES can form the basis for planning and impact assessment defining site-specific targets for the protection, enhancement and restoration of biodiversity and ES. Defining these targets is usually an expert-led endeavour, during which general environmental objectives from EU, national or federal state level are concretized to each specific site. In this process, some priorities for protection, enhancement and restoration based on particularly high importance of individual biodiversity or ES are considered, but also the overlaying of different sectoral maps is considered to identify areas of particular importance for multiple uses. All these maps help identification of spatial impacts of planning options. The expert-led process is guided primarily by existing laws, directives and strategies for the protection of nature, among others the German Nature Conservation Act, which calls for the protection of biodiversity, several essential ecosystem functions with meaning for society, as well as the visual landscape for recreation.

The final output is referred to as a 'target concept'. The degree to which the objectives set out in this target concept can actually be met depends on the weighing process within regional planning procedures that lead up to the development of the final regional plan, which integrates the interests of nature conservation and various other sectors. However, even if the nature conservation sector has achieved inclusion of nature conservation considerations in the regional plan, the nature conservation targets still need to be implemented, which is often challenging due to the need for cooperation and coordination of various stakeholders and land users.

The assessment of impacts in regional planning is implemented in two ways. First, the landscape framework plan provides information on the sensitivity of nature and landscape to specific pressures. For example, the landscape plan provides information on where the siting of wind energy plants should be avoided due to specific sensitivities of species populations (e.g., bats) or specific sensitivities against impacts on the visual landscape (e.g., in areas of particular aesthetic value such as the border between hilly landscapes and the German northern floodplains). The second and even more important way the landscape framework plan contributes to the incorporation of biodiversity and ES in regional planning is that the landscape plan functions as a Strategic Environmental Assessment (SEA) for the regional plan. As such, the landscape framework plan provides relevant information and enables appropriate consideration of issues of biodiversity and ES in the regional plan and decision-making.

10.6 A CHECKLIST FOR ECOSYSTEM SERVICES AND BIODIVERSITY IMPACT ASSESSMENT OF SPATIAL PLANNING

1. *Identify green infrastructure and benefits derived from ecosystem services:* Identify the various components of green infrastructure including water areas. Protected areas form the core of green infrastructure. Analyse the spatial variation in the capability of ES provision (such as in the Helsinki-Uusimaa regional plan or the Region Hannover case). Concentrate especially on areas of particularly high significance for biodiversity or single ES (and often legal protection), multi-functional

ES 'oases', and areas where trade-offs in ES provision are present. Use methods based on scientific knowledge in carrying out analyses.

2. *Identify demand for ecosystem services:* Assess the demand for ES from the region itself and from outside (e.g., in the form of nature tourism). Food and clean drinking water are basics but remember communities' need for flood protection or microclimate regulation. Demands can be derived from existing legislation (as implicit demands according to democratic decisions by society, cf. Baró et al. 2015) as well as empirically from potential users and beneficiaries, as in the case of the Helsinki-Uusimaa region through an open PPGIS survey.
3. *Find out people's needs and engage them (co-creation of knowledge):* The real meanings of green and blue areas cannot be known without asking people. Use various methods of collaborative planning and map surveys to engage the residents of the area. For example, Region Hannover engaged stakeholders and citizens in a two-year participation effort to jointly develop a 'Leitbild' for the future development of the region, including thematic guided tours and discussions, workshops, online consultation and even an online mapping effort where citizens could identify places in need for recreation enhancement measures.
4. *Remember the importance of cooperation between different administrative sectors:* Planning of green infrastructure as well as assessing impacts of spatial planning on ES demands cooperation from different sectoral administrations and experts, which was done in the Helsinki-Uusimaa regional plan process. The Regional Council had gathered wide groups of stakeholders and experts for each of the regional plan's themes, one of which was green infrastructure. Natural flood protection requires a different type of expertise from that for sustainable food production or living environment planning that supports people's physical and mental well-being. Co-generating knowledge and participatory plan-making that systematically involves stakeholders and users from the very beginning are thus key preconditions for success.
5. *Preserve important green areas and connections between them:* Make spatial planning decisions that help to preserve robust and functional ecosystems. With strategic visions and spatial plan markers and instructions it is possible to support the sustainable flow of ES. Link with and harness available planning and governance instruments for implementing planning results in practice. For example, in the City of Tampere, core nature areas and connectivity between them were assessed both for current and future situations based on implementation of statutory general and detailed land use plans, to assess what

impact the plans would have for the most important green space entities in the city. If the analysis results had shown that the land use plans would have led to breaking of ecological connections, there would still have been the opportunity to modify them.

6. *Define alternative spatial plans and assess their impacts on ecosystem services:* Secure the effectiveness of the spatial plan by assessing its (and the alternative plans') impacts on ES before approving the plan officially. Identify alternative solutions and the means to moderate unavoidable detriments. In Hannover the Regional Landscape Plan was designed for this purpose: it helps regional planners to assess how different spatial planning alternatives would impact on biodiversity and ES.
7. *Monitor the realization of the spatial plan in subordinate plans and decision-making:* In order to ensure a functional green infrastructure and a flow of vital ES, strategic visions must be reflected at all scales of spatial planning resolutions. Take care that the understanding of the justification for the delineated green infrastructure in spatial plans at a certain level moves on to all other scales of planning. Ensure that the key infrastructure remains unbroken and continuous in finer-scale spatial planning. Deliver planning results in ways that are accessible and useful to a variety of potential users. Integrate realization options in plan-making from the very outset. In the Helsinki-Uusimaa region this was done through continuous interaction with municipal planners and policy-makers along the regional plan development process. Municipal planners were also taught how to understand the methods used in assessing the natural values and ES provided by green infrastructure. This way the significance of taking into account impacts caused by spatial planning decisions is concretely passed to the next planning level.

NOTE

1. Examples of Internet-based map survey tools can be found at the following websites: Maptionnaire (http://maptionnaire.com/en/); Landscape Values & PPGIS Institute (http://www.landscapevalues.org/); Harava – Survey Tool for Smart Planning (https://www.eharava.fi/en/).

REFERENCES

Albert, C. and C. von Haaren (2014), 'Implications of applying the green infrastructure concept in landscape planning for ecosystem services in peri-urban areas: an expert survey

and case study', *Planning Practice & Research* [online], accessed 23 January 2016 at http://www.tandfonline.com/doi/abs/10.1080/02697459.2014.973683?journalCode=cppr20.
Albert, C., J. Aronson and C. Fürst et al. (2014), 'Integrating ecosystem services in landscape planning: requirements, approaches, and impacts', *Landscape Ecology*, **29**(8), 1277–85.
Albert, C., C. von Haaren and C. Galler (2012), 'Ökosystemdienstleistungen: Alter Wein in neuen Schläuchen oder ein Impuls für die Landschaftsplanung?', *Naturschutz und Landschaftsplanung*, **44**, 142–8.
ARL (ed.) (2003), *Planungsbegriffe in Europa, Deutsch-Niederländisches Handbuch der Planungsbegriffe*, Hannover: Akademie für Raumforschung und Landesplanung – ARL.
ARL (ed.) (2005), *Handwörterbuch der Raumordnung*, Hannover: Akademie für Raumforschung und Landesplanung – ARL.
ARL (ed.) (2011), *Grundriss der Raumordnung und Raumentwicklung*, Hannover: Verlag der ARL.
Baró, F., D. Haase, E. Gómez-Baggethun and N. Frantzeskaki (2015), 'Mismatches between ecosystem services supply and demand in urban areas: A quantitative assessment in five European cities', *Ecological Indicators*, **55**, 146–58.
Becker, G. and M. Richard (1990), *The Biotope Area Factor as an Ecological Parameter – Principles for its Determination and Identification of the Target*, Berlin: Landschaft Planen & Bauen.
Bergsten, A., O. Bodin and F. Ecke (2013), 'Protected areas in a landscape dominated by logging – a connectivity analysis that integrates varying protection levels with competition–colonization tradeoffs', *Biological Conservation*, **160**, 279–88.
Brown, G. (2013), 'Relationships between spatial and non-spatial preferences and place-based values in national forests', *Applied Geography*, **44**, 1–11.
Brown, G., M. Kelly and D. Whitall (2013), 'Which "public"? Sampling effects in public participation GIS (PPGIS) and volunteered geographic information (VGI) systems for public lands management', *Journal of Environmental Planning and Management*, **57**(2), 190–214.
Calabrese, J.M. and W.F. Fagan (2004), 'A comparison-shopper's guide to connectivity metrics', *Frontiers in Ecology and the Environment*, **2**(10), 529–36.
Chan, K.M.A., A.D. Guerry and P. Balvanera et al. (2012), 'Where are cultural and social in ecosystem services? A framework for constructive engagement', *Bioscience*, **62**(8), 744–56.
City of Tampere (2014), '*Kantakaupungin yleiskaava 2040. Osallistumis- ja arviointisuunnitelma 26.3.2014*' [Local master plan for the inner city 2040 – participation and assessment scheme, 26 March 2014], accessed 26 January 2016 at http://www.tampere.fi/liitteet/o/tYcUE45eO/OsallistumisJaArviointisuunnitelma.pdf [in Finnish].
Emanuelsson, K. and J. Persson (2014), *En kontextanpassad grönytefaktormodell. Sveriges lantbruksuniversitet* [A Context-adapted Green Factor Model], Alnarp: Fakulteten för landskapsarkitektur, Trädgårds- och växtproduktionsvetenskap [in Swedish].
European Commission (2011), *The EU Biodiversity Strategy to 2020*, Luxembourg: Publications Office of the European Union.
European Environment Agency (2014), *Spatial Analysis of Green Infrastructure in Europe, EEA Technical Report No. 2/2014*, Luxembourg: European Environment Agency.
European Environment Agency (2015), CICES: towards a common classification of ecosystem services [website], accessed 6 August 2015 at http://cices.eu/.
García-Nieto, A.P, C. Quintas-Soriano and M. García-Llorente et al. (2015), 'Collaborative mapping of ecosystem services: the role of stakeholders' profiles', *Ecosystem Services*, **13**, 141–52.
Geneletti, D. (2011), 'Reasons and options for integrating ecosystem services in strategic environmental assessment of spatial planning', *International Journal of Biodiversity Science, Ecosystem Services & Management*, **7**(3), 143–9.
Geneletti, D. (2013), 'Assessing the impact of alternative land-use zoning policies on future ecosystem services', *Environmental Impact Assessment Review*, **40**, 25–35.
Geneletti, D. (2014), *Integrating Ecosystem Services in Strategic Environmental Assessment: A Guide for Practitioners*, a ProEcoServ report, Nairobi: United Nations Environment Program (UNEP).

Hatton Macdonald, D., R. Bark and A. Coggan (2014), 'Is ecosystem service research used by decision-makers? A case study of the Murray-Darling Basin, Australia', *Landscape Ecology*, **29**(8), 1447–60.

Honrado, J.P., C. Vieira and C. Soares et al. (2013), 'Can we infer about ecosystem services from EIA and SEA practice? A framework for analysis and examples from Portugal', *Environmental Impact Assessment Review*, **40**, 14–24.

Itkonen, P., L. Kopperoinen and A. Viinikka et al. (2015), 'Case: mapping green infrastructure and ecosystem services in the Helsinki-Uusimaa Region', in J.-P. Jäppinen and J. Heliölä (eds), *Towards a Sustainable and Genuinely Green Economy. The Value and Social Significance of Ecosystem Services in Finland (TEEB for Finland). Synthesis and Roadmap, The Finnish Environment*, Helsinki: Ministry of the Environment, pp. 46–55.

Järvelä, E. and T. Tiihonen (2014), *Climate-proof City. Developing a Green Factor Tool for the City of Helsinki*, Helsinki: City of Helsinki and Eero Paloheimo Ecocity Ltd.

Kelemen, E., M. García-Llorente and G. Pataki et al. (2014), 'Non-monetary techniques for the valuation of ecosystem services', in M. Potschin and K. Jax (eds), *Ecosystem Services Reference Book*, accessed 23 January 2016 at www.openness-project.eu/library/reference-book.

Klain, S.C. and K.M.A. Chan (2012), 'Navigating coastal values: participatory mapping of ecosystem services for spatial planning', *Ecological Economics*, **82**, 104–13.

Kopperoinen, L., P. Itkonen and J. Niemelä (2014), 'Using expert knowledge in combining green infrastructure and ecosystem services in land use planning: an insight into a new place-based methodology', *Landscape Ecology*, **29**(8), 1361–75.

Kruuse, A. (2011), *GRaBS Expert Paper 6: The Green Space Factor and the Green Points System*, London: Town and Country Planning Association.

Kyttä, M. and M. Kahila (2011), 'SoftGIS methodology – building bridges in urban planning', *GIM International (The Global Magazine for Geomatics)*, **25**(3), 37–41.

Kyttä, M., A. Broberg and T. Tzoulas et al. (2013), 'Towards contextually sensitive urban densification: location-based softGIS knowledge revealing perceived residential environmental quality', *Landscape and Urban Planning*, **113**, 30–46.

Laita, A., J. Kotiaho and M. Mönkkönen (2011), 'Graph-theoretic connectivity measures: what do they tell us about connectivity?', *Landscape Ecology*, **26**(7), 951–67.

Maes, J., A. Teller and M. Erhard et al. (2013), *Mapping and Assessment of Ecosystems and their Services. An Analytical Framework for Ecosystem Assessments Under Action 5 of the EU Biodiversity Strategy to 2020*, Luxembourg: Publications Office of the European Union.

Milcu, A., J. Hanspach and D. Abson et al. (2013), 'Cultural ecosystem services: a literature review and prospects for future research', *Ecology and Society*, **18**(3), 565–98.

Ministry of the Environment (2002), *Finland's National Land Use Guidelines*, Helsinki: Ministry of the Environment, Land Use Department.

Moreno, J., I. Palomo and J. Escalera et al. (2014), 'Incorporating ecosystem services into ecosystem-based management to deal with complexity: a participative mental model approach', *Landscape Ecology*, **29**(8), 1407–21.

Nadin, V. (2006), *The Role and Scope of Spatial Planning. Literature Review. Spatial Plans in Practice: Supporting the Reform of Spatial Planning*, Wetherby, UK: Communities and Local Government, accessed 20 January 2016 at http://webarchive.nationalarchives.gov.uk/20120919132719/http://communities.gov.uk/publications/planningandbuilding/spatialplans4.

Partidário, M.R. and R.C. Gomes (2013), 'Ecosystem services inclusive strategic environmental assessment', *Environmental Impact Assessment Review*, **40**, 36–46.

Pascual-Hortal, L. and S. Saura (2006), 'Comparison and development of new graph-based landscape connectivity indices: towards the prioritization of habitat patches and corridors for conservation', *Landscape Ecology*, **21**(7), 959–67.

Peifer, H. (2009), *CORILIS Version 3 – A Short Technical Description*, European Environment Agency, accessed 17 February 2015 at http://www.eea.europa.eu/data-and-maps/data/corilis-2000-2#tab-methodology.

Potschin, M. and R. Haines-Young (2011), 'Ecosystem services: exploring a geographical perspective', *Progress in Physical Geography*, **35**(5), 575–94.

Raymond, C.M., B.A. Bryan and D. Hatton MacDonald et al. (2009), 'Mapping community values for natural capital and ecosystem services', *Ecological Economics*, **68**(5), 1301–15.

Region Hannover (2009), *Regional Planning in the Hannover Region*, Hannover: Region Hannover.

Region Hannover (2013), *Landschaftsrahmenplan der Region Hannover* [Regional Landscape Plan of Region Hannover], Hannover: Region Hannover.

Saura, S., Ö. Bodin and M.-J. Fortin (2014), 'Stepping stones are crucial for species' long-distance dispersal and range expansion through habitat networks', *Journal of Applied Ecology*, **51**(1), 171–82.

Sindell, M. and Z. Thomas (2011), *Green Factor: Redux. Leveraging Seattle's Green Code to Create Urban Spaces*, accessed 21 January 2016 at http://www.gglo.com/wp-content/uploads/2011-08-31_UW-Green-Factor-Final.pdf.

Söderman, T. and S.-R. Saarela (eds) (2011), *Kestävät kaupunkiseudut. Kriteereitä ja mittareita suunnittelun työvälineiksi* [Sustainable urban regions – criteria and indicators for land use planning], *Finnish Environment 25/2011*, Helsinki: Finnish Environment Institute [in Finnish; English abstract].

Söderman, T., P. Itkonen and J. Rinne et al. (2014), *Ekosysteemipalvelut ja viherrakenne Tampereella – EVITA-hankkeen loppuraportti* [Ecosystem Services and Green Infrastructure in Tampere – The Final Report of the EVITA project], Tampere: City of Tampere, accessed 17 February 2015 at http://www.tampere.fi/material/attachments/e/rnJW3eIn6/EVITA_Ekosysteemipalvelut_ja_viherrakenne.pdf [in Finnish].

Söderman, T., L. Kopperoinen and V. Yli-Pelkonen et al. (2012), 'Ecosystem services criteria for sustainable development in urban regions', *Journal of Environmental Assessment Policy and Management*, **14**(2), 1250008–48.

Soille, P. and P. Vogt (2009), 'Morphological segmentation of binary patterns', *Pattern Recognition Letters*, **30**(4), 456–9.

Spangenberg, J.H., C. von Haaren and J. Settele (2014), 'The ecosystem service cascade: further developing the metaphor. Integrating societal processes to accommodate social processes and planning, and the case of bioenergy', *Ecological Economics*, **104**, 22–32.

Statistics Finland (2014), 'National accounts', accessed 21 January 2016 at http://www.tilastokeskus.fi/til/kan_en.html.

Stenning, E. (2008), 'An assessment of the Seattle green factor: increasing and improving the quality of urban green infrastructure', Master of Urban Planning thesis, Department of Urban Design and Planning. University of Washington.

Van Assche, K., R. Beunen and M. Duineveld et al. (2013), 'Co-evolutions of planning and design: risks and benefits of design perspectives in planning systems', *Planning Theory*, **12**(2), 177–98, accessed 29 January 2016 at http://plt.sagepub.com/content/12/2/177.short.

Von Haaren, C., C. Albert and J. Barkmann et al. (2014), 'From explanation to application: introducing a practice-oriented ecosystem services evaluation (PRESET) model adapted to the context of landscape planning and management', *Landscape Ecology*, **29**(8), 1335–46.

11. Ecosystem services in marine Environmental Impact Assessment: tools to support marine planning at project and strategic scales*

Tara Hooper, Olivia Langmead and Matthew Ashley

11.1 INTRODUCTION

It has been reported that the total economic value per unit area of marine and coastal habitats exceeds that of terrestrial systems by an order of magnitude (De Groot et al., 2012). Whatever the limitations of such monetary valuation, it serves to illustrate the importance of the ecosystem services (ES) provided by estuaries, seagrass beds, mangroves, coral reefs, salt marshes, and wider coastal and marine areas. The human activities taking advantage of these ES – which include fisheries, recreation, energy, ports and shipping, communication infrastructure and aggregate extraction – result in an increasingly crowded coastal zone, conflicts between sectors, and environmental impact.

The development of offshore renewable energy is a case in point. Seventy-five per cent of the installed capacity of offshore wind in Europe has been deployed since 2010, with 19 per cent of the total capacity installed in 2014 (EWEA, 2015). In the UK alone 1500 turbines had been installed or were in construction by the end of 2014 in wind farms covering nearly 800 km^2, and plans for a further 2700 turbines have been given consent or formally submitted for consent (4COffshore, 2014). Germany, the Netherlands, Belgium and Denmark are also developing offshore wind zones in the North Sea. In 2013 Northern Europe had 90 per cent of offshore wind capacity but countries including China, Japan and the United States are actively pursuing opportunities in this sector (GWEC, 2014).

The need to understand the economic, social and environmental impacts of this rapidly expanding sector is self-evident, but it has been questioned whether the existing Environmental Impact Assessment (EIA) process is adequate for renewable energy. The EIA framework leads to an emphasis on local and predominantly negative impacts, and is less effective at evaluating positive and non-local benefits, such as climate change mitigation

(Smart et al., 2014). ES approaches move beyond the unidirectional perspective of harm caused by human activity (Loomis and Paterson, 2014a), and thus offer a broader framework with which to consider the additional benefits of developments such as marine and offshore energy, and to evaluate the trade-offs that arise.

In seeking to mitigate existing environmental impacts and conserve biodiversity, the establishment of marine protected areas (MPAs) is another example of policies influencing the use of coastal space. In 2011, Natura 2000 sites (which include Special Protection Areas and Special Areas of Conservation under the European Union Birds and Habitats Directives) covered 4.3 per cent of European Union waters, and an ongoing process of additional national and regional initiatives has augmented this network (EEA, 2012). Globally the number of countries and territories that have created MPAs in more than 10 per cent of their marine areas (the goal outlined in the Aichi Biodiversity Target 11 within the Convention on Biological Diversity) more than doubled between 2010 and 2012 (Spalding et al., 2013).

Here again an ES approach has the potential to support MPA development, as a framework for defining objectives and monitoring impacts. The traditional focus of MPAs has been on biodiversity conservation, but describing this focus in terms of maintaining the delivery of wider ES has been advocated as a means of increasing support and improving compliance (Cárcamo et al., 2014). Such an approach allows the human benefits of MPAs to be better articulated and understood – as well as demonstrating how biodiversity is fundamental to ES generally (Palumbi et al., 2009) studies have considered MPAs in terms of specific ES such as their value for recreation and tourism (Kenter et al., 2013) and in climate change mitigation (McLeod et al., 2009).

The expansion of offshore renewable energy and MPAs does not occur in isolation; there is the potential for interaction within and between these sectors and with other marine activities. Conflicts between management objectives, and between stakeholder groups, arise from failure to adequately balance calls for increasing access and use with the need for enhanced preservation and protection (Loomis and Paterson, 2014a). Wider marine planning is therefore required to explicitly consider trade-offs under different management approaches (Lester et al., 2013) and hence minimize conflicts and maximize the benefits that can be obtained from marine ecosystems.

Capturing information about the flows of ES and their beneficiaries can support the movement of marine planning beyond descriptive assessments and towards approaches that better capture the dynamic interactions within the socio-ecological system (Le Cornu et al., 2014). UNESCO

(2014) gives examples of more than 20 countries attempting to implement marine spatial planning and there is a growing tendency for such policies to include an explicit focus on ES, for example in the European Union and USA (under the draft Directive for Marine Spatial Planning and Integrated Coastal Management, and Executive Order 13547 respectively).

Despite such examples of support for ES approaches at a high level within policy, examples of actual practice remain limited. This is due in part to a lack of practical guidance on methods for implementation (Werner et al., 2014). The acceptance and application of ES approaches amongst practitioners is growing, however. In a study of 44 practitioners involved in marine planning worldwide, 95 per cent had included some ES information in the development of their marine plan (Le Cornu et al., 2014). This was most likely related to cultural and provisioning services, with less than half of the practitioners interviewed using information about regulating or supporting services (ibid.). Practitioners also tend to rely on readily available biophysical and social data, neglecting assessment of the linkages between social and ecological systems, which is required to further understanding of the interactions between people and their environment (ibid.).

The marine environment poses particular challenges because the complexity of marine ecosystems is a barrier to understanding the goods and services they deliver (Lopes and Videira, 2013), to the extent that it has been suggested that our understanding of these systems will never be complete (Shucksmith and Kelly, 2014). There is little empirical evidence to demonstrate the exact role of ecosystem processes and biodiversity in the supply of ES, leading to a reliance on largely theoretical models of these relationships (Börger et al., 2014). Spatial assessment of marine ES is further complicated by the mobile nature of marine resources, such as fish, mammals and the plankton that support them, which move between locations during different seasons and at different stages in their life cycle (ibid.).

Despite these challenges, progress continues to be made in identifying, measuring and mapping marine ES such that the impacts of developments and the implications of management strategies can be better understood.

This chapter explores the process required for adopting an ES approach to Environmental Impact Assessment in marine systems, including the role of stakeholders in identifying ES and selecting assessment priorities. The application, and adaptation, of classification frameworks in practice is considered, and the need for appropriate indicators that permit the manageable measurement and reporting of ES is discussed. Modelling and spatial mapping are key tools for evaluating and communicating potential change, and methods for these are demonstrated using two case studies: (1) broad-scale mapping to identify key areas of ES delivery in a context of regional energy planning; and (2) the use of Bayesian belief network modelling to consider how ES delivery might change following the establishment of MPAs.

11.2 IDENTIFYING AND EVALUATING MARINE ECOSYSTEM SERVICES

EIAs are a core component of the planning process for infrastructure developments. ES approaches are conceptually well aligned with EIAs as they seek to evaluate the ecological, social and economic implications of change. The advantage of ES approaches is in providing a common language and clear framework for assessing trade-offs (Granek et al., 2010) and as such they have the potential to provide a more systematic and transparent approach to impact assessment and hence support decision-making.

One of the factors that restricts the use of ES information in impact assessment and marine management is that existing research has had a tendency to focus on a single or small suite of services in a small number of habitats. In a review of 145 studies, 48 per cent had considered just one service, while only 13 per cent assessed six or more (Liquete et al., 2013). The most studied marine ES highlighted by the review were food provision (particularly fisheries), followed by water purification, coastal protection, and recreation and tourism, with the research focusing on mangroves, coastal wetlands and coral reefs. However, there are a growing number of studies that suggest practical methods for considering a greater number of ES, hence supporting assessments that seek to assess a broad range of impacts.

Hooper et al. (2014) outline a methodology for applying an ES approach to assess the impacts of a specific infrastructure development, using a hypothetical tidal barrage as a case study. Their approach focuses on evaluating changes to the delivery of environmental benefits, which are defined as the point at which a direct gain in human welfare provided by environmental services is realized. The approach thus facilitates monetary valuation and cost–benefit analysis, as this is often an aspiration of ES approaches. However, valuation is not always the desired endpoint of ES assessments, and the steps of the process (Figure 11.1) are also applicable to an evaluation of the underlying ES. This latter approach may be appropriate if a more biophysical output, and/or the consideration of supporting services, is required.

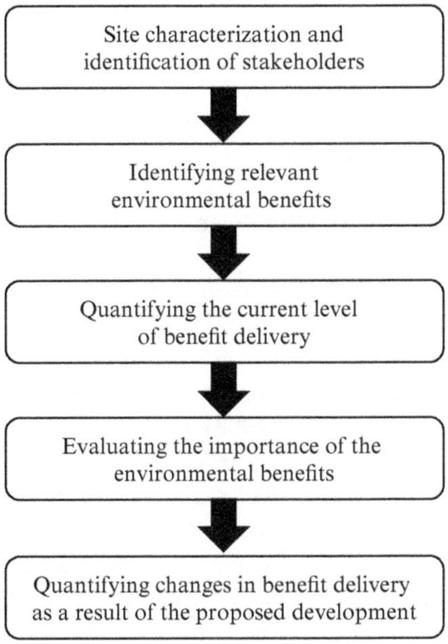

Source: As proposed by Hooper et al. (2014).

Figure 11.1 The key steps in using an ES approach for impact assessment

11.2.1 Setting the Scene

The first step in assessing potential impacts is to understand the area that will be affected by the proposed development in terms of the habitats, species, conservation designations, social and economic uses and cultural importance. As ES approaches are by nature anthropocentric, key to any assessment is identifying the people who benefit from the affected environment. A number of tools exist (such as content analysis, institutional analysis, interviews, and social network analysis) that can be used to formalize the process of conceptualizing institutional context and identifying stakeholders (Lopes and Videira, 2013). This process should look beyond the immediate locale, as there may be regional, national or even international benefits provided by the area.

Following this broad characterization of the area, a detailed inventory is required that identifies the specific benefits provided by the ecosystem, the compilation of which should again involve stakeholders. For a comprehensive assessment, the scope of this inventory should be as broad as possible, and Hooper et al. (2014) use the term 'environmental' (as opposed to 'ecosystem') to allow for the inclusion of benefits that rely solely on abiotic components of the system. This approach is particularly applicable to energy, as a service that does not relate to the biota, and also remains in line with the existing EIA process by permitting the inclusion of benefits such as transport and aggregate extraction.

11.2.2 The Role of Classification Frameworks

The inventory should follow a defined framework that provides a logical classification and grouping of the benefits, and so aids data collection and interpretation of the outputs. Many authors have followed the Millennium Ecosystem Assessment (2003) framework in classifying marine ES – for example, Beaumont et al. (2006, 2008) whose work has been adopted and adapted by authors including Everard et al. (2010), Hussain et al. (2010) and Remoundou et al. (2009). More recently, attention has focused on frameworks that seek to distinguish more clearly between services and benefits, and so facilitate monetary valuation, for example the UK National Ecosystem Assessment (2011) and, specifically for the marine environment, Saunders et al. (2010). There is also ongoing work to develop common standards, particularly through development of the Common International Classification of Ecosystem Services (CICES, 2013; Haines-Young and Potchin, 2013), which seeks to link to the System of Environmental-Economic Accounting (SEEA; United Nations et al., 2014).

However, there is little evidence for the consistent use of any one framework in existing empirical assessments. In a systematic review of 145 papers considering marine ES, Liquete et al. (2013) found that 68 per cent of the studies did not follow a particular classification. This lack of consistency may be undesirable as it can reduce comparability between assessments (Fletcher et al., 2011), but in practice it is often not possible to match empirical assessments with theoretical classifications (Liquete et al., 2013) to the extent that it has been questioned whether there will ever be a universally applicable framework (Fisher et al., 2009).

The priority remains to select the best approach for a given assessment, while considerations such as how its outcome may be compared to other studies are secondary. In the absence of a universal classification, existing frameworks serve as a useful guide, but are likely to require adaptation to suit the specific situation. A description of the priorities for, and assumptions made in defining, the classification framework for a particular assessment will support understanding of its outputs and limitations. For example, Hooper et al. (2014), whose assessment was intended to support valuation, explain the process by which benefits were categorized and justify the omission of supporting services.

11.2.3 Identifying Priorities for Assessment

The purpose of an inventory is to identify the ES or benefits that are currently delivered, and hence might be affected by a new coastal development, but marine ecosystems are highly complex and support a large and diverse range of services. A common recommendation in tackling this complexity in practice is to focus on those ES that are of the highest priority (Werner et al., 2014). Identifying these requires understanding of the social values that define the services provided (Loomis and Paterson, 2014a), and so wide stakeholder involvement is once again essential.

Participatory approaches including focus groups, systems mapping, citizens' juries, and multi-criteria analysis are useful methods for determining the importance of specific coastal ES and benefits (Lopes and Videira, 2013). Examples of the application of these methods include simple multi-criteria analysis to determine how stakeholders rank the importance of ES supplied by a marine protected area and how they perceive threats to them (Cárcamo et al., 2014). This approach of identifying priorities based on the vulnerability as well as the importance of marine ES is also used by Werner et al. (2014) who developed a matrix to link the perceived societal and financial value of marine ES to the level of stress, as a tool for rapidly identifying management priorities. A 'triage' method has also been proposed to support strategic decision-making by identifying the aim and scope of ES assessments to ensure that they are meaningful, useful and feasible (Pendleton et al., 2015).

11.2.4 Choosing Indicators for Ecosystem Services and Benefits

Having identified and prioritized the ES and benefits provided at the site, the next stage is to determine the level at which they are supplied, to provide a baseline against which potential impacts can be evaluated. ES impact assessments usually seek to quantify, and potentially to determine a monetary value for, the affected services or benefits, although meaningful quantification is not always possible particularly in the case of cultural services (Hooper et al., 2014). However, as UK policy guidance notes, it is imperative that an impact assessment identifies all potential costs and benefits, whether these are monetized, quantified or described in qualitative terms (HM Treasury, 2013).

In order to assess services and benefits, and the potential changes to their supply, it must first be established what needs to be measured. Some direct measures exist, such as the quantity of fish caught, the number of people taking part in a recreational activity, or the rate of carbon uptake by a particular habitat. However, it is often impossible or impractical to measure marine ES directly, and so indirect or proxy measures must be used instead (Liquete et al., 2013; Loomis and Paterson, 2014b). The process is further complicated because there are neither agreed indicators for marine ES (most established indicators for the condition of the marine environment are not specific to ES) nor even a consistent approach to indicator selection (Hattam et al., 2014).

There are examples of studies proposing indicators for multiple marine ES or benefits, such as the generic approach taken by Boehnke-Henrichs et al. (2013) as well as specific case studies of a macrotidal estuary (Hooper et al., 2014) and a large offshore sandbank (Hattam et al., 2014). The indicators defined in existing studies have been characterized as falling into one of three categories: (1) biophysical structures and processes that underpin the *capacity* of an ecosystem to provide services; (2) the *flow* of ES from functions that ultimately contribute to human well-being; and (3) the specific societal *benefit* that results from that flow (Liquete et al., 2013). That review also demonstrated that researchers have proposed a high number of indicators for the benefits resulting from provisioning and cultural services but fewer for the underlying capacity and flow, while the converse is true for most regulating services, with a high number of indicators for ES capacity compared to those for the ultimate benefits received (ibid.).

What is considered a service, and hence what constitutes a relevant and useful indicator, is likely to be context specific (Hattam et al., 2014). As in the case of classification frameworks, there may be no suitable list of indicators already available, requiring this to be defined for each assessment. Literature reviews are a useful starting point for identifying indicators (Boehnke-Henrichs et al., 2013), with expert deliberation also a suitable approach for defining the parameters to be measured in the assessment (Hattam et al., 2014).

Indicators of ecosystem functions and services are by nature ecological, while those for benefits must capture the human use or enjoyment resulting from service delivery (ibid.). The applicability of any proposed indicator should be assessed against three key criteria: the availability of, or potential to collect, data (measurability), the ability to detect change over time (sensitivity), and the potential to respond to the specific change or impact (specificity) (Hattam et al., 2014; Loomis and Paterson, 2014b). Two further criteria should also be considered in the selection process: the possibilities for aggregating or disaggregating the indicator for use at different spatial scales (scalability) and the potential usefulness of the indicator in other locations and situations (transferability) (Hattam et al., 2014). Not all indicators that have been suggested in the existing literature fulfil these criteria, as they may be insufficiently temporally sensitive, or lack a demonstrable connection to specific environmental parameters that may be impacted (Boehnke-Henrichs et al., 2013).

Poorly chosen indicators are perhaps indicative of the particular challenges presented in the marine environment. These challenges (as described in Hattam et al., 2014) include the mobility of marine species that use different habitats at different times of the year, and a lack of data, particularly for regulating services. There may also be a spatial disconnect between the supply of services and their beneficiaries, and potentially a small number of direct beneficiaries (in particularly remote or offshore locations), making changes difficult to detect. It can also be challenging to link cultural services to specific, measurable ecosystem components, and the confounding effects of multiple stressors (e.g., climate change and fishing impacts) mean many indicators lack specificity.

Even where appropriate indicators are selected, simply focusing on one level in the ecosystems services cascade (i.e., considering solely benefits or services or functions) may not provide enough information for adequate impact assessment. Measuring only benefits, for example, provides no indication of the security of their supply or how this may be impacted by developments (Hooper et al., 2014). Ideally, therefore, a suite of indicators is required for each ES that can capture changes in the service itself, the underlying ecosystem functions, and the resulting human benefits (Hattam et al., 2014). While it may be desirable to take a fully comprehensive approach to evaluating ES, any impact assessment must be manageable in practice, creating the further challenge of producing a suite of indicators that is small enough to be workable but does not oversimplify the complexity of the ecosystem (Loomis et al., 2014).

It is also important that the outputs of any impact assessment are useful to decision-makers. The development of indices has been suggested as one approach for consolidating the results derived from evaluation of an extensive suite of variables for ES and functions, as a means of facilitating communication (Loomis and Paterson, 2014b). The hierarchical approach proposed by Loomis et al. (2014a, 2014b) involves the rank ordering and aggregation of indicators in progressive steps (Figure 11.2) enabling the status of the highest-order indices (such as 'Water column' in Figure 11.2) to be presented as a 'report card', which can be easily interpreted. However, caution should be applied to such an approach, as it may be an oversimplification of complex problems, and can be misleading if the indices are poorly constructed or misinterpreted (Loomis and Paterson, 2014b).

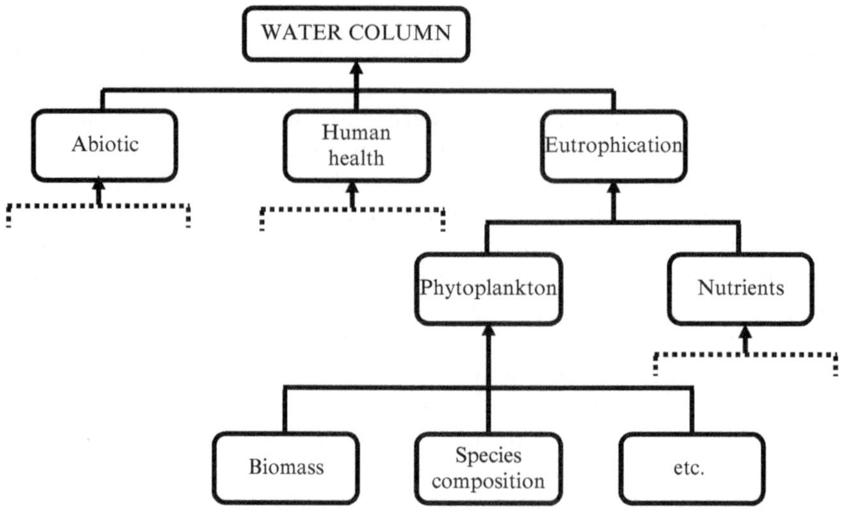

Source: Loomis et al. (2014a).

Figure 11.2 An example of the hierarchical aggregation of indicators to construct a water column index

11.2.5 Evaluating Change

Having determined the relevant ES, functions and benefits, and evaluated their baseline level, the next step of an impact assessment is to predict how they are going to respond to the changes in pressures brought about by the development or management intervention. There are some examples of assessments that have considered specific impacts on specific ES, such as loss of salt marsh on coastal protection (Brisson et al., 2014), recreational benefits of artificial reefs (Chen et al., 2013), carbon capture by restored seagrass (Duarte et al., 2013) and the impacts of offshore wind farms (Mangi, 2013). However, such examples are not extensive within the literature, and published research tends to focus on topics such as the implications of climate change (e.g., Nelson et al., 2013), ocean acidification (e.g., Garrard and Beaumont, 2014) and invasive species (e.g., McLaughlan et al., 2014) as opposed to stressors that can be managed at a local scale.

Unless it is possible to conduct empirical research as part of the impact assessment, expert judgement is likely to be needed in the absence of existing information on how a particular development impacts on particular services. A larger body of literature exists on the responses and sensitivity of habitats and species to disturbances. This knowledge can be used to infer the likely response of the ES that depend on those habitats and species. Alternatively, expert deliberation can be used to directly predict the likely changes in ES delivery. However, the absence of empirical evidence will decrease confidence and increase uncertainty in predictions of ES response. The need to adequately evaluate and communicate uncertainty has already been highlighted for impact assessment in general (Leung et al., 2015).

11.3 MODELLING AND SPATIAL MAPPING APPROACHES

Where resources permit, modelling can be a useful tool in predicting the responses of ES to impacts. Understanding the spatial distribution of services is also often desirable, and mapping aids visualization of competing demands and the impacts of new initiatives (Shucksmith and Kelly, 2014). The growing interest in ES approaches has seen a corresponding growth in the availability of modelling and spatial mapping tools.

Although effort has been focused on terrestrial habitats, there are examples of applications for the marine environment. These include the web-based Artificial Intelligence for Ecosystem Services (ARIES) (Villa et al., 2014), which has been used to map the supply of key ES in the Puget Sound region of the northwest United States (Bagstad et al., 2014). The Integrated Valuation of Ecosystem Services and Tradeoffs (InVEST) tool (Natural Capital Project, 2015) contains a number of models relevant to the marine environment, including blue carbon, coastal protection, habitat risk assessment, marine water quality, recreation, and scenic quality. These tools have been applied in practice to situations including marine spatial planning in British Columbia and coastal zone management in Belize (Ruckelshaus et al., 2013).

The two case studies presented below illustrate the application of mapping and modelling approaches at two different scales.

11.3.1 Broad-scale Evaluation: Identifying Key Areas of Service Delivery

An important objective of spatially mapping baseline information is to identify key areas of service provision, and so inform future protection of these areas against pressures that could result in service loss (Shucksmith and Kelly, 2014). The Bristol Channel has been identified as a strategic area for marine and offshore energy (RegenSW and Marine Energy Matters, 2012), which prompted a similar broad-scale assessment to rapidly determine important areas for particular ES, and hence identify potential conflicts and sites that might be sensitive to impacts from energy development (Ashley, 2014). Key aspects of Ashley's research are summarized for this case study.

The Bristol Channel and Severn Estuary area of the UK contains a diverse array of habitats, landscapes and seascapes, containing habitats and features of international conservation importance, as well as some of the UK's largest ports and trade links. To provide a rapid assessment, six key ES (those that were both widespread across the study site and of high importance to the region) were selected: carbon storage, flood alleviation, wild food and fish, ports and shipping, coastal/estuarine archaeology, and sense of place.

The aim of the study was to relate the habitats and ecosystem features present to the level of supply of each key benefit. Habitat information for the study site was obtained from the Mapping European Seabed Habitats project (MESH). Geographical information system software (ESRI ARC GIS 10) was utilized to spatially represent environmental benefit supply-level data across the region. The mapping approach divided the study region into 10 km^2 hexagonal planning cells. 'Supply levels' between one (low) and five (high) were attributed to each planning cell, for each individual ES or environmental benefit. Confidence maps relating to mapped results were also created to indicate the level of detail provided by data resources.

Where quantified data were available, such as biomass or number of individuals of a species (such as wild salmon within a tributary river), supply levels were ascertained by calculating five 'Jenks breaks' across the range of the dataset. Jenks breaks cluster data in a given number of categories through calculations that ensure highest statistical differences are present between each category. Where ES or benefits relied on the presence of particular habitat or environmental feature (for instance, the relative role of mud flats and salt marsh habitat in storing carbon) data provided in the literature or expert assessment were utilized to define supply levels. As the study was identifying broad-scale hotspots of a given ES or benefit supply, a planning cell was assigned the value corresponding to the habitat or feature of the greatest benefit present in that cell.

To take carbon storage as an example, published datasets on carbon sequestration values (Andrews et al., 2006; Alonso et al., 2012) and supply levels designated by Liekens et al. (2013), Jacobs et al. (2013) and Potts et al. (2013), were used to allocate supply levels to the habitats within the Severn Estuary and Bristol Channel area, based on broad predicted habitat type and additional data such as flora species present. The mapping process (Figure 11.3) clearly displayed the importance for carbon storage of the inner Severn and smaller estuaries feeding into the

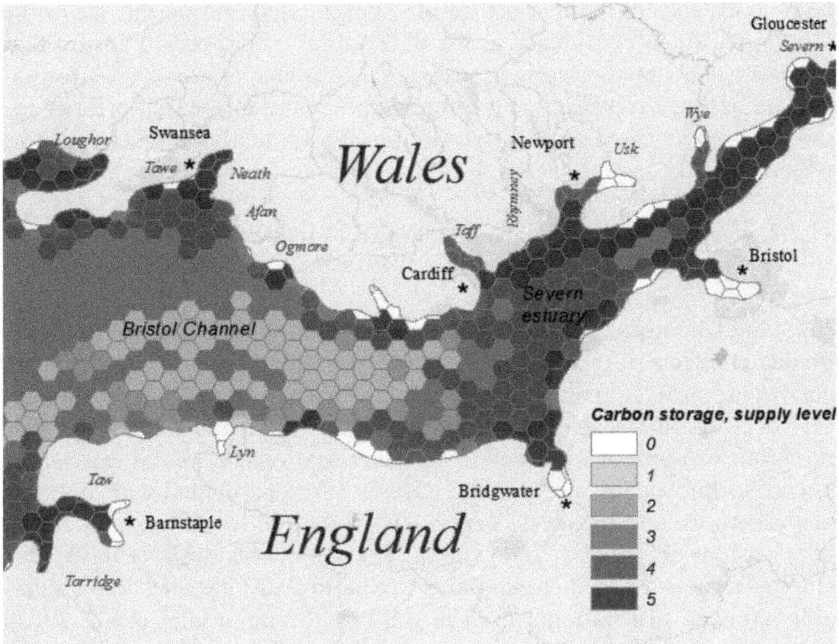

Note: The level of carbon stored ranges from 1 (lowest) to 5 (highest), 0 = no habitat data present.

Figure 11.3 Carbon storage levels based on detailed habitat type (substratum and species)

wider Bristol Channel. The wider study also demonstrated the importance of the region's intertidal and nearshore areas for flood risk management, archaeological resources and wild food and fisheries, as well as the presence of internationally significant marine transport links and ports, the presence of which support regional, national and international trade and industry (Ashley, 2014).

Certain challenges were identified during the process. In particular, it was noted that assigning across the entire planning cell the highest benefit supply level provided by a single habitat present within that cell may overestimate the supply of the benefit across the cell. However, adopting a precautionary principle is appropriate in studies of this nature, which aim to identify key areas that may require further research or protection. The study also exposed a shortage of evidence on the levels of specific ES benefits provided by particular habitat types, for instance a lack of studies quantifying carbon storage by intertidal mud flats. Such weaknesses limit confidence in the mapped results, potentially limiting the use of the results, although confidence in the data was also mapped to ensure that this issue was appropriately reported. The absence of recorded information on cultural services was also highlighted, and a research priority for the development and application of appropriate methods to assess these was identified.

11.3.2 Modelling and Mapping the Potential Impacts of Specific Interventions

Mapping techniques can also be combined with modelling to display the predicted effects of changing human pressures, and these approaches can also be applied at finer scales. Bayesian belief network (BBN) models are becoming increasingly widespread in environmental modelling. This probabilistic approach fits well to real-life situations because it can handle uncertain knowledge and missing data, is easily combined with decision analysis tools, and facilitates expert elicitation and learning about causal relationships (Uusitalo, 2007). In the marine context, BBNs have been employed in studies including the evaluation of pressures on vulnerable habitats at a national level in the UK (Stelzenmüller et al., 2010), implications of climate change on coral reef and hard substrates in South Florida (Fletcher et al., 2014), outcomes for society following a number of plausible alternative social and economic pathways on Black Sea ecosystems (Langmead et al., 2009), and in considering ES trade-offs in the development of coastal shrimp aquaculture in Thailand (Schmitt and Brugere, 2013).

In this case study (which summarizes work by Langmead et al., 2015), an approach was developed to explore the implications of MPA designation on the provision of ES by subtidal sedimentary habitats in the North Devon UNESCO Biosphere Reserve (NDBR). Little is known about the importance of sedimentary habitats in supporting service provision and within the NDBR marine area up to four new MPAs may be designated, with unknown consequences on service provision. The approach involved developing a BBN to represent relationships between human activities, subtidal seabed habitats and their ES and linking it with a geospatial database.

A habitat map was generated using multiple data sources (e.g., McMahon and Golding, 2010) and modelled layers (UKSeaMap; McBreen et al., 2011), together with a corresponding confidence layer. Habitats were grouped into broad habitat types with similar sensitivity to fishing pressure (after Hall et al., 2008). This was gridded to 1 km^2 and used, together with water depth, to provide the geospatial context for the model. A fishing pressure map was constructed using demersal mobile fishing activity as a proxy indicator for seabed abrasion, benchmarked to broad levels of fishing intensity. This was adapted to create a second pressure map representative of an MPA designation scenario. Our assumption was that demersal towed gear would be prohibited within four recommended MPA sites (Lieberknecht et al., 2011). Fishing effort would be lost from some MPA sites, while for others it would be displaced outside MPA boundaries (based on consultation with resident fishers).

Relationships between habitats and ES provision were described by assigning levels of provision to each habitat for (1) nursery provision for commercial fish species (literature review of juvenile sediment type and depth preferences); (2) carbon sequestration (sediment mud content) and (3) waste remediation (bioturbation, evaluated using empirical data). An assumption was made that the level of ES provision was proportional to habitat sensitivity to fishing pressure, thus service provision by sensitive habitat types would be lower than resistant/tolerant habitat types under equivalent fishing pressure exposure due to greater functional disruption. This was used to construct conditional probability distributions to underpin the causal relationships within the model.

The model was run for each 1 km^2 of the NDBR using geospatial data from habitat map and water depth and was conditioned with (1) the distribution of fishing pressure and (2) the altered fishing pressure representing the effect of MPA designation. Service provision maps by service type were generated representative of the current provision of ES and likely changes to service provision with MPA designation. The three ES types were aggregated into a single map using stakeholder preferences, and differences between the current estimation of service provision and the MPA scenario were assessed.

The scenario representing the designation of recommended MPAs shows that changes in provision of the different ES types is variable. Nursery provision (Figure 11.4) increased in some areas (especially in the North of Lundy site) and decreased in others (due to pressure increases from fishing displacement), while both carbon sequestration and waste remediation increased in the protected areas but did not decrease in the areas subject to fisheries displacement as these had negligible service provision under the current scenario.

The case study clearly demonstrated to stakeholders the potential effects of management on service provision, with important implications for spatial planning, and wider marine governance. It showed that both the direct implications of MPA designation could be assessed (at least for some services), and that secondary implications of management interventions can also be captured, such as fisheries displacement in this case.

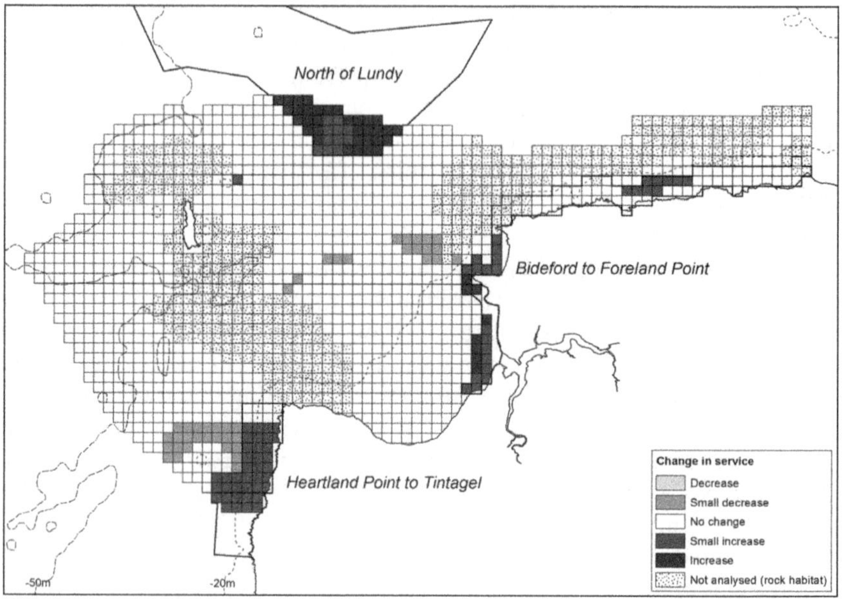

Figure 11.4 *An example model-derived scenario outcome showing changes in nursery provision with recommended MPA designation (relative to current provision)*

11.4 CONCLUSIONS

The potential for ES approaches is well understood: the drivers for this are not just academic, as evidenced by the explicit references in, for example, marine spatial planning policy. Recent work to develop indicators for marine ES has been an important step in operationalizing the approach, as it provides guidance on what can, and should, be measured. Research more generally is moving beyond the development of conceptual frameworks, and practical assessments demonstrate that there is no universal framework or suite of indicators, and these must be derived within the specific context of each assessment. Comprehensive assessments should consider underlying processes as well as services and benefits, to ensure that potential impacts on each level of the cascade are considered.

Stakeholders have vital roles at different stages of an ES-based impact assessment: the anthropocentric nature of the approach requires a full understanding of who benefits and how. Local knowledge is also a key component of any expert judgement on how services may respond to changes in pressures, and stakeholder involvement is required to support prioritization of the components to be assessed. This engagement of stakeholders, and the ultimate use of the outputs by decision-makers, requires methods to rationalize and communicate complex information, including the communication of uncertainty.

Modelling and mapping approaches are effective analysis and communication tools that enable important areas for ES provision to be identified, and potentially prioritized for protection. They also assist in identifying possible trade-offs between areas of high intensities of human activities and service provision. The main limitations though are the information requirements in terms of habitat maps, understanding of the relationships between marine habitats and ES provision, and how human pressures may reduce service flows.

It remains the case that further empirical research would support a more rapid uptake of ES approaches to impact assessment. Particular areas in which future research should focus include understanding of the linkages between ecological processes and ES, and the simultaneous assessment of multiple ES, particularly those beyond fisheries, recreation, water purification and coastal protection, in habitats other than mangroves, wetlands and coral reefs. Further empirical evaluation is required to determine the response of marine ES to pressures, and to establish thresholds and tipping points. There are also spatial issues that require additional research, in terms of how to account for the mobility and distinct life cycle stages of marine species, and the disconnect between services and beneficiaries. Although understanding of discrete impacts is not yet comprehensive, methods to assess cumulative impacts effectively are urgently required, and would relate more closely to the needs of project and strategic planning in an increasingly crowded coastal zone.

NOTE

* The work reported for the Bristol Channel case study was funded by the Natural Environment Research Council Marine Renewable Energy Knowledge Exchange programme. Information to support the mapping was derived from data made available under the European Marine Observation Data Network (EMODnet) Seabed Habitats project (www.emodnet-seabedhabitats.eu), funded by the European Commission's Directorate General for Maritime Affairs and Fisheries (DG MARE). The Data Owner and EMODnet Seabed Habitats consortium accept no liability for the use of this data or for any further analysis or interpretation of the data. The North Devon case study was carried out as part of the VALMER project, which was supported by the European cross-border cooperation programme INTERREG IV A France (Channel) England, co-funded by the ERDF. The authors are also very grateful to Charly Griffiths of the Marine Biological Association, UK, for producing the maps for the North Devon case study.

REFERENCES

4COffshore (2014), 'Global offshore wind farms database', accessed 5 December 2014 at http://www.4coffshore.com/windfarms/.

Alonso, I., K. Weston and R. Gregg et al. (2012), *Carbon Storage by Habitat – Review of the Evidence of the Impacts of Management Decisions and Condition of Carbon Stores and Sources*, Natural England Research Reports No. NERR043.

Andrews, J.E., D. Burgess and R.R. Cave et al. (2006), 'Biogeochemical value of managed realignment, Humber estuary, UK', *Science of the Total Environment*, **371**(1–3), 19–30.

Ashley, M.C. (2014), *Ecosystem Service Mapping in the Severn Estuary and Inner Bristol Channel*, report for NERC Marine Renewable Energy Knowledge Exchange Project, RSPB and Plymouth Marine Laboratory, Plymouth.

Bagstad, K.J., F. Villa and D. Batker et al. (2014), 'From theoretical to actual ecosystem services: mapping beneficiaries and spatial flows in ecosystem service assessments', *Ecology and Society*, **19**(2), 64.

Beaumont, N., M. Austen and S. Mangi et al. (2008), 'Economic valuation for the conservation of marine biodiversity', *Marine Pollution Bulletin*, **56**(3), 386–96.

Beaumont, N., M. Townsend and S. Mangi et al. (2006), *Marine Biodiversity: An Economic Valuation*, report prepared for Defra.

Boehnke-Henrichs, A., C. Baulcomb and R. Koss et al. (2013), 'Typology and indicators of ecosystem services for marine spatial planning and management', *Journal of Environmental Management*, **130**, 135–45.

Börger, T., N.J. Beaumont and L. Pendleton et al. (2014), 'Incorporating ecosystem services in marine planning: the role of valuation', *Marine Policy*, **46**, 161–70.

Brisson, C.P., T.C. Coverdale and M.D. Bertness (2014), 'Salt marsh die-off and recovery reveal disparity between the recovery of ecosystem structure and service provision', *Biological Conservation*, **179**, 1–5.

Cárcamo, P., R. Garay-Flühmann and F.A. Squeo et al. (2014), 'Using stakeholders' perspective of ecosystem services and biodiversity features to plan a marine protected area', *Environmental Science and Policy*, **40**, 116–31.

Chen, J.L., C.T. Chuang and R.Q. Jan et al. (2013), 'Recreational benefits of ecosystem services on and around artificial reefs: a case study in Penghu, Taiwan', *Ocean and Coastal Management*, **85**(A), 58–64.

CICES (2013), *CICES Version 4.3*, accessed 5 February 2015 at http://cices.eu/.

De Groot, R., L. Brander and S. van der Ploeg et al. (2012), 'Global estimates of the value of ecosystems and their services in monetary units', *Ecosystem Services*, **1**(1), 50–61.

Duarte, C.M., T. Sintes and N. Marbà (2013), 'Assessing the CO_2 capture potential of seagrass restoration projects', *Journal of Applied Ecology*, **50**(6), 1341–9.

European Environment Agency (EEA) (2012), *Protected Areas in Europe – An Overview*, European Environment Agency Report No. 5/2012, Luxembourg: Publications Office of the European Union.

European Wind Energy Association (EWEA) (2015), *Wind in Power. 2014 European Statistics*, accessed 21 January 2016 at http://www.ewea.org/fileadmin/files/library/publications/statistics/EWEA-Annual-Statistics-2014.pdf.

Everard, M., L. Jones and B. Watts (2010), 'Have we neglected the societal importance of sand dunes? An ecosystem services perspective', *Aquatic Conservation: Marine and Freshwater Ecosystems*, **20**(4), 476–87.

Fisher, B., R.K. Turner and P. Morling (2009), 'Defining and classifying ecosystem services for decision making', *Ecological Economics*, **68**(3), 643–53.

Fletcher, P.J., C.R. Kelble and W.K. Nuttle et al. (2014), 'Using the integrated ecosystem assessment framework to build consensus and transfer information to managers', *Ecological Indicators*, **44**, 11–25.

Fletcher, S., J. Saunders and R.J.H. Herbert (2011), 'A review of the ecosystem services provided by broad-scale marine habitats in England's MPA network', *Journal of Coastal Research*, **SI 64**, 378–83.

Garrard, S.L. and N.J. Beaumont (2014), 'The effect of ocean acidification on carbon storage and sequestration in seagrass beds; a global and UK context', *Marine Pollution Bulletin*, **86**(1), 138–46.

Granek, E.F., S. Polasky and C.V. Kappel et al. (2010), 'Ecosystem services as a common language for coastal ecosystem-based management', *Conservation Biology*, **24**(1), 207–16.

GWEC (2014), *Global Wind Report Annual Market Update 2013*, Global Wind Energy Council, accessed 12 December 2014 at http://www.gwec.net/publications/global-wind-report-2/global-wind-report-2013/.

Haines-Young, R. and M. Potschin (2013), *Common International Classification of Ecosystem Services (CICES): Consultation on Version 4, August–December 2012*, EEA Framework Contract No. EEA/IEA/09/003.

Hall, K., O.A.L. Paramor and L.A. Robinson et al. (2008), *Mapping the Sensitivity of Benthic Habitats to Fishing in Welsh Waters – Development of a Protocol*, CCW Report No. 8/12.

Hattam, C., J.P. Atkins and N. Beaumont et al. (2014), 'Marine ecosystem services: linking indicators to their classification', *Ecological Indicators*, **49**, 61–75.

HM Treasury (2013), *The Green Book, Appraisal and Evaluation in Central Government*, London: TSO.

Hooper, T., P. Cooper and A. Hunt et al. (2014), 'A methodology for the assessment of local-scale changes in marine environmental benefits and its application', *Ecosystem Services*, **8**, 65–74.

Hussain, S.S., A. Winrow-Giffin and D. Moran et al. (2010), 'An ex ante ecological economic assessment of the benefits arising from marine protected areas designation in the UK', *Ecological Economics*, **69**(4), 828–38.

Jacobs, S.W., D. Vandenbruwaene and O. Vrebos et al. (2013), *Ecosystem Service Assessment of TIDE Estuaries*, study report in the framework of the Interreg IVB project TIDE, ECOBE, UA, Antwerp, Belgium, accessed May 2014 at http://www.tide-toolbox.eu/reports/.

Kenter, J.O., R. Bryce and A. Davies et al. (2013), *The Value of Potential Marine Protected Areas in the UK to Divers and Sea Anglers*, report for the UK National Ecosystem Assessment, Cambridge, UK: UNEP-WCMC.

Langmead, O., T. Hooper and C. Griffiths et al. (2015), 'Case study process: North Devon', in W. Dodds and L.A. Friedrich et al. (eds), *The Potential Role of Ecosystem Service Assessment in Marine Governance in the Western Channel*, VALMER Work Package 4 evidence base report, VALMER project, pp. 25–44.

Langmead, O., A. McQuatters-Gollop and L.D. Mee et al. (2009), 'Recovery or decline of the northwestern Black Sea: a societal choice revealed by socio-ecological modelling', *Ecological Modelling*, **220**(21), 2927–39.

Le Cornu, E., J.N. Kittinger and J.Z. Koehn et al. (2014), 'Current practice and future prospects for social data in coastal and ocean planning', *Conservation Biology*, **28**(4), 902–11.

Lester, S.E., C. Costello and B.S. Halpern et al. (2013), 'Evaluating tradeoffs among ecosystem services to inform marine spatial planning', *Marine Policy*, **38**, 80–89.

Leung, W., B. Noble and J. Gunn et al. (2015), 'A review of uncertainty research in impact assessment', *Environmental Impact Assessment Review*, **50**, 116–23.

Lieberknecht, L.M., T.E.J. Hooper and T.M. Mullier et al. (2011), *Finding Sanctuary Final Report and Recommendations*, a report submitted by the Finding Sanctuary stakeholder project to Defra, the Joint Nature Conservation Committee, and Natural England.

Liekens, I., S. Broekx and L. de Nocker (2013), *Manual for the Valuation of Ecosystem Services in Estuaries*, a study commissioned by Antwerp Port Authority (APA), accessed May 2014 at http://www.tide-toolbox.eu/reports.

Liquete, C., C. Piroddi and E.G. Drakou et al. (2013), 'Current status and future prospects for the assessment of marine and coastal ecosystem services: a systematic review', *PLoS ONE*, **8**(7), e67737.

Loomis, D.K. and S.K. Paterson (2014a), 'The human dimensions of coastal ecosystem services: managing for social values', *Ecological Indicators*, **44**, 6–10.

Loomis, D.K. and S.K. Paterson (2014b), 'Human dimensions indicators of coastal ecosystem services: a hierarchical perspective', *Ecological Indicators*, **44**, 63–8.

Loomis, D.K., P.B. Ortner and C.R. Kelble et al. (2014), 'Developing integrated ecosystem indices', *Ecological Indicators*, **44**, 57–62.

Lopes, R. and N. Videira (2013), 'Valuing marine and coastal ecosystem services: an integrated participatory framework', *Ocean and Coastal Management*, **84**, 153–62.

Mangi, S.C. (2013), 'The impact of offshore wind farms on marine ecosystems: a review taking an ecosystem services perspective', *Proceedings of the IEEE*, **101**(4), 999–1009.

McBreen, F., N. Askew and A. Cameron et al. (2011), *UK SeaMap 2010: Predictive Mapping of Seabed Habitats in UK Waters*, Joint Nature Conservation Committee Report No. 446.

McLaughlan, C., B. Gallardo and D.C. Aldridge (2014), 'How complete is our knowledge of the ecosystem services impacts of Europe's top 10 invasive species?', *Acta Oecologica*, **54**, 119–30.

McLeod, E., R. Salm and A. Green et al. (2009), 'Designing marine protected area networks to address the impacts of climate change', *Frontiers in Ecology and the Environment*, **7**(7), 362–70.

McMahon, J. and T. Golding (2010), *Atlantic Array. Environmental Impact Assessment Scoping Report*, Swindon, UK: Bristol Channel Zone Limited.

Millennium Ecosystem Assessment (2003), *A Framework for Assessment*, Washington, DC: Island Press.

Natural Capital Project (2015), *Integrated Valuation of Ecosystem Services and Tradeoffs (InVEST)*, accessed 26 February 2015 at http://www.naturalcapitalproject.org/InVEST.html.

Nelson, E.J., P. Kareiva and M. Ruckelshaus et al. (2013), 'Climate change's impact on key ecosystem services and the human well-being they support in the US', *Frontiers in Ecology and the Environment*, **11**(9), 483–893.

Palumbi, S., P. Sandifer and J. Allan et al. (2009), 'Managing for ocean biodiversity to sustain marine ecosystem services', *Frontiers in Ecology and the Environment*, **7**(4), 204–11.

Pendleton, L., R. Mongreul and N. Beaumont et al. (2015), 'A triage approach to improve the relevance of marine ecosystem services assessments', *Marine Ecology Progress Series*, **530**, 183–93.

Potts, T., D. Burdon and E. Jackson et al. (2013), 'Do marine protected areas deliver flows of ecosystem services to support human welfare?', *Marine Policy*, **44**, 139–48.

RegenSW and Marine Energy Matters (2012), *Bristol Channel Energy – A Balanced Technology Approach Discussion Document November 2012*, accessed 26 February 2015 at http://regensw.s3.amazonaws.com/bristol_channel_energy_balanced_technology_approach_20121127_c541010d0b3719f8.pdf.

Remoundou, K., P. Koundouri and A. Kontogianni et al. (2009), 'Valuation of natural marine ecosystems: an economic perspective', *Environmental Science and Policy*, **12**(7), 1040–52.

Ruckelshaus, M., E. McKenzie and H. Tallis et al. (2013), 'Notes from the field: lessons learned from using ecosystem service approaches to inform real-world decisions', *Ecological Economics*, **115**(C), 1–21.

Saunders, J., R. Tinch and S. Hull (2010), *Valuing the Marine Estate and UK Seas: An Ecosystem Services Framework*, report for the Crown Estate.

Schmitt, L.H.M. and C. Brugere (2013), 'Capturing ecosystem services, stakeholders' preferences and trade-offs in coastal aquaculture decisions: a Bayesian belief network application', *PloS ONE*, **8**(10), e75956.

Shucksmith, R.J. and C. Kelly (2014), 'Data collection and mapping – principles, processes and application in marine spatial planning', *Marine Policy*, **50**, 27–33.

Smart, D.E., T.A. Stojanovic and C.R. Warren (2014), 'Is EIA part of the wind power planning problem?', *Environmental Impact Assessment Review*, **49**, 13–23.

Spalding, M.D., I. Meliane and A. Milam et al. (2013), 'Protecting marine spaces: global targets and changing approaches', *Ocean Yearbook*, **27**(1), 213–48.

Stelzenmüller, V., J. Lee and E. Garnacho et al. (2010), 'Assessment of a Bayesian belief

network – GIS framework as a practical tool to support marine planning', *Marine Pollution Bulletin*, **60**(10), 1743–54.
UK National Ecosystem Assessment (2011), *The UK National Ecosystem Assessment Technical Report*. Cambridge: UNEP-WCMC.
UNESCO (2014), 'Marine spatial planning around the world', accessed 3 December 2014 at http://www.unesco-ioc-marinesp.be/msp_around_the_world.
United Nations, European Commission, Food and Agriculture Organization, International Monetary Fund, Organisation for Economic Co-operation and Development and The World Bank (2014), *System of Environmental-Economic Accounting 2012 Central Framework*, New York: United Nations.
Uusitalo, L. (2007), 'Advantages and challenges of Bayesian networks in environmental modelling', *Ecological Modelling*, **203**(3), 312–18.
Villa, F., K.J. Bagstad and B. Voigt et al. (2014), 'A methodology for adaptable and robust ecosystem services assessment', *PLoS ONE*, **9**(3), e91001.
Werner, S.R., J.P. Spurgeon and G.H. Isaksen et al. (2014), 'Rapid prioritization of marine ecosystem services and ecosystem indicators', *Marine Policy*, **50**(Part A), 178–89.

12. Understanding the impacts of ecotourism on biodiversity: a multi-scale, cumulative issue influenced by perceptions and politics*

David Newsome and Michael Hughes

12.1 INTRODUCTION AND CONTEXT

In this chapter, we explore the complex character of ecotourism as a phenomenon and how this relates to understanding the potential impacts on biodiversity. We consider the importance of biodiversity in identifying natural places as ecotourism attractions and the role ecotourism plays in the conservation of biodiversity. Ecotourism, however, is not without risks in regard to disturbing wildlife and damaging the environment. Positive and negative impacts are therefore identified but the nature of these impacts varies according to the type of ecotourism activities and how the tourism operators meet the tourists' expectations. We will take the view that ecotourism is a specific type of tourism and that, perhaps unlike many other forms of tourism, has an overall positive impact on biodiversity. This is despite that fact that there may be real and recognizable deleterious impacts occurring as a result of tourism development and activities. We are, however, of the view that in comparison to many of the existing threats to biodiversity, ecotourism has the potential for positive outcomes.

The type and severity of environmental impacts are also influenced by political and socio-economic factors that apply in the areas in which the biodiversity occurs. Protected areas, such as national parks, play a vital role in conserving biodiversity and tourism is considered an encouragement for public engagement with conservation that is regarded as central to the role of parks themselves. Protected area managers have an active role to play in understanding what ecotourism is and in promoting sustainable tourism. This is achieved according to the application of different management strategies such as the provision and maintenance of visitor facilities, controlling where tourists go and what they do and in the provision of educational programmes. The potential negative impacts of tourism on biodiversity often interact with wider landscape-level impacts. Such impacting factors that are derived from the landscape matrix include the presence of pest animals, weeds and human-related activities that compromise biodiversity conservation. Acknowledging that impact assessment can include wider environmental and social issues this chapter focuses on the biodiversity aspects of impact assessment. Accordingly, we conclude this chapter with an exploration of the implications for impact assessment in regard to the complex characteristics of ecotourism and their interaction with biodiversity. This will be achieved by examining the case of the ecotourism–conservation nexus associated with iconic species such as gorillas in Africa. Overall we consider the need to understand the positive and negative impacts of ecotourism on biodiversity at various scales and levels of socio-political complexity. In doing so we unravel the role of people's and institutional perceptions of what is regarded as an impact and the cumulative effects of a number of land uses and human activities, of which tourism is only one, on biodiversity.

12.2 WHAT IS ECOTOURISM?

The complexity of what ecotourism comprises is reflected in the number of definitions that have been developed over the years. Indeed, generating definitions for ecotourism has been something of an academic 'cottage industry', with more than 80 definitions published in recent decades (e.g., Hughes, 2013). The multitude of definitions, and associated understandings, can lead to some confusion as to what actually constitutes an ecotourism product (e.g., Newsome et al., 2005; Newsome and Rodger, 2013; Newsome, 2013). Despite the plethora of definitions, most authors have identified a set of common principles across variants of ecotourism. These principles include that ecotourism:

- is a nature-based activity;
- includes environmental education as a core component of the experience;
- adheres to sustainability principles;
- benefits the region or area in which it takes place.

Further to this, ecotourism is considered to occupy a spectrum of products and experiences based on the types of nature-based locations visited, how the experience is managed and where the emphasis lies in relation to the core principles listed above. Ecotourism is thus complex as it encompasses a range of activities from niche market to mass forms of tourism in a variety of geographic locations and scales of focus (Fennell, 2001; Fennell and Weaver, 2005; Weaver, 2005; Beaumont, 2011; Buckley,

2013; Newsome, 2013). This can lead to a range of terminology that spans a spectrum from mass to niche ecotourism (Newsome et al., 2013), soft to hard ecotourism (Orams, 2001) or 'light' to stringent ecotourism (Buckley, 2013).

Various authors have discussed the core features of ecotourism listed above (e.g., Fennell, 2001; Weaver, 2008; Krider et al., 2010; Buckley, 2013; Hughes, 2013; Newsome et al., 2013). Buckley (2013) noted that while these principles may be common across various manifestations of ecotourism, there is inconsistency in how they are understood and implemented. For example, while most definitions include references to sustainability, few clearly define what this means in practice. Thus the application of sustainability will depend on where one may be in the world and on the tourism agency and tour operator understanding of what ecotourism is. The nature-based setting as a fundamental aspect of ecotourism can also include a range of understandings depending on how nature is defined (ibid.). Thus while there is a broad consensus on the general attributes of ecotourism, the devil is in the detail of the priority of each criterion and how each is applied (Buckley, 2013). Variations in the implementation of ecotourism principles leads to categorizations of ecotourism practice that can include products that are 'lax' or 'greenwashed' through to more stringent forms (ibid., p. 13). How ecotourism is interpreted and hence put into practice has implications for the potential impacts on biodiversity.

Respective conceptualizations of the ecotourism spectrum differ slightly in emphasis suggesting that the spectrum is actually multidimensional (Buckley, 2013). For example, Weaver's (2005) mass to niche ecotourism is based on the numbers of tourists involved, the extent and character of facilities provided and the degree to which the experience is tailored to specific demands of the individual tourist. Mass ecotourism provides a generic product and experience based on facilities designed to cater for large numbers of tourists simultaneously while niche ecotourism affords a tailored experience for small numbers of tourists. In comparison, Orams's (2001) soft to hard ecotourism focuses on the type of natural area accessed in terms of remoteness and ecological integrity. Soft ecotourism occurs in easily accessible locations that are generally ecologically compromised, have visitor amenities and facilities affording modern comforts and a highly structured educational component, while hard ecotourism includes locations that are remote, ecologically pristine, and difficult to access with no or few visitor facilities. From yet another perspective, Buckley's (2013) lax to stringent ecotourism seems mainly based on the extent to which sustainable practices are effectively implemented. Lax ecotourism is essentially greenwash. That is, it purports to be an ecotourism product (perhaps for marketing purposes) but fails to effectively implement any of the core principles associated with sustainable practice. In contrast, stringent ecotourism adheres strictly to the complete set of ecotourism principles. Wherever it falls along the various spectra, ecotourism is generally focused on sites that are nature based, scenic, contain charismatic or iconic wildlife, exceptional biodiversity or unique natural phenomena. Owing to their uniqueness or diversity and endemism, ecotourism sites are also often biodiversity 'hotspots' identified in part by being rich in species, habitats or ecosystems that may also be threatened with degradation, loss and even extinction (e.g., Myers et al., 2000).

12.3 BIODIVERSITY AND ECOTOURISM

As with ecotourism, biodiversity is a concept with a range of definitions and applications. In its literal sense, biodiversity refers to the variety of life but what this means both as a concept and in practice is harder to define consistently (Maier, 2012). Biodiversity is generally conceived as variety of the biotic elements of an environment at a range of scales. Biodiversity may be considered at a local, regional or global scale and involves a variety of discrete measures that can include abundance, richness and evenness (Myers et al., 2000). For example, biodiversity can refer to species quantity and diversity (including genetic diversity) within a habitat, habitat quantity and diversity within an ecosystem and so forth (Magurran and McGill, 2010). Depending on how it is defined will determine how impact is measured and hence, assessment of the influence of activities such as ecotourism on biodiversity. Measuring and assessing biodiversity is problematic due to the difficulty of precise definition, its fluctuating nature, difficulties in comprehensively measuring all biotic components at a given scale and specificity to geographical, biological and temporal context. As such, there are no benchmarks for what is 'good biodiversity' that can be generally applied (Maier, 2012, p. 501). This means that in many cases ecotourism impacts on biodiversity can be challenging to assess.

Aside from scientific measures of biodiversity, tourist perceptions play an important role (such as ascribing a subjective value) in determining the usefulness and quality of an environment as suggested by notions such as the 'tourist gaze' (Urry, 1992). Tourism, including ecotourism, at a destination is an experiential product co-constructed by the tourist as a consumer of the experience together with the operator or tourism manager as the supplier of the experience. The various classifications of ecotourism reflect this dynamic where mass ecotourism at a highly managed location may be construed as a real or valuable nature experience by some individuals while others may only consider a niche-market, remote-area experience as true ecotourism (Rolston, 1998; Waitt et al., 2003; Hughes, 2013). Along a similar vein, a non-expert individual may view a highly degraded natural area as pristine nature despite its poor measures of biodiversity, and obtain just as satisfactory an experience and associated benefit as would an expert in a remote relatively pristine natural area with 'good biodiversity' (Hughes, 2013, 2014). That is, most tourists understand nature based on the experiential appreciation of biota that represent the broader concept of nature. For example, a tree or a wild animal is a symbolic representation of the broader idea of nature (metonymic cues). Therefore, for some people a satisfactory and valued nature experience can be achieved merely by viewing individual biota. In contrast, biodiversity is a scientific concept that in itself cannot be experienced in the same way, but is measured according to a range of rational assumptions (Maier, 2012; Hughes, 2013). This highlights a difference between conceptions of nature and how it relates to biodiversity, which are important when considering the range of ecotourism and the potential impacts it can have.

Despite the complexities, there are reported measureable relationships between ecotourism and biodiversity. For example, Siikamäki et al. (2015) found that the annual number of visits to specific national parks in Finland was positively correlated with biodiversity measures. Biodiversity was quantified according to the number of Natura 2000 habitat types and occurrence of species on the IUCN Red List within respective Finnish national parks. Further, recreation within the national parks tended to focus on what could be termed 'hot spots' with high biodiversity values. Similar to the biodiversity measure described by Siikamäki et al. (2015), Dawson et al. (2015, p. 133) established a link between biodiversity and 'last-chance tourism' based on a positive correlation between tourism demand and the threatened status of a range of natural phenomena at varying scales. They identify this phenomenon as tourism demand based on the perceived last chance to see threatened species or habitats before they disappear, or unique ecosystems and landscapes undergoing irreversible changes. This may include mass ecotourism-type experiences such as polar bear viewing in Manitoba, Canada. It may also include niche ecotourism experiences such as the 'Climate Change Challenge, Mt Kilimanjaro Mission' (p. 140).

The Tree Top Walk in the Southwest of Western Australia is a specific example of mass ecotourism discussed by Hughes (2013). This ecotourism experience is located in a recognized global biodiversity hotspot within a remnant area of degraded forest consisting of the endemic giant red tingle tree (*Eucalyptus jacksonii*). An extensive agricultural area surrounds the remnant forest. The site is accessible and has purpose-built facilities and amenities that cater for up to 200 000 visitors per year. The site is designed with a strong emphasis on environmental education and sustainable practice intended to minimize visitor impact on the forest environment. Local environmental benefits include a significant reduction in littering, trampling, soil compaction and tree damage compared to the original site configuration as well as some socio-economic benefits through employment of local residents for operations and maintenance of the site. These attributes arguably place the Tree Top Walk as an ecotourism experience that occurs on a mass tourist scale but that is managed based on the core principles of ecotourism to minimize negative impacts and promote positive impacts on the site and region.

12.4 UNDERSTANDING THE IMPACTS OF ECOTOURISM ON BIODIVERSITY

Ecotourism, as with any form of tourism, will inevitably have impacts on the locations where it occurs and these may be both negative and positive. The extent and direction of impacts are likely to vary in direct relation to where on the various spectra an ecotourism activity falls. That is, impacts can occur at different temporal and geographic scales depending on the type of ecotourism and when and where it takes place. In other words, the impacts of ecotourism are complex to measure because they are context specific (Hughes and Carlsen, 2008). For example, the case of viewing gorillas in Uganda, described in this chapter, is focused on a specific species that resides in a small geographic location while each encounter is tightly controlled in terms of duration and proximity between tourists and gorillas. Most biodiversity-related impacts in this case would be confined according to the restricted area and time spent with the gorillas. In contrast, wildflower viewing in Western Australia is focused on a wide range of species, occurs across a large geographical area with the season lasting several months. In this case impacts of tourism activity dispersed across such a wide range of parameters are likely to be more widespread geographically and temporally (Mason et al., 2015).

While specific to context, there is a range of potential negative impacts that can include disturbance to various wildlife behaviours such as foraging, nesting or resting, habituation of wildlife to humans through provisioning, desensitization of wildlife to humans and other potential threats (Hughes and Carlsen, 2008). For example, Lewis and Newsome (2003) noted that poorly regulated ecotourism activity centred on stingrays was associated with a risk of injury to the stingrays – an interesting example of trampling animals rather than vegetation! Muellner et al. (2004) and Biggs (2013) noted that an activity as apparently benign as bird watching can result in disturbance to nesting and other behaviours if occurring during sensitive breeding times or if playback calls are used to attract birds. Poorly managed ecotourism focused on turtles can lead to nest damage (Tisdell and Wilson, 2002). Adverse impacts may also occur to habitats and broader ecosystems through littering, removal or trampling of flora, soil compaction and erosion, and spreading or introduction of pathogens (such as the soil-borne water mould *phytophthora*), unless there is proper management (see, e.g., Ballantyne and Pickering, 2012; Ramchurjee, 2013; Newsome et al., 2013; Fennell, 2015; Mason et al., 2015).

Positive impacts cited in the literature include increased public environmental awareness through education, economic contributions to local areas through visitor expenditure and ecotourism enterprise and motivation for conservation of natural phenomena based on value as an ecotourism focal point. Along these lines, authors such as Siikamäki et al. (2015) and Dawson et al. (2015) argue that the positive correlations they identify between ecotourism and biodiversity measures highlight the role ecotourism can play in promoting biodiversity conservation. For example, while Dawson et al. (2015) point out the potential negative impacts of last-chance tourism in ecologically sensitive locations, they also highlight the opportunity to harness ecotourism demand to provide an incentive to support the endangered system. They note that such ecotourism affords an opportunity to raise public awareness through education, generate local economic benefits through visitor expenditure and related business and hence promote conservation of threatened species and systems. In Torres del Paine Biosphere reserve in Chile, ecotourism has been considered to potentially promote conservation of a rare orchid species (Vidal et al., 2012). Other authors also identify specific examples where development of ecotourism could serve to encourage conservation of unique biodiversity and natural phenomena (see, e.g., Munn, 1992 on macaws; Schwitzer et al., 2006 on lemurs; O'Connor et al., 2009 on whales; Anderson et al., 2011 on manta rays; Cisneros-Montemayor et al., 2013 on sharks; Newsome and Hassell, 2014 on biodiversity). The ecotourism demand for rare and threatened examples of biodiversity fits closely with the notion of biodiversity 'hotspots' as defined by Myers et al. (2000) and could thus assist with conservation of endangered species in such areas. Although ecotourism impacts can include various negative aspects, the primary difference between ecotourism (which adheres to specific core principles) and other forms of tourism and recreation lies in ecotourism's emphasis on managing impacts for the benefit of the nature-based setting in which it occurs. This chapter further details a specific case in Africa where ecotourism can function to promote biodiversity conservation as part of a broader regional management context.

12.5 AN EXAMPLE OF THE SOCIO-POLITICAL AND ECONOMIC CONTEXT OF ECOTOURISM AND ITS RELATIONSHIP WITH BIODIVERSITY

12.5.1 Psycho-social Touristic Context

A driving force in regard to the relationship, and therefore importance, of ecotourism to biodiversity is the focus that is often placed on rare, iconic, beautiful and interesting species. This is exemplified by the words of Nigel Collar (BirdLife International) who wrote: 'The immeasurable diversity of wildlife is one of the richest sources of curiosity, stimulus and emotional satisfaction we possess: we cannot value its wealth in dollars, and we cannot calculate its power in terms of utility' (Collar, 1986, p. 18).

To a certain extent, a dollar value can be assigned to the wildlife viewing experience and compared with other land uses that may not be as sustainable (e.g., logging) or remunerative (e.g., agriculture on nutrient poor soils). Collar (2003) went on to write that according to the 'present' trend, and into the future, the continued attitudes and activities of humans will result in large areas of the Earth being devoid of wild areas and that the quality of all human life will be compromised as a result of this degradation. Many concerned scientists and specialists have argued that Collar's reflections on the value of biodiversity as an expression of humans experiencing a 'freedom of the mind' can be realized through tourism. People have a deep interest in wildlife and wild places (e.g., Newsome et al., 2005, 2013) that is often expressed through a strong interest in conservation (e.g., Buckley, 2009). Such ecotourism development (as described by many, e.g., Page and Dowling, 2002; Buckley, 2009) is also important in engaging local communities, such as those that live in the vicinity of biodiversity, in conserving rather than overly exploiting their natural resources (e.g., see Newsome and Hassell, 2014).

Perhaps no greater example of the sentiment highlighted by Collar (1986) is that expressed by tourists who undertake ecotours to Africa to experience great apes (bonobos, chimpanzees and gorillas) in the wild. An indication of the emotional interest and impact of many ecotourism ventures that focus on spectacular and interesting wildlife is exemplified by tourist statements on TripAdvisor (Box 12.1).

12.5.2 The East African Afromontane Biodiversity Hotspot and Tourism

The East African Afromontane Biodiversity Hotspot comprises an arc of mountains and rift valley that extends from Saudi Arabia to Mozambique. According to BirdLife International (2012) the area contains 2350 endemic

BOX 12.1 THREE TRIPADVISOR REVIEWS OF TOURIST EXPERIENCES OF VIEWING ICONIC WILDLIFE IN THE EAST AFRICAN AFROMONTANE BIODIVERSITY HOTSPOT

'There were about 16 gorillas in our group and they ended up in a big circle around us. The silverback did the usual chest beating, we saw a couple mating, there were baby gorillas swinging in the trees and one of the mid-sized gorillas actually came up to me and played around with my leg. Very scary at the time but an experience I will never forget. This was the highlight of our trip for sure'.

(http://www.tripadvisor.com.au/ShowUserReviews-g673829-d1908641-r121232793-Bwindi_Impenetrable_National_Park-Kabale_Western_Region.html#)

'It is fascinating (and a little scary) to see how human-like the mountain gorillas are. Their brown eyes hold a great deal of expression, and their ears and fingers could easily be mistaken for humans (except for the colour). Visiting the mountain gorillas and watching them interact with each other, in their home, is a truly magical, once in a lifetime experience'.

(http://www.tripadvisor.com.au/ShowUserReviews-g673829-d1908641-r248963230-Bwindi_Impenetrable_National_Park-Kabale_Western_Region.html#)

'You have to trek them to believe them. The gorillas are a breath-taking experience; you can't really describe how magnificent they are. The guides were wonderful and looked after the whole group and never rushed anyone. It is a trek but makes it all worth it. My friend struggled a bit but made it; just be prepared for some serious heavy breathing, the altitude makes it tough to start but you soon get used to it. Seeing these gorillas should be on everyone's bucket list'.

(http://www.tripadvisor.com.au/ShowUserReviews-g673829-d1908641-r248963230-Bwindi_Impenetrable_National_Park-Kabale_Western_Region.html#)

plants and 1300 bird species, 110 of which are endemic to the region. There are also 100 endemic mammals out of a total of around 500 species. Iconic species include the great apes such as the eastern chimpanzee and Grauer's gorilla. Moreover, BirdLife International (2012) also note that only about 10 per cent of original vegetation remains intact and only 15 per cent of that total land area is officially protected. Important conservation efforts include the development of policies focusing on the co-benefits of biodiversity conservation, improving the protection and management of key areas and the sustainable financing of conservation corridors and key conservation areas.

The Albertine Rift is part of the western Rift Valley of Africa but also comprises the Rwenzori Massif and Virunga volcanoes and, because of significant endemic and endangered species, is recognized as one of the most important conservation sites in Africa (Plumptre et al., 2007). Important protected areas include the Virunga and Kahuzi Biega National Parks in the Democratic Republic of Congo and the Bwindi Impenetrable and Kibale National Parks in Uganda.

In East Africa, Uganda for example, a range of natural features and biodiversity are advertised and marketed in the tourism promotion literature and related websites for trips to protected areas. Some key Ugandan ecotourism products include open plains safaris, large herds of African wildlife, tree-climbing lions, butterflies, tropical forests, iconic mountain scenery, bird watching (especially for iconic birds such as the shoebill and Albertine endemics) and treks to see gorillas and chimpanzees in the wild (e.g., see Naturetrek, 2013; Natural World Safaris, 2016). Two key protected areas, namely Bwindi and Kibale National Parks, both of which contain significant plant and animal biodiversity, (Figures 12.1 and 12.2) are highly significant ecotourism destinations in Uganda.

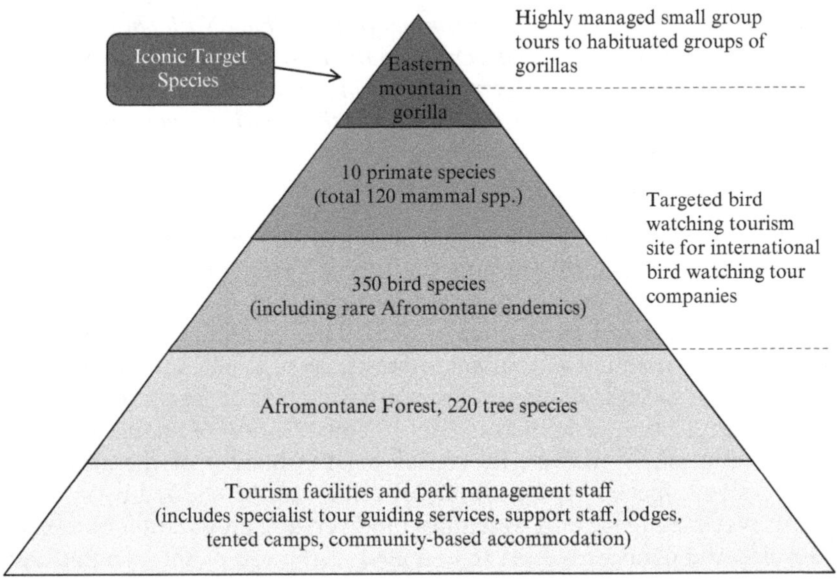

Figure 12.1 *Pyramid of ecotourism interest in Bwindi National Park, a protected area occurring in the East African Afromontane Biodiversity Hotspot. Tourism facilities form the base of the pyramid to illustrate the point that tourism can support the biodiversity on which it depends*

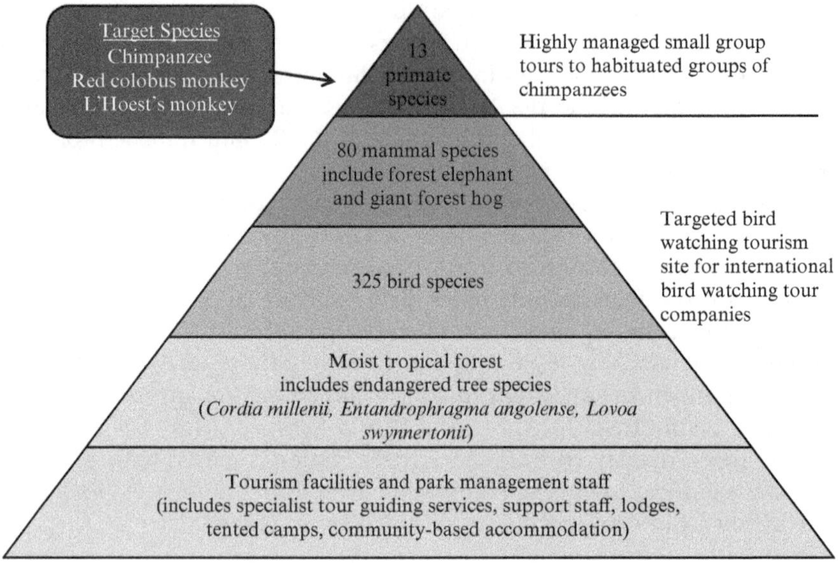

Figure 12.2 Pyramid of ecotourism interest in Kibale National Park, a protected area occurring in the East African Afromontane Biodiversity Hotspot. Tourism facilities form the base of the pyramid to illustrate the point that tourism can support the biodiversity on which it depends

Bwindi National Park is where tourists go to experience mountain forest and especially the gorilla groups that occur there (Figure 12.1). Kibale National Park is a middle altitude forest site with the main focus being on chimpanzees and birds (Figure 12.2). The conceptual diagrams, represented as a pyramid of tourism interests, serve to illustrate that tourist interest in a specific iconic species can have wider conservation implications as protection of habitat is vital to conservation of species of major tourist interest. Moreover, by conserving the habitat of the aforementioned iconic species other ecotourism values can be realized by virtue of tourism-related (e.g., bird watching) biodiversity conservation outcomes. Gorillas and chimpanzees as represented at the top of the pyramid thus comprise highly valued ecosystem components of the total biodiversity of these aspects of the East African Afromontane Biodiversity Hotspot.

In Uganda there are 11 habituated gorilla groups that can be seen by tourists. There are four habituated groups located at four different locations in Bwindi National Park. Park management (Figure 12.1) issues 32 permits daily for small-group (six to eight people) excursions to view the different groups of wild mountain gorillas. To ensure minimal impact on the gorillas, and as progress towards sustainable tourism, only a limited number of permits are issued each day. There are very strict rules of conduct in order to ensure safe viewing and minimal disturbance to the gorillas (e.g., see Bwindi National Park, n.d.). The rules focus on aspects such as one visit to a particular group a day, stays are limited to 1 hour and there is guide-controlled maintenance of safe observation distances.

The habituation programme and rules for tourist interaction are designed to minimize the negative impacts of tourism on target species. However, various workers have reported problems associated with gorilla tourism and close contact between humans with gorillas such as the risk of human-carried disease especially affecting gorilla groups (e.g., Bermejo et al., 2006; Rwego et al., 2008; Mittermeier et al., 2013). Such problems have also been reported in regard to ecotourism that is centred on chimpanzees (e.g., Johns, 1996; Goldberg et al., 2007; Humle, 2011).

Despite such potential problems there is a strong opinion that ecotourism primarily benefits great apes in Africa. For example, Lepp (2007) found that at Bigodi (a bird watching and primate site for international tourists) in Uganda, local people benefited from ecotourism in terms of income and community development (positive environmental impact). At Bwindi National Park there are lodges directly associated with the gorilla trekking activities and additional accommodation and services available for ecotourists beyond the borders of the park. Many of the 'high end' accommodation providers in Uganda pride themselves on their sustainable approaches to tourism development. Claims of sustainability relate to the design of ecotourism accommodation facilities as well as use of local produce and materials.

Tourist spend in Uganda confers, and has the capacity to increase, significant positive economic impacts (Moyini, 2012). Having recognized the positive economic impacts of wildlife-based ecotourism in Uganda in particular, some authors have reported the economic situation to be complex and value driven. Adams and Infield (2003) reported on the complex relationships between multiple stakeholders surrounding the economic benefits of gorilla tourism. Of particular interest is the work of Laudati (2010) where the role of ecotourism as a positive socio-economic benefit is explored in the context of poverty alleviation. Laudati (2010) asserts that the expectation of foreigners (ecotourists) to experience 'untouched and pristine wilderness' may impact on some local communities, for example by excluding them from natural areas, because they were traditionally (or are) dependent on natural forest products for their livelihoods. This last point provides a useful segue into the final part of this chapter where the complex interaction between ecotourism and biodiversity and the context of Environmental Impact Assessment can be considered in regard to a complex mix of socio-economic and environmental factors.

12.6 WHAT CAN BE LEARNT IN REGARD TO THE IMPACT ASSESSMENT OF ECOTOURISM ON BIODIVERSITY?

In understanding the environmental impacts of ecotourism on biodiversity it is important to consider the broader socio-political context in which this issue sits. The Eastern Afromontane Biodiversity Hotspot population of eastern gorillas has been and still is insecure due to intense pressure from armed rebels. Benz and Benz-Schwarzburg (2010) note that all countries where great apes live have had some kind of war since 1946 and that resources are damaged and lost during war due to the absence of effective controls and poor law enforcement. A significant issue has been the proliferation of firearms together with a breakdown of law and order and increase in illegal poaching and logging (Mittermeier et al., 2013).

Another important problem relating to the conservation of great apes in Africa is deforestation (e.g., Walsh et al., 2003). Mittermeier et al. (2013) recognize the rapid removal of forest outside protected areas as a significant threat. In West Africa 80 per cent of original forest cover is gone and one national park in the Ivory Coast lost more than 90 per cent of its forest cover in less than ten years. There is significant loss of forest where the great apes occur, much of it associated with human population growth (ibid.). This problem is slower in the Congo Basin than in other areas but is likely to increase with future expansion of industrial agriculture (ibid.). While forest clearing at mine sites completely destroys natural habitat this is usually quite localized and it is the associated infrastructure such as service roads that facilitate penetration into remote areas that exacerbate other detrimental factors (e.g., Figure 12.3). Poaching is a major problem in Africa and logging plus mining roads allow rapid access to areas that would otherwise require foot journeys that could take weeks. Hence, the penetration of roads for logging and mining tends to speed up the poaching problem while newly created roads also attract settlers who clear forest for subsistence farming.

There is a long history of the hunting of great apes in Africa especially if it is/has been part of cultural behaviour of particular cultural groups (ibid.). In this regard Leverington et al. (2010) reported that hunting, killing and collecting animals, logging and wood harvesting were the most significant threats to the conservation of species and protected area management effectiveness at the global scale. According to Mittermeier

Source: Photograph by Sue Littleton.

Figure 12.3 *Logging access road traversing lowland tropical rainforest Cameroon, West Africa*

et al. (2013) the western lowland gorilla population in Gabon has declined by 50 per cent from 1980 to 2000 due to a combination of poaching and deaths due to gorillas contracting the Ebola virus. The virus is a major threat to both the western lowland gorilla and human populations in Central Africa (Rizkalla et al., 2007) and while there are health programmes aimed at helping the human populations there are no such programmes for the apes. As stated previously, international tourists, as well as locals, are potential carriers of disease such as measles and influenza to gorilla populations. However, Humle and Kormos (2011) reported that the main cause of death in eastern and western populations of chimpanzees was Ebola associated with ape–human contact, probably via contact with animals killed for the bush meat trade, and because the great apes are highly tactile, the ready spread of the virus from contact with dead/diseased animals (e.g., see Rouquet et al., 2005). Fragmentation of landscapes has also been implicated in the spread of the Ebola virus in Africa (Laporta, 2014).

It is thus evident that the conservation of great apes in Africa serves as an example of how ecotourism can function by adding economic and employment values to local communities. Mittermeier et al. (2013) note that the mountain gorilla population in Uganda is increasing and this is likely to be strongly connected with the ecotourism value of these animals and their habitat. However, problems, even with the supposed benign character of ecotourism, such as the risk of human-carried disease transmission to target species, are apparent. Despite this, and compared to other pressures, effectively managed ecotourism can be a positive activity in regard to the conservation of biodiversity. Most of the great apes in Africa reside in protected areas but even in these places there can be problems of hunting and poaching for food and traditional medicines. These latter issues are regarded as significant problems where there is inadequate protection and park management. This has been identified as a problem by Leverington et al. (2010), who have reported on poor management effectiveness in many of the world's protected areas. These problems occur at various scales such as landscape-level impacts such as pollution, agricultural encroachment, invasion of pest species, illegal logging and inadequate resources to manage a particular park or other type of protected area.

At the same time the future of great ape conservation beyond existing protected area networks (which are potential ecotourism resources) due to deforestation is especially problematic where there is a combination of existing agriculture, rapid human population growth, a lack of enforcement controls on illegal logging (either due to corruption or inadequate park management effectiveness) and planned industrial agricultural projects. Newsome et al. (2013, pp. 195–9) discuss the problem of potential loss of important ecotourism resources and wildlife connectivity. There is a major problem when there are dominant government policies that emphasize agriculture, mining (e.g., Figure 12.4) and timber rather than wildlife-based ecotourism!

If there are no specific reasons for tourists to visit areas that have high biodiversity, governments will not easily be convinced to protect them especially in the absence of financial incentives. In this regard, Environmental and Social Impact Assessment especially needs to be considered in the context of places that are high in biodiversity that have no ecotourism. For example, almost a decade ago, Plumptre et al. (2007) made an important point in regard to the biodiversity of the Albertine Rift. They noted that those areas that remained unrecognized and not properly protected or designated as protected areas are lost opportunities for positive impact ecotourism development. However, this additional ecotourism potential is also dependent on constructive political views and social stability.

Note: Madagascar is one of the world's most significant biodiversity hotspots. Many plants and animals are at low population levels and existing protected areas are at risk from poor management.

Source: Photograph by David Newsome.

Figure 12.4 Photograph of currency in Madagascar reflecting dominant government policy in regard to environmental assets

12.7 AN ENVIRONMENTAL IMPACT ASSESSMENT FRAMEWORK IN REGARD TO THE ECOTOURISM–BIODIVERSITY NEXUS

Environmental Impact Assessment (EIA) has been conceptualized as a process that directs decision-making toward sustainability (Hughes and Morrison-Saunders, 2015). In this sense it shares common ground with ecotourism in having sustainable practice as a core ideal. Despite its sustainability claims, EIA is traditionally focused at the specific project level and generally emphasizes minimization of negative impacts (Gibson, 2013). Ecotourism's core principles include contributions of site and regional benefit in addition to minimizing negative impacts. Furthermore, the impact of ecotourism on biodiversity can be understood at a range of geographical and temporal scales. While these are ideals, in practice ecotourism can vary widely in application as indicated by the multidimensional spectra discussed earlier in this chapter. In line with this, Hughes and Morrison-Saunders (2015) note a trend in EIA towards consideration of impacts in the context of broader scales and provision of benefits rather than simply mitigating costs, referred to as sustainability assessment. This trend in EIA recognizes that individual activities and impacts at a specific time and place have implications at the wider scale, resulting in incremental cumulative effects (Gibson, 2013). It also recognizes that the impacts of a specific activity, such as ecotourism, is one impact among many, for example poaching, logging and mining. The recognition of system-wide regional contexts in EIA complements a need to move toward consideration of the broader context in regard to ecotourism impacts within the wider socio-political and geographical context as exemplified by great apes as an ecotourism attraction in Uganda.

The complex, nuanced character of ecotourism and biodiversity and how they relate to each other presents challenges in terms of applying EIA methods. Effective application of EIA requires a clear set of criteria and specific objectives (indicators) in order to determine the type and extent of impact of a specific activity at a given location and to identify actions to minimize any negative impacts (Dahlitz and Morrison-Saunders, 2015). EIA indicators are usually context specific, determined by the type of project being assessed, its site location and time frame. The UNWTO (2004) developed some very general concepts that tourism impact indicators should address that basically reflect the core principles of ecotourism. These include:

- reduction of negative environmental impacts;
- generation of positive impacts;
- information and environmental interpretation;
- contributions toward nature conservation.

Spencely and Bien (2013, p.409) take these into consideration, along with other international ecotourism-related criteria when outlining a set of ecotourism standards that are also quite general, perhaps due to the complex character of ecotourism and its impacts on biodiversity. These include the presence in an operation of:

- a focus on personal experiences of nature;
- interpretation and education that raises awareness of nature;
- an active contribution to biodiversity conservation;
- positive economic and social contributions to local communities;
- local community involvement;
- locally appropriate scale for tours, accommodation and attractions.

It is evident from past research cited in this chapter that the impacts of individual ecotourism operations and developments are case specific and rely on a series of conditions associated with the type of activity and when and where it takes place. With this in mind, and in consideration of the ecotourism multidimensional aspects, EIA in relation to ecotourism might consider indicators based on generally applicable criteria at the wider scale. Drawing on the evidence presented in this chapter, and the trend in EIA toward consideration of the broader scale, Table 12.1 outlines a set of 'big picture' principles that apply to ecotourism and the broader context in which it occurs. What is important here is that the extent to which ecotourism principles can be effectively implemented in a given location is reliant

Table 12.1 Big picture principles for ecotourism EIA at a given location

Scale	Category	Attributes
Broad	Regional socio-economic factors	Government natural resource policy Implementation of international conservation strategies, plans and agreements Political stability Economic stability Community well-being (e.g., HDI index)
	Ecotourism resources	Biota/biodiversity Iconic species Landscape diversity Protected area status Ecological sensitivity Remoteness and accessibility Funding for management Management capacity Management effectiveness Pressures from other land uses
	Ecotourism development	Scale of development (temporal and geographic) Sustainable accommodation design Local community engagement Locally sourced employment and suppliers Conservation-focused benefits Economic benefits to local communities Wildlife interaction plans Monitoring
Specific	Ecotourism activities associated with development	Codes of conduct Group size limits Presence of guides Education and interpretation Proactive conservation-based activities Monitoring

on the broader context in which it occurs. As a consequence this influences the character and scale of impacts that may result from ecotourism developments and activities.

As suggested by the case of ecotourism focused on great apes in Uganda, another possible approach could be consideration of impacts that may occur in the absence of ecotourism. The case study highlights the circumstance whereby the presence of ecotourism has provided the rationale for biodiversity conservation. Conserving the great apes requires conservation of the broader ecosystem in which they reside and depend on for survival. While ecotourism may pose some threat as noted in the literature, these threats should be considered relative to the potential impacts that may occur in the absence of ecotourism in this region. It is strongly evident that if there were no demand by tourists to see great apes in Uganda, the area in which they reside would be exposed to other land use pressures that are currently held in check by the higher priority given to ecotourism. Governments and land managers will prioritize land uses according to the perceived relative benefit respective land uses afford (Hughes et al., 2013). Land uses that are considered of greater importance or benefit will displace those of lesser status. If great ape-based ecotourism did not exist or did not afford significant benefits, other land uses would encroach on the area, precipitating a potentially greater threat to the apes, their habitat and the associated biodiversity. Therefore, if EIA was to be conducted on an ecotourism operation, consideration of the broader political, social and geographical context is vital to establish the relative impacts of ecotourism in terms of the alternative land uses that could occur in its absence.

12.8 CONCLUSION

Ecotourism spans a multidimensional spectrum based on attributes such as organism identity, location and accessibility, scale as well as implementation (Fennell, 2001; Weaver, 2005; Fennell and Weaver, 2005; Beaumont, 2011; Buckley, 2013; Newsome, 2013). The negative and positive impacts of ecotourism can vary depending on where on these various spectra a particular development or activity resides. While negative impacts of ecotourism have been widely reported, particularly in relation to sensitive natural areas, there are also positive impacts primarily focused on facilitating conservation of biodiversity (e.g., Schwitzer et al., 2006; Anderson et al., 2011; Cisneros-Montemayor et al., 2013; Hughes, 2013; Dawson et al., 2015; Siikamäki et al., 2015). This is based on the effective application of fundamental ecotourism principles such as minimizing negative impacts, providing local benefits, promoting biodiversity conservation and adhering to sustainability principles more generally.

Given that ecotourism is generally focused on biodiversity hotspots, and reliant on the establishment of protected areas, it may be construed that the absence of ecotourism could result in biodiversity being lost. Ecotourism thus values biodiversity in situ, whereas other pressures and land uses may generally involve extractive or destructive practices that result in fragmentation and loss of biodiversity. Accordingly, ecotourism promotes ecosystem integrity that is essential for the conservation of biodiversity. Moreover, EIA in theory should determine the most appropriate land use for a given area or location according to the environmental values that have been identified as important. Such values, however, depend on various political, regional and management contexts. As evidenced by the Ugandan great ape case, the absence of ecotourism could translate into an absence of local and regional benefits (economic and social) afforded to the community. Ecotourism benefits result in a high value being placed on the region's biodiversity and hence its conservation. It is strongly evident that the great ape-based ecotourism operations in Uganda have displaced (or held at bay) other pressures such as mining, logging, agriculture and hunting. It is highly probable that the absence of ecotourism in this context may facilitate the expansion of other land uses that pose a significantly greater threat to biodiversity values.

NOTE

* The authors would like to thank Susan Moore for early ideas in the development of this chapter and Phil Ladd for reviewing the manuscript and helping them to clarify some important points in the preparation of this chapter.

REFERENCES

Adams, W.M. and M. Infield (2003), 'Who is on the gorilla's payroll? Claims on tourist revenue from a Ugandan National Park', *World Development*, **31**(1), 177–90.

Anderson, R.C., A.M. Shiham and A.M. Kitchen-Wheeler et al. (2011), 'Extent and economic value of manta ray watching in Maldives', *Journal of Tourism in Marine Environments*, **7**(1), 15–27.

Ballantyne, M. and C. Pickering (2012), 'Ecotourism as a threatening process for wild orchids', *Journal of Ecotourism*, **11**(1), 34–47.

Beaumont, N. (2011), 'The third criterion of ecotourism: are ecotourists more concerned about sustainability than other tourists?', *Journal of Ecotourism*, **10**(2), 135–48.

Benz, S. and J. Benz-Schwarzburg (2010), 'Great apes and new wars', *Civil Wars*, **12**(4), 395–430.

Bermejo, M., J.D. Rodríguez-Teijeiro and G. Illera et al. (2006), 'Ebola outbreak killed 5000 gorillas', *Science*, **314**(5805), 1564.
Biggs, D. (2013), 'Birding, sustainability and ecotourism', in R. Ballantyne and J. Packer (eds), *International Handbook on Ecotourism*, Cheltenham, UK and Northampton, MA, USA: Edward Elgar Publishing, pp. 394–403.
BirdLife International (2012), *Eastern Afromontane Biodiversity Hotspot*, accessed July 2012 at http://www.cepf.net/Documents/Eastern_Afromontane_Ecosystem_Profile_Final.pdf.
Buckley, R. (ed.) (2009), *Ecotourism: Principles and Practices*, Wallingford, UK: CABI Publishing.
Buckley, R. (2013), 'Defining ecotourism: consensus on core, disagreement on detail', in R. Ballantyne and J. Packer (eds), *International Handbook on Ecotourism*, Cheltenham, UK and Northampton, MA, USA: Edward Elgar Publishing, pp. 9–14.
Bwindi National Park (n.d.), 'Uganda safaris', accessed 29 January 2016 at http://www.bwindiforestnationalpark.com/uganda-safaris.html.
Cisneros-Montemayor, A.M., M. Barnes-Mauthe and D.A. Abdulrazzak et al. (2013), 'Global economic value of shark ecotourism: implications for conservation', *Oryx*, **47**(3), 381–8.
Collar, N. (1986), 'Species are a measure of man's freedom: reflections after writing a Red Data book on African birds', *Oryx*, **20**(1), 15–19.
Collar, N. (2003), 'Beyond value: biodiversity and the freedom on the mind', *Global Ecology and Biogeography*, **12**(4), 265–9.
Dahlitz, V. and A. Morrison-Saunders (2015), 'Assessing the utility of environmental factors and objectives in environmental impact assessment practice: Western Australian insights', *Impact Assessment and Project Appraisal*, **33**(2), 142–7.
Dawson, J., R.H. Lemelin and E. Stewart et al. (2015), 'Last chance tourism: a race to be last?', in M. Hughes, D. Weaver and C. Pforr (eds), *The Practice of Sustainable Tourism: Resolving the Paradox*, London: Routledge, pp. 133–45.
Fennell, D. (2001), 'A content analysis of ecotourism definitions', *Current Issues in Tourism*, **4**(5), 403–21.
Fennell, D. (2015), *Ecotourism*, 4th edition, London: Routledge.
Fennell, D. and D. Weaver (2005), 'The ecotourism concept and tourism–conservation symbiosis', *Journal of Sustainable Tourism*, **13**(4), 373–90.
Gibson, R. (2013), 'Avoiding sustainability trade-offs in environmental assessment', *Impact Assessment and Project Appraisal*, **31**(1), 2–12.
Goldberg, T.L., T.R. Gillespie and T.R. Rwego et al. (2007), 'Patterns of gastrointestinal bacterial exchange between chimpanzees and humans involved in research and tourism in western Uganda', *Biological Conservation*, **135**(4), 511–17.
Hughes, M. (2013), 'Ecotourists and views of nature', in R. Ballantyne and J. Packer (eds), *International Handbook on Ecotourism*, Cheltenham, UK and Northampton, MA, USA: Edward Elgar Publishing, pp. 108–16.
Hughes, M. (2014), 'Researching the links between parklands and health', in C. Voigt and C. Pforr (eds), *Wellness Tourism: A Destination Perspective*, London: Routledge, pp. 147–60.
Hughes, M. and J. Carlsen (2008), 'Human–wildlife interaction guidelines in Western Australia', *Journal of Ecotourism*, **7**(2/3), 142–54.
Hughes, M. and A. Morrison-Saunders (2015), 'Promoting interdisciplinary sustainable tourism', in M. Hughes, D. Weaver and C. Pforr (eds), *The Practice of Sustainable Tourism: Resolving the Paradox*, London: Routledge, pp. 133–45.
Hughes, M., M. Tye and R. Jones (2013), 'Whose land is it anyway? Contesting urban fringe nature-based tourism and recreation in Western Australia', *Tourism Recreation Research*, **38**(1), 29–42.
Humle, T. (2011), 'The 2003 epidemic of a flu-like respiratory disease at Bossou', in T. Matsuzawa, T. Humle and Y. Sugiyama (eds), *The Chimpanzees of Bossou and Nimba*, Tokyo: Springer, pp. 325–33.

Humle, T. and R. Kormos (2011), 'Chimpanzees in Guinea and West Africa', in T. Matsuzawa, T. Humle and Y. Sugiyama (eds), *The Chimpanzees of Bossou and Nimba*, Tokyo: Springer, pp. 393–401.
Johns, B.G. (1996), 'Responses of chimpanzees to habituation and tourism in the Kibale Forest, Uganda', *Biological Conservation*, **78**(3), 257–62.
Krider, R.E., A. Arguello and C. Campbell et al. (2010), 'Trait and image interaction: in ecotourism preference', *Annals of Tourism Research*, **37**(3), 779–801.
Laporta, G.Z. (2014), 'Landscape fragmentation and Ebola outbreaks', *Memórias do Instituto Oswaldo Cru*, **109**(8), 1088.
Laudati, A. (2010), 'Ecotourism: the modern predator? Implications of gorilla tourism on local livelihoods in Bwindi Impenetrable National Park, Uganda', *Environment and Planning D* [abstract], **28**(4), 726–43.
Lepp, A. (2007), 'Residents' attitudes towards tourism in Bigodi village, Uganda', *Tourism Management*, **28**, 876–85.
Leverington, F., K.L. Costa and H. Pavese et al. (2010), 'A global analysis of protected area management effectiveness', *Environmental Management*, **46**(5), 685–98.
Lewis, A. and D. Newsome (2003), 'Planning for stingray tourism at Hamelin Bay, Western Australia: the importance of stakeholder perspectives', *International Journal of Tourism Research*, **5**(5), 331–46.
Magurran, A.E. and B.J. McGill (2010), *Biological Diversity: Frontiers in Measurement and Assessment*, New York: Oxford University Press.
Maier, D. (2012), *What's So Good About Biodiversity? A Call for Better Reasoning About Nature's Value*, New York: Springer Science and Business Media.
Mason, S., D. Newsome and S. Moore et al. (2015), 'Recreational trampling negatively impacts vegetation structure of a global biodiversity hotspot', *Biodiversity and Conservation*, **24**(11), 2685–707.
Mittermeier, R.A., A.B. Rylands and D.E. Wilson (2013), *Handbook of the Mammals of the World. Vol. 3. Primates*, Barcelona: Lynx Edicions.
Moyini, Y. (2012), *Analysis of the Economic Significance of Gorilla Tourism in Uganda. International Gorilla Conservation Programme*, accessed 22 January 2016 at http://www.igcp.org/wp-content/themes/igcp/docs/pdf/MoyiniUganda.pdf.
Muellner, A., K. Linsenmair and M. Wikelski (2004), 'Exposure to ecotourism reduces survival and affects stress response in hoatzin chicks *(Opisthocomus hoazin)*', *Biological Conservation*, **118**(4), 549–58.
Munn, C.A. (1992), 'Macaw biology and ecotourism, or when a bird in the bush is worth two in the hand', in S.R. Beissinger and N.F.R. Snyder (eds), *New World Parrots in Crisis*, Washington, DC: Smithsonian Institution Press, pp. 47–72.
Myers, N., R.A. Mittermeier and C.G. Mittermeier et al. (2000), 'Biodiversity hotspots for conservation priorities', *Nature*, **403**(6772), 853–8.
Natural World Safaris (2016), 'Africa & Indian Ocean', accessed 29 January 2016 at http://www.naturalworldsafaris.com/destinations/africa-and-the-indian-ocean.aspx.
Naturetrek Wildlife Holidays Worldwide (2013), 'Uganda – shoebills to gorillas'; 'Uganda mammals and mountains', accessed 29 January 2016 at https://www.naturetrek.co.uk/website/searchpage.aspx?max=154207&country=UGA,&region=AF.
Newsome, D. (2013), 'An ecotourist's recent experience in Sri Lanka', *Journal of Ecotourism*, **12**(3), 210–20.
Newsome, D. and S. Hassell (2014), 'Tourism and conservation in Madagascar: the importance of Andasibe National Park', *Koedoe*, **56**(2), 1–8.
Newsome, D. and K. Rodger (2013), 'Wildlife tourism', in A. Holden and D. Fennell (eds), *A Handbook of Tourism and the Environment*, London and New York: Routledge, pp. 345–58.
Newsome, D., R. Dowling and S. Moore (2005), *Wildlife Tourism*, Clevedon, UK: Channel View Publications.
Newsome, D., S. Moore and R. Dowling (2013), *Natural Area Tourism: Ecology, Impacts and Management*, 2nd edition, Clevedon, UK: Channel View Publications.

O'Connor, S., R. Campbell and H. Cortez et al. (2009), *Whale Watching Worldwide: Tourism Numbers, Expenditures and Expanding Economic Benefits*, Yarmouth MA: International Fund for Animal Welfare (IFAW).
Orams, M. (2001), 'Types of ecotourism', in D. Weaver (ed.), *The Encyclopedia of Ecotourism*, Wallingford, UK: CABI, pp. 23–36.
Page, S. and R. Dowling (2002), *Ecotourism*, London: Pearson Education.
Plumptre, A.J., T.R. Davenport and M. Behangana et al. (2007), 'The biodiversity of the Albertine Rift', *Biological Conservation*, **134**(2), 178–94.
Ramchurjee, N. (2013), 'Impacts of ecotourism to Rajiv Gandhi National Park (Nagarhole), Karnataka', *Environmental Development and Sustainability*, **15**(6), 1517–25.
Rizkalla, C., F. Blanco-Silva and S. Gruver (2007), 'Modeling the impact of Ebola and bushmeat hunting on Western Lowland Gorillas', *EcoHealth*, **4**(2), 151–5.
Rolston, H. (1998), 'Aesthetic experience in forests', *Journal of Aesthetics and Art Criticism*, **56**(2), 157–66.
Rouquet, P., J.M. Froment and M. Bermejo et al. (2005), 'Wild animal mortality monitoring and human Ebola outbreaks, Gabon and Republic of Congo, 2001–2003', *Emerging Infectious Diseases*, **11**(2), 283–90.
Rwego, I.B., G. Isabirye-Basuta and T.R. Gillespie et al. (2008), 'Gastrointestinal bacterial transmission among humans, mountain gorillas, and livestock in Bwindi Impenetrable National Park, Uganda', *Conservation Biology*, **22**(6), 1600–607.
Schwitzer, C., N. Schwitzer and G.H. Randriatahina et al. (2006), '"Programme Sahamalaza": New perspectives for the in situ and ex situ study and conservation of the blue-eyed black lemur (*Eulemur macaco flavifrons*) in a fragmented habitat', in *Proceedings of the German-Malagasy Research Cooperation in Life and Earth Sciences*, pp. 135–49.
Siikamäki, P., K. Kangas and A. Paasivaara et al. (2015), 'Biodiversity attracts visitors to national parks', *Biodiversity and Conservation*, **24**(10), 2521–35.
Spencely, A. and A. Bien (2013), 'Ecotourism standards: international accreditation and local certification', in R. Ballantyne and J. Packer (eds), *International Handbook on Ecotourism*, Cheltenham, UK and Northampton, MA, USA: Edward Elgar Publishing, pp. 404–17.
Urry, J. (1992), 'The tourist gaze and the "environment"', *Theory Culture and Society*, **9**(3), 1–26.
Tisdell, C. and C. Wilson (2002), 'Ecotourism for the survival of sea turtles and other wildlife', *Biodiversity and Conservation*, **11**(9), 1521–38.
United Nations World Tourism Organization (UNWTO) (2004), *The Guidebook on Indicators of Sustainable Development for Tourism Destinations*, Madrid: UNWTO.
Vidal, O.J., C. San Martin and S. Mardones et al. (2012), 'The orchids of Torres del Paine Biosphere Reserve: the need for species monitoring and ecotourism planning for biodiversity conservation', *Gayana Botánica*, **69**(1), 136–46.
Waitt, G., R. Lane and L. Head (2003), 'The boundaries of nature tourism', *Annals of Tourism Research*, **30**(3), 523–45.
Walsh, P.D., K.A. Abernethy and M. Bermejo et al. (2003), 'Catastrophic ape decline in western equatorial Africa', *Nature*, **422**(6932), 611–14.
Weaver, D. (2005), 'Mass and urban ecotourism: new manifestations of an old concept', *Tourism Recreation Research*, **30**(1), 19–26.
Weaver, D. (2008), *Ecotourism*, Milton, QLD: John Wiley and Sons, Ltd.

13. Exploring the trade-offs between wind energy and biodiversity conservation*

Lea Bulling and Johann Köppel

13.1 INTRODUCTION

Environmental policies designed to trigger renewable energy and those to protect biodiversity are not necessarily aligned, creating a dilemma as far as different ecosystem services are concerned. Climate change and particularly global warming affect biodiversity and can lead to a higher extinction risk (Bellard et al., 2012), compromising goals of biodiversity conservation. Renewable energy productions emit smaller amounts of greenhouse gases compared to carbon-based energy generation and are therefore an adequate solution to help slow down global warming (Arent et al., 2011). Mitigation measures for climate change are promoted through the transformation of energy systems (towards renewables, i.e., 'Energiewende'), with ambitious plans of forerunners such as Germany and the USA. The development of wind and solar energy (Geissler et al., 2013) and also the development of hydroenergy within the past few years have contributed to more sustainable energy production. Simultaneously, they have led to changes in landscape and ecosystems due to new forms of land use causing environmental impacts; that is, effects on biodiversity and wildlife (Drewitt and Langston, 2006; Kuvlesky et al., 2007). In addition, climate mitigation efforts can trigger renewable energy development but face challenges from biodiversity protection instruments and practices.

Even though many questions on ecosystem functions and potential trade-offs remain – for example, temporal and spatial effects of changes (De Groot et al., 2010) – the ecological scale of effects (from global to local) needs to be considered in decision-making processes and biodiversity trade-offs have to be balanced (Allison et al., 2014). Management is possible in multiple planning and governance levels and in project development (Geissler et al., 2013; Köppel et al., 2014). Essentially, planning and siting of renewable energy is guided by the precautionary principle in an attempt to carefully address biodiversity challenges. Yet this impact assessment attitude creates limitations, struggling to negotiate the aforementioned trade-offs.

In this chapter, an overview over current conflicts of biodiversity and energy generation is given for species groups mostly affected by development (e.g., birds and selected mammals). This is followed by potential solutions and exemplary case studies showing good practice in order to show a more effective future in wind energy development with wildlife considerations.

13.2 WIND ENERGY'S EFFECTS ON BIODIVERSITY

Trade-offs between renewable energies and biodiversity-focused ecosystem functions and services are complex and produce positive as well as negative effects (Katzner et al., 2013; Allison et al., 2014). Therefore, impact assessments are essential for understanding and responding to those effects (Schuster et al., 2015).

Within the past decades a variety of research has been done on solar energy (e.g., Tsoutsos et al., 2005; Lovich and Ennen, 2011; Stoms et al., 2013; Hernandez et al., 2014) but focus on the assessment of renewable energy trade-offs has been on wind energy and potential impacts on wildlife. The well-being of bats, birds and marine mammals has been mostly examined. Within the last couple of years, research focusing on offshore wind energy increased substantially, catching up on the vast amount of onshore studies particularly within Europe (Schuster et al., 2015). Research on impacts on terrestrial mammals (e.g., Colman et al., 2012; Helldin et al., 2012), tortoises (Lovich and Ennen, 2013), fish and benthos (e.g., Wilhelmsson et al., 2006; Vaissière et al., 2014) have only been marginally conducted; however, it will not be addressed here.

The trade-offs can be classified into direct effects on the species and indirect effects (Kunz et al., 2007; National Wind Coordinating Collaborative (NWCC), 2010; Katzner et al., 2013). Direct effects cover all collisions with the facility, for example, rotor blades or turbine tower. Indirect effects consist of loss of habitat, habitat fragmentation, and displacement of individuals by the construction, facility operation, and/or decommissioning of wind farms. The effects are distributed differently among animal groups and species (Schuster et al., 2015).

13.2.1 Birds

Collisions leading to mortality can occur on the tower, at the nacelle, or with the rotor blades with the collision risk dependent on flight height and flight behavior aspects. Some birds show avoidance behavior around wind turbines and consequently are exposed to a lower risk of colliding with the wind farm (May, 2015). Collision risk is essentially influenced by bird species, abundance, behavior (depending on sex, age and season), weather, topography, wind farm layout, and lighting (Drewitt and Langston, 2006; Marques et al., 2014; Schuster et al., 2015).

Due to minimal space consumption of wind turbines and its associated infrastructure, they do not cause extensive loss of habitat, yet can cause indirect habitat loss during construction and operation by disturbance or displacement of birds. For instance, during construction, species such as curlews (*Numenius arquata*) were presumably displaced by construction activity within the wind farm area and relocated to breed at nearby sites (Pearce-Higgins et al., 2012). But other species such as the stonechat (*Saxicola rubicola*) or skylark (*Alauda arvensis*) were able to benefit temporarily from the construction due to an enhancement of habitat quality by vegetation disturbances (ibid.). Yet during operations, the majority of wintering farmland birds (Devereux et al., 2008) and raptors (Madders and Whitfield, 2006) most likely do not suffer from displacement by wind turbines. For some species, displacement can lead to a disturbance during foraging (Rees, 2012), breeding, and other comfort behavior (Steinborn et al., 2011).

Impacts on birds caused by wind farms are highly species specific and therefore a differentiated consideration of impacts based on behavior is necessary. Impacts on resting and wintering birds, farmland-breeding birds, raptors, and migrating birds are analyzed further.

Resting birds are perceived as more sensitive to wind farms than breeding birds, by showing greater avoidance behavior near turbines (Hötker et al., 2005; Reichenbach and Steinborn, 2006). In addition, habitat quality appears to influence the extent of effects, with wind turbines in natural habitats causing higher negative impacts (Larsen and Madsen, 2000; Pearce-Higgins et al., 2009; Steinborn et al., 2011). For both breeding and resting birds, the effect of displacement is better explored, but impacts by collisions are not to be neglected as some species or in general, local common species (e.g., gulls), show behavior that causes high collision risks (Grünkorn et al., 2009; Zimmerling et al., 2013). Further studies on collision rates of breeding and resting birds are needed as the potential effects emphasize the importance of informed turbine siting for this species group.

Raptors are particularly susceptible to collisions with wind turbines (Madders and Whitfield, 2006; Lucas et al., 2008; Dahl et al., 2013; Pagel et al., 2013; Marques et al., 2014) and in order to understand effects of wind energy development on them it is particularly important to consider site characteristics like topography and wind currents (Barrios and Rodríguez, 2004; Hoover and Morrison, 2005), as well as nearby breeding and foraging grounds (Dürr, 2009; Camiña, 2011). The selection of wind farm locations including the siting of single turbines, as well as their design, seem to influence flight response and therefore fatality rates (Smallwood and Thelander, 2004). Nonetheless, to some extent it remains debatable whether raptor abundance and avoidance response can be used to significantly predict collision risk (Schuster et al., 2015). Even though raptors are likely aware of the turbine, they might be inattentive during searching flights (Smallwood and Thelander, 2004; Dahl et al., 2013) or their season-specific interaction behavior (Barrios and Rodríguez, 2004; Hull and Muir, 2013), increasing the collision risk. Moreover, while mortality rates from collision are comparatively well documented for raptors at numerous wind farms, additional mortality including indirect mortality from increased adult mortality, reproductive skipping, or displacement, remains unknown (Schuster et al., 2015). In this context, long-term effects including potential habituation need to be further observed.

Assessing the effects on migratory birds remains particularly challenging as our knowledge on migration patterns, especially over the sea, is still limited as well as our technical capabilities to monitor and quantify collisions (ibid.). Yet considerable progress has been made during the last years within this field (e.g., Hill et al., 2014). So far, at-sea collision events have only been observed at other offshore structures (e.g., research platforms, gas and oil rigs), which differ in structure and lighting; thus the transferability of the findings to offshore wind farms remains debatable (Hüppop et al., 2006). It is known that weather conditions not only at sea but also at departure points strongly influence migration intensity and timing as well as avoidance responses that seem relatively consolidated for most species in favorable weather conditions (ibid.; Plonczkier and Simms, 2012). Times of high migration activity and therefore high potential collision risk occur on only a few occasions during the year, especially during nighttime, and peaking in spring and autumn. Birds decrease their flight altitude under adverse weather conditions and in addition are attracted to turbine lighting, thus exposing them to a higher risk of collision (Hüppop et al., 2006; Aumüller et al., 2011).

13.2.2 Bats

The impact assessment of wind farms on bats has revealed that bats primarily suffer from wind turbine operation-caused fatality. Several studies (Kunz et al., 2007; Arnett et al., 2008; Doty and Martin, 2012) have observed fatalities on an alarming scale; however, they are highly variable depending on the site and period (Arnett et al., 2008). The number of bat fatalities can even exceed the number of bird fatalities (Doty and

Martin, 2012). The question of why bats collide is explained by the high rotation speed of the blade tips leaving an insufficient amount of time for bats to react and alter their flight direction. However, other hypotheses regarding bat–turbine interaction remain debatable. The reasons why some bat species are attracted to operating wind turbines, causing them to fly close to moving blades, are not yet fully understood. Methodological approaches for impact assessment on bats at wind farms are complex and can lead to an under-representation of fatalities if not executed correctly and should include parameters like searcher efficiency or scavenger removal (Arnett et al., 2008).

For bats, collision risk with the turbine increases with an increase in activity (Arnett et al., 2008). Activity is especially high at night (as they are nocturnal) and correlates with meteorological factors such as wind speed, temperature and precipitation (Arnett et al., 2008; Behr et al., 2011). Most fatalities occur at nights with low wind speeds. Furthermore, collision risk increases during migration in late summer and fall (ibid.) and during hunting activities (Arnett et al., 2008; Rydell et al., 2010; Brinkmann et al., 2011). Additional to local species, some authors (Voigt et al., 2012; Baerwald et al., 2014) point out the impact on migrating species and far-reaching ecological and population impacts (Voigt et al., 2015). Moreover, including forested areas for wind energy development, as recently done in Germany, and increasing turbine height due to repowering might lead to an alteration of the described potential effects and needs to be further investigated.

13.2.3 Marine Mammals

Noise caused by the construction and operation of offshore wind farms can affect marine wildlife. While different techniques are used for the installation of wind turbines offshore, the use of monopiles has become most utilized. To install monopiles, they are driven into the ground, which causes underwater noise (Danish Energy Agency, 2013). Observing potential effects on marine mammals revealed that underwater noise emission during piling activities requires substantial mitigation efforts. Noise effects from operating wind turbines have been considered unlikely so far (Madsen et al., 2006). These effects, being highly variable dependent on numerous parameters, are challenging to assess as animal displacement and sound levels alone account for commonly used measurable parameters. Studies (e.g., Nedwell et al., 2007; Lucke et al., 2009) on species-specific temporary threshold shift (TTS) and permanent threshold shift (PTS), accounting for real-life conditions (e.g., background noise), are therefore indispensable to assessing potential effects of underwater noise emission on marine mammals. Species particularly impacted by those noises are those that show high acoustic sensitivity such as seals (*Halichoerus grypus*, *Phoca vitulina*) and cetaceans (Gordon et al., 2007). Currently, however, most offshore wind farms are located in relatively shallow waters and therefore most assessments of impacts have been done for inshore species such as harbor porpoises (*Phocoena phocoena*) and bottlenose dolphins (*Tursiops truncatus*).

Over time, even temporal effects such as the interruption of species behavior could have an effect on population level, which is virtually impossible to quantify. A potential long-term positive effect (that, of course, needs better understanding) is the exclusion of fisheries in the area, bringing an increase in prey availability inside the wind farm and thus possibly serving as refuge for marine mammals and ascertaining a habitat gain (Bergström et al., 2014).

13.2.4 Remaining Challenges in Assessing the Impacts

Currently it is not possible to quantify long-term effects on population levels. The process of decommissioning and its potential impacts caused by the deconstruction and removal of piles on marine mammals is unknown, yet some authors (Diederichs et al., 2008) are apprehensive about the variety of negative effects. Assessing cumulative and long-term effects will become even more crucial in the future with further wind energy development, posing a great challenge to Environmental Impact Assessment (Schuster et al., 2015). The magnitude of cumulative effects further depends on the remaining habitat availability and quality outside the impacted area and we must be aware that cumulative effects do not only result from the collectivity of several wind farms. The negative trade-offs for biodiversity caused by wind energy production also need to be assessed in relation to trade-offs caused by other anthropogenic activities like, for instance, other forms of energy production (e.g., Erickson et al., 2005; Sovacool, 2013).

In order to assess the significance of potential impacts and the effectiveness of mitigation measures, it is important to increase our understanding of population sizes and fatality rates in general (Voigt et al., 2015). Still, this might not be possible in every case due to, for example, inter-annual and seasonal population dynamics or migratory behavior. Pre-construction observations might not correlate with post-construction conditions due to the introduction of the turbine structures, which possibly could result in an alteration of species abundance and composition. Additionally, an unknown 'dark figure' of wind farm-induced mortality poses a great challenge to mortality estimations and therefore impact assessment involving mortality rates from off-site deaths, indirect mortality due to reproductive skipping, decreased fitness, or a decrease in habitat quality, all of which seem virtually impossible to determine. Yet through growing experience in impact assessment for wind farms, several issues have been reduced and improvements of fatality rate estimations through elaborated research in the field of scavenger removal (e.g., by Hull and Muir, 2010; Smallwood et al., 2010, 2013; Paula et al., 2011; Huso and Erickson, 2013; Mathews et al., 2013; Bastos et al., 2013; Huso et al., 2014) have been possible and innovative approaches like the use of detection dogs for carcass searches (Paula et al., 2011; Mathews et al., 2013) have been developed and tested.

Thoroughly assessing the potential effects of renewable energy systems and substantially pursuing the mitigation sequence (avoid, minimize, reduce & compensate) provide the potential to limit unintended trade-offs – when implemented successfully. This includes coping with local impacts in order to limit cumulative effects on a landscape scale as well.

13.3 AVOIDANCE: LIMITING IMPACTS THROUGH SITING AND ENVIRONMENTAL PLANNING

Based on the gained knowledge from impact assessment and about the behavior of species, a wide variety of avoidance and minimization measures have been developed at all planning and project levels. Comprehensive avoidance of effects on particular populations or individuals can best be achieved by well-informed designation of wind energy areas. Pre-construction mitigation measures (May et al., 2015) such as macro-avoidance strategies, when empowered by impact assessments, are meant to circumnavigate biodiversity hotspots and protected sites. In terms of spatial planning, macro-siting consists of siting in areas with low ecological spatial resistance like agricultural lands or other pre-stressed areas, consequently avoiding intact habitats (Arizona Game and Fish Department, 2008) or even protected areas (Reichenbach and Steinborn, 2006). Not only can bundling with other infrastructure projects or existing wind turbines lead to a minimization of negative impacts, but also selected landscape elements that tend to attract sensitive species can be avoided to limit collision risk – for example, linear structures like rivers, hedges or forest edges for bats (Baerwald and Barclay, 2009; Kelm et al., 2014).

If detailed information about the presence of wind energy-sensitive species is available, siting decisions are more effective via better incorporation of distances to valuable habitats (Drewitt and Langston, 2006; Rodrigues et al., 2006; Piorkowski and O'Connell, 2010; Arnett and Baerwald, 2013), for example, nesting and roosting sites or flight corridors. Some approaches suggest completely avoiding siting wind farms within potential high-risk areas for sensitive species (Obermeyer et al., 2011; Jakle, 2012; Länderarbeitsgemeinschaft der Staatlichen Vogelschutzwarten, 2015). Commonly suitable instruments for implementation are predetermined buffer zones (e.g., Obermeyer et al., 2011; Landesamt für Umwelt, Gesundheit und Verbraucherschutz, 2012; Länderarbeitsgemeinschaft der Staatlichen Vogelschutzwarten, 2015), which are placed around protected areas and bird nesting sites that are to be excluded from further planning. Determining species-specific buffer zones requires essential knowledge about the species and its behavior (e.g., home range) and furthermore, applying buffer zones requires data on the location of nesting sites. Yet this highly preventive approach has crucial weaknesses as it is mostly applicable at a local planning scale or in permitting processes (Köppel et al., 2014) due to lack of nesting site data at larger scales. This approach is less useful in tiered planning systems (e.g., Germany), which aim to eliminate conflicts early. Furthermore, it can lead to an extensive exclusion of suitable sites, resulting in a limited availability of potential wind energy areas (Nagel et al., 2014). Depending on the home range of a species, buffer zones can stretch out up to a radius of 10 km surrounding the nest site of, for instance, black storks (*Ciconia nigra*) (Länderarbeitsgemeinschaft der Staatlichen Vogelschutzwarten, 2015). Particular species that move between nests cannot be considered for buffer zones (Nagel et al., 2014). Finally, planning strictly in accordance with buffer zones does not completely eliminate the possibility of fatality or disturbance, only leaving the planner with a residual risk.

A rather new approach of dealing with mitigating negative impacts in planning and siting of wind farms on a wider scale is the use of local zoning plans. They can be helpful tools with which to make informed macro-siting decisions. Local zoning plans can be developed for one individual species and can categorize an area into several zones reaching from high risk to low/no risk. The several forms of land use within habitats of local populations place high pressure on many species; therefore an integrated approach appears to be suitable.

The definition of high-density needs to be adapted based on species- and regional-specific factors, as also seen in the case study (Box 13.1). High density can have different minimum densities depending on the area, the commonness of a species, and the assigned responsibility for that species.

Both presented approaches of macro-siting by excluding high-density areas and nesting sites have advantages and drawbacks. If species

BOX 13.1 MACRO-AVOIDANCE TOOL FOR WESTERN CAPERCAILLIE, BLACK FOREST (GERMANY)

Currently there are around 600 western capercaillie (*Tetrao urogallus*) individuals in the Black Forest in an area of 50 000 ha. Survivability of the capercaillie population is only possible with a minimum long-term population size of 500 individuals. Consequently, all measures introduced in the capercaillie Action Plan should be implemented, including a wind energy zoning map (Forstliche Versuchs- und Forschungsanstalt, 2013). An assessment of impacts was made for all potential wind energy sites (areas with a minimum wind speed of 5.25 m/s 100 m above ground), classed into four categories (see Table 13.1) and displayed on a zoning map (Forstliche Versuchs- und Forschungsanstalt and ForstBW, 2012; Forstliche Versuchs- und Forschungsanstalt, 2013). The definition of valuable habitats was based on population size, breeding, and nursery areas collected with relevant stakeholders. In order to guarantee genetic exchange between sub-populations, stepping stones were identified and displayed as well, according to the categories shown in Table 13.1. The maps operate as the guidance document for wind energy planning without forming a binding legal framework. For the planning processes in this region, further research for capercaillie is not necessary, which can speed up planning processes and therefore pose an adequate solution for both wind farm developers and nature conservation.

Table 13.1 Categories used in the wind energy zoning map and evaluation tool for western capercaillie

	Definition	Implication for planning
1	Main habitats Breeding sites or courtship display on those sites or in close proximity Stepping stones between two essential habitats (if there is only one connection) Minimum size of 1 km	Permit not possible (exclusion from wind energy)
2	Habitats Populated areas with medium habitat potential Non-populated/formerly populated areas with high habitat potential Stepping stones and corridors	Very problematic (Cat. 2) and less problematic (Cat. 3) Species protection assessment is necessary Implementation of mitigation measures (micro-siting, minimization or measures for continuous ecological functionality)
3	Minor habitats Currently or potentially	

Table 13.1 (continued)

	Definition	Implication for planning
	marginally used by capercaillie (low or medium habitat potential) Stepping stones of secondary priority	
4	Other Currently and most possibly in the future not used as habitats Areas not relevant as corridors	No restrictions for wind energy

Source: Based on Forstliche Versuchs- und Forschungsanstalt (2013).

protection should be included early on when siting wind farms, the approach of only considering nest sites is very limited. Changes in nest sites are likely for some species such as the red kite (*Milvus milvus*) (Scheller et al., 2013) and cannot be considered in the approach of using buffer zones around nest sites. Excluding high-density areas, on the other hand, allows species to move their nests within the high-density area without losing protection. Particularly for long-term protection of species, this approach is suitable. This approach is useful for species with a limited distribution area, whereas for common nationwide species this approach is less helpful because it does not balance the trade-offs (biodiversity vs loss of earnings) well, leading to an extremely high exclusion zone as soon as high-risk exclusion zones of several species are combined. Consequently, not all conflicts can be eliminated during the siting process but rather have to be minimized with other mitigation measures.

In the development stage of site-specific project planning, there are also several opportunities to limit negative trade-offs. Precise turbine siting and wind farm layout, defined as micro-siting, provides the opportunity to decrease impacts by arranging turbines linearly or as small clusters in order to increase visibility (Larsen and Madsen, 2000; Drewitt and Langston, 2006). Furthermore, the choice of wind facility design (i.e., tower height, rotor size, tower width) can affect the collision risk for birds and bats because the perception of an obstacle is different. Increasing tower visibility via higher contrast between tower and sky (e.g., towers with a green-colored bottom gradient fading into white or grey upwards) can lead to a minimization of collisions for low-flying bird species (Dürr, 2011).

13.4 MINIMIZATION: LIMITING IMPACTS BY MANAGEMENT

Under the circumstances that negative impacts on biodiversity cannot be avoided through siting, there are several other measures to minimize through land and facility management. Besides the facility itself, the wind turbine's immediate surroundings influence the attractiveness for animals and thereby the possibility of presence. Raptors particularly – for example, red kite (*Milvus milvus*) – forage over good visible ground and consequently show increased presence on recently mowed grasslands (Mammen et al., 2014). Deriving a mitigation measure from this behavior observation is to make the immediate surrounding of the turbine unattractive for sensitive species to minimize their collision risk. Increasing habitat quality off-site is often considered a compensatory measure either for habitat loss caused by construction or for the incidental take of species (cf. Habitat Conservation Plan). It can, however, distract sensitive bird species and lure them away from the risk zone directly surrounding the turbines. It is a less invasive approach than, for instance, deterrence measures. Finally, post-construction mitigation measures such as temporary shutdowns during times of high abundance (e.g., migration) or increased risk (e.g., seasonal flight behavior) can minimize trade-offs that could not be eliminated during the planning processes. For example, peak bat activity and collision risk can be predicted and applied in temporary turbine curtailments, including factors of wind speed, temperature, precipitation, time of the year, and daytime into algorithms. Most measures have already – to a greater or lesser extend successfully – been implemented.

Integrated approaches can also be seen in Habitat Conservation Plans (HCP) as implemented in the US Section 10(a) of the Endangered Species Act, stating that projects, for example, wind farms, can be permitted even if they might result in the take (including capture, harm, or killing) of a threatened or endangered species. The main requirement for issuing a permit in this case is that the take be incidental and that the applicant minimizes and mitigates the impacts of such takings. Other countries such as Germany have similar approaches, with special authorizations in the incident of an accidental death varying in their degree of takes. The overall strategy of managing trade-offs by compensating them from the very beginning is specified in an HCP. This approach has been successfully implemented with increasing interest (US Fish & Wildlife Service, 2005; Langpap and Kerkvliet, 2012) in many fields and as the case study in Box 13.2 also shows for renewable energy projects. An HCP includes an assessment of likely impacts as well as measures of minimization, mitigation, and monitoring that the proponent will implement

BOX 13.2 HABITAT CONSERVATION PLAN (HCP) FOR INDIANA BATS AT BEECH RIDGE WIND ENERGY PROJECT IN WEST VIRGINIA, USA

The Beech Ridge Wind Energy Project faced challenges during the permitting process due to the occurrence of Indiana bats (*Myotis sodalis*) and Virginia big-eared bats (*Corynorhinus townsendii virginianus*), both listed as endangered species under the Endangered Species Act. A Habitat Conservation Plan (HCP) was developed to identify and mitigate potential impacts on the endangered bats. Impact assessment showed no potential loss of habitat; however, take of both species was predicted due to migration through the facility or hibernation in the area (US Fish & Wildlife Service, 2013b).

In order to minimize potential take of bats, thorough on-site mitigation in the project design and planning was adapted. As a macro-avoidance measure, the number of turbines was reduced from 124 to 100 turbines and the location was moved farther away from hibernacula to lower the collision risk. Additionally, project operation was altered in order to reduce take by 60 percent for the Indiana bat. A curtailment plan was developed that included a raise of cut-in speeds (from 3.5 m/s to 4.8 m/s) between 15 July and 15 October starting at sunset and lasting for five hours during times of expected high bat activity. Based on monitoring results during the first years, the curtailment plan can be adapted (e.g., adaptive management) for an improvement of results (Beech Ridge Energy LLC, 2013).

The expected remaining risk was estimated to a maximum possible take of 53 Indiana bats over the entire project duration of 25 years. For the incidental take of those 53 individuals (approx. 0.712 percent of local population/year), the issuance of an Incidental Take Permit was required. This remaining risk was then considered an unavoidable impact and mitigated through off-site measures improving habitat conditions and supporting recovery of endangered species (see Table 13.2). In this case, conservation projects to protect priority habitat such as winter hibernacula or summer maternity colonies or roosts were stated in the HCP (Beech Ridge Energy LLC, 2013).

Table 13.2 Biological mitigation goals and objectives to decrease risk for Indiana bats at Beech Ridge Wind Energy Project

	HCP Action	Measure (Example)	Risk of Taking
No Mitigation	–	–	Initial risk ■■■
Avoidance	Adapted project design and planning	Reduction of project size from 124 to 100 turbines	Lowered risk ■■□
Minimization	Curtailment plan and adaptive management	Turbine shutoff during times of high bat activity	Remaining risk ■□□

Table 13.2 (continued)

	HCP Action	Measure (Example)	Risk of Taking
Compensation	Compensated by off-site mitigation	Protection of Indiana bat hibernacula off-site	Remaining risk is compensated ☐☐☐

Source: Own compilation based on Beech Ridge Energy LLC (2013).

(US Fish & Wildlife Service, 2005). In wind energy projects, typical measures include protecting or restoring habitats within a certain distance to the project location in compensation for the forthcoming destruction of habitat (Wilhere, 2009; Gartman, 2014) in order to foster the entire population. The extent of this habitat development should at least guarantee a 'no net loss' of habitat in order to trade off negative effects for wildlife. A challenge of this is the identification of a species-specific take threshold, which the Fish and Wildlife Service (FWS) and the proponent can agree upon. The HCP has to appoint actions to be taken in case the threshold is exceeded.

The major advantage of this approach is the possible project implementation benefitting renewable energy development whilst pushing through potential endangerment of a certain species. When executed properly they can be an effective overall approach, providing habitat for the entire population and consequently strengthening species long term while allowing wind energy development to continue.

While the use of project conservation plans could prove beneficial, compensation payments (in-lieu fees) might also contribute to stabilize populations of species of concern on a landscape level (Davis et al., 2009; Jakle, 2012), however Cole (2011) encourages use of avoidance and minimization measures over compensation. Offsetting impacts to wildlife is possible by replacing or restoring habitat (on- or off-site) directly or by paying in-lieu fees for a third party to do so. In addition, investments in mitigation or conservation banks is possible (Davis et al., 2009; Jakle, 2012). A first step needs to be the assessment of direct and indirect impacts and a deduction of mitigation needs.

Even though knowledge of wind energy's wildlife impacts has been growing within the past decade, we still have to face the ever-remaining uncertainties in Biodiversity Impact Assessments (Köppel et al., 2014). Adaptive management (Walters, 2001; Williams et al., 2009) and planning (Kato and Ahern, 2008) is a concept to help cope with this difficulty. The

National Research Council (2004, p. 2) has defined adaptive management as a decision process that 'promotes flexible decision making that can be adjusted in the face of uncertainties as outcomes from management actions and other events become better understood'. An adaptive management process has a specific goal in mind, defined in an initial step during the process. All subsequent management decisions are made based on the available information (Williams, 2011). In particular, monitoring, however, can advance the understanding of remaining knowledge gaps and therefore completes the mitigation hierarchy (Köppel et al., 2014). Based on monitoring results, this 'learning process' is triggered by comparing predicted and actual results and adapting management actions (Williams et al., 2009; Williams, 2011). In Köppel et al. (2014) the point is made that the possible degree of adaption is limited after construction of wind turbines, but examples of how modifications with adaptive measures can be implemented are given. Commonly known for a couple of years are curtailment plans for local and migratory bat populations (Brinkmann et al., 2011). In order to reduce risk for those species, algorithms based on activity patterns (Behr et al., 2011; Lagrange et al., 2013) can be adapted throughout operation of turbines. Monitoring of bat activity at the nacelle level can indicate times with increased collision risk and hence identify potential shutdown periods (Brinkmann et al., 2011). Those particular periods are almost impossible to predict before project commissioning, therefore they remain an uncertainty. The challenge in this approach is balancing trade-offs for biodiversity conservation and economic interests, as extensive shutdown times could compromise cost effectiveness as well as increase risk for financial predictability and thereby endanger development of renewable energies. Consequently, it becomes necessary to develop a case-specific (potentially legally binding) concept including possible monitoring results and connecting them to project adaptation measures. This approach allows the proponent and investors better economic statements due to a relatively calculable risk. Accordingly, many cases including adaptive management are projects with a very profitable site, ensuring profit even in times of supplementary adjustments. Adaptive management is currently implemented for several species including birds as seen in the case study using the example of the California condor (Box 13.3). However, in comparing a theoretical framework with accomplished projects there is always room for improvement. Often the process of knowledge gain is not included in the ongoing project but rather in the subsequent project, preventing improved management in the current project. Provided that the learning process leads to adaptions in the ongoing management of a wind farm they are often only imposed during the first years of operation and do not last the entire life span of

BOX 13.3 ADAPTIVE MANAGEMENT STRATEGY FOR CALIFORNIA CONDOR AT ALTA EAST WIND ENERGY FACILITY IN CALIFORNIA, USA

In the Alta East Wind Project, there are remaining uncertainties about potential negative impacts on the California condor (*Gymnogyps californianus*), which is currently recovering from severe population declines through statewide recovery programs. In order to limit negative impacts, an Adaptive Management Strategy (US Fish & Wildlife Service, 2013a) has been developed by the project proponent Alta Wind Power Development, LLC in cooperation with the Bureau of Land Management (BLM) and FWS. In the area of the wind farm no individuals of the California condor have been observed so far; however, with the increasing population size and the wind farm being located in a potentially suitable habitat, it is possible that use of this area will occur during the 20-year operation phase. The 'Condor Monitoring and Avoidance Plan' was developed to react to this potentially increasing collision risk caused by increasing population size as well as new flight patterns. Several mitigation measures were developed (see Table 13.3); for example, surveillance with a very high frequency (VHF) detection system, employing a full-time biologist on-site to monitor the presence of condors (currently all wild condors have a VHF transmitter) and establish curtailment measures if one ventured near the facility (e.g., implement a Condor Monitoring and Avoidance Plan). The case of the Alta East Wind Farm is one example of how the trade-off between biodiversity and renewable energy can be limited using the adaptive management approach. Wildlife organizations, however, criticize that the legal take of one California condor is accepted within the project (Barringer, 2013; Sahagun, 2013).

Table 13.3 Tiers and actions developed in the Adaptive Management Strategy for Alta East Wind Project

Tier	Action
Detection of California condor < 16 mile detection perimeter, but > 8 miles away	FWS, BLM and Alta Windpower discuss details of detection
Detection of California condor < 8 mile detection perimeter > One observation in 30 days or > Two observations in 60 days or > One observation of one individual within 2 miles in one year	FWS, BLM and Alta Windpower discuss details of detection and determine appropriate action Continued full-time observation
Group of California condors moves within 2 miles of the project boundary	Real-time curtailment to ensure California condors are not killed

Table 13.3 (continued)

Tier	Action
Change in percentage of population wearing VHF units (> 30% not equipped with VHF transmitters)	Nighttime-only operations (no further action if an alternative detection system is deployed)
California condor take	Nighttime-only operation

Source: Based on US Fish & Wildlife Service (2013a).

a wind farm (e.g., Windpark Ellern as introduced in Köppel et al., 2014). This limits the chances of success of Adaptive Management.

13.5 CONCLUSION

Some authors (Allison et al., 2014) have apprehended a deadlock in renewable energy development due to remaining uncertainties about the risks for biodiversity in low-carbon energy production. However, understanding the effects of renewable energies on wildlife and avoiding them by adequate impact assessment proves crucial to preventing a major green vs green dilemma. Environmental planning and management strategies can be designed to supplement the ecosystem services concept in biodiversity conservation for green energy. As current knowledge gaps will only be partially overcome in the future, some uncertainties will always remain, particularly due to constant technical development and dynamic ecosystem development.

After more than a decade spent on comprehensive research on wind energy's wildlife implications and mitigation strategies, we are substantially better supplied with methods, measures, and promising tools to tackle unintended trade-offs. Nevertheless, critical methodological issues have been the assessment of long-term impacts (e.g., habituation of birds), cumulative impacts (e.g., on marine mammals), uncertainties (e.g., population assessments of bats) and the spatial scope of impacts (e.g., migratory species). At the same time, such issues have also triggered innovations such as adaptive management, planning and zoning approaches, and state-of-the-art micro-avoidance concepts. Such efforts contribute to reconcile ecosystem services' conflicts between reducing carbon loads and safeguarding biodiversity.

NOTE

* This chapter evolved from research carried out by the Environmental Assessment and Planning Research Group of the Berlin Institute of Technology (TU Berlin), funded by the German Federal Ministry for Economic Affairs and Energy (BMWi) on the basis of a decision by the German Bundestag. The authors would like to thank Eva Schuster and Victoria D. Gartman for their comments, which helped to improve the manuscript.

REFERENCES

Allison, T.D., T.L. Root and P.C. Frumhoff (2014), 'Thinking globally and siting locally – renewable energy and biodiversity in a rapidly warming world', *Climatic Change*, **126**(1–2), 1–6.

Arent, D.J., A. Wise and R. Gelman (2011), 'The status and prospects of renewable energy for combating global warming', *Energy Economics*, **33**(4), 584–93.

Arizona Game and Fish Department (2008), *Guidelines for Reducing Impacts to Wildlife from Wind Energy Development in Arizona*, revised 23 November 2009: Phoenix, AZ: Arizona Game and Fish Department.

Arnett, E.B. and E.F. Baerwald (2013), 'Impacts of wind energy development on bats: implications for conservation', in R.A. Adams and S.C. Pedersen (eds), *Bat Evolution, Ecology, and Conservation*: New York: Springer, pp. 435–56.

Arnett, E.B., W.K. Brown and W.P. Erickson et al. (2008), 'Patterns of bat fatalities at wind energy facilities in North America', *Journal of Wildlife Management*, **72**(1), 61–78.

Aumüller, R., K. Boos and S. Freienstein et al. (2011), 'Beschreibung eines Vo-gel-schlag-er-eig-nisses und seiner Ursachen an einer Forschungsplattform in der Deutschen Bucht' [Description of a bird strike event and its causes at a research platform in the German Bight], *Vogelwarte*, **49**, 9–16.

Baerwald, E.F. and R. Barclay (2009), 'Geographic variation in activity and fatality of migratory bats at wind energy facilities', *Journal of Mammalogy*, **90**(6), 1341–9.

Baerwald, E.F., W.P. Patterson and R.M.R. Barclay (2014), 'Origins and migratory patterns of bats killed by wind turbines in southern Alberta: evidence from stable isotopes', *Ecosphere*, **5**(9), 118.

Barringer, F. (2013), 'Turbine plans unnerve fans of condors in California', *New York Times*, 25 May.

Barrios, L. and A. Rodríguez (2004), 'Behavioural and environmental correlates of soaring-bird mortality at on-shore wind turbines', *Journal of Applied Ecology*, **41**(1), 72–81.

Bastos, R., M. Santos and J.A. Cabral (2013), 'A new stochastic dynamic tool to improve the accuracy of mortality estimates for bats killed at wind farms', *Ecological Indicators*, **34**, 428–40.

Beech Ridge Energy LLC (2013), *Beech Ridge Wind Energy Project Habitat Conservation Plan Greenbrier and Nicholas Counties, West Virginia*, Chicago, IL: Beech Ridge Energy LLC.

Behr, O., R. Brinkmann and I. Niermann et al. (2011), 'Vorhersage der Fledermausaktivität an Windenergieanlagen' [Prediction of bat activity at wind turbines], in R. Brinkmann, O. Behr and I. Niermann et al. (eds) *Entwicklung von Methoden zur Untersuchung und Reduktion des Kollisionsrisikos von Fledermäusen an Onshore-Windenergieanlagen: Ergebnisse eines Forschungsvorhabens*, Göttingen: Cuvillier, pp. 287–322.

Bellard, C., C. Bertelsmeier and P. Leadley et al. (2012), 'Impacts of climate change on the future of biodiversity', *Ecology Letters*, **15**(4), 365–77.

Bergström, L., L. Kautsky and T. Malm et al. (2014), 'Effects of offshore wind farms on marine wildlife – a generalized impact assessment', *Environmental Research Letters*, **9**(3), 034012.

Brinkmann, R., O. Behr and I. Niermann et al. (2011), *Entwicklung von Methoden zur Untersuchung und Reduktion des Kollisionsrisikos von Fledermäusen an Onshore-Wind-ener-gie-an-la-gen: Ergebnisse eines Forschungsvorhabens* [Development of Methods for Investigating and Reducing the Risk of Collision of Bats with Onshore Wind Turbines: Results of a Research Project], Göttingen: Cuvillier.

Camiña, Á (2011), 'Ein vermeidbarer Konflikt: Geier und Windenergieanlagen' [An avoidable conflict: vultures and wind turbines], *Der Falke*, **58**(12), 504–7.

Cole, S.G. (2011), 'Wind power compensation is not for the birds: an opinion from an environmental economist', *Restoration Ecology*, **19**(2), 147–53.

Colman, J.E., S. Eftestøl and D. Tsegaye et al. (2012), 'Is a wind-power plant acting as a barrier for reindeer *Rangifer tarandus tarandus* movements?', *Wildlife Biology*, **18**(4), 439–45.

Dahl, E.L., R. May and P.L. Hoel et al. (2013), 'White-tailed eagles (*Haliaeetus albicilla*) at the Smøla wind-power plant, Central Norway, lack behavioral flight responses to wind turbines', *Wildlife Society Bulletin*, **37**(1), 66–74.

Danish Energy Agency (2013), *Danish Offshore Wind. Key Environmental Issues – A Follow-up*, Vattenfall: The Danish Nature Agency/DONG Energy, accessed 24 January 2016 at http://www.orbicon.dk/media/Havvindm%C3%B8llebog.pdf.

Davis, A., T. Weis and K. Halsey et al. (2009), *Enabling Progress Compensatory Mitigation Scenarios for Wind Energy Projects in the US*, report for American Wind Wildlife Institute.

De Groot, R.S., R. Alkemade and L. Braat et al. (2010), 'Challenges in integrating the concept of ecosystem services and values in landscape planning, management and decision making', *Ecosystem Services – Bridging Ecology, Economy and Social Sciences*, **7**(3), 260–72.

Devereux, C.L., M.J. Denny and M.J. Whittingham (2008), 'Minimal effects of wind turbines on the distribution of wintering farmland birds', *Journal of Applied Ecology*, **45**(6), 1689–94.

Diederichs, A., G. Nehls and M. Dähne et al. (2008), *Methodologies for Measuring and Assessing Potential Changes in Marine Mammal Behaviour, Abundance or Distribution Arising from the Construction, Operation and Decommissioning of Offshore Windfarms*, report commissioned by COWRIE Ltd.

Doty, A.C. and A.P. Martin (2012), 'Assessment of bat and avian mortality at a pilot wind turbine at Coega, Port Elizabeth, Eastern Cape, South Africa', *New Zealand Journal of Zoology*, **40**(1), 75–80.

Drewitt, A.L. and R. Langston (2006), 'Assessing the impacts of wind farms on birds', *Ibis*, **148**(S1), S29–S42.

Dürr, T. (2009), 'Zur Gefährdung des Rotmilans *Milvus milvus* durch Windenergieanlagen in Deutschland' [The endangerment of the red kite *Milvus milvus* by wind turbines in Germany], *Informationsdienst Naturschutz Niedersachsen*, **29**(3), 185–91.

Dürr, T. (2011), 'Dunkler Anstrich könnte Kollisionen verhindern: Vogelunfälle an Windradmasten' [Dark paint could prevent collisions: bird accidents at wind turbine masts], *Der Falke*, **58**(12), 499–501.

Erickson, W.P., G.D. Johnson and D.P. Young (2005), 'A summary and comparison of bird mortality from anthropogenic causes with an emphasis on collisions', *USDA Forest Service General Technical Reports Pacific Southwest Research Station*, **191**, 1029–42.

Forstliche Versuchs- und Forschungsanstalt and ForstBW (FVA) (2012), *Planungsgrundlage 'Windkraft und Auerhuhn'* [Planning Guide 'Wind Energy and Capercaillie], accessed 27 January 2015 at http://www.fva-bw.de/forschung/bui/windenergie_auerhuhn/Windenergie_und_Auerhuhn_525_100.pdf.

Forstliche Versuchs- und Forschungsanstalt (FVA) (2013), *Bewertungshilfe 'Auerhuhn und Windenergie im Schwarzwald': Basis: Karten- und Planungsgrundlagen der FVA* [Evaluation Tool 'Capercaillie and Wind Energy in the Black Forest': Basis: Maps and Planning Guide by the Forest Research Institute], accessed 21 July 2015 at http://www.fva-bw.de/publikationen/sonstiges/131007auerhuhn_und_windkraft_bewertung.pdf.

Gartman, V. (2014), 'Policies and measures for protected species in wind energy: an assessment between the US and Germany', Master's thesis, Texas Tech University.
Geissler, G., J. Köppel and P. Gunther (2013), 'Wind energy and environmental assessments – a hard look at two forerunners' approaches: Germany and the United States', *Renewable Energy*, **51**, 71–8.
Gordon, J., D. Thompson and D. Gillespie et al. (2007), 'Assessment of the potential for acoustic deterrents to mitigate the impact on marine mammals of underwater noise arising from the construction of offshore windfarms', St Andrews, Scotland: COWRIE Ltd.
Grünkorn, T., A. Diederichs and D. Poszig et al. (2009), 'Wie viele Vogel kollidieren mit Windenergieanlagen? How many birds collide with wind turbines?', *Natur und Landschaft*, **84**(7), 309–14.
Helldin, J.O., J. Jung and W. Neumann et al. (2012), *The Impacts of Wind Power on Terrestrial Mammals: A Synthesis*, Report No. 6510, Stockholm: Swedish Environmental Protection Agency.
Hernandez, R.R., S.B. Easter and M.L. Murphy-Mariscal et al. (2014), 'Environmental impacts of utility-scale solar energy', *Renewable and Sustainable Energy Reviews*, **29**, 766–79.
Hill, R., K. Hill and R. Aumüller et al. (2014), 'Of birds, blades and barriers: detecting and analysing mass migration events at Alpha Ventus', in Federal Maritime and Hydrographic Agency and Federal Ministry for the Environment, Nature Conservation and Nuclear Safety (eds), *Ecological Research at the Offshore Windfarm Alpha Ventus*: Wiesbaden: Springer Fachmedien, pp. 111–31.
Hoover, S. and M.L. Morrison (2005), 'Behavior of red-tailed hawks in a wind turbine development', *Journal of Wildlife Management*, **69**(1), 150–59.
Hötker, H., K. Thomsen and H. Köster (2005), 'Auswirkungen regenerativer Ener-gie-ge-win-nung auf die biologische Vielfalt: am Beispiel der Vögel und der Fledermäuse' [Impacts of renewable energy production on biodiversity: the example of birds and bats], *BfN-Skripten No. 142*, Bonn: Naturschutzbund.
Hull, C.L. and S. Muir (2010), 'Search areas for monitoring bird and bat carcasses at wind farms using a Monte-Carlo model', *Australasian Journal of Environmental Management*, **17**(2), 77–87.
Hull, C.L. and S.C. Muir (2013), 'Behavior and turbine avoidance rates of eagles at two wind farms in Tasmania, Australia', *Wildlife Society Bulletin*, **37**(1), 49–58.
Hüppop, O., J. Dierschke and K. Exo et al. (2006), 'Bird migration studies and potential collision risk with offshore wind turbines', *Ibis*, **148**(S1), S90–S109.
Huso, M.M.P. and W.P. Erickson (2013), 'A comment on "Novel scavenger removal trials increase wind turbine-caused avian fatality estimates"', *The Journal of Wildlife Management*, **77**(2), 213–15.
Huso, M.M.P., D.H. Dalthorp and D.A. Dail et al. (2014), 'Estimating wind-turbine caused bird and bat fatality when zero carcasses are observed', *Ecological Applications*, **25**, 1213–25.
Jakle, A. (2012), *Wind Development and Wildlife Mitigation in Wyoming: A Primer*, Laramie, WY: Ruckelshaus Institute of Environment and Natural Resources.
Kato, S. and J. Ahern (2008), '"Learning by doing": adaptive planning as a strategy to address uncertainty in planning', *Journal of Environmental Planning and Management*, **51**(4), 543–59.
Katzner, T., J.A. Johnson and D.M. Evans et al. (2013), 'Challenges and opportunities for animal conservation from renewable energy development', *Animal Conservation*, **16**(4), 367–9.
Kelm, D.H., J. Lenski and V. Kelm et al. (2014), 'Seasonal bat activity in relation to distance to hedgerows in an agricultural landscape in Central Europe and implications for wind energy development', *Acta Chiropterologica*, **16**(1), 65–73.
Köppel, J., M. Dahmen and J. Helfrich et al. (2014), 'Cautious but committed: moving toward adaptive planning and operation strategies for renewable energy's wildlife implications', *Environmental Management*, **54**(4), 744–55.

Kunz, T.H., E.B. Arnett and B. Cooper et al. (2007), 'Assessing impacts of wind-energy development on nocturnally active birds and bats: a guidance document', *Journal of Wildlife Management*, **71**(8), 2449–86.

Kuvlesky, W.P., L.A. Brennan and M.L. Morrison et al. (2007), 'Wind energy development and wildlife conservation: challenges and opportunities', *Journal of Wildlife Management*, **71**(8), 2487–98.

Lagrange, H., P. Rico and Y. Bas et al. (2013), 'Mitigating bat fatalities from wind-power plants through targeted curtailment: results from 4 years of testing of CHIROTECH©' in Naturvårdsverket (ed.), *Book of Abstracts: Conference on Wind Power and Environmental Impacts Stockholm 5–7 February, Conference on Wind Power and Environmental Impacts (CWE), Stockholm.*

Länderarbeitsgemeinschaft der Staatlichen Vogelschutzwarten (LAG VSW) (2015), Ab-stands-emp-feh-lung-en für Windenergieanlagen zu bedeutsamen Vogellebensräumen sowie Brutplätzen ausgewählter Vogelarten' [Buffer zone recommendations for wind turbines near important bird habitat and breeding grounds of selected species], revision of 15 April, Neschwitz: Länderarbeitsgemeinschaft der Staatlichen Vogelschutzwarten.

Landesamt für Umwelt, Gesundheit und Verbraucherschutz (LUGV) (2012), 'Tierökologische Abstandskriterien für die Errichtung von Windenergieanlagen in Brandenburg (TAK)' [Avian buffer zones for wind energy development in Brandenburg (TAK)]: Landesamt für Umwelt, Gesundheit und Verbraucherschutz.

Langpap, C. and J. Kerkvliet (2012), 'Endangered species conservation on private land: assessing the effectiveness of habitat conservation plans', *Journal of Environmental Economics and Management*, **64**(1), 1–15.

Larsen, J.K. and J. Madsen (2000), 'Effects of wind turbines and other physical elements on field utilization by pink-footed geese (*Anser brachyrhynchus*): a landscape perspective', *Landscape Ecology*, **15**(8), 755–64.

Lovich, J.E. and J.R. Ennen (2011), 'Wildlife conservation and solar energy development in the desert southwest, United States', *BioScience*, **61**(12), 982–92.

Lovich, J.E. and J.R. Ennen (2013), 'Assessing the state of knowledge of utility-scale wind energy development and operation on non-volant terrestrial and marine wildlife', *Applied Energy*, **103**, 52–60.

Lucas, M., G.F.E. Janss and D.P. Whitfield et al. (2008), 'Collision fatality of raptors in wind farms does not depend on raptor abundance', *Journal of Applied Ecology*, **45**(6), 1695–703.

Lucke, K., U. Siebert and P.A. Lepper et al. (2009), 'Temporary shift in masked hearing thresholds in a harbor porpoise (*Phocoena phocoena*) after exposure to seismic airgun stimuli', *The Journal of the Acoustical Society of America*, **125**(6), 4060–70.

Madders, M. and D.P. Whitfield (2006), 'Upland raptors and the assessment of wind farm impacts', *Ibis*, **148**(S1), S43–S56.

Madsen, P.T., M. Wahlberg and J. Tougaard et al. (2006), 'Wind turbine underwater noise and marine mammals: implications of current knowledge and data needs', *Marine Ecology Progress Series*, **309**, 279–95.

Mammen, K., U. Mammen and A. Resetaritz (2014), 'Rotmilan' [Red kite], in *Greifvögel und Windkraftanlagen: Problemanalyse und Lösungsvorschläge: Schlussbericht für das Bundesministerium für Umwelt, Naturschutz und Reaktorsicherheit*, Bergenhusen, Berlin: Husum, pp. 13–100.

Marques, A.T., H. Batalha and S. Rodrigues et al. (2014), 'Understanding bird collisions at wind farms: an updated review on the causes and possible mitigation strategies', *Biological Conservation*, **179**, 40–52.

Mathews, F., M. Swindells and R. Goodhead et al. (2013), 'Effectiveness of search dogs compared with human observers in locating bat carcasses at wind-turbine sites: a blinded randomized trial', *Wildlife Society Bulletin*, **37**(1), 34–40.

May, R.F. (2015), 'A unifying framework for the underlying mechanisms of avian avoidance of wind turbines', *Biological Conservation*, **190**, 179–87.

May, R., O. Reitan and K. Bevanger et al. (2015), 'Mitigating wind-turbine induced avian

mortality: sensory, aerodynamic and cognitive constraints and options', *Renewable and Sustainable Energy Reviews*, **42**, 170–81.
Nagel, P.B., T. Schwarz and J. Köppel (2014), 'Ausbau der Windenergie – Anforderungen aus der Rechtsprechung und fachliche Vorgaben für die planerische Steuerung' [Expansion of wind power – requirements from case law and technical guidelines for planning control], *UPR*, **10**, 371–82.
National Research Council (NRC) (2004), *Adaptive Management for Water Resources Project Planning*, Washington, DC: NRC.
National Wind Coordinating Collaborative (NWCC) (2010), 'Wind turbine interactions with birds, bats, and their habitats: a summary of research results and priority questions', factsheet, accessed 23 January 2016 at https://www1.eere.energy.gov/wind/pdfs/birds_and_bats_fact_sheet.pdf.
Nedwell, J., S. Parvin and B. Edwards et al. (2007), *Measurement and Interpretation of Underwater Noise During Construction and Operation of Offshore Windfarms in UK Waters*, Subacoustech Report No. 544R0738 to COWRIE Ltd.
Obermeyer, B., R. Manes and J. Kiesecker et al. (2011), 'Development by design: mitigating wind development's impacts on wildlife in Kansas', *PloS ONE*, **6**(10), e26698.
Pagel, J.E., K.J. Kritz and B.A. Millsap et al. (2013), 'Bald eagle and golden eagle mortalities at wind energy facilities in the contiguous United States', *Journal of Raptor Research*, **47**(3), 311–15.
Paula, J., M.C. Leal and M.J. Silva et al. (2011), 'Dogs as a tool to improve bird-strike mortality estimates at wind farms', *Journal for Nature Conservation*, **19**(4), 202–8.
Pearce-Higgins, J.W., L. Stephen and A. Douse et al. (2012), 'Greater impacts of wind farms on bird populations during construction than subsequent operation: results of a multi-site and multi-species analysis', *Journal of Applied Ecology*, **49**(2), 386–94.
Pearce-Higgins, J.W., L. Stephen and R.H.W. Langston et al. (2009), 'The distribution of breeding birds around upland wind farms', *Journal of Applied Ecology*, **46**(6), 1323–31.
Piorkowski, M.D. and T.J. O'Connell (2010), 'Spatial pattern of summer bat mortality from collisions with wind turbines in mixed-grass prairie', *The American Midland Naturalist*, **164**(2), 260–69.
Plonczkier, P. and I.C. Simms (2012), 'Radar monitoring of migrating pink-footed geese: behavioural responses to offshore wind farm development', *Journal of Applied Ecology*, **49**(5), 1187–94.
Rees, E.C. (2012), 'Impacts of wind farms on swans and geese: a review', *Wildfowl*, **62**, Slimbridge, UK: Wildfowl & Wetlands Trust, 37–72.
Reichenbach, M. and H. Steinborn (2006), 'Windkraft, Vögel, Lebensräume – Ergebnisse einer fünfjährigen BACI-Studie zum Einfluss von Windkraftanlagen und Habitatparametern auf Wiesenvögel' [Wind power, bird, habitats – results of a five-year BACI study on the influence of wind turbines and habitat parameters on grassland birds], *Osnabrücker Naturwissenschaftliche Mitteilungen*, **32**, 243–59.
Rodrigues, L., L. Bach and L. Biraschi et al. (2006), *Wind Turbines and Bats: Guidelines for the Planning Process and Impact Assessments Version 1.0*, Advisory Committee of the EUROBATS Agreement, Ljubljana, Slovenia.
Rydell, J., L. Bach and M. Dubourg-Savage et al. (2010), 'Bat mortality at wind turbines in Northwestern Europe', *Acta Chiropterologica*, **12**(2), 261–74.
Sahagun, L. (2013), 'Terra-Gen gets OK on wind farm in wake of condor decision', *Los Angeles Times*, 24 May.
Scheller, W., F. Vökler and A. Güttner (2013), 'Rotmilankartierung 2011/2012 in Meck-lenburg – Vorpommern' [Red kite mapping 2011/2012 in Mecklenburg-Western Pomerania]. SALIX – Kooperationsbüro für Umwelt- und Landschaftsplanung.
Schuster, E., L. Bulling and J. Köppel (2015), 'Consolidating the state of knowledge: a synoptical review of wind energy's wildlife effects', *Environmental Management*, **56**(2), 300–331.
Smallwood, K.S. and C.G. Thelander (2004), *Developing Methods to Reduce Bird Mortality in the Altamont Pass Wind Resource Area*, Ojai, CA: BioResource Consultants.

Smallwood, K.S., D.A. Bell and B. Karas et al. (2013), 'Response to Huso and Erickson's comments on novel scavenger removal trials', *The Journal of Wildlife Management*, **77**(2), 216–25.
Smallwood, K.S., D.A. Bell and S.A. Snyder et al. (2010), 'Novel scavenger removal trials increase wind turbine – caused avian fatality estimates', *The Journal of Wildlife Management*, **74**(5), 1089–96.
Sovacool, B.K. (2013), 'The avian benefits of wind energy: a 2009 update', *Selected Papers from World Renewable Energy Congress XI*, **49**, 19–24.
Steinborn, H., M. Reichenbach and H. Timmermann (2011), '*Windkraft – Vögel – Lebensräume': Ergebnisse einer siebenjährigen Studie zum Einfluss von Windkraftanlagen und Habitatparametern auf Wiesenvögel* ['*Wind Power – Birds – Habitats': Results of a Seven Year Study on the Influence of Win Turbines and Habitat Parameters on Grassland Birds*], Oldenburg: ARSU GmbH.
Stoms, D.M., S.L. Dashiell and F.W. Davis (2013), 'Siting solar energy development to minimize biological impacts', *Renewable Energy*, **57**, 289–98.
Tsoutsos, T., N. Frantzeskaki and V. Gekas (2005), 'Environmental impacts from the solar energy technologies', *Energy Policy*, **33**(3), 289–96.
US Fish & Wildlife Service (FWS) (2005), 'Habitat Conservation Plans', Section 10 of the Endangered Species Act, 2005, Arlington, VA: US Fish & Wildlife Service, accessed 24 January 2016 at https://www.fws.gov/endangered/esa-library/pdf/HCP_Incidental_Take.pdf.
US Fish & Wildlife Service (FWS) (2013a), *Biological Opinion for the Alta East Wind Project, Kern County, California*, Ventura, CA: US Fish & Wildlife Service.
US Fish & Wildlife Service (FWS) (2013b), *Biological Opinion on an Application for an Incidental Take Permit Submitted by Beech Ridge Energy LLC*, Ventura, CA: US Fish & Wildlife Service.
Vaissière, A., H. Levrel and S. Pioch et al. (2014), 'Biodiversity offsets for offshore wind farm projects: the current situation in Europe', *Marine Policy*, **48**, 172–83.
Voigt, C.C., L.S. Lehnert and G. Petersons et al. (2015), 'Wildlife and renewable energy: German politics cross migratory bats', *European Journal of Wildlife Research*, **61**(2), 213–19.
Voigt, C.C., A.G. Popa-Lisseanu and I. Niermann et al. (2012), 'The catchment area of wind farms for European bats: a plea for international regulations', *Biological Conservation*, **153**, 80–86.
Walters, C.J. (2001), *Adaptive Management of Renewable Resources*, Caldwell, NJ: Blackburn Press.
Wilhelmsson, D., T. Malm and M.C. Öhman (2006), 'The influence of offshore windpower on demersal fish', *ICES Journal of Marine Science: Journal du Conseil*, **63**(5), 775–84.
Wilhere, G.F. (2009), 'Three paradoxes of habitat conservation plans', *Environmental Management*, **44**(6), 1089–98.
Williams, B.K. (2011), 'Passive and active adaptive management: approaches and an example', *Adaptive Management for Natural Resources*, **92**(5), 1371–8.
Williams, B.K., R.C. Szaro and C.D. Shapiro (2009), *Adaptive Management: The US Department of the Interior Technical Guide*, Washington, DC: Adaptive Management Working Group.
Zimmerling, R.J., A.C. Pomeroy and M.V. d'Entremont et al. (2013), 'Canadian estimate of bird mortality due to collisions and direct habitat loss associated with wind turbine developments', *Avian Conservation and Ecology*, **8**(2), 10.

14. Cumulative effects of dams on biodiversity*

Asha Rajvanshi

14.1 INTRODUCTION

Escalation in human population and increased water usage are placing greater demands on the world's freshwater supplies (Postel, 2000). Consequently, river systems that support the highest biological diversity in any landscape have become radically transformed by hydrological structures throughout the world. Global estimates indicate that nearly 50 000 dams (Berga et al., 2006) have regulated 60 percent of the world's river flows in over 150 countries (Schelle et al., 2004) and have created reservoirs, spreading over 500 000 km^2. With the hydropower dams and other water development infrastructure becoming a dominant feature of many freshwater systems worldwide (WCD, 2000; Gleik, 2003), it is estimated that another 1500 dams are currently under construction (IJHD, 2004). Although this absolute number of dams around the world may further change, Asia's proportionate share of the global total of dams would continue to remain relatively high (Dudgeon, 2000).

The environmental impact assessments of the dams from around the world have brought about a radical realization that large dams that were thought to be overwhelmingly contributing to huge economic benefits are actually doing so at the expense of the natural assets as most dams are concentrated in species-rich areas, in sites that overlap with ecologically sensitive areas or in critical habitats designated for conserving rare and threatened species. Among such sites identified for construction of dams are protected habitats in many national parks, World Heritage Sites (WHS) that according to UNESCO represent 'outstanding cultural or natural importance to the common heritage of humanity', and critical habitats outside designated conservation areas supporting endemic and other biodiversity elements that command significant conservation values.

According to International Rivers, the Three Parallel Rivers site in Yunnan Province of China, which is known to be the epicenter of Chinese biodiversity, is threatened by the recently revived proposal of building a 13-dam cascade along the main body of the Nu River flowing through the site. The proposed Huay Samong Dam on the Huay Samong River is located in the World Natural Heritage site in Prajeen Buri Province, between the Pang Sida and Tab Lan National Parks. Likewise, the 222 MW Chan 75 Dam was built in the La Amistad UNESCO Biosphere Reserve Buffer Zone in Panama. The proposed hydropower projects (e.g., Rishi-Ganga I – 70 MW and Rishi-Ganga II – 35 MW) are located in the Rishi Ganga River catchment within the Nanda Devi National Park World Heritage Site in India, which offers critical habitats for several endangered mammals, especially the snow leopard, Himalayan musk deer and bharal or the Himalayan blue sheep (Rajvanshi et al., 2012).

The pace of construction of large dams as a means of energy expansion (Ginneken, 2015) is relatively higher in the mega-diverse countries of the world (Figure 14.1), which include middle-income countries: Brazil, Russia, India, China and South Africa (BRICS).

The most publicized construction is occurring in the Mekong Basin in China (Huang and Yan, 2009; Ziv et al., 2012), the Amazon in Brazil (Da Silva et al., 2010; Soito and Freitas, 2011), and the Himalayan region in India (Grumbine and Pandit, 2013). Almost all of these river basins support both high levels of biodiversity and a large human population dependent upon fisheries or agriculture maintained by free-flowing river floodplain processes (Dudgeon, 2000; Richter et al., 2010).

The needs for combining economic benefits from hydropower development and protecting ecosystems goods and services that sustain societies and biological wealth on earth are, however, far greater than those associated with developments in other sectors. At the same time, the challenges of achieving the synergies between the economic objective of power generation and the ecological incentives of conserving the biological diversity are often too complicated as the combined impacts of multiple dams are invariably much greater than the simple sum of their direct impacts. A range of planning tools – for example, sectoral environmental assessments (EA), basin-wide EAs, regional EAs, and cumulative EA – have evolved. All of these approaches can be used to innovate and improve impact assessment outcomes for aiding hydropower development planning.

This chapter puts forward the typology of cumulative impacts and presents the merits of conducting Cumulative Impact Assessment (CIA) of hydropower projects for identification of basin-wide impacts that may represent aggregate, multiple and altogether new impacts on ecological systems. Section 14.2 reviews the potential impacts of dams on biodiversity. Section 14.3 defines the central properties of CIA and expands on the typology of cumulative impacts as it applies to dams and how these are distinguishable based on contextual, spatial and temporal dimensions of impacts and multiplicity of drivers of change. Section 14.4 draws the linkages of CIA with Strategic Environmental Assessment (SEA) and presents examples that

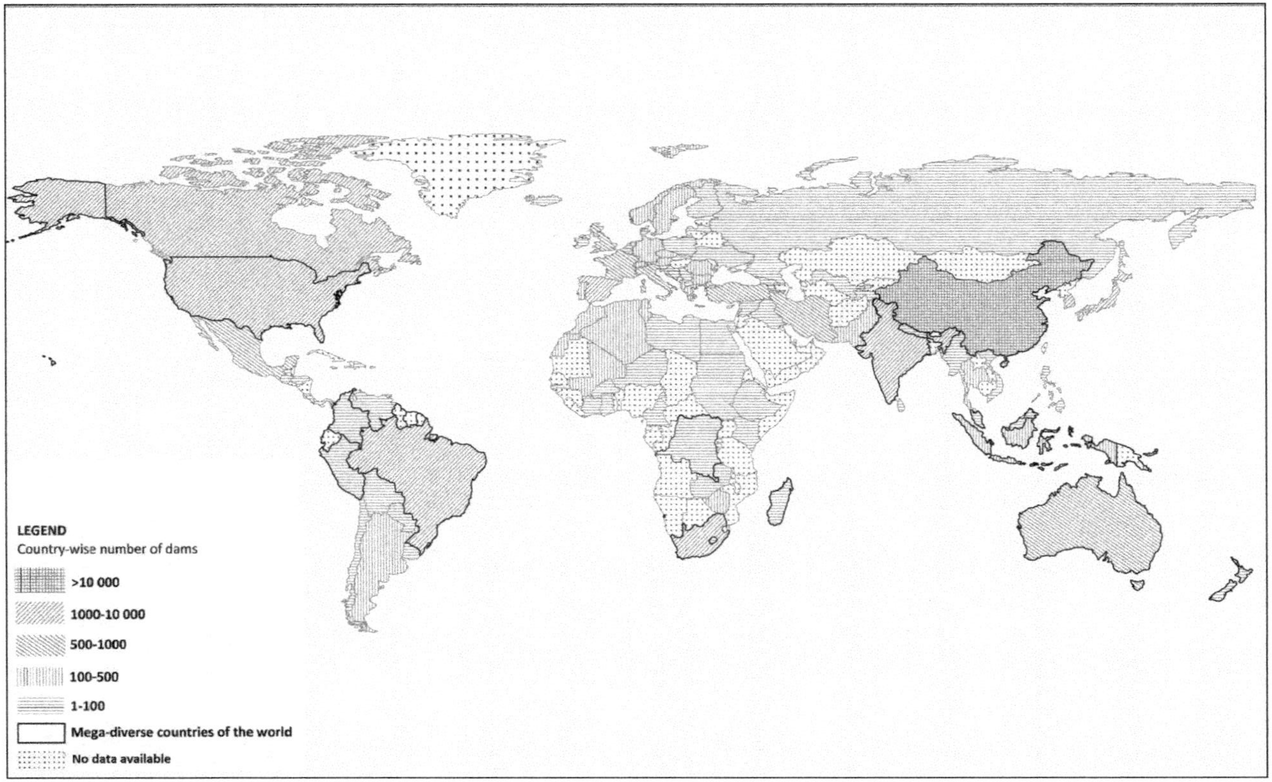

Figure 14.1 *Most dams are located in mega-diverse regions of the world*

reflect the contributions of CIA in SEAs for planning energy projects in locations that are least disruptive to key ecological processes and operating them in ways that protect biodiversity and maintain key ecological services. The conclusions in Section 14.5 highlight the merits of applying CIA and discuss the way forward for achieving the successful outcome from the application of CIA framework in the context of environmental assessment of dams.

14.2 POTENTIAL EFFECTS OF DAMS ON BIODIVERSITY

Many of the major catchments in the world now contain multiple dams. Damming a river has the potential to generate multi-pronged impacts on floral and faunal species and their habitats and ecosystem(s) goods and services by submerging forests and fragmenting the river ecosystems. Large-scale hydrological alterations lead to a suite of environmental impacts. Several researchers have elaborated on these impacts (Brooker, 1981; Petts, 1984; Casado et al., 1989; Liu and Yu, 1992; Rosenberg et al., 1997; Bergkamp et al., 2000; Brismar, 2002, 2004; Friedl and Wüest, 2002; Kibler and Tullos, 2013).

Dams have submerged large tracts of land creating 16.7 million reservoirs with a combined storage capacity of approximately 8070 km^3 and have increased the earth's terrestrial surface water area by more than 305 000 km^2 (Lehner et al., 2011). Within a basin, the greater the number of dams, the greater is the fragmentation of river ecosystems. An assessment of 227 of the major river basins in the world showed that 37 percent of the large rivers are strongly affected by fragmentation and altered flows, 23 percent are moderately affected, and 40 percent are unaffected (Revenga et al., 2000; WCD, 2000). Disruption of hydrologic connectivity that is vital for the water-mediated transport of matter, energy and organisms within or between elements of the hydrologic cycle (Freeman et al., 2007) is a major impact of dams that has far-reaching challenges of conserving biodiversity and protection of ecosystem services. Dams bring about changes in the natural hydrograph of a river, the flow regime, the functions of the tributaries and in the characteristics of the riparian and aquatic habitats, wetland and floodplains vegetation, delta mangroves, coastal habitats and the estuarine ecosystems. Dams can negatively impact downstream ecosystems by degrading water quality, obstructing and/or delaying fish passage, and eliminating critical spawning and rearing habitat for migratory species (Ligon et al., 1995; Bunn and Arthington, 2002). These losses in ecosystem functions and services can elevate social costs because of loss of important fisheries (Dudgeon, 2000) and also eliminate numerous other important societal and ecological benefits (e.g., fertility of flood plains) provided by free-flowing rivers (Postel and Richter, 2003). Most direct and indirect impacts of dams are captured in Figure 14.2.

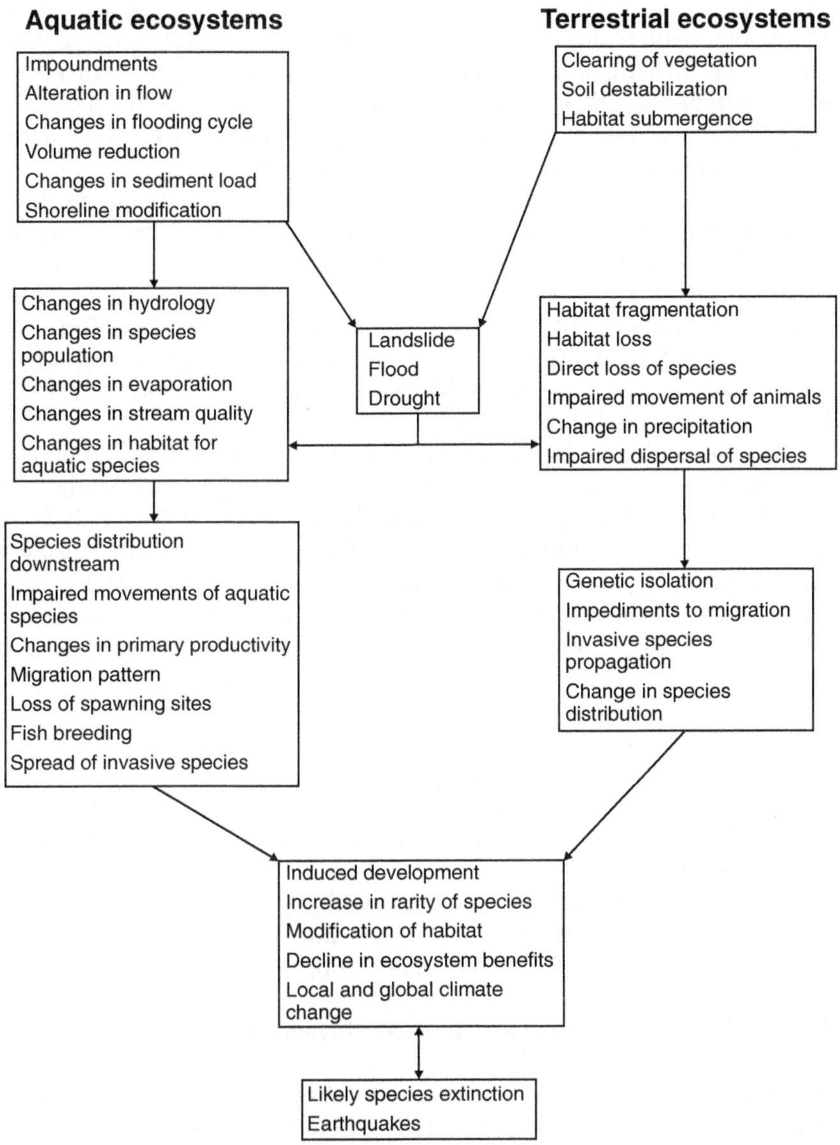

Figure 14.2 Direct and indirect ecological impacts of dams

14.3 CUMULATIVE IMPACT ASSESSMENT OF DAMS

Impacts in nature are not experienced sequentially, from one dam to another. It is also almost impossible to unscramble the impacts from those of individual or multiple projects or from those resulting from other indirect perturbations in a landscape. Anticipating impacts is therefore not synonymous with accurately predicting the nature and magnitude of impacts from several projects having multiple sources and complex pathways, seen together on ecosystems and biodiversity.

Addressing cumulative impacts or effects of large development projects is therefore increasingly gaining importance. Such effects refer to 'the net result of environmental impact from a number of projects and activities' (Sadler, 1996, p. 161). These impacts can also be explained as spatially or temporally accumulated changes in the environment resulting from the perturbations originating from one or more resource sectors and activities over time and space (Contant and Wiggins, 1991; Spaling and Smit, 1993; Krzyzanowski and Almuedo, 2010).

Assessment of cumulative impacts is guided by a CIA process that can be explained as a systematic procedure for identifying and evaluating the significance of effects from multiple activities based on the analysis of the causes, pathways and consequences of these impacts (Cooper, 2004). According to Hegmann et al. (1999), the CIA is environmental assessment as it should always have been: an Environmental Impact Assessment (EIA) done well. This captures the notion that cumulative impacts can also result from singular developments at a project level.

No specific definition of cumulative impacts has yet been agreed upon (Cooper and Sheate, 2002; Wärnbäck and Hilding-Rydevik, 2009), and various interpretations and typologies have been proposed. Some central properties of cumulative impacts that have been recognized are: (1) cumulative effects may be experienced by different types of environmental impact receptors and at varying scales; (2) are triggered by multiple causes or impacting factors; and (3) are generated by multiple impact pathways, generally involving multiple root causes and lower and higher order effects, interlinked by cause–effect relationships.

14.3.1 Typology of Cumulative Impacts of Dams on Biodiversity

Attempts to study the cumulative effects of hydrological alterations are relatively recent compared to the study of individual dams and reservoir developments (e.g., Hall, 1971; Hecky et al., 1984). Cumulative impacts are contextual and encompass a broad spectrum of dam-induced impacts on biodiversity on different targets and on varying spatial and temporal scales associated with dams. Several studies have illustrated how forms of cumulative impacts may be differentiated (Peterson et al., 1987; Cada and Hunsaker, 1990). For discussing the impacts in the context of dams as potential drivers of change in Valued Ecosystem Components (VECs) over time, Figure 14.3 provides the typology of cumulative effects that are associated with construction and operation of hydropower and irrigation dams.

These different types of impacts are further explained here and are substantiated with examples:

1. Cumulative impacts could represent several accumulated impacts resulting from multiple lower-order effects, induced by a single hydropower project that may cause persistent environmental effects that become either interactive or additive to result in a more significant impact or a major catastrophe. Examples of such cumulative effects can be seen in land clearing for dams that may represent a visible effect that may bring about alteration in habitat used by multiple species across taxonomic groups and finally the extinction of the sensitive, Rare, Threatened and Endangered (RET) species.
2. Cumulative impacts could also be a manifestation of the multiple effects induced by a single dam when triggering of several actions may lead to multiple impacts that are sufficiently large. Examples of such impacts could be seen in a dam-induced fragmentation of forest and riverine habitats, where decline in numbers and diversity of species begin to show and thereafter may become a continuous process resulting in small isolated populations surviving in disconnected habitats (see Box 14.1).
3. Some cumulative impacts on biodiversity and ecosystem services may not result from a single stand-alone project but are a product of the ecological, socio-economic and cultural impacts from existing projects in combination with the incremental negative impacts over the long term (see Box 14.2).
4. Some cumulative effects can result from a series of hydroelectric projects constructed or planned on the same river or within the same watershed causing compounding effects on multiple receptors or ecosystem benefits that could be synergistic and irreversible over time. Examples of such effects can be best visualized in a scenario of multiple dams on a single river in a larger basin that trigger a range of impacts (e.g., conversion of the river into several lotic ecosystems, reduced volume and flows, changes in channel morphology, decline in

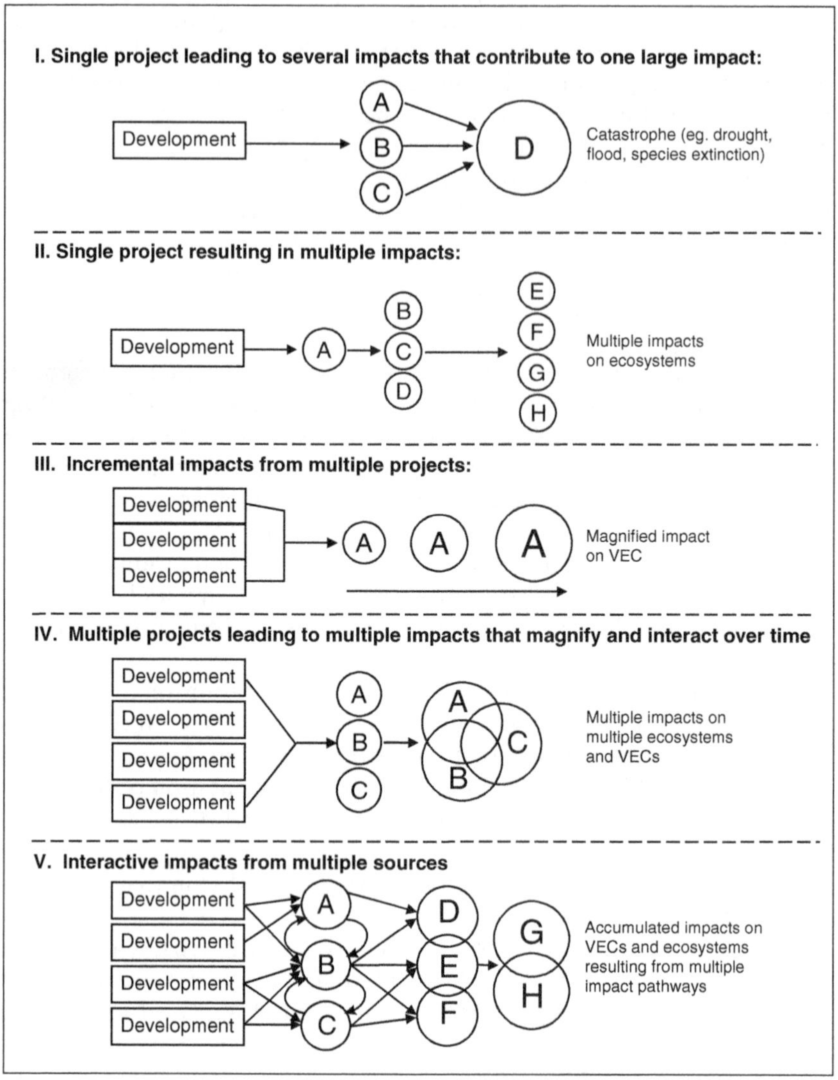

Figure 14.3 *Typology of cumulative impacts of hydropower and irrigation dams on biodiversity*

nutrient loads for aquatic species) that singly or in combination may influence multiple receptors such as fishes, aquatic mammals, terrestrial species and human communities dependent on ecosystem benefits of fisheries and fertile flood plains (see Boxes 14.3 and 14.4).

BOX 14.1 CUMULATIVE IMPACTS OF INDIVIDUAL DAMS

The Three Gorges Dam (TGD), built in China in 2006, is 1.4 miles long and 604 feet tall and represents the world's largest dam. TGD is fed by the Yangtze River and its 26 turbines to provide 18 000 megawatts of power. The dam, which has significantly altered the river system downstream because of decreases in freshwater flow and more saltwater creeping up the Yangtze, has endangered the fish populations that are already threatened by overfishing (Hvinstendahl, 2008). The dam has reduced the endangered baiji or Yangtze River dolphin's habitat and facilitated an increase in ship traffic, finally driving the species to functional extinction (Carew-Reid et al., 2010).

Since the operations of the original Theun-Hinboun Project in Lao PDR began in 1998, villagers have consistently reported that erosion occurring downstream of the power plant along the Hai and Hinboun continues to accelerate and intensify sedimentation. By 2005, it was estimated that after eight years of operations, the Hai River had widened by up to 45 m, with 68 ha of land (or between 8.5 and 14 million tons of soil) lost to erosion. Some of the land along the Hai that has collapsed into the river over the years was fertile agricultural soil previously relied upon by villagers for farming. Over the years, the area of land lost increased as the flow of water discharged downstream exacerbated bank erosion. Theun-Hinboun Hydropower Project, which became operational nearly two decades ago, resulted in the continuous decline in the quality of riparian habitat and forest and reduction in fish and other aquatic species (International Rivers, 2014 and FIVAS, 2007). Concessions for two additional hydropower dams on the Hinboun River have been granted with little consideration for the additional and cumulative impacts on the local people or on the ecosystems.

Dong Nai 2 Dam in Vietnam, which is under construction, would inundate and radically alter the ecology of the Dong Nai River for approximately 22.5 km upstream of the dam. The zone of influence (ZOI) of Dong Nai 2 covers 6601 ha of Ta Dung Nature Reserve and 1168 ha of South West Lam Dong Nature Reserve. The ZOI of Dong Nai 2 extends to approximately 4 km within Deo Nui San key biodiversity areas (KBAs), which is separated from South West Lam Dong Nature Reserve by Highway 20. It is feared that road construction (to link with the dam site) is likely to become yet another cause of disturbance inside the KBA from construction activities and pollution, thus extending the ZOI (Carew-Reid et al., 2010).

5. Some cumulative impacts associated with dams are generated through complex impact pathways, involving multiple root causes and lower- and higher-order effects, interlinked by cause–effect relationships. Several workers have highlighted the importance of considering multiple pathways and root causes in determining cumulative effects in EIA (Hyder Consulting, 1999; Modak and Biswas, 1999; Brismar, 2004). Such aggregated impacts may appear in the immediate time frame or may occur in the long term as a consequence of linked effects that become manifested in the foreseeable future (see Box 14.5). This

BOX 14.2 CUMULATIVE IMPACT RESULTS FROM INCREMENTAL CHANGES IN ECOLOGICAL CONDITIONS OVER TIME

Syr Darya and Amu Darya Rivers are central Asia's lifeline and flow into the Aral Sea and form the Aral Sea basin. In the 1960s, the Soviet government decided to divert those rivers by construction of a dam in Kazakhstan between the northern and southern parts of Aral Sea so that they could irrigate the desert region surrounding the sea for growing cotton rather than supplying water to the Aral Sea basin. Not only was all the water being diverted into canals at the expense of the Aral Sea supply, but 25–75 percent of it (depending on the time period) was being soaked up by the desert and blatantly wasted. Consequently, the water level in the Aral Sea started drastically decreasing from the 1960s onward. By the 1980s, the water from these rivers was completely utilized with little flow reaching the Aral Sea basin. The aggregated impacts of these projects over 30 years resulted in the splitting of the Aral Sea into two water bodies, the so-called 'Southern Aral Sea' and 'Northern Aral Sea'. By 2002, the level of the Southern Aral Sea had fallen by 22 m. The smaller lakes within the southern Aral Sea that were stopped from being fed by river flows became highly saline due to evaporation, causing some or all fishes reintroduced in the 1990s to die. Even re-watering the Southern Aral could not compensate for the increased salinity over the years. By 2005 finally, the inflow from the Syr Darya could be retained only in the Northern Aral Sea. The major cumulative socio-economic impacts included the desertification of Aral Sea that was once the world's fourth largest freshwater lake covering 68 300 km^2, decimation of once abundant ports and fishing communities and complete halt of fishery and the increasing salinity of the rivers that have become polluted with fertilizer and pesticides and contaminated with agricultural chemicals.

Source: ADB (2010).

category of impacts is best explained in the analysis of the influence of dams on climate change. As dams divert water from rivers, they may have significant additional climate change impacts because they drain and dry up downstream wetlands that are 'carbon sinks' holding vast amounts of greenhouse gases in soils. The draining and drying-up effects cause carbon and methane to be released and emitted into the air.

14.4 RELEVANCE OF CIA IN INFORMING DECISIONS AND ITS RELATIONSHIP WITH SEA

CIA can be undertaken at different planning levels: project and strategic levels. At the project level, CIA can help enhance the environmental performance of individual projects, improve the foci of regional and

BOX 14.3 SYNERGISTIC IMPACTS OF MULTIPLE DAMS ON THE SAME RIVER

The Siang River originates in the Chemayungdung mountain ranges, which are nearly 60 miles south-east of Mansarovar Lake in the Mount Kailash Range in Southern Tibet at an elevation of 5300 m. A spring called Tamchok Khambab spills from the glacier, which later gathers breadth and volume to become the Tsangpo, the highest river in the world. The river enters Arunachal Pradesh near Gelling from where it is known as Siang and flows for 294 km before its confluence with the Dibang and Lohit Rivers. The Siang Basin displays extraordinary richness of floral and faunal species in more than 15 000 km^2 of forest representing 82.26 percent of the entire Siang Basin. The basin has more than 100 species of orchids, 16 species of rhododendron, 14 species of bamboo, 14 species of cane, and overall 27 rare, endangered and threatened (RET) species and 46 endemic plant species. Approximately 25 (18 percent) of mammalian species found are protected species listed in India's Wildlife (Protection) Act, 1972. Out of 447 species of birds, 31 are Schedule I species. The single basin consists of five Important Bird and Biodiversity Areas (IBAs). On this river alone, 44 dams that have been planned along its entire length would result in the conversion of free-flowing river into a 208.5 km lentic ecosystem posing significant challenges by altering the entire Siang River aquatic system, which in turn would impact the aquatic biodiversity, adversely affecting fish populations and their migration behavior in the long term (SANDRP, 2014).

integrated assessment of impacts of several projects operating in the region and can inform strategic-level assessment. SEA allows these cumulative impacts to be addressed at an aggregate level.

CIA at strategic level addresses the overall effects of policies, plan and programs. It can also evaluate cumulative effects from gradual, piecemeal environmental change caused by a number of individually small projects. It identifies resources that are at risk from a cumulative change and recognizes the regional perspective of impacts.

On the other hand, the outcome of SEA can provide a revised scope for CIA (Figure 14.5). Thus SEA and CIA draw selectively from each other for improving decision outcomes (Boxes 14.5 and 14.6). By focusing on how a given receptor is affected by the totality of plans, projects and activities, rather than on the effects of a particular plan or project, CIA links the different scales of environmental assessment and provides a critical insight for conflict resolution, reforms in policy and fresh arguments for decision-making. Depending on whether the assessment of cumulative effects is built upon environmental assessments or will feed into strategic decisions, the CIA is an ideal, region-based approach that is closely linked to the concepts of 'limits of acceptable change' and 'thresholds of significance', which appear to be strongly emerging tools in SEA.

BOX 14.4 CUMULATIVE IMPACTS OF MULTIPLE DAMS ON THE INDIAN HIMALAYAN RIVER SYSTEMS

The state of Uttarakhand, which is uniquely endowed with glaciers and rain-fed monsoonal rivers, has a hydropower potential of the order of 20 000 MW against which only about 3164 MW (approx. 16 percent) has been harnessed. The hydropower development in the region is being recognized as having critical significance in meeting the state's energy demand for sustainable development. Based on the current energy plan of the state, as many as 70 hydroelectric projects are largely to be concentrated in two river basins with regard to Alaknanda and Bhagirathi. Among the various allotted hydroelectric projects in these two basins, 17 are commissioned hydroelectric projects with total installed capacity of 1851 MW; 14 projects of 2538 MW capacity are in the advanced stages of construction and 39 projects with installed capacity of 4644 MW are in various stages of planning.

The courses of Bhagirathi and Alaknanda support a number of forest formations that support unique biodiversity values and high densities of species, particularly Rare, Threatened and Endangered (RET) species of flora and fauna. Among the fishes are included several threatened species such as the golden mahseer (*Tor putitora*) and snow trout (*Schizothorax richardsonii*) that breed in the Himalayan Alaknanda and Bhagirathi river basins. The Gangotri National Park (NP) is located in the upper catchment of Bhagirathi. About 15 species of mammals and 150 bird species have been documented from Gangotri NP. This includes some of the rare and charismatic species such as snow leopard, black bear, brown bear, musk deer, blue sheep, Himalayan tahr, Himalayan monal, koklass pheasant, and Himalayan snowcock. Within the Alaknanda Basin, the Nanda Devi NP and Valley of Flowers NP are two areas with rich biodiversity values that have together been designated as a UNESCO World Heritage Site.

In the event of development proceeding as planned, of the 1121 km long stretch of rivers that flow in the entire Alaknanda-Bhagirathi basins, a minimum of 526.8-km-long river stretch would be affected due to changes in the flows. Similarly, 9494.69 ha of forest would be lost on account of land take for infrastructure and submergence if all planned developments are implemented. Changes in length of free-flowing rivers and loss of terrestrial habitats would be the two key factors that would become the root causes of multiple impacts that would have compounding effects on a range of receptors including aquatic and terrestrial species, ecosystems benefits and the far-reaching consequences for biodiversity conservation and well-being of dependent human communities. Dam-induced cumulative effects would also have social and economic consequences. For instance, these rivers that command great religious sentiment and value would have reduced flow, which will constrain its use for cultural and religious activities of the Hindu community of India.

Potential cumulative effects of dams on two of the Himalayan rivers on major receptors from multiple impact pathways are presented in Figure 14.4.

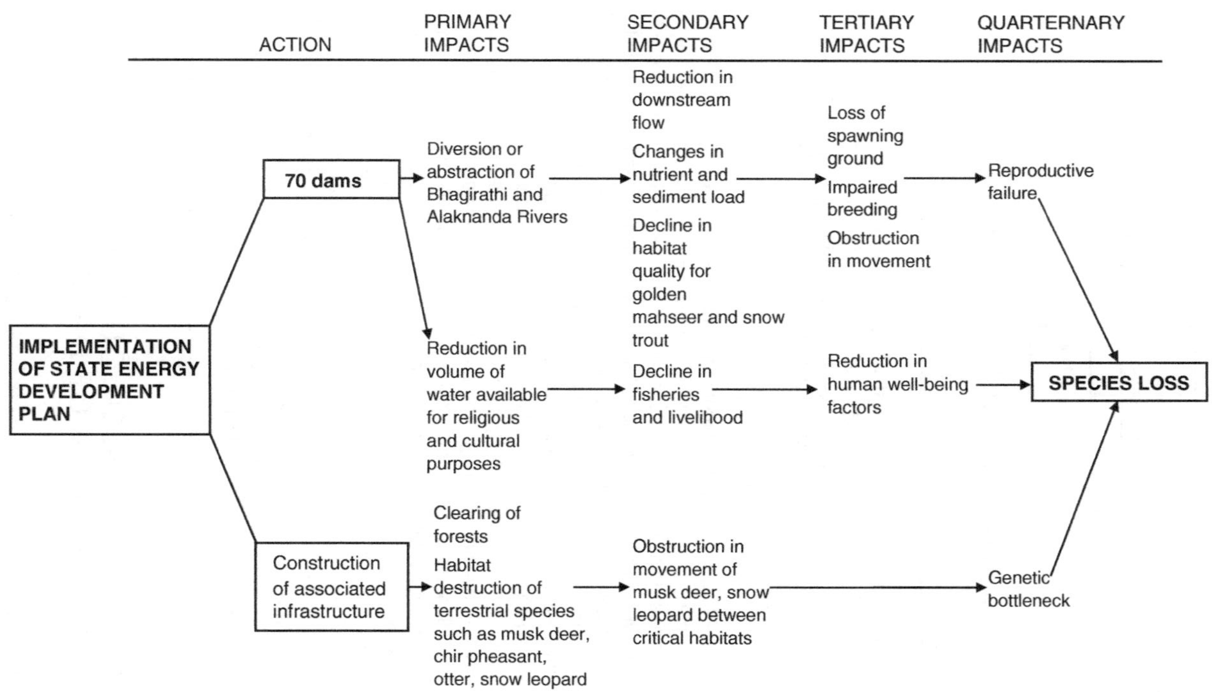

Source: Rajvanshi et al. (2012).

Figure 14.4 Potential cumulative effects of dams on two of the main Himalayan rivers

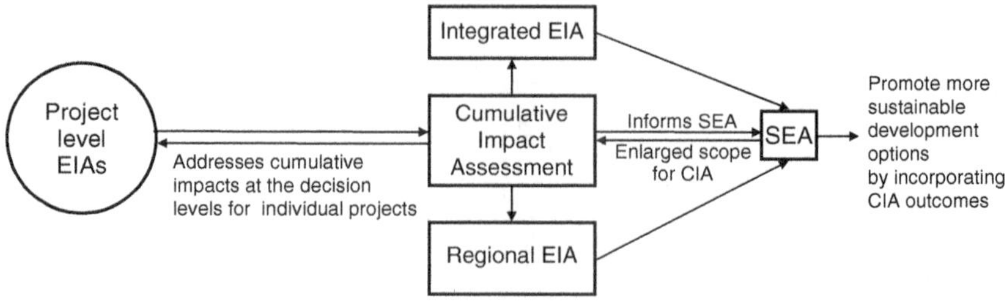

Figure 14.5 Relationship of CIA with project-level EIA and SEA

BOX 14.5 RESULTS OF CUMULATIVE IMPACT ASSESSMENT CAN INFORM STRATEGIC ENVIRONMENTAL ASSESSMENT (SEA) AND GENERATE MORE OPTIONS FOR DECISION-MAKING

The Mekong River basin, site of the biggest inland fishery in the world and home to 65 million people in six countries (China, Myanmar, Laos, Thailand, Vietnam and Cambodia), is undergoing massive hydropower development within a watershed of 800 000 km^2. Driven by increasing demand for electricity and a desire for economic development, 11 dams are being planned on the main river: 41 on the tributaries that are expected to be completed within the next four years and yet another 10–37 tributary dams likely to be built between 2015 and 2030. Ziv et al. (2012) modeled fish migration in the event that all of the proposed dams are constructed, and predicted that the fish productivity would reduce by 51 percent and endanger 100 migratory fish species. Although the fate of 27 tributary dams is yet to be determined, studies assessed that the cumulative losses in fish biodiversity and production would be greater than from the proposed dams on the upper reaches of the Lower Mekong River. The study also suggested that the cumulative impacts of all 27 projects would be far more serious, not only because of the total area that will be blocked for fish migration, but also because some regions have more important fish than others (ibid.). For the reasons that these dams that were not subjected to SEA and could have catastrophic effects on fish productivity and biodiversity, the cumulative assessments argued the need for reconsideration of hydropower development in the entire Mekong River basin.

14.5 KEY TO SUCCESSFUL OUTCOME FROM CONDUCTING CIA OF DAMS

Since the objective of CIAs of dams is invariably to provide an empirical basis for identifying key immediate and future impacts, including those on biodiversity and ecological systems for informing decisions, these must

BOX 14.6 CONTRIBUTION OF CUMULATIVE IMPACT ASSESSMENT IN PROVIDING OPTIONS FOR ENERGY PLANNING TO SAFEGUARD CONSERVATION VALUES OF INDIAN HIMALAYAN LANDSCAPE

The state of Uttarakhand in the northwestern region of India has an estimated hydropower potential of 20236 MW, against which only about 16 percent (3164 MW) has been harnessed so far. From the recent energy plans prepared for the developments in hydropower sector of the state, it became evident that as many as 70 hydroelectric projects would be concentrated on the River Ganges and its tributaries in the two river basins with regard to Alaknanda and Bhagirathi. Of these, 17 dams are already commissioned, 14 projects are under various stages of construction and 39 are at the proposal stage.

The Ministry of Environment, Forest and Climate Change, Government of India, called for Cumulative Impact Assessment (CIA) of all 70 dams to estimate the extent to which hydropower potential identified in the basins should be developed without risking stability of landforms and environment in the Himalayan region and without endangering the biodiversity of the region.

The CIA study was undertaken by the Wildlife Institute of India to assess the combined effects of multiple developments or activities on a range of receptors with an objective to generate 'alerts' for safeguarding priority areas for conservation from impacts of hydropower schemes (existing, under construction and proposed) in the two basins; provide a 'risk forecast' for Rare, Threatened and Endangered (RET) species and critical habitats in the event of developments proceeding as planned; and to finally present a menu of scenarios for decision-makers to approve options that best help in aligning energy planning with biodiversity conservation for sustainable developments in the energy sector in the state. This assessment also aimed to contribute to strategic planning for promoting 'green decisions' that are more environmentally sustainable, especially in terms of leading to greater conservation of biodiversity while meeting goals of energy planning.

Four scenarios were developed to explore various trajectories of change that may lead to a broadening range of plausible alternatives for securing and safeguarding priority biodiversity values. These were:

- Scenario 1: Presenting the exclusive impacts of only the commissioned projects to assess their influence on the true biodiversity baseline of the basins in the 'no dam' scenario.
- Scenario 2: Assessing the combined impacts of all commissioned projects and those under different stages of construction on biodiversity values.
- Scenario 3: Evaluating the cumulative effects of all projects including commissioned projects, those under progressive stages of construction and those that are still in the form of proposals for consideration.
- Scenario 4: Presenting the incremental impacts of the proposed projects by targeting only the proposed projects.

Based on the significance of existing and future impact potential of all dams, three alternative development options were presented for making decisions with respect to energy planning in the state (Table 14.1). These alternatives provided the estimates of overall gains and losses for biodiversity and power production in the event of developments proceeding as planned or when regulated by proposing exclusion of some dams to optimize benefits for conservation and power development.

Table 14.1 Estimates of overall gains and losses for biodiversity and power production

Recommend Commissioned Projects Only	Recommend Commissioned Projects and those Under Construction	Regulate Power Production by Excluding select 24 Projects from the Plan
Commissioned projects have already impacted the biodiversity Prospects of conserving biodiversity likely to be further compromised by projects under construction Decline in population of golden mahseer, an endangered fish in upstream stretches of Bhagirathi River, has already occurred due to construction of a major dam, the Tehri Dam	47% river stretch would be additionally affected by all projects under different phases of construction 87% fish species would be affected by changes in environmental flow of rivers already influenced by commissioned projects and by development of dam in progress Diversion of 1700 ha of forest area for dams would lead to a loss of critical habitats of many RET species	37% reduction in river length that would be affected 21.71% decrease in the total forest land required (9494.68 ha) 27% reduction in planned power generation capacity Power deficit by improving efficiency of existing projects can be effectively compensated Reduction in transmission loss that currently accounts for 40% of total power generated could reduce the need for projects

The Government of India has greatly benefitted from this cumulative assessment in understanding the dimensions of impacts linked to planned and unplanned developments and has accordingly looked at the options of exclusion of some of the proposed dams from the state's energy plan.

Source: Rajvanshi et al. (2012).

be designed to predict the spatial and temporal aspects of impacts against the logically determined ecological thresholds. Adequate data, a suite of appropriate assessment tools and a legally supported framework become important requisites for achieving successful CIA outcomes.

A number of practical lessons for overcoming the challenges of conducting CIA and improving the practice and process of CIA have emerged. One of the big challenges that constrain CIA of dams is either the lack of ecological data or the fact that biodiversity-relevant data remains usually limited to descriptive information that cannot be used as a basis for devising a scorecard for numerically presenting the biodiversity values of the site. As the assignment of biodiversity values to land units is a key component of the CIA process, there is a strong need to develop biodiversity criteria and measurable standards against which the biodiversity values of the site can be evaluated. Biodiversity criteria must at least capture the importance of the ecological/biological characteristics of these areas, especially in terms of rarity, vulnerability or irreplaceability of the values. While no standard set of criteria has emerged, multitudes have been used in practice (Smith and Theberge, 1986; Treweek, 1999; Regan et al., 2007). Some of the important and more often used criteria for valuing biodiversity for the baseline stage in CIAs of dams (ICEM, 2007; Rajvanshi et al., 2012) are summarized in Table 14.2.

Other recommended criteria include diversity of habitats affected, catchment connectivity at the sub-basin scale and direct and indirect influence to lands designated as conservation and biodiversity priorities.

The priorities and targets set in the National Biodiversity Strategy and Action Plan process should also be additionally considered in developing biodiversity criteria that becomes equally relevant in the given context of biodiversity conservation priorities of different countries.

Failure to delineate precise boundaries of study area is another factor that further compounds the difficulty of conducting ecological assessment for CIA. For most CIA of dams, wildlife habitats forming parts of the submergence areas of individual dams, areas of land diverted for associated infrastructure and construction of utilities and receiving areas for people and wildlife to be displaced by dams become discernible areas for defining impact zones. The intervening distance between dam site and areas that provide extended buffers for the protection of the habitats of many species from adverse impacts of construction activities are seldom included in the influence zone for ecological assessments. A commonly held notion that species found in national parks and sanctuaries are not likely to be impacted by the projects in the vicinity can make CIA illogical because of the assumption that species confined within protected areas would not use other suitable habitats within the zones of submergence of dam. The

Table 14.2 Criteria for assessing biodiversity values

Criteria	Description
RET (rare, endangered and threatened) species, as per IUCN and other global criteria	Number of RET species present in the sub-basin
Protected species	Number of species that are listed in the national legislations reflecting statutory provisions for protection of species
Species richness	Number of different species represented in a set or collection of individuals
Medicinal species	Number of species having medicinal properties hence valuable ecologically and economically
Endemic species	Number of endemic species present in the sub-basin, reflecting the irreplaceable values and the national importance that the species command
Habitat diversity	Number of habitat types as a surrogate for habitat heterogeneity and biodiversity richness
Species richness	Number of different species present in given land units
Breeding/congregation	Presence/absence of breeding sites and congregation opportunities for the target taxonomic group in a sub-basin
Migratory pathways/ corridor	Presence/absence of migratory pathways/corridor for aquatic biodiversity in the sub-basins

assessment of impacts of Debang Multipurpose Project in Arunachal Pradesh in India on wildlife species of Mehao Wildlife Sanctuary, located about 11 km from reservoir periphery followed similar assumptions and arguments in portraying the impacts of the project (Chernaik, 2007).

CIA must consider that many endangered mammalian species with large individual ranges are not capable of determining when they would remain within the bounds of a wildlife sanctuary and that there would be likely impacts on faunal elements of the ecosystem if the area under project(s) provides habitat for the animal species prior to development of a dam.

Although scholarly research has been limited to developing process frameworks or practical methodologies for cumulative impact analysis of dams, available literature justifies that a number of tools for predicting cumulative effects of dams are being successfully employed. These include the use of matrices, impact network models or systems models for detailed assessment of cause–effect relationships between actions and valued resources and impact interactions and the trends analysis for assessing the status of resources and changes in the occurrence or intensity of stresses over time (Cooper, 2004). Also, spatial analysis using geographical information systems (GIS) to analyze changes in landscape features in response to cumulative impacts of development actions in the context of dams are being well employed (Mennis, 2002). Similarly, scenario building is being increasingly encouraged in CIAs as a 'futuring' approach for predicting outcomes of various scenarios to improve the information for decision-makers (Duinker and Greig, 2007).

Approaches that are especially relevant in the context of assessment of impacts on biodiversity need to be further improvised and new or more innovative tools must be evolved for more significant contributions to address impacts on biodiversity and ecosystem benefits. Among such methods, the most appropriate method is the one that best provides an assessment of the effects on the Valued Ecosystem Component (VEC) being examined. VEC commands sufficient merit and should therefore be increasingly encouraged for good and meaningful outcomes of CIA of dams. Canter and Ross (2010) strongly argue that VECs should form the fulcrum of CIA studies, as they provide a means to adequately integrate project effects at local, regional, and strategic spatial areas. Focusing on VECs in CIA can provide an opportunity for taking a closer look at key environmental attributes and their sustainability implications than may be necessary in project EAs (Bérubé, 2007; Senner, 2011).

Approaches employed for assessing cumulative impact of dams that recognize river systems as providers of goods and services are also equally relevant for portraying undesirable impacts of large dam projects and must be encouraged. More recent CIAs of dams (Rajvanshi et al., 2012; World Bank, 2012; Kibler and Tullos, 2013) have recommended lengths of river channel inundated or dewatered by dams as more reliable criteria for assessing cumulative impacts on ecosystem services offered by flowing rivers. Aggregated impacts on a VEC (e.g., aquatic habitat, fish species) may be more significant than the individual project-specific impacts on the same VEC. Conclusions drawn from CIA in such cases can provide a convincing basis for reviewing policy on values of minimum flow to be maintained in the river by individual project owners or in defining the limits of diversion of water for harnessing hydroenergy based on water budget of a river basin or a watershed.

This benefit of CIA is best illustrated through an example of the CIA of 70 dams in Indian Himalaya (Rajvanshi et al., 2012). Based on the CIA, it could be concluded that to sustain riverine ecology with special reference to fishes in the Alaknanda and Bhagirathi, 20 percent of monthly average of flow would have to be maintained as a Minimum Environmental Flow required (MEFR) (cumec/day) during dry season (November to March), 25 percent during winter (October and April) and 30 percent of monthly average of high flow season (May to September). The suggested MEFR would provide the necessary environmental cues to trigger the breeding and migration behavior of Himalayan fishes. Based on this assessment, the Ministry of Environment, Forests and Climate Change, Government of India, is working on minimum flow requirement in these basins as a condition for clearance of projects.

Supportive political and policy contexts for taking CIA forward are other key requisites for successful practice. While development of frameworks for actual understanding of how to evaluate cumulative impacts has largely taken place in some countries such as the USA and Canada – partly as a result of community pressures on science and government and partly in response to strong legislative support – many countries still lack the legal backing to encourage CIA as a mandatory approach for upscaling biodiversity. The situation is far bleaker in the context of countries that are leading in dam building. The support for propelling CIA as a promising decision support tool would require continued development of new frameworks, exchange of experience and lessons learnt among impact assessment professionals, academia and all other beneficiaries of CIA outcomes.

NOTE

* The author is grateful to faculty colleagues whose valuable contributions in the study of cumulative impacts of dams in Alaknanda and Bhagirathi basins in India provided the backdrop for Box 14.4. The author also sincerely thanks Akanksha Saxena, Debojyoti Mukherjee and Narinder Bist of the EIA Cell of the Wildlife Institute of India who provided significant inputs in illustrations for this chapter.

REFERENCES

Asian Development Bank (ADB) (2010), *Central Asia Atlas of Natural Resources*, Manila: Asian Development Bank.

Berga, L., J.M. Buil and E. Bofill et al. (2006), 'Dams and reservoirs, societies and environment in the 21st century', in *Proceedings of the International Symposium on Dams in Societies of the 21*st *Century, Barcelona*, 18 June 2006.

Bergkamp, G., M. McCartney and P. Dugan et al. (2000), *Dams, Ecosystem Functions and Environmental Restoration, Thematic Review II.1*, prepared for the World Commission on Dams (WCD).

Bérubé, M. (2007), 'Cumulative effects assessment at Hydro-Quebec: what have we learned?', *Impact Assessment and Project Appraisal*, **25**(2), 101–9.

Brismar, A. (2002), 'River systems as providers of goods and services: a basis for comparing desired and undesired effects of large dam projects', *Journal of Environmental Management*, **29**(5), 598–609.

Brismar, A. (2004), 'Attention to impact pathways in EIAs of large dam projects', *Environmental Impact Assessment Review*, **24**(1), 59–87.

Brooker, M.P. (1981), 'The impact of impoundments on the downstream fisheries and general ecology of rivers', in M.P. Brooker (ed.), *Advances in Applied Biology*, London: Academic Press, pp. 91–152.

Bunn, S.E. and A.H. Arthington (2002), 'Basic principles and ecological consequences of altered flow regimes for aquatic biodiversity', *Environmental Management*, **30**(4), 492–507.

Cada, G.F. and C.T. Hunsaker (1990), 'Cumulative impacts of hydropower development: reaching a watershed in impact assessment', *Environmental Professional*, **12**(1), 2–9.

Canter, L. and B. Ross (2010), 'State and practices of cumulative effects assessment and management: the good, the bad and the ugly', *Impact Assessment and Project Appraisal*, **28**(4), 261–8.

Carew-Reid, J. J. Kempinski and A. Clausen (2010), *Biodiversity and Development of the Hydropower Sector: Lessons from the Vietnamese Experience – Volume I: Review of the Effects of Hydropower Development on Biodiversity in Vietnam*, Hanoi: ICEM – International Centre for Environmental Management for the Critical Ecosystem Partnership Fund.

Casado, C., D. Garcia de Jalon and C.M. Del Olmo et al. (1989), 'The effect of an irrigation and hydroelectric reservoir on its downstream communities', *Regulated Rivers: Research and Management*, **4**(1), 275–84.

Chernaik, M. (2007), *Evaluation of the Environmental Impact Assessment Report and Environmental Management Plan for the Dibang Multipurpose Project (3000 MW)*, St. Eugene, OR: Environmental Law Alliance Worldwide.

Contant, C.K. and L.L. Wiggins (1991), 'Defining and analyzing cumulative environmental impacts', *Environmental Impact Assessment Review*, **11**(4), 297–309.

Cooper, L.M. (2004), 'Guidelines for cumulative effects assessment in SEA of plans', *EPMG Occasional Paper No. 04/LMC/CEA*, London: Imperial College.

Cooper, L.M. and W. Sheate (2002), 'Cumulative effects assessment: a review of UK environmental impact statements', *Environmental Impact Assessment Review*, **22**(4), 415–43.

Da Silva, J.J.L.S., M. Marques and J.M. Damasco (2010), 'Impactos do desenvolvimento do potencial hidroelectrico sobre os ecossistemas aquaticos do Rio Tocantins' [Impacts of the development of the hydroelectric potential on the aquatic ecosystems of the Rio Tocantins], *Revista Ambiente & Agua*, **5**(1), 189–203.

Dudgeon, D. (2000), 'Large-scale hydrological changes in tropical Asia: prospects for riverine biodiversity', *Bioscience*, **50**(9), 793–806.

Duinker, P.N. and L.A. Greig (2007), 'Scenario analysis in environmental impact assessment: improving explorations of the future', *Environmental Impact Assessment Review*, **27**(3), 206–19.

FIVAS (2007), *Ruined River, Damaged Lives: Impacts of the Theun-Hinboun Hydropower Project on Downstream Communities in Lao PDR*, Oslo: The Association for International Water Studies.

Freeman, M.C., C.M. Pringle and C.R. Jackson (2007), 'Hydrologic connectivity and the contribution of stream headwaters to ecological integrity at regional scales', *Journal of the American Water Resources Association*, **43**(1), 5–14.

Friedl, G. and A. Wüest (2002), 'Disrupting biogeochemical cycles: consequences of damming', *Aquatic Science*, **64**(1), 55–65.

Ginneken, V.M. (2015), 'A decade of sustainable hydropower development – what have we learned?', accessed 17 February 2015 at http://www.hydroworld.com/articles/print/volume-23/issue-1/articles/a-decade-of-sustainable-hydropower-development-what-have-we-learned.html.

Gleik, P.H. (2003), 'Global freshwater resource: soft-path solutions for the 21st century', *Science*, **302**(5650), 1524–8.

Grumbine, R.E. and M.K. Pandit (2013), 'Threats from India's Himalaya dams', *Science*, **339**(6115), 36–7.

Hall, G.E. (ed.) (1971), *Reservoir Fisheries and Limnology*, Special Publication No. 8, Washington, DC: American Fisheries Society.

Hecky, R.E., R.W. Newbury and R.A. Bodaly et al. (1984), 'Environmental impact prediction and assessment: the Southern Indian Lake experience', *Canadian Journal of Fisheries and Aquatic Sciences*, **41**(4), 720–32.

Hegmann, G., C. Cocklin and R. Creasy et al. (1999), *Cumulative Effects Assessment Practitioners Guide*, Hull, Quebec, prepared by AXYS Environmental Consulting Ltd. and the CEA Working Group for the Canadian Environmental Assessment Agency.

Huang, H. and Z. Yan (2009), 'Present situation and future prospect of hydropower in China', *Renewable and Sustainable Energy Review*, **13**(6), 1652–6.

Hvinstendahl, M. (2008), 'China's three gorges dam: an environmental catastrophe?' *Scientific American*, 25 March, accessed 3 October 2011 at http://www.scientificamerican.com/article.cfm? id=chinas-three-gorges-dam-disaster.

Hyder Consulting (1999), *Study on the Assessment of Indirect and Cumulative Impacts as well as Impact Interactions*, EC DGXI Environment, Nuclear Safety and Civil Protection, Brussels, accessed 25 January 2016 at http://ec.europa.eu/environment/archives/eia/eia-studies-and-reports/pdf/volume1.pdf.

International Centre for Environmental Management (ICEM) (2007), *Pilot Strategic Environmental Assessment in the Hydropower Sub-sector, Final Report: Risks to Biodiversity from Hydropower in the 6th Power Development Plan*, Hanoi: ICEM for World Bank, MONRE, MOI and EVN.

International Journal of Hydropower and Dams (IJHD) (2004), 'World Atlas and Industry Guide', *Impact Assessment and Project Appraisal*, **25**(2), 101–9.

International Rivers (2014), *Failure to Restore: An Assessment of the Impacts of the Theun-Hinboun Hydropower Dam Projects on Downstream Communities in Laos*, accessed 24 January 2016 at http://www.banktrack.org/manage/ems_files/download/thhp_and_thxp_downstream_impacts_report_2014_pdf/thhp_and_thxp_downstream_impacts_report_2014.pdf.

Kibler, K.M. and D.D. Tullos (2013), 'Cumulative biophysical impact of small and large hydropower development in Nu River, China', *Water Resources Research*, **49**(6), 3104–18.

Krzyzanowski, J. and P.L. Almuedo (2010), *Cumulative Impacts of Natural Resource Development on Ecosystems and Wildlife: An Annotated Bibliography for British Columbia*, Forrex Forum for Research and Extension in Natural Resources, Kamloops, BC, accessed 13 March 2015 at http://www.forrex. org/sites/default/files/forrex_series /FS26.pdf.

Lehner, B., C.R. Liermann and C. Revenga et al. (2011), 'High-resolution mapping of the world's reservoirs and dams for sustainable river-flow management', *Frontiers in Ecology and the Environment*, **9**(9), 494–502.

Ligon, F.K., W.E. Dietrich and W.J. Trush (1995), 'Downstream ecological effects of dams', *Bioscience*, **45**(3), 183–91.

Liu, J.K. and Z.T. Yu (1992), 'Water quality changes and effects on fish populations in the Hanjiang River, China, following hydroelectric dam construction', *Regulation Rivers: Resource Management*, **7**(4), 359–68.

Mennis, J. (2002), 'Using geographic information systems to create and analyze statistical surfaces of population and risk for environmental justice analysis', *Social Science Quarterly*, **83**(1), 281–97.

Modak, P. and A.K. Biswas (1999), *Conducting Environmental Impact Assessment for Developing Countries*, Tokyo: United Nations University Press.

Peterson, E.B., Y.-H. Chan and N.M. Peterson et al. (1987), *Cumulative Effects Assessment in Canada: An Agenda for Action and Research*, Hull, Quebec: Canadian Environmental Assessment Research Council.

Petts, G.E. (1984), *Impounded Rivers: Perspectives for Ecological Management*, Chichester, UK: Wiley.

Postel, S.L. (2000), 'Entering an era of water scarcity: the challenges ahead', *Ecological Applications*, **10**(4), 941–8.

Postel, S. and B. Richter (2003), *Rivers for Life: Managing Water for People and Nature*, Washington, DC: Island Press.

Rajvanshi, A., R. Arora and V.B. Mathur et al. (2012), *Assessment of Cumulative Impacts of Hydroelectric Projects on Aquatic and Terrestrial Biodiversity in Alaknanda and Bhahirathi Basins, Uttarakhand*, Wildlife Institute of India, Technical Report.

Regan, H.M., F.W. Davis and S.J. Andelman et al. (2007), 'Comprehensive criteria for biodiversity evaluation in conservation planning', *Biodiversity and Conservation*, **16**(9), 2715–28.

Revenga, C., J. Brunner and N. Henninger et al. (2000), *Pilot Analysis of Global Ecosystems: Freshwater Systems*, Washington, DC: World Resources Institute.

Richter, B.D., S. Postel and C. Revenga et al. (2010), 'Lost in development's shadow: the downstream human consequences of dams', *Water Alternatives*, **3**(2), 14–42.

Rosenberg, D.M., F. Berkes and R.A. Bodaly et al. (1997), 'Large-scale impacts of hydroelectric development', *Environmental Review*, **5**(1), 27–54.

Sadler, B. (1996), *Environmental Assessment in a Changing World: Evaluating Practice to Improve Performance*, International Study of the Effectiveness of Environmental Assessment Final Report, Canada: International Association for Impact Assessment and Canadian Environment Assessment Agency.

SANDRP (2014), 'Cumulative impact assessment study of Siang basin in Arunachal Pradesh: serious shortcomings; pro large hydro bias', *South Asia Network on Dams, Rivers and People* [blog], accessed 25 January 2016 at https://sandrp.wordpress.com/2014/02/18/cumulative-impact-assessment-study-of-siang-basin-in-arunachal-needs-urgent-improvement/.

Schelle, P., U. Collier and J. Pittock (2004), 'Rivers at risk – dams and the future of freshwater ecosystems', paper presented in the 7th International River Symposium, 4 September, Brisbane: World Wildlife Fund (WWF), accessed 14 March 2015 at htpp://assets.panda.org/downloads/riversatriskriversymposiumpapercollier.doc.

Senner, R. (2011), 'Appraising the sustainability of project alternatives: an increasing role for cumulative effects assessment', *Environmental Impact Assessment Review*, **31**(5), 502–5.

Smith, P.G.R. and J.B. Theberge (1986), 'A review of criteria for evaluating natural areas', *Environmental Management*, **10**(6), 715–34.

Soito, J.L.D.S. and M.A.V. Freitas (2011), 'Amazon and the expansion of hydropower in Brazil: vulnerability, impacts and possibilities for adaptation to global climate change', *Renewable and Sustainable Energy Reviews*, **15**(6), 3165–77.

Spaling, H. and B. Smit (1993), 'Cumulative environmental change: conceptual frameworks, evaluation approaches, and institutional perspectives', *Environmental Management*, **17**(5), 587–600.

Treweek, J. (1999), *Ecological Impact Assessment*, Hoboken, NJ: Wiley-Blackwell.

Wärnbäck, A. and T. Hilding-Rydevik (2009), 'Cumulative effects in Swedish EIA practice – difficulties and obstacles', *Environmental Impact Assessment Review*, **29**(2), 107–15.

World Bank (2012), 'Sample guidelines: cumulative environmental impact assessment for hydropower projects in Turkey', *Working Paper No. 76998*, Washington, DC: World Bank.

World Commission on Dams (WCD) (2000), *Dams and Development: A New Framework for Decision Making*, London: Earthscan.

Ziv, G., E. Baran and S. Nam et al. (2012), 'Trading-off fish biodiversity, food security, and hydropower in the Mekong river basin', *Proceedings of the National Academy of Sciences of the United States of America*, **109**(15), 5609–14.

PART III

CURRENT ISSUES AND CHALLENGES

15. Addressing the interactions between biodiversity conservation and poverty alleviation in impact assessment

Dilys Roe and Davide Geneletti

15.1 INTRODUCTION

Throughout the years, biodiversity has received increasing attention in impact assessment, as shown by the many guidance documents (Rajvanshi et al., 2007; EC, 2013), scientific publications (Treweek et al., 2005; Scolozzi and Geneletti, 2011, 2012) and reviews of practice (Geneletti, 2006; Gontier et al., 2006) that have been produced to foster biodiversity-inclusive impact assessment processes. The Convention on Biological Diversity (CBD) explicitly recognizes that impact assessment processes, such as Environmental Impact Assessment (EIA) and Strategic Environmental Assessment (SEA), are important tools to support decision-making for development proposals that have potentially significant impacts on biodiversity. Voluntary guidelines on 'biodiversity-inclusive impact assessment' were first adopted by parties to the CBD in 2002 and then further refined in 2006 (SCBD, 2006). The conceptual framework of these guidelines explores how physical and social (and economic) interventions lead to biophysical and social changes, which in turn influence different aspects of biodiversity. If these impacts are significant, this has an impact on the ecosystem services (ES) provided by biodiversity, which can then affect human well-being.

In line with the three objectives of the Convention, the CBD guidance is intended to explore the impacts of external interventions on biodiversity conservation, on the potential for sustainable use, and on access to, and/or rights over biological resources (ibid.). What is still overlooked, however, are the implications of biodiversity impacts for poverty alleviation. The Strategic Plan for Biodiversity 2011–2020 emphasizes that biodiversity 'contributes to local livelihoods, and economic development, and is essential for the achievement of the Millennium Development Goals, including poverty reduction' (SCBD, 2010). There is, however, no straightforward relationship between conserving biodiversity and reducing poverty and thus no straightforward path between the impacts on ES and impacts on human well-being in the CBD conceptual framework.

First, the links between ES and biodiversity are complex in themselves (Cardinale et al., 2012), and second, different groups of poor people use and benefit from different components or attributes of biodiversity and so some bits of biodiversity may be more important than others when it comes to assessing their contribution to poverty alleviation (Roe et al., 2014). Understanding how poor people interact with biodiversity and how this interaction contributes to their livelihoods can provide a more nuanced understanding of the social consequences of biodiversity impacts and allow more insights to be derived from Biodiversity Impact Assessments.

Clearly the poverty implications of biodiversity loss are not going to be relevant in every context in which a Biodiversity Impact Assessment might be carried out. In poor countries, however, poverty alleviation is often a far higher political priority than biodiversity conservation. Making the case that minimizing the negative impacts of external interventions on biodiversity can also help avoid any unexpected consequences for poverty can play a crucial role in attracting the attention of policy-makers and ensuring political buy-in to the mitigation actions that are proposed. The Strategic Plan for Biodiversity (SCBD, 2010) highlights the need to 'mainstream biodiversity across government and society' (Strategic Goal A), including integrating biodiversity into 'national and local development and poverty reduction strategies and planning processes' (Aichi Target 2).

This chapter starts by presenting a conceptual framework to guide thinking about the interactions between different components or attributes of biodiversity and different dimensions of poverty. The framework recognizes that both biodiversity and poverty are complex multi-dimensional concepts and, furthermore, that the interactions between the two are mediated by a wide range of factors. In particular, cross-cutting determinants such as governance, policies on poverty and biodiversity protection, and population growth and density that are associated with the socio-economic context are critical in determining whether or not biodiversity leads to actual poverty reduction. The chapter then continues by reviewing the existing evidence base on biodiversity–poverty linkages, drawing on a systematic mapping of the published and grey literature (Section 15.3). The findings of the review are used in Section 15.4 to formulate key distinctions that help to understand whether and how biodiversity helps in alleviating poverty, and in Section 15.5 to discuss their implications for impact assessment practice.

15.2 LINKING BIODIVERSITY AND POVERTY ALLEVIATION – A CONCEPTUAL FRAMEWORK

Biodiversity is defined by the CBD as 'the variability among living organisms from all sources including, inter alia, terrestrial, marine and other aquatic ecosystems and the ecological complexes of which they are part; this includes diversity within species, between species and of ecosystems'. But the term is often used without any particular focus on variability, and often simply as a synonym for nature in general. The term biodiversity is often used to refer to the *amount* (in terms of abundance or biomass) of a particular species or type of resource, or the *extent* of habitat rather than diversity per se (Leisher et al., 2013; Vira and Kontoleon, 2013). When thinking about how biodiversity contributes to poverty alleviation it is useful to consider whether it is indeed variability that is important or whether some other aspect of biodiversity is more critical.

Poverty is another term with many different definitions. The simplest usually relate to some level of material wealth – for example, the Millennium Development Goal to 'eradicate extreme poverty' refers to the billion-plus people whose income is less that US$1 a day. However, poor people often do not define themselves in cash income terms – indeed the concept of cash is completely meaningless for some indigenous communities who live outside of the cash economy. It has therefore become increasingly recognized that poverty is multi-dimensional. The World Bank, for example, describes poverty as 'to be hungry, to lack shelter and clothing, to be sick and not cared for, to be illiterate and not schooled' (World Bank, 2001) while Sen discusses poverty in terms of 'capabilities' to fulfil a productive life including good health, access to education and political freedom (Sen, 1999). This conceptualization of poverty is similar to that of human well-being, and indeed, poverty can be thought of as the opposite – or absence – of well-being (Millennium Ecosystem Assessment, 2005).

At some levels the relationship between biodiversity and poverty is absolute – biodiversity underpins the delivery of essential ES on which the whole of humanity is dependent. But there is certainly no linear relationship. For example, the Millennium Ecosystem Assessment (MA) suggested that while many millions of people have benefited from the transformation of ecosystems and exploitation of natural resources, the benefits have not been evenly or equitably distributed, with the poor being the biggest losers (ibid.). Other commentators have noted the dynamic and context-specific nature of the biodiversity conservation–poverty alleviation relationship (Kepe et al., 2004) and have suggested that factors such as individual access to, and control over, resources, policies on poverty and biodiversity protection, population growth and density are critical in determining whether or not the existence, or use, of biodiversity leads to poverty reduction (Tekelenburg et al., 2009).

Understanding how biodiversity contributes to poverty alleviation/ human well-being or livelihood support thus requires thinking about the relationship in a number of stages. First, it requires unpacking the term biodiversity to understand which particular components or attributes of biodiversity are important in supporting poor people's livelihoods. Is diversity important or is it more about the availability of certain key species? Are particular species important or is it rather the general type of resource, for example, fish, or function, for example, pollinators? Second, it requires unpacking the term poverty to understand which particular dimensions are influenced by biodiversity. Is it all about the income that can be derived or does biodiversity also affect other dimensions of poverty/ well-being such as health, food security, energy security and so on?

It is also important to understand the mechanism by which value from biodiversity is extracted. Is it through harvesting and consuming wild products? Is it from using genetic diversity to improve traditional crop varieties? Is it from setting up biodiversity-based tourism enterprises? Mechanisms can be categorized into direct use, indirect use, non-use, and 'biodiversity dis-services'. In this context, the terms 'use' and 'non-use' are employed slightly differently to that commonly used in the concept of total economic value (TEV). Consistent with TEV, 'direct use' refers to the consumptive use of certain species or resources such as non-timber forest products (NTFPs) for subsistence or tradable purposes. In this case species or resources are removed from nature. Indirect use refers to the benefits derived from biodiversity being left in situ rather than being removed and includes, for example, use of biodiversity for tourism or for soil nutrient cycling. Non-use refers to the co-benefits that can arise from biodiversity management efforts such as jobs. These first three categories all refer to the benefits that people get from biodiversity. The final category – 'biodiversity dis-services' – therefore refers to the negative impacts that components of biodiversity can have on people such as the health impacts of parasites and diseases, or the loss of crops and livestock as a result of wildlife incursions.

Finally, some understanding is needed of how these mechanisms actually work to contribute to poverty alleviation. Do they generate income? Do they provide a form of insurance or risk management? Do they act as a safety net and support basic needs in the absence of other options? Understanding the role of biodiversity requires differentiating between use of biodiversity for subsistence purposes only or to generate a tradable surplus or income; between immediate use and contributions to longer-term resilience; between using biodiversity as a regular matter of course or as a last resort in times of emergency; and between supporting livelihood security or undermining it.

The actual impact that biodiversity has on poverty will be mediated by the environmental, political and institutional context that regulates the nature and severity of poverty, the natural availability of biodiversity, and the ability of poor people to access and use biodiversity. The impact could be positive, neutral or negative (e.g., if human wildlife conflict or vector-borne disease exacerbates poverty). The use of biodiversity by poor people will also have an impact on biodiversity itself. The nature of this impact (biodiversity enhancement or depletion) will determine the long-term sustainability of use and ongoing ability of biodiversity to provide poverty alleviation services. Figure 15.1 provides a visual summary of the conceptual framework.

15.3 THE EVIDENCE BASE ON BIODIVERSITY–POVERTY LINKAGES

This section summarizes the results of a systematic mapping of the published and grey literature on the contribution of biodiversity to poverty alleviation/well-being and/or supporting local livelihoods (Roe et al., 2014). From a starting point of over 10000 studies – identified through an extensive keyword search – we identified 387 relevant studies of which 249 included some kind of measure of the contribution of biodiversity to poverty alleviation or well-being. Relevant studies were those that met our criteria of identifying a link between biodiversity use/non-use and an effect on poverty/local livelihoods in medium- or low-income countries (so, for example, we rejected any that were concerned with the effect of poverty/poor people on biodiversity, or those that we located in high-income countries, or those that were about how conservation *interventions* influence poverty as opposed to how biodiversity itself influences poverty).

15.3.1 Defining Biodiversity

Our interpretation of the term 'biodiversity' in the context of this review merits some discussion. Although the CBD definition of biodiversity encompasses 'living organisms from all sources', we were predominantly interested in natural habitats and wild species – what Balmford et al. (2002) describe as 'wild nature' – rather than all living organisms. However, the line between what is wild and what is not is very fuzzy. For example, we did not include mainstream agricultural crops but we did

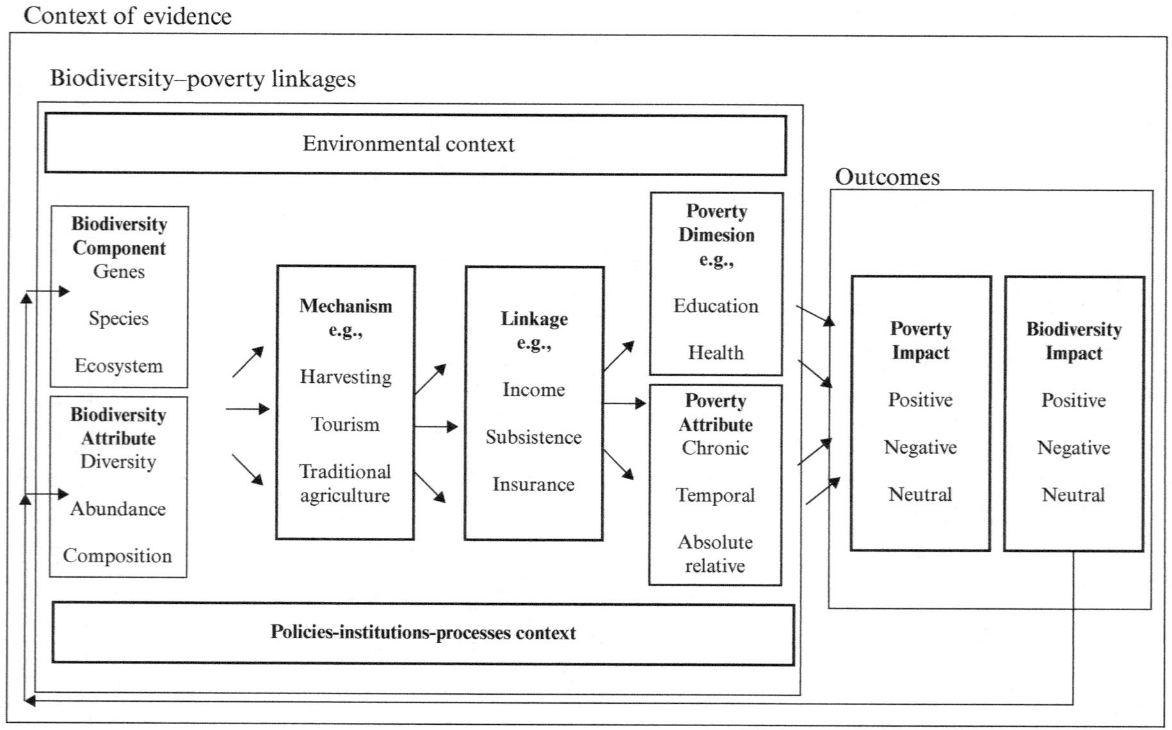

Source: Roe et al. (2014).

Figure 15.1 Conceptual framework for reviewing biodiversity–poverty linkages

include indigenous varieties of crops and crop wild relatives, or locally domesticated wild species. Similarly, we did not include modern livestock as a component of biodiversity but we did include traditional breeds or landraces. We also omitted micro-organisms, parasite and disease vectors such as mosquitoes. We recognize, however, that the impact of such living organisms on poverty as a result of their health impacts is probably far more significant than of any other component of biodiversity.

Finally, we did not cover the broad ES literature unless a study specifically linked the provision of a specific ES with particular components of biodiversity and then made a link to poverty. Much has been written about the varied and complex inter-linkages between biodiversity and ES and the evidence requires careful interpretation. For example, focusing on provisioning and regulating services, Cardinale et al. (2012) found that there is evidence that biodiversity either directly influences, or is strongly correlated with, some provisioning and regulating services but not all. The services most closely linked to ecosystem functions were those where the evidence for a link to diversity was also strongest. For many of the ES reviewed, the evidence for effects of biodiversity was mixed, unclear, or there was insufficient data to evaluate the link. They also found cases where increased biodiversity may actually be deleterious to ES – for example, where diverse natural enemy communities sometimes inhibit biocontrol because they attack each other. Overall, Cardinale et al. 'caution against making sweeping statements that biodiversity always brings benefits to society' (p. 63).

Mace et al. (2012) identify three different roles for biodiversity in ES: as a regulator of ecosystem processes, as a final ES, and as a good. Our review encompassed studies on biodiversity in each of these roles but within that has predominantly adopted what Mace et al. would refer to as a 'conservation perspective' where the focus is on 'a subset of biodiversity that includes charismatic species and those on threatened species lists' (p. 21). Specifically, we focused on a subset of biodiversity that is predominantly wild (or at least not domesticated to an industrial scale) and of a predominantly larger than micro-organism scale (but not ignoring wild genetic resources). The sections below describe the evidence, following the structure of the conceptual framework discussed above.

15.3.2 Description of the Evidence Base

Components and attributes of biodiversity

The evidence addressed many different components of biodiversity – from genes to ecosystems (Figure 15.2). In addition to the standard 'genes, species, ecosystems' typology, we added the category 'guilds' to capture

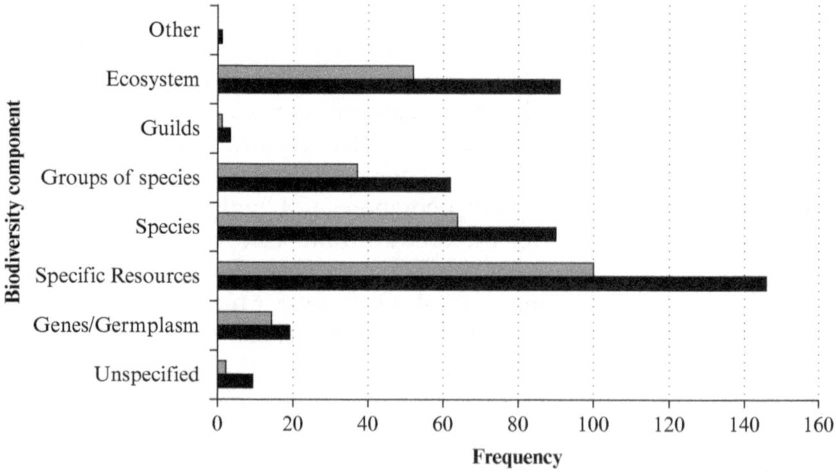

Note: Dark grey columns represent total dataset meeting primary inclusion criteria; light grey columns represent subset with a measure of poverty impact.

Source: Roe et al. (2014).

Figure 15.2 Components of biodiversity studied

studies on, for example, the role of pollinators. We made a distinction between studies that focused on individual species and those that focused on groups of species (e.g., African plains game, or 'the big five' – African lion, African elephant, Cape buffalo, African leopard and white/black rhinoceros). We also distinguished studies that were focused on particular types of resources – rather than specific species – for example, NTFPs. We found that the resources category accounted for the largest number of studies that we reviewed (38 per cent of all papers) within which the most commonly studied type of resource was NTFPs. We found only a very small evidence base making a link between guilds such as pollinators and poverty, or genetic resources and poverty. Figure 15.2 summarizes the results.

In keeping with our finding about NTFPs we found that forests were by far the most studied ecosystem. Artificial terrestrial habitats (e.g., agricultural land) were also reasonably well studied but very little evidence was found relating to mountains, deserts and introduced/exotic habitats.

In most cases, the abundance or extent of biodiversity was the attribute that made it important for poverty alleviation. Diversity itself was the least frequently noted attribute (Figure 15.3).

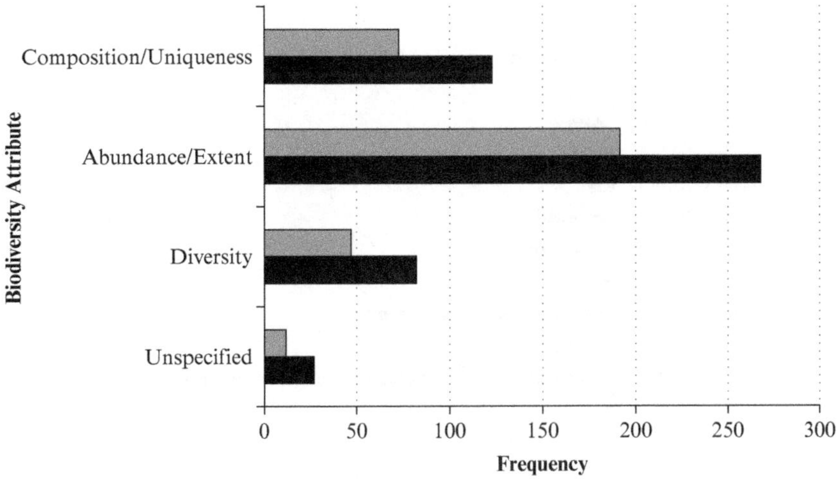

Note: Dark grey columns represent total dataset meeting primary inclusion criteria; light grey columns represent subset with a measure of poverty impact.

Source: Roe et al. (2014).

Figure 15.3 Is diversity important per se? Attributes of biodiversity studied

Dimensions of poverty

The Ecosystem Services for Poverty Alleviation (ESPA) programme has produced a conceptual framework for understanding poverty (Suich, 2012). This identifies – from a review of the literature – 19 commonly used dimensions of poverty. Our analysis only identified studies that addressed 11 of these, plus an additional dimension – energy security. Although most studies examined more than one dimension of poverty, the most commonly studied was income (70 per cent of all papers). Other commonly studied dimensions were food security (32 per cent of all papers) and asset accumulation (23 per cent of all papers) while the least commonly studied were energy, shelter and safe water. Figure 15.4 describes the frequency of study of each dimension.

Mechanisms linking biodiversity and poverty

Not surprisingly, the evidence reveals that the most common mechanism for linking biodiversity and poverty is through direct use of different components of biodiversity. Biodiversity disservices have received little attention in the literature – although our review did not attempt to address the literature on the links between poverty and human and livestock

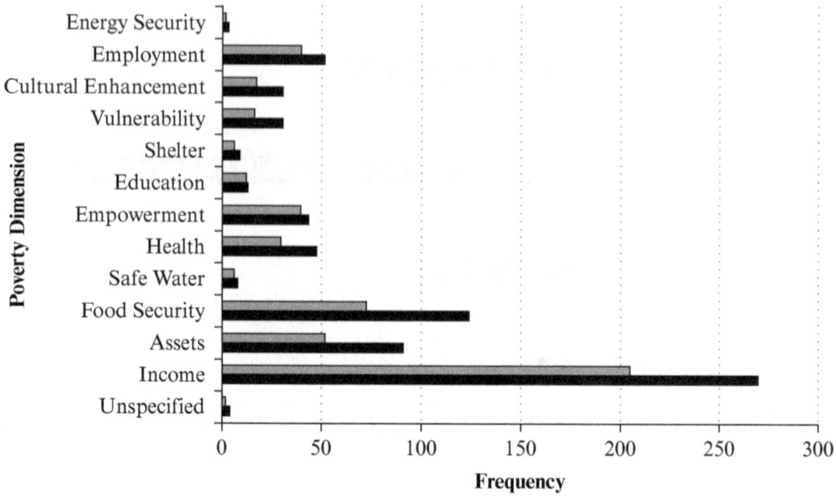

Note: Dark grey columns represent total dataset meeting primary inclusion criteria; light grey columns represent subset with a measure of poverty impact.

Source: Roe et al. (2014).

Figure 15.4 Dimensions of poverty studied

health – which may contain some useful insights into the significance of biodiversity. Figure 15.5 illustrates the diversity of mechanisms covered in the literature, the most common being NTFP harvesting (noted in over half of the studies reviewed).

In most cases, biodiversity influences poverty alleviation by acting as a source of income generation for poor people and/or through supporting basic needs. Less commonly identified processes were contributing to longer-term resilience and acting as a safety net. For most of the mechanisms reviewed biodiversity influences multiple dimensions of poverty – way beyond the income effect. This was particularly pronounced in studies of fishing, NTFPs, tourism and wild plant cultivation where almost all dimensions of poverty were considered in one or more study.

Impact of biodiversity use on poverty – and on biodiversity

The overwhelming majority of studies reviewed described a positive effect of biodiversity on poverty – although only two-thirds of these actually included any *measure* of that effect. Measures related to almost all the dimensions of poverty were used although the most common type of measure was income related. Non-income measures included increases

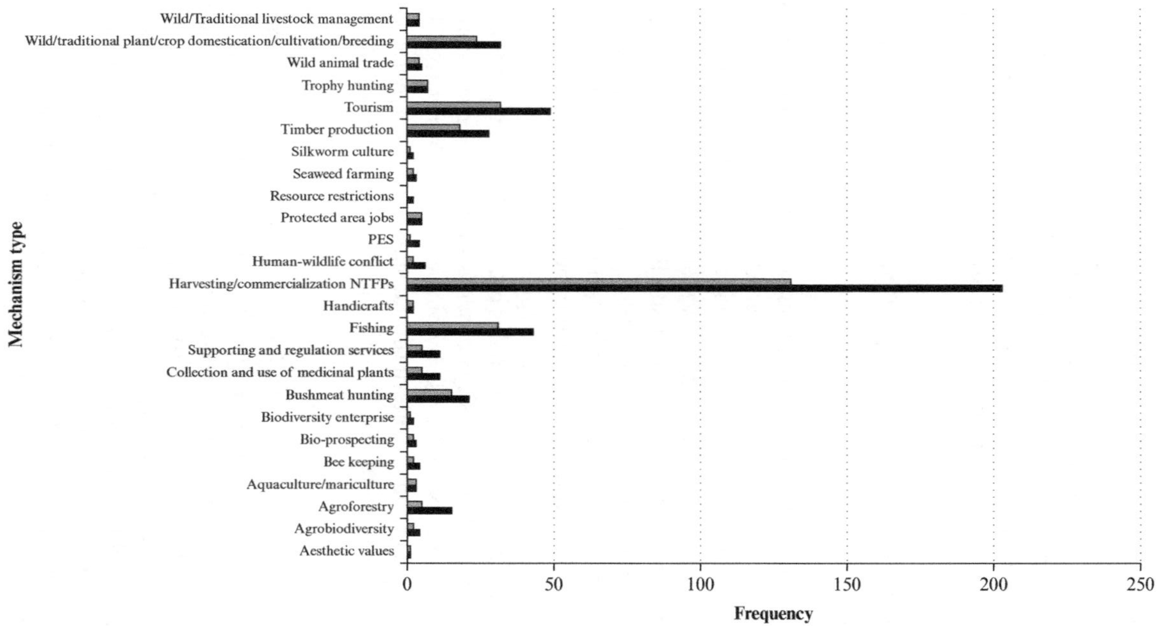

Note: Dark grey columns represent total dataset meeting primary inclusion criteria; light grey columns represent subset with a measure of poverty impact.

Source: Roe et al. (2014).

Figure 15.5 How do poor people interact with biodiversity? Different mechanisms by which people experience costs and benefits of biodiversity

in food availability and intake; numbers of jobs created; improvements in health; improvements in asset productivity. But only just over half of the studies included any consideration of whether the use of biodiversity was sustainable. Of these, use was considered sustainable in 65 per cent of cases. When this finding was compared with the type of use of biodiversity it can be seen – not surprisingly – that most doubts about the sustainability of use were associated with direct (consumptive) use mechanisms. Biodiversity use was considered to be unsustainable in 21 per cent of direct use studies compared to 13 per cent for indirect use studies.

15.4 DOES BIODIVERSITY HELP IN ALLEVIATING POVERTY?

The evidence base described above would suggest that the answer to this question is yes: poor people use – or interact with – biodiversity in a variety of different ways and the body of evidence suggests an overall positive effect. The reality, however, is that few studies have actually demonstrated a clear link between use of biodiversity and a change in state of poverty. Claims about the poverty reduction role of biodiversity thus need to be treated with some circumspection. Similarly, claims may be made that biodiversity has delivered poverty benefits when in fact it was a single species (e.g., trophy hunting revenues from elephants) or it was the collective biomass of a number of species (e.g., fisheries, NTFPs). Furthermore, many of the beneficiaries of biodiversity conservation/use are rural communities who happen to be located close by but who may not necessarily be defined as 'poor' by national or international standards. It is therefore necessary to make a number of clear distinctions in any discussion of biodiversity–poverty links:

- *Poverty reduction vs poverty prevention:* It is important to differentiate between: (1) the safety net role that biodiversity plays in supporting rural people's livelihoods in developing countries (e.g., see Vira and Kontoleon, 2013 for a meta-analysis of studies) and (2) the role that biodiversity can play in actually lifting people out of poverty (e.g., WRI, 2005). Sunderlin et al. (2005) make this same point with respect to the link between forests and poverty alleviation.
- *Biodiversity vs biomass vs selected species:* Species abundance or biomass, or the economic value of a single species or habitat can often make a far more significant contribution to poverty reduction – at least in the short term – than variability. Abundance, biomass, individual ecosystems, species and genes are all components or

attributes of biodiversity but often the term biodiversity is used to refer to just one of these components. Vira and Kontoleon (2013, p. 53) reviewing the evidence for the dependence of the poor on biodiversity note: 'the term "nature's resources" better captures the generic categories of resources that have been studied in this literature. These include forests, both in terms of wood-based and NTFPs, mangroves, fish, wild animals (bushmeat) and wild plants (including herbs), and common pool resources (CPRs) more generally'. Nevertheless, while only a limited number of species may be in direct use, they themselves are dependent on biodiversity to maintain the ecological systems of which they are part. Diversity is also important for food security, risk avoidance, livelihood insurance and the delivery of critical ES.

- *Biodiversity use vs biodiversity conservation:* Some uses of biodiversity may deliver significant poverty benefits in the short term but not all uses of biodiversity are sustainable in the long term (e.g., felling of Brazil nut trees for timber).

Despite all the problems identified above, and despite caveats about the state of the evidence base it is clear that that billions of poor people living in rural areas of developing countries are directly dependent on 'biodiversity' to support their livelihoods (disservices notwithstanding) whether it be individual components, or whether it really is the variability among living organisms.

15.5 IMPLICATIONS FOR BIODIVERSITY IMPACT ASSESSMENT

It seems clear from the discussion above that efforts should be made to encourage conservation or restoration of biodiversity when impact assessment of a planned intervention indicates potentially negative consequences. Because the rural poor depend on biodiversity for their day-to-day livelihoods, it is logical that if it is conserved or restored it can continue to provide livelihood support functions.

In response to the Strategic Plan for Biodiversity 2011–2020, parties to the CBD are revising their National Biodiversity Strategies and Action Plans (NBSAPs) to better emphasize the links between biodiversity, poverty reduction and national development – including the need to recognize the impact of development interventions on biodiversity and the subsequent knock-on impacts on poor people. In Botswana, for example, the revised NBSAP highlights the importance of biodiversity to people's livelihoods and when threats

BOX 15.1 LINKING BIODIVERSITY AND POVERTY ALLEVIATION IN THE UGANDA NBSAP

The Ugandan second NBSAP (NEMA, 2015) emphasizes throughout the strategy the link between biodiversity and national economic development, poverty reduction and people's livelihoods. This includes attention to the role both biodiversity and ecosystem services play in supporting some of the poorest and most vulnerable people and in particular the rural poor, landless and women. It also recognizes that biodiversity underpins large portions of the country's economy (e.g., fishing, tourism, livestock, agriculture, forestry, energy).

The strategy also discusses how current development policies and political processes are a threat to biodiversity, and in particular the following two issues: (1) The emphasis within the National Development Plan (NDP) on supporting sectors that can provide quick economic returns. The environment and natural resource sector is not prioritized within the NDP because it does not yield quick returns. This has meant that there has been pressure to convert landscape meant for biodiversity into other forms. (2) The recent push by some politicians to degazette protected areas so that they can be used for agriculture and industrial expansion.

In response to this the NBSAP includes targets to influence the NDP and improving management practices of agricultural lands. The strategy also identifies new and emerging issues into which biodiversity concerns will need to be integrated including responses to climate change, discovery of oil in the Albertine Rift and production of biofuels.

to biodiversity are discussed, the impact of these threats is discussed both in terms of the impact on biodiversity and on people's livelihoods (MEWT, 2014). The Uganda revised NBSAP provides more insights into the links between biodiversity, development and poverty and the need to ensure that development interventions – such as oil exploration – do not impact on biodiversity to the detriment of local livelihoods (Box 15.1)

However, the intervention that is employed to reach biodiversity conservation/restoration outcomes may itself have a negative poverty impact. For example, strict enforcement of protected area and other exclusionary land uses may actually increase local incidence of poverty through the loss of resource access. The way conservation interventions are designed and implemented is a key determinant of their poverty impacts – whether they are protected areas (e.g., West et al., 2006), Payments for Ecosystem Services (PES) schemes (e.g., Wunder, 2008) or any other mechanisms (e.g., Leisher et al., 2013). Incorporating poverty alleviation into Biodiversity Impact Assessment therefore requires attention to not just how impacts on biodiversity from external intervention might have knock-on impacts on poor people, but also how efforts to mitigate those impacts on biodiversity may also have unintended consequences for poor people.

This chapter sheds light on some of the issues that should be addressed in Biodiversity Impact Assessment of any planned development interventions if the subsequent implications for poverty are to be taken into account. Specifically, there is a need to:

- Unpack 'biodiversity', and clarify which are the most important components of biodiversity for poor people (e.g., the availability of some key species; the availability of a particular type of resource such as bushmeat or firewood). This would help to steer the scoping stage of impact assessment towards the issues that matter the most to poor people.
- Unpack 'poverty', by identifying its different dimensions and disaggregating the potential impacts of the planned intervention on each of them. In this way, the impact chain can be unveiled and the drivers that are likely to cause changes in the individual components of poverty (or well-being), can be properly identified and modelled. This approach is suggested by the Millennium Ecosystem Assessment framework (MA, 2005), but is rarely applied in impact assessment practice, where studies tend to produce evaluations of the overall impact of proposed developments. This type of analysis could also help to disaggregate impacts for different groups of beneficiaries. These could be, for example, people with different income levels, or different degree of dependency from a given biodiversity resource or service (due to livelihood systems, culture, lifestyle, etc.) (UNEP, 2014). In this way, clear predictions of winners and losers associated with planned interventions can be produced (Geneletti, 2013, 2015), and used to inform decision-making, but also to enhance public engagement, which is central to impact assessment processes.
- Understand the way in which poor people extract values from biodiversity, so as to predict the effects on these processes that the planned intervention is likely to cause, for example, in terms of reducing the quality or quantity of a given resource or generating biodiversity 'dis-services'. This would provide much more context-specific information on which to base possible alternative strategies that minimize negative interferences and/or enhance positive ones. In this way, impact assessment could be used not only in a reactive way to assess the effects of proposals, but also in a more proactive way, to improve the design and implementation of such proposals.

We recognize that the consideration of poverty issues within Biodiversity Impact Assessment expands the scope of such exercises – which typically operate under tight time and resource constraints. Hence, it is important to keep the additional analysis and information load to a minimum, in order for the impact assessment process to be effective in actually supporting decision-makers. Nevertheless, just keeping in mind some of the issues discussed in this chapter can contribute to making Biodiversity Impact Assessment much more effective in poor countries and can help to design interventions that not only minimize their impact on biodiversity but also minimize unintended consequences for the poor.

REFERENCES

Balmford, A., A. Bruner and P. Cooper et al. (2002), 'Economic reasons for conserving wild nature', *Science*, **297**(5583), 950–53.

Cardinale, B.J., E. Duffy and A. Gonzalez et al. (2012), 'Biodiversity loss and its impact on humanity', *Nature*, **486**(7401), 59–67.

European Commission (EC) (2013), *Guidance on Integrating Climate Change and Biodiversity into Strategic Environmental Assessment*, accessed 25 January 2016 at http://ec.europa.eu/environment/eia/pdf/EIA%20Guidance.pdf.

Geneletti, D. (2006), 'Some common shortcomings in the treatment of impacts of linear infrastructures on natural habitat', *Environmental Impact Assessment Review*, **26**(3), 257–67.

Geneletti, D. (2013), 'Assessing the impact of alternative land-use zoning policies on future ecosystem services', *Environmental Impact Assessment Review*, **40**(1), 25–35.

Geneletti, D. (2015), 'A conceptual approach to promote the integration of ecosystem services in strategic environmental assessment', *Journal of Environmental Assessment Policy and Management*, **17**(4), 1550035.

Gontier, M., B. Balfors and U. Mortberg (2006), 'Biodiversity in environmental assessment – current practice and tools for prediction', *Environmental Impact Assessment Review*, **26**(3), 268–86.

Kepe, T., M. Saruchera and W.J. Webster (2004), 'Poverty alleviation and biodiversity conservation: a South African perspective', *Oryx*, **38**(2), 143–5.

Leisher, C., M. Sanjayan and J. Blockhus et al. (2013), 'Does conserving biodiversity work to reduce poverty? A state of knowledge review', in D. Roe, J. Elliott and C. Sandbrook et al. (eds), *Biodiversity Conservation and Poverty Alleviation: Exploring the Evidence for a Link*, Oxford: Wiley-Blackwell.

Mace, G.M., K. Norris and A.H. Fitter (2012), 'Biodiversity and ecosystem services: a multilayered relationship', *TREE*, **27**(1), 19–26.

Millennium Ecosystem Assessment (MA) (2005), *Ecosystems and Human Wellbeing – Biodiversity Synthesis*, Washington, DC: World Resources Institute.

Ministry of Environment, Wildlife and Tourism (MEWT) (2014), *Revised National Biodiversity Strategy and Action Plan – Draft Unpublished Report*, Gaborone, Botswana: MEWT.

National Environmental Management Authority (NEMA) (2015), *National Biodiversity Strategy and Action Plan*, Kampala: NEMA.

Rajvanshi, A., V.B. Mathur and U.A. Iftikhar (2007), *Best Practice Guidance for Biodiversity-inclusive Impact Assessment: A Manual for Practitioners and Reviewers in South Asia* (CBBIA-IAIA Guidance Series), Capacity Building in Biodiversity and Impact Assessment (CBBIA) Project, International Association for Impact Assessment (IAIA), accessed 25 January 2016 at http://people.exeter.ac.uk/rwfm201/cbbia/downloads/asi/Manual_Guidance_final.pdf.

Roe, D., M. Fancourt and C. Sandbrook et al. (2014), 'Which components or attributes of biodiversity influence which dimensions of poverty?', *Environmental Evidence*, **3**(1), 3.
Scolozzi, R. and D. Geneletti (2011), 'Spatial rule-based assessment of habitat potential to predict impact of land use change on biodiversity at municipal scale', *Environmental Management*, **47**(3), 368–83.
Scolozzi, R. and D. Geneletti (2012), 'A multi-scale qualitative approach to assess the impact of urbanization on natural habitats and their connectivity', *Environmental Impact Assessment Review*, **36**, 9–22.
Secretariat of the Convention on Biological Diversity (SCBD) (2006), *Voluntary Guidelines on Biodiversity-inclusive Impact Assessment*, Montreal: SCBD.
Secretariat of the Convention on Biological Diversity (SCBD) (2010), *Decision X/2: Strategic Plan for Biodiversity 2011–2020*, Montreal: SCBD.
Sen, A. (1999), *Development as Freedom*, London: Zed Books.
Suich, H. (2012), 'Conceptual framework: poverty', ESPA, accessed 26 January 2015 at http://www.espa.ac.uk/files/espa/Povertyframework.pdf.
Sunderlin, W.D., A. Angelsen and B. Belcher et al. (2005), 'Livelihoods, forests, and conservation in developing countries: an overview', *World Development*, **33**(9), 1383–402.
Tekelenburg, A., B.J.E. ten Brink and M.C.H. Witmer (2009), *How do Biodiversity and Poverty Relate? An Explorative Study*, Bilthoven: Netherlands Environmental Assessment Agency.
Treweek, J., R. Therivel and S. Thompson et al. (2005), 'Principles for the use of strategic environmental assessment as a tool for promoting the conservation and sustainable use of biodiversity', *Journal of Environmental Assessment Policy & Management*, **7**(2), 173–99.
UNEP (2014), *Integrating Ecosystem Services in Strategic Environmental Assessment: A Guide for Practitioners*, a Proecoserv report by D. Geneletti, accessed 26 January 2016 at http://www.ing.unitn.it/~genelab/documents/GuidelineESintoSEA.pdf.
Vira, B. and A. Kontoleon (2013), 'Dependence of the poor on biodiversity: which poor, what biodiversity? A state of knowledge review', in D. Roe, J. Elliott and C. Sandbrook et al. (eds), *Biodiversity Conservation and Poverty Alleviation: Exploring the Evidence for a Link*, Oxford, UK: Wiley-Blackwell.
West, P., J. Igoe and D. Brockington (2006), 'Parks and peoples: the social impact of protected areas', *Annual Review of Anthropology*, **35**(1), 251–77.
World Bank (2001), *World Development Report 2000/2001: Attacking Poverty*, Washington, DC: World Bank.
World Resource Institute et al. (WRI) (2005), *World Resources 2005 – The Wealth of the Poor: Managing Ecosystems to Fight Poverty*, Washington, DC: UNEP, UNDP, World Bank and World Resources Institute.
Wunder, S. (2008), 'Payments for environmental services and the poor: concepts and preliminary evidence', Environment and Development *Economics*, **13**(3), 279–97.

16. Biodiversity offsets for 'no net loss' through impact assessment

Susie Brownlie and Jo Treweek

16.1 INTRODUCTION

Simply stated, biodiversity offsets are intended to balance loss of biodiversity due to development with commensurate gain. Loss of biodiversity is escalating, and climate change is likely to accelerate that loss. The current rate of species extinction is estimated to be 100–1000 times more than would be expected to occur naturally (Rockström et al., 2009). This is an issue of deep concern, not only because biodiversity has intrinsic value, but also because it underpins a wide range of ecosystem services (ES) on which humans depend. If current rates of loss continue, the capacity of ecosystems to provide for the needs of present and future generations is uncertain (CBD, 2010).

It is against this background that interest in biodiversity offsets has strengthened. Environmental policies and regulations in developed countries have increasingly incorporated 'no net loss' and preferably 'net gain' (NNL/NG) concepts (Ten Kate and Crowe, 2014) in their goals for biodiversity. Impact assessment (IA) is seen as a key implementation mechanism (e.g., Morandeau and Vilaysack, 2012) because it is endorsed by several international conventions as a tool for mainstreaming biodiversity into planning and decision-making, is widely promoted as a tool for corporate social responsibility and is seen by financial institutions as a key tool to manage environmental and social risks of the developments in which they invest. One of its advantages is that it assigns liability to developers to compensate for environmental damage, according to the 'polluter pays principle', as in Europe, for example (Darbi et al., 2009). Residual losses of biodiversity are accepted in many IA systems, provided that reasonable efforts are made to avoid or minimize impacts. This approach is one of the main reasons why addition of an offset step to the mitigation hierarchy is considered necessary if there is genuine intent to achieve NNL/NG.

Recognizing the role that offsets could play in strengthening IA through the mitigation hierarchy, more than 60 countries have adopted, or are developing, policies or laws requiring compensation or offsets (Ten Kate

Table 16.1 Legal requirements for biodiversity offsets or compensation

Laws/Regulations	Species/Habitat Legislation	EIA	Offset/ Compensation
Existing	USA, Canada, European Union, Germany, Switzerland, Australia	Mexico, Chile, South Africa, China, Pakistan	USA, Mexico, Brazil, Colombia, Paraguay, Netherlands, Australia
Planned or potential	UK, France, Vietnam, New Zealand	Argentina, Uganda, Madagascar, Ghana, Egypt, EU, India, Japan, Malaysia, Mongolia, Pakistan, Russia, South Korea, Thailand, Peru	Costa Rica, Panama, South Africa, Uganda, Madagascar, Indonesia, Mongolia

Source: Table 1 in Doswald et al. (2012), reproduced with permission from the authors and publishers.

and Crowe, 2014). These provisions are included in regulatory frameworks for development planning and IA (Doswald et al., 2012) or in conservation policies or legislation.

Biodiversity offsets may be developed to comply with financial lending or legal requirements (Table 16.1), or on a voluntary basis by businesses as part of their risk management policies to reduce reputational risks or to develop a social licence to operate (e.g., ICMM IUCN, 2013). Increasingly, companies are undertaking voluntary offsets as part of their corporate strategies (e.g., Rio Tinto, 2004). These are largely in the extractive industries (oil, gas and mining) but there are also examples from other sectors, such as water and urban development (e.g., Ten Kate et al., 2004; Darbi et al., 2009; Madsen et al., 2011). Even when offsets are implemented outside any formal IA context, however, procedures analogous to those used in IA are likely to be needed to obtain the information and understanding about biodiversity risks and impacts that are required to design and implement offsets in accordance with good practice.

In contrast with views highlighting the potential benefits of incorporating biodiversity offsets in IA, strong concerns have been raised that offsets might legitimize activities that would otherwise not be approved (Hayes and Morrison-Saunders, 2007) and serve as a 'licence to trash' nature. Some authors believe that the challenges in trying to counterbalance biodiversity losses with gains may be insurmountable, and that offsets might 'perversely yield worse biodiversity outcomes' than weakly enforced prohibitions (Walker et al., 2009, p. 155). According to Quigley and Harper (2006), offsets are at best slowing down, not halting, the rate of habitat loss; Walker et al. concur, suggesting that biodiversity offsets to date have 'facilitated development while perpetuating biodiversity loss' (2009, p. 149). Legitimate concerns exist that an early focus on biodiversity offsets, rather than their use as a 'last resort', diverts attention from strict application of the mitigation hierarchy in planning, IA and decision-making (Walker et al., 2009; Villarroya et al., 2014). Civil society organizations have expressed concern that consideration of biodiversity offsets as part of the initial project 'go' or 'no go' decision may lead to authorization of projects that are inappropriate because of the severity of their biodiversity impacts (BBOP, 2012a).

In Western Australia, where several offsets have been implemented and there are established policies and requirements, it has been suggested that a project that is not acceptable without offsets should not be acceptable with them: offsets should not be used to enable environmental trade-offs (Hayes and Morrison-Saunders, 2007). Echoing this stance, regulators in the USA state that offsets may not be used as a way of reducing environmental impacts 'to make a potentially avoidable project appear more acceptable' (McKenney and Kiesecker, 2010, p. 167).

The rest of this chapter explores the issues associated with incorporating biodiversity offsets in IA in line with good practice principles and standards. It then goes on to consider some of the aspects that need further consideration to allay fears that use of offsets might reduce the effectiveness of IA and accelerate loss of biodiversity due to development.

16.2 BIODIVERSITY OFFSETS: DEFINITIONS, PRINCIPLES AND 'NO NET LOSS'

There are numerous definitions of biodiversity offsets, spanning the regulatory, business and scientific sectors. The Business and Biodiversity Offsets Programme (BBOP) provides the most widely used definition:

> Biodiversity offsets are measurable conservation outcomes resulting from actions designed to compensate for significant residual adverse biodiversity impacts arising from project development after appropriate prevention and mitigation measures have been taken. The goal of offsets is to achieve NNL/NG of biodiversity on the ground with respect to species composition, habitat structure, ecosystem function and people's use and cultural values associated with biodiversity (BBOP, 2012b, p. 13).

According to this definition, offsets are a category of 'compensation' distinguished by the use of explicit accounting for losses and gains using a single 'currency' or 'metric'. In the USA, however, 'compensatory mitigation' is used synonymously with 'offset' and in some other applications of offsetting it is considered acceptable to use beneficial activities or compensation to counterbalance negative impacts on biodiversity without any strict accounting for NNL/NG. According to the definition given above, compensation provided through payment for training, research or other activities that will not result in measurable conservation outcomes on the ground does not constitute an offset because tangible gains are not assured and the gains that result may not be commensurate in type or amount with the losses that occurred.

An offset strives to balance losses of biodiversity in time and space with equivalent gains. The measures used in this exchange and offset design must capture what we care about (Salzman and Ruhl, 2000). These significant values, referred to as 'key biodiversity components', may be populations or species, habitats, communities, ecosystems, ecological or evolutionary processes, or arguably ES.

The ability to design offsets hinges on a clear and practical definition of 'NNL/NG'. Although offsets are mainly intended to achieve NNL/NG, precise objectives of offset policies are not always clearly defined (e.g., Villarroya et al., 2014) and may have different emphasis. For example, the International Finance Corporation (IFC) Performance Standard (PS) 6 requires NNL/NG to be demonstrated only for certain biodiversity features, based on criteria and thresholds. Interpretations of NNL/NG vary according to stakeholder values, by decisions on how biodiversity is measured, the scope of development impacts considered, and the temporal and spatial scale at which the NNL/NG goal is applied (Gardner et al., 2013). Since the Convention on Biological Diversity affords intrinsic value to all biodiversity components, the goal of NNL/NG means no net reduction in the diversity within and among species and vegetation types, their long-term viability, and functioning of species assemblages and ecosystems, including ecological and evolutionary processes (ibid.). The extent to which human benefits from biodiversity are addressed in offsetting systems varies between countries/systems and the majority address them independently.

The mitigation hierarchy is fundamental to IA best practice for biodiversity (ICMM IUCN, 2013). In line with the 'prevention principle' (that encourages measures to protect the environment rather than repairing damage), and the 'precautionary principle' (favouring caution when scientific knowledge is incomplete, and consequences of actions are uncertain and potentially serious), the mitigation hierarchy prioritizes avoidance of damage at source, followed by efforts to minimize impacts and then to repair damage. Compensation or offsets form a logical final step in this sequence. Existing legislation and IA approaches often fail to mitigate impacts on biodiversity sufficiently for NNL/NG to be achieved (Rainey et al., 2014) and this failure is due in part to the lack of requirements to address residual impacts in order to demonstrate NNL. The EU EIA Directive, for example, requires only measures to avoid, reduce or restore impacts, and does not require any ecological compensation unless developments are proposed within its Natura 2000 network.

Recognizing this deficiency, the IFC's PS6, 'Biodiversity Conservation and Sustainable Management of Living Natural Resources' makes explicit reference to use of biodiversity offsets as part of the IA mitigation hierarchy as one of a suite of measures to achieve 'net gain' where development would otherwise have residual impacts on areas of 'critical biodiversity' (IFC, 2012). The Performance Standards for the operational basis of the Equator Principles[1] and the 80 banks from 34 countries that subscribe to these principles must thus apply the full mitigation hierarchy in their project finance due diligence procedures (Doswald et al., 2012). As presented in PS6, appropriate application of the mitigation hierarchy is intended to ensure that risks to biodiversity are appropriately avoided and managed before any decision is made to offset them (Doswald et al., 2012; IFC, 2012; ICMM IUCN, 2013). Underlining the importance of avoidance in the mitigation hierarchy is recognition that not all negative impacts on biodiversity can be offset – some biodiversity may be irreplaceable if lost. 'Good practice' use of offsets relies on IA to apply the mitigation hierarchy faithfully, to ensure good integration between specialist studies within the impact assessment process, to engage with stakeholders, and to measure residual negative impacts as a reliable basis for designing and planning implementation of offsets.

The BBOP, an international multi-stakeholder group, developed ten 'good practice' principles for biodiversity offsets (BBOP, 2012b) to provide a basis for rigorous approaches (Box 16.1) and to underpin a global standard. The following section considers the challenges of integrating offsets with IA in accordance with these principles and requirements.

16.3 INTEGRATING OFFSET PLANNING WITH THE IA PROCESS

Biodiversity offsets can be considered for residual effects of plans, programmes or for individual projects, but have predominantly been applied to projects. Impact assessment can contribute to offset design and implementation by providing a structured approach to gathering information,

BOX 16.1 BIODIVERSITY OFFSET PRINCIPLES

1. *Adherence to the mitigation hierarchy:* A biodiversity offset is a commitment to compensate for significant residual adverse impacts on biodiversity identified after appropriate avoidance, minimization and on-site rehabilitation measures have been taken according to the mitigation hierarchy.
2. *Limits to what can be offset:* There are situations where residual impacts cannot be fully compensated for by a biodiversity offset because of the irreplaceability or vulnerability of the biodiversity affected.
3. *Landscape context:* A biodiversity offset should be designed and implemented in a landscape context to achieve the expected measurable conservation outcomes taking into account available information on the full range of biological, social and cultural values of biodiversity and supporting an ecosystem approach.
4. *No net loss:* A biodiversity offset should be designed and implemented to achieve in situ, measurable conservation outcomes that can reasonably be expected to result in no net loss and preferably a net gain of biodiversity.
5. *Additional conservation outcomes:* A biodiversity offset should achieve conservation outcomes above and beyond results that would have occurred if the offset had not taken place. Offset design and implementation should avoid displacing activities harmful to biodiversity to other locations.
6. *Stakeholder participation:* In areas affected by the project and by the biodiversity offset, the effective participation of stakeholders should be ensured in decision-making about biodiversity offsets, including their evaluation, selection, design, implementation and monitoring.
7. *Equity:* A biodiversity offset should be designed and implemented in an equitable manner, which means the sharing among stakeholders of the rights and responsibilities, risks and rewards associated with a project and offset in a fair and balanced way, respecting legal and customary arrangements. Special consideration should be given to respecting both internationally and nationally recognized rights of indigenous peoples and local communities.
8. *Long-term outcomes:* The design and implementation of a biodiversity offset should be based on an adaptive management approach, incorporating monitoring and evaluation, with the objective of securing outcomes that last at least as long as the project's impacts and preferably in perpetuity.
9. *Transparency:* The design and implementation of a biodiversity offset, and communication of its results to the public, should be undertaken in a transparent and timely manner.
10. *Science and traditional knowledge:* The design and implementation of a biodiversity offset should be a documented process informed by sound science, including an appropriate consideration of traditional knowledge.

Source: BBOP (2012b, p. 10).

assessing and evaluating impacts on biodiversity, applying the mitigation hierarchy and engaging a diverse range of stakeholders to understand their issues and values (BBOP, 2009a). In practice, addressing offsets effectively within IA depends on many factors. Offsets may require longer lead times, more detailed or different data, different specialist skills, and/or a wider spatial scope than is typically provided (ibid.). The extent to which offset studies are completed within the IA process prior to authorization of a project is likely to be case and context specific, but should at least be sufficient to give assurance that 'non-offsetable' impacts on biodiversity would be avoided, that an offset could and would be successfully implemented, and that adequate financial provision for implementation has been made.

16.3.1 Scoping and Baseline

Early identification of critical biodiversity, its sensitivity to impacts, and the likely ability to offset any residual impacts, is crucial. This enables the appropriateness and acceptability of offsets to be considered when there is still time to review alternatives, which is essential if there is any possibility that impacts might be difficult or impossible to offset. The potential need for offsets should thus be considered at the scoping stage, especially if IA is being undertaken for proposals in biodiversity-rich or -sensitive environments such as in a biodiversity hotspot and centre of endemism in the Northern Cape, South Africa (Box 16.2). In this way, the necessary

BOX 16.2 CASE STUDY: MINE IN NORTHERN CAPE PROVINCE, SOUTH AFRICA

A zinc mining company plans to expand its mine in the semi-desert 'Bushmanland' part of the Succulent Karoo Global Biodiversity Hotspot; also a centre of endemism (Hughes et al., 2015). This region is home to unique succulent flora and is a national priority area for protected area expansion. A botanist identified and mapped sensitive areas on the mining site, comprising communities of plant species of concern and important ecological corridors across the landscape. The mapping was informed by systematic biodiversity plans for the region; several features had already been identified as 'Critical Biodiversity Areas'.

Early specialist input enabled the review of layout plans for the mine and changes in the siting of waste dumps and the tailings dam, as well as roads and mining methodology, to avoid or prevent, and minimize potentially significant impacts. The availability of regional spatial data of significant habitats and species provided the basis for a well-informed evaluation of biodiversity impacts in a regional context, and the identification of optimal, contiguous offset sites that contributed to securing critical habitat in this unique region.

information can be obtained without duplication of effort or delay, and adequate financial provision for the implementation of offsets can be built into project budgets.

As well as supporting rigorous assessment, evaluation and mitigation of impacts, identifying the intrinsic, use and cultural values of biodiversity during the early screening and scoping phases of IA, through stakeholder engagement and gathering of baseline information, provides a sound basis for deciding whether avoidance is appropriate.

IA processes typically accommodate specialist studies to assess impacts on biodiversity and recommend mitigation, but may fall short of providing measures of residual negative impacts that could form the basis for detailed offset design, and planning and budgeting for offset implementation. To provide this information, it may be necessary to expand the scope of baseline studies, for example, to include information at the species, community and ecosystem level, and potential offset locations. As indicated in the BBOP principles, a landscape-scale perspective is needed to understand the threats and pressures affecting biodiversity and how these may be exacerbated by a development proposal (see also Pilgrim et al., 2013).

16.3.2 Assessing and Evaluating Impacts on Biodiversity

Assessing impacts typically involves analysis of their nature, magnitude/severity, extent and duration. Pilgrim et al. (2013) refer to the severity of the impact as being its intensity at a defined (usually spatial) scale; its extent as the proportion of the population/range of a given biodiversity feature (inversely related to the viability of the remainder of that feature) and its duration as the time period for which impacts can be expected to continue.

Evaluating impacts involves a judgement of the impact's significance, that is, whether the impacts are acceptable to stakeholders, require mitigation and/or compensation, or are unacceptable (CBD and NCEIA, 2006). Significance lies at the core of IA; it is not the same thing as 'impact magnitude', which is the prerogative of scientists and technical experts, and more amenable to objective and quantitative procedures (Lawrence, 2007). 'Significance' takes into account the values that affected communities attach to biodiversity, and broad societal values expressed through laws or policy related to environmental and biodiversity conservation, conservation priorities and strategies.

Essential elements in assessing and evaluating impacts on the intrinsic value of biodiversity include (BBOP, 2012a):

- the conservation or threat status of affected ecosystems, habitats and species at different spatial scales;

- the proportion of overall ranges and population distribution that might be affected;
- known priorities, for example, populations of threatened or endemic species, significant concentrations and source populations, or unique communities and habitats in the area; and
- sufficient information about biodiversity composition, structure and function to understand ecological dependencies, processes and relationships.

The significance of impacts on use and cultural values would take into account levels of dependence on, and importance of, biodiversity to affected communities.

Significant impacts on intrinsic biodiversity values that could be 'non-offsetable' must be identified at the earliest possible stage of the IA. Likewise, significant impacts on use and cultural values (ES) that may be difficult to compensate must be identified, engaging stakeholders to ensure that their values and concerns are taken into consideration. These impacts must preferably be avoided/prevented or, if this is not possible and they are of sufficient magnitude, they should determine that the 'no go' alternative is selected (Brownlie et al., 2013). A precautionary approach should be taken where there are insufficient scientific data, uncertainty about predictions or outcomes, or a risk of significant harm to biodiversity; higher risks and/or greater potential harm to biodiversity require greater reliability and certainty of information (CBD and NCEIA, 2006) to support identification of mitigation measures including offsets.

Offset design relies on a quantitative measure of the residual negative impacts that need to be compensated by the offset. To date IA generally has not involved measurement of impacts; importantly, consideration of offsets in the IA process introduces the need to quantify likely biodiversity loss using a suitable metric. The transition from significant residual impacts to offset design can be made smoother by using the same metrics to predict habitat losses or species population declines in the IA process. Losses can be tracked and quantified throughout the IA process, for example in the pre-impact situation and the post-impact situation with or without mitigation, to give a transparent account of efforts to avoid and minimize impacts.

16.3.3 Applying the Mitigation Hierarchy

Strict adherence to the mitigation hierarchy of first avoiding or preventing impacts, then minimizing them, rehabilitating or restoring damage, and only then considering compensation, means that the use of offsets to achieve NNL/NG must only be confirmed when specialist studies have been undertaken, alternatives have been fully investigated, and options have been exhausted. While offsets have potential to improve the status quo, they are challenging to implement and will always carry a greater risk of failure than avoidance of damage in the first place. Strict adherence to the mitigation hierarchy's sequence of steps helps to ensure appropriate emphasis on avoidance. There is broad agreement that this first and most important step in the hierarchy 'is ignored more often than it is implemented' (Clare et al., 2011, p. 165), but equally it is possible that early consideration of scope for offsets, and recognition of the potential challenges and costs, might improve application of earlier steps in the mitigation hierarchy – it is usually cheaper to avoid impacts on priority biodiversity than to offset them.

Although the mitigation hierarchy must be followed and avoidance of impacts prioritized, it is important to think about the full range of mitigation options, including offsets, from the start of an IA process.

16.3.4 Planning for Implementation

Mitigation measures for negative impacts stipulated in the IA are the basis for preparing an Environmental Management Plan (EMP) or, when focused on biodiversity, a Biodiversity Management Plan (BMP). Typically, the EMP or BMP focuses on on-site mitigation measures. Biodiversity offsets may require a number of off-site interventions, potentially managed by different implementers. For this reason it often makes sense to prepare a separate Biodiversity Offset Management Plan (BOMP).

Planning for offset implementation involves providing the necessary resources and financial security to establish, secure and manage offset areas for at least as long as the project's impacts and preferably in perpetuity (BBOP, 2012b). Without adequate planning, the risks of offsets underperforming or failing due to poor management may undermine their long-term contribution (Gardner et al., 2013). Funds must be provided to cover the costs of establishing and managing the offset in the long term, including staffing and infrastructure requirements, monitoring and auditing needs, adjusted for inflation. Possible financing vehicles comprise an endowment, a sinking or revolving fund, or a combination of options. The extent to which planning for financial provision occurs within the IA process, rather than after a conditional authorization, will depend on the specific case and context. Trusts have become increasingly popular as vehicles for offset financing (BBOP, 2009d). Other ways to fund the operational offset include nature-based tourism, 'payment for ES' (PES), or sale of products from the sustainable harvest of resources. Offset management invariably requires a project manager who coordinates and oversees the implementation of a range of actions.

Offset implementation activities may be broader in scope and require more detailed and longer-term roles, responsibilities and legal, institutional and financial arrangements than on-site management. For implementation to be effective, roles and responsibilities, actions and performance targets must be explicit.

Permanence is a key requirement in many offset policies but a major failure in practice; there is consensus that protection of offset areas in perpetuity is preferable, but companies may be reluctant to fund offsets for any longer than the duration of residual project impacts (McKenney and Kiesecker, 2010). It is thus also important to consider who would take responsibility for managing offset sites when a developer's obligations end, and to plan 'handover' to an appropriate, capacitated institution or organization (ICMM IUCN, 2013).

Strong arguments for considering offset requirements early in the IA process are the long lead time often needed to secure access to land on which to implement offsets and the need for early costing of offset implementation to ensure adequate financial provision in project budgets.

16.4 OFFSET DESIGN WITHIN IA

It is crucial to design offsets to achieve required or clearly defined objectives. Whether or not impacts can be offset is 'fundamentally defined by what offsetting is intended to achieve' (Pilgrim et al., 2013, p. 377). 'Choosing the most appropriate calculation method for an offset scheme. . .requires a clear decision as to the fundamental objective of the offset policy' (Bull et al., 2014b, p. 10).

Choosing the most appropriate type and location of offset, and measuring biodiversity losses and gains, are essential steps in designing offsets and planning their implementation. The gains that can be achieved need to be both theoretically and practically achievable, given the options that exist to provide offsets in the landscape. The extent to which these issues are addressed in detail during the IA process varies considerably. In most cases, the IA process stops at the point where residual impacts on biodiversity are confirmed, the requirements for offsets are established, there is reasonable assurance that they would be implemented and sufficient resources have been allocated to implement them successfully.

It is not appropriate to rely on offsets to achieve NNL/NG if the plans for implementing them are vague. The IA practitioner and/or biodiversity specialist should at least be able to confirm that there are suitable sites and established techniques available that can be relied upon to deliver the required gains. An overview of the various issues that need to be considered in offset design is included in this section.

16.4.1 Balancing Biodiversity Losses and Gains

Suitable 'metrics' need to be chosen to compare losses due to impacts with gains through an offset. Loss due to development can be balanced by gains in the same or different type of biodiversity. These considerations are addressed below.

Like for like or 'in kind'

Rules to govern the acceptability of offsets, given the particular losses of biodiversity that will occur, may be stipulated in government policy or relevant standards and lender requirements. The BBOP principles also provide a basis for appropriate exchanges. As a general rule offsets targeting the same type of biodiversity, or involving the same biodiversity components, are preferred to 'out of kind' offsets. The greater the level of uncertainty or risk, vulnerability or irreplaceability of affected biodiversity, the stronger the requirement for a 'like for like' exchange (assuming that impacts are offsetable). Otherwise there is a risk that inappropriate trade-offs will occur and NNL/NG will not be achieved. For priority biodiversity, a bespoke equivalency analysis is therefore generally required. For widespread/common species and habitat types, 'ratios of acceptable loss or minimum levels of conservation' could be appropriate (Quétier and Lavorel, 2011, p. 7). The irreplaceability and vulnerability of biodiversity inform whether offsets would be acceptable and feasible, and the most appropriate type of offset to use (Figure 16.1).

'Out of kind' and 'trading up'

In some jurisdictions, 'trading up' exchanges are allowed, whereby the offset components are different from those lost through an impact (out of kind) but are of higher conservation priority.

The workability of rigid 'like for like' offsets is questioned. They may force offsets to focus on features that may not represent regional conservation priorities (Hayes and Morrison-Saunders, 2007) and may not deliver optimum conservation outcomes: 'conservation gains from out of kind offsets may outweigh those from strict in kind NNL interpretations' (Bull et al., 2014b, p. 2). Priority-focused offsets rather than strict 'like for like' exchanges may offer benefits (Habib et al., 2013) and in some countries (e.g., Germany), a requirement for strict functional equivalence

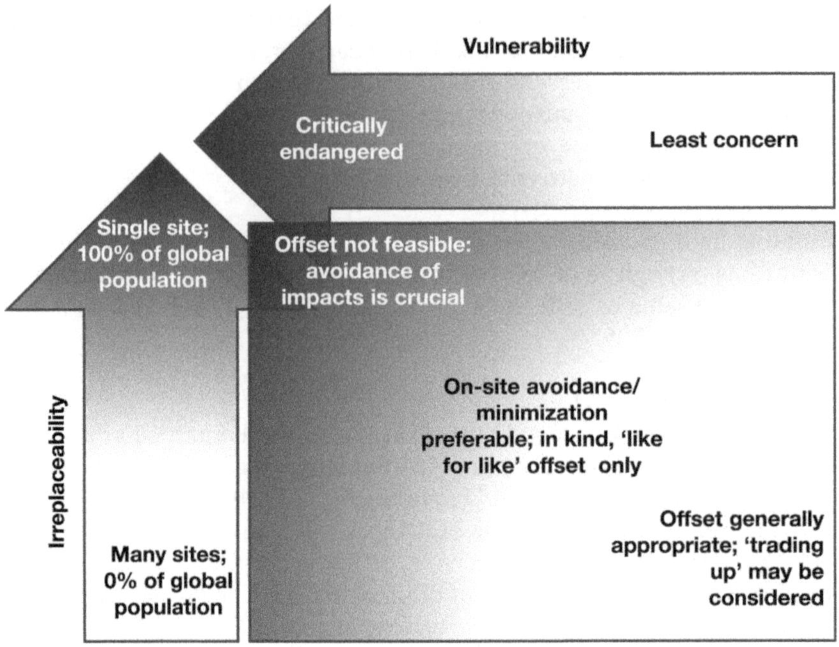

Source: Adapted from Figures 5 and 6, BBOP (2012b).

Figure 16.1 Irreplaceability and vulnerability considerations in deciding on an appropriate type of offset

between impact and offset is no longer imposed; the offset must be similar to the impact in geographic and functional terms but may relate to a different type of habitat (Morandeau and Vilaysack, 2012). This requirement is also found in the UK's pilot offset system for biodiversity at lower levels of conservation priority. Although addressed at policy level, these provisions have a bearing on IA practice. In some countries, developers have to develop their own interpretations of what constitutes an acceptable offset.

Despite some acceptance of 'out of kind' offsets (e.g., in the United States and Australia), there is little support for 'very out of kind' compensation such as funding for education and training (McKenney and Kiesecker, 2010). In Australia, a distinction is made between 'direct' and 'indirect' ('secondary'/'contributing') offsets; the latter may constitute a defined proportion of the offset. A direct offset provides biodiversity values, while indirect offsets support their intended outcome and may include, for example, new research (Hayes and Morrison-Saunders, 2007).

Measuring losses and gains

According to Bull et al. (2013, p. 5) 'ideally all biodiversity losses addressed through offsets should be reversible'. A good offset will account for loss and gain of biodiversity at all levels of biological organization, as well as the role of changes in structure, composition and function of biodiversity in influencing provision of ES to different stakeholders (BBOP, 2012a). A good metric should thus capture the type, amount and condition of the key biodiversity components that are being lost and gained, and should take into account the underlying ecological processes/function that sustain these components. Since it is generally impossible to measure every form of biodiversity, 'surrogates' or 'proxies' are often used to represent key biodiversity components.

A wide range of metrics has been developed over the last 40 years. Most use some measurement of habitat area as a basic proxy unit for calculating biodiversity losses and the conservation gain that must be achieved, others use a combination of area and 'condition' or 'quality' of habitat (BBOP, 2012a). Area alone is not an adequate metric; such proxy measures of biodiversity need to be scaled against changes in the components of conservation concern (e.g., presence of a particular species); doubling of habitat area is not likely to lead to a doubling in species (BBOP, 2012d).

A number of complementary measures may be used to account for distinct biodiversity components and functions (e.g., rare species, wetland functions); one method uses a 'biodiversity distinctiveness' score as a measure of the exchange (Treweek et al., 2010); others take into account levels of landscape connectivity, are designed to make an explicit contribution to national conservation and sustainable use objectives (BBOP, 2009c, 2012a), or determine 'ecosystem and species credits' at prospective offset sites (e.g., NSW, 2014). Populations, or other indicators of species diversity and abundance, may be measurable in some cases. Alternatively, the availability of suitable habitat and its capacity to support viable populations can serve as a proxy (BBOP, 2012a).

There is increasing interest in the use of these types of metric to quantify biodiversity impacts in IA so that losses and gains can be tracked throughout the mitigation hierarchy. This approach underpins Rio Tinto's Biodiversity Strategy, for example, and its stated intention to demonstrate a net positive impact on biodiversity (Rio Tinto, 2004).

Some examples of methodologies used in different countries are:

- The United States's Habitat Evaluation Procedure and Habitat Suitability Indices (BBOP, 2009c);
- Australia's State of Victoria Habitat Hectares method (BBOP, 2009c);

- State of South Australia's Significant Environmental Benefit (BBOP, 2009c);
- Colombia's 'offsets at landscape scale' framework and use of offset ratios (Saenz et al., 2013); and
- South Africa's 'basic offset ratio' adjusted to take into account a range of considerations (Brownlie and Botha, 2009).

Different countries advocate different approaches and metrics, specify the use of one or more indicators, and may require consideration of benchmarks and landscape context, time lags and/or uncertainties (Quétier and Lavorel, 2011). Budgets, time frames, information needs and the availability of information, as well as the specific socioeconomic context of the project site, have a significant influence on the choice of approach (BBOP, 2009b).

16.4.2 Actions to Deliver Offset Gains

The types of conservation action proposed to achieve required gains need to be realistic, given the range of locations available and the capacity that exists to implement them.

'Restoration' offsets

Where proven methods exist, habitat may be re-constructed/re-created. Activities may incorporate the reduction/removal of current threats by, for example, introducing sustainable livelihoods or substitute materials.

Restoration has numerous limitations (Middle and Middle, 2010). To date, re-creation of ecosystems with all component species and functions has proved prohibitively expensive or impossible (Bekessy et al., 2010). Many attempts to restore habitats fail, either in terms of structure or function when compared with reference ecosystems. Newly created or restored habitat may thus end up being unsuitable for a target species or communities at the end of the day (e.g., Moreno-Mateos et al., 2012; Pickett et al., 2013). Curran et al. (2014) found little support that ecological restoration leads to a NNL/NG of biodiversity. According to Maron et al. (2012), many expectations of restoration are not supported by evidence; the promise of full restoration may increase the chance that damage to biodiversity is permitted. However, increased use of offsets within IA may improve understanding of actions needed for successful restoration and raise awareness of the need to plan the resources needed earlier in the process.

Uncertainties associated with restoration favour the use of habitat banks for offsets (Box 16.3). These banks constitute areas in which biodiversity gains have already been achieved through habitat creation,

BOX 16.3 BIODIVERSITY BANKING

Mitigation banks for wetland offsets in the USA were started in the 1990s, followed by conservation banks for endangered species (Doswald et al., 2012). In Germany, 'land pools' aim to provide or concentrate conservation measures (Darbi et al., 2009). In Australia, BioBanking (New South Wales) and BushBroker programmes (Victoria) are market-based banking mechanisms involving ecosystem and species credits; France and the Netherlands are trialling compensation banks (Morandeau and Vilaysack, 2012); a number of South American countries (e.g., Brazil, Costa Rica, Chile and Mexico) have good potential for a banking scheme (Bovarnick et al., 2010).

restoration and/or protection, to be converted into credits. Credits can be sold to developers to meet their offset obligations. Habitat banks can demonstrate that the requisite biodiversity gains have been secured before development begins. In some cases, companies (e.g., Kennecott) may establish their own habitat banks to meet anticipated offset requirements (BBOP, 2009b).

Advantages of habitat banks include (Hayes and Morrison-Saunders, 2007; Darbi et al., 2009) greater geographic scales of mitigation and flexibility in offset location, reduced delay and risk of offset failure, and generally more cost-effective conservation. However, there are constraints relating to the set-up costs of habitat banks and the uncertainty of markets. Assessments in the USA and Australia show that their ecological objectives have not always been reached (Morandeau and Vilaysack, 2012). Nevertheless they offer some advantages to developers, allowing them to transfer responsibility for offset delivery on completion of the IA process, without longer-term commitment or the need to build 'in house' capacity for conservation.

'Protection' or 'averted loss' offsets

Where there is real, imminent or projected loss of biodiversity, its protection can generate 'gains' for biodiversity: the difference between background rates of loss 'without' and 'with' protection of biodiversity in an area constitutes these gains. This offset type is appropriate where background rates of biodiversity loss are high and relatively large intact areas can be protected from transformation, as in Madagascar (Temple et al., 2012), and/or in areas where restoration is not considered to be achievable. Protection may present lower risks than restoration. However, oversimplifications in determining the appropriate size and spatial allocation of offsets to landscapes currently limit confidence in their success (Sonter et al., 2014).

BOX 16.4 FINANCIAL COMPENSATION AS OFFSETS

Programmes in some countries (e.g., Brazil, Russia, India) accept financial compensation as a substitute for in situ conservation (Morandeau and Vilaysack, 2012). Payment is made to various entities depending on the country: dedicated funds managed by government or an approved not-for-profit conservation body, foundation or association. In other countries, financial transfers are tolerated but inferior to 'in kind' compensation (e.g., Netherlands, Morocco, Australia). Financial compensation is only accepted in exceptional cases as a 'last resort' in some countries, when it is impossible to implement 'in kind' measures (e.g., Austria and Switzerland), and when the project is deemed to be of overriding public interest or there is no irreversible destruction of protected species' habitats (e.g., Germany).

Compensation packages or financial compensation

In some countries financial compensation (Box 16.4) to a dedicated offsets fund may be an option instead of in situ measures. For relatively common biodiversity, predetermined checklists could be used to calculate debits, which could be compensated through an in lieu fee, for example (Eftec et al., 2010). However, there is a risk that offsets might become functionally and spatially disconnected from the impact (Darbi et al., 2009) and/or funds may be diverted away from conservation.

Local stakeholders affected by a project and/or its offset may be compensated 'in kind' or paid for their losses. Financial compensation is distinct from financial guarantee mechanisms for an offset, and financial provision to establish and manage 'in situ' offset activities.

Whichever type is used, the equivalence of losses and gains must be demonstrated. Different countries may favour different interventions. In the USA, the priority for delivery of wetland offsets is restoration, creation, enhancement and protection. As a final step, the proponent may choose between paying financial compensation, using the services of a mitigation bank (Darbi et al., 2009), or paying an in lieu fee to a green fund or government agency that is expected to accomplish the desired offset (Bovarnick et al., 2010).

16.4.3 Identifying and Securing Locations to Deliver Offsets

There may be a number of suitable options in the landscape that would satisfy offset requirements. It is best to try to meet all requirements in one area to generate simultaneous gains for several targets, concentrating management effort and reducing overall costs. It is advisable to check any existing or pending rights or claims on candidate sites (e.g., mining rights), as well as customary rights (e.g., to grazing land).

As a general rule, there are higher chances that there would be continuity of ecological processes if offsets were located as close to impacted areas as possible, in the same landscape unit or watershed. On-site offsets have advantages in this regard, but may offer small, disconnected fragments of habitat inconsistent with conservation priorities in the landscape. In that case it is preferable to seek off-site offsets.

Offsets must deliver benefits to those communities whose use and cultural values attached to biodiversity are affected by development; the biodiversity in the offset should provide acceptable substitute values to those lost. Although separate offsets for intrinsic, use and cultural values are feasible, it is preferable to prioritize offsets near impacted areas that could deliver the biological and socioeconomic outcomes needed, and widen the areas of search as necessary (BBOP, 2012a). Involvement of local communities in offset implementation may also be sought and their willingness to participate needs to be explored. This is an aspect of offset design that has been relatively neglected and requires close collaboration between social and biodiversity practitioners. The IA process needs to be designed to foster this collaboration.

Legal options and approaches to securing offsets in the long term may differ in different countries but need to ensure that offset obligations are passed on to new land/resource owners (BBOP, 2009d). There is a range of options to 'secure' habitat in the long term, including formal protection, or entering into contracts or covenants with individuals or communities in which they conserve habitat in return for payment or other benefits. Covenants may be preferable since they avoid the need to acquire land (e.g., Hayes and Morrison-Saunders, 2007), reducing costs and land pressures associated with offsets (Morandeau and Vilaysack, 2012).

Optimizing offsets in the landscape

Offsets should preferably be located in a priority area for conservation identified in biodiversity plans, connecting priority areas and/or consolidating ecological corridors (Brownlie and Botha, 2009). Numerous authors (e.g., from USA, Australia, South Africa and Colombia) raise the need for long-term regional planning and a landscape-scale perspective to ensure that offsets are optimally located in a strategic context. A landscape vision 'ensures that the biologically and ecologically important features remain the core conservation targets throughout the process' (Saenz et al., 2013, p. 4977). Harnessing landscape planning can improve offset site selection and conservation returns (McKenney and Kiesecker, 2010).

Systematic conservation planning approaches (e.g., Kiesecker et al., 2009; Saenz et al., 2013) blend landscape-level planning with application of the mitigation hierarchy to arrive at a 'conservation portfolio' from which offset sites can be selected. In this approach, impacts are quantified based on conservation targets identified in the landscape-level plan, and areas most similar to impacted areas could be selected as offset sites. 'By first avoiding or minimizing impacts to occurrences of biological targets with high irreplaceability and/or vulnerability, then ensuring that impacts are restored onsite. . .and finally offsetting any remaining residual impacts, we can provide a framework truly consistent with sustainable development' (Saenz et al., 2013, p. 4977). There is some interest in the use of Strategic Environmental Assessment (SEA) to provide a framework to optimize location of offsets in the landscape according to conservation priority; SEA may be particularly beneficial in contexts where environmental planning systems are relatively weak or under-resourced.

When selecting offset sites, consideration must be given to possible 'leakage' that occurs when an offset displaces pressure on biodiversity (e.g., deforestation or poaching) to another area. Instead of remedying a project's residual negative impacts, therefore, an offset may simply shift a threat to another location, thwarting NNL/NG.

16.5 CHALLENGES IN OFFSET DESIGN

The use of offsets in IA fundamentally poses the problem of balancing certain biodiversity loss with uncertain gains. Drawing on numerous sources, the following challenges in designing a biodiversity offset can be identified, which must be addressed to make IA an effective tool for NNL/NG:

- using suitable metrics;
- dealing with 'non-offsetable' impacts;
- ensuring equivalence of losses and gains;
- ensuring 'additionality' and reliable baselines; and
- integrating ES in offset design.

These challenges are addressed below.

16.5.1 Metrics

The quantification of biodiversity losses and gains poses significant challenges because of the inherent complexity of biodiversity, and the variety of ways in which it can be measured. Different methods for calculating the required offset activities result in divergent outcomes for biodiversity; 'the method used to quantify losses and gains strongly influences the biodiversity outcomes of offsetting, implying that offsets generated using different methodologies are not transferable between jurisdictions' (Bull et al., 2014b, p. 2).

More specificity in the choice of metrics might be more conducive to achieving NNL/NG, but it also makes offsets less likely to be implemented and may complicate the use of trading with mitigation banks (Quétier and Lavorel, 2011). Using additional metrics results in additional complication and expense and beyond some point will not justifiably reduce uncertainty in quantifying biodiversity further (Bull et al., 2013).

Despite the numerous methods available, all are subject to criticism. Many fail to consider all the relevant factors/functions for a site, ignore the dynamic nature of ecosystems, compare 'apples and oranges', rely too heavily on untested weighting/scoring mechanisms, or produce highly inaccurate results (McKenney, 2005). Quigley and Harper (2006) found that at least 63 per cent of projects designed to offset fish habitat loss in Canada failed to achieve the stated target of NNL despite meeting legal requirements. The use of crude metrics, poor biodiversity surrogates and over-simplified metrics also fails to account for significant environmental and social welfare values across space, type and time (e.g., Walker et al., 2009). In Italy the lack of established methods appears to hinder offsets; no metrics were provided in 19 of 35 cases (Rega, 2013).

These shortcomings, together with the frequent failure of restoration as a core offset mechanism, and time lag effects between offsets and impacts on threatened biodiversity (Bekessy et al., 2010), means that there is a high risk of losing biodiversity in each exchange and failing to achieve NNL/NG.

16.5.2 Limits to Offsets

'Biodiversity offsets are not appropriate for "critical" or "non-substitutable" biodiversity assets which cannot be replaced using known techniques or within reasonable time frames' (Treweek, 2009, p. 32). 'Some things, such as critical habitat for listed threatened species, are not tradeable' (Bekessy et al., 2010, p. 154). Major financial institutions (e.g., World Bank, 2004) recognize the principle that, due to the high value of some biodiversity resources, there are – effectively – 'no go zones' for new extractive industry investments. The challenge lies in determining where those limits lie. Factors that indicate a high risk that impacts would be 'non-offsetable' (BBOP, 2012c), comprise:

- a high proportion and condition of biodiversity (population, species or ecosystem) to be residually impacted;
- a high level of irreplaceability or uniqueness of features and sites (e.g., habitats for highly specialized or uniquely adapted species, restricted in distribution, relatively immobile and/or slow to regenerate affected biodiversity) and probable scarcity of suitable offset options in the landscape; and
- a high vulnerability of affected biodiversity (highly threatened ecosystems/species), especially where the cause of decline is unknown, or not tractable.

There are currently no globally accepted thresholds that define when impacts would be 'non-offsetable'. However, indicative thresholds (BBOP, 2012c) include the following:

- The IFC (2012) differentiates between 'Natural Habitat' and 'Critical Habitat'. Offsets for the former must achieve 'no net loss'; the latter must achieve a 'net gain' outcome. 'Critical Habitat' is identified on the basis of three species-level criteria; impacts on Tier 1 Critical Habitat are likely to be non-offsetable.
- In South Africa, impacts on 'Critically Endangered' ecosystems are regarded as non-offsetable.
- In New South Wales, Australia, there are areas where clearing of native vegetation is not permitted and impacts cannot be compensated unless their condition is low or they are not viable.

Thresholds indicating conservation priorities can help to identify situations where there is a high risk that impacts may be non-offsetable. Examples include (BBOP, 2012c):

- IUCN Red List criteria used to track the extinction risk of species. The IFC draws on these thresholds to indicate likely non-offsetability.
- Systems proposed to be valid at the international level for identifying ecosystems at risk of irreversible loss, using IUCN-type categories with a set of explicit criteria and thresholds.
- The EU Habitats and Birds Directives and the EU Water Framework Directive, which set out guidelines with indicators and thresholds for determining the conservation status of listed habitats and species.
- Various methods and systems to identify conservation priorities according to vulnerability- and irreplaceability-based criteria (e.g., systematic conservation planning).

- Some national interpretations of 'High Conservation Value' areas with indicative thresholds for irreplaceable resources.

16.5.3 Equivalence in the Exchange

Ensuring equivalence between losses and gains necessitates accounting for uncertainties. There are three key areas of risk and uncertainty, namely time lags, uncertainties about the measure of residual negative impacts, and uncertainties about the outcome of offset activities. There are other complex issues such as climate change, with shifts in ecological functioning and dynamics, which pose significant uncertainties when designing offsets.

Time lags

Development results in immediate biodiversity losses. Protection offsets may not be immediate, and positive management actions take time to mature. A time lag is expected between the creation of habitat and habitation by the target species, as some resources may only become available in later successional stages (Pickett et al., 2013). These delays may have significant consequences for biodiversity conservation (Maron et al., 2012), and even if biodiversity levels were eventually restored, the ES lost in the interim could necessitate additional compensation (Bull et al., 2014a).

Time lags between habitat loss and offset habitat becoming available to a species are critical. According to Maron et al. (2010), many animal species need old-growth habitat (e.g., tree hollows) that takes decades or centuries to form; a reduction in habitat availability of that duration may lead to extinction of highly threatened species. The success of habitat offsets where there is a significant delay between habitat loss and gain for a threatened species is likely to be low, because resource bottlenecks become a significant threat to the species' persistence. For this reason, some countries require offsets to be established before development proceeds (Bekessy et al., 2010).

Uncertainties about residual negative impacts

There may be uncertainties in predicting negative impacts on biodiversity due, amongst others, to the quality of available data and/or inherent uncertainty regarding ecosystem dynamics. The effectiveness of measures proposed to avoid, minimize or repair impacts may be unpredictable or overly optimistic, resulting in uncertainty about – or underestimates of – residual negative impacts.

Uncertainties about offset outcomes

The risk of an unsuccessful offset is high where there are no tried and tested techniques to achieve the required biodiversity outcomes (BBOP, 2012c). For example, where re-creation of habitat constitutes the core offset activity, there may be significant uncertainties about the result. There are also implementation challenges unrelated to biodiversity that could jeopardize the success of an offset. Where these risks are material, offset planners need to consider whether it is realistic to pursue offsets (BBOP, 2012a).

16.5.4 Additionality and Baselines

An offset must deliver conservation gains over and above what is already taking place or planned, existing obligations and/or legal requirements. The additional benefit attributable to an offset activity is the difference between the scenario with the activity and the alternative scenario without it; calculations of loss and gain need to consider the biodiversity baseline and trends. Unfortunately, assumptions about both scenarios are frequently not explicit, wrong or both. For example, it is often assumed that the offset would remove all threats; protection may prevent habitat loss, but it may be less effective at ameliorating threats like invasive species (Maron et al., 2013).

The current drivers of biodiversity loss (e.g., deforestation), and the background (without the offset) rates of biodiversity loss, must be determined. Bull et al. (2014a) note that the perception of whether NNL/NG has been achieved changes significantly when outcomes are measured against different baselines.

The scope for additionality differs in different countries and contexts. Protection of existing habitat is not recognized as an offset in Germany because there is no additional benefit (Darbi et al., 2009). In some countries, the removal of invasive alien species may be 'additional', while in others, laws require clearing of these species and their eradication could not count as an offset. Resources for managing protected areas, albeit contentious, may be seen as 'additional' when the capacity needed for their management is lacking and expected to deteriorate in the long term.

16.5.5 Ecosystem Services

In some definitions the scope of offsets includes the use and cultural values of biodiversity, which are akin to ES. The importance of biodiversity in ecosystem resilience and helping to maintain multiple ES is widely recognized. In rural areas of developing countries specifically, projects frequently impact on communities who rely on natural resources.

Understanding the socioeconomic and cultural context is crucial for understanding the affected biodiversity when designing and implementing offsets (BBOP, 2009b). Do local communities rely on the biodiversity of the region for their subsistence, livelihoods, health, or safety from natural hazards? Do they hold certain biodiversity features as culturally or spiritually significant?

There are a number of approaches to addressing ES in IA (Landsberg et al., 2013; WBCSD, 2013). There are differences in the focus of 'ecosystem services'; some use the Millennium Ecosystem Assessment's categories (MA, 2005), while others (Landers and Nahlik, 2013) limit consideration to 'final' rather than 'intermediate' ES.

Offsets targeting ES exist, but may not be labelled as such: for example, water quality trading credits and fisheries compensation (ICMM IUCN, 2013). Depending on their timing and location, biodiversity offsets may be sufficient to provide compensation to affected parties for impacted ES. In other situations, particular services may need to be offset separately; water purification and provision of fuel wood are common examples.

Economic valuation and financial compensation (e.g., for impacts on forest livelihoods) and measures of loss and gain of ES over time (e.g., months of lost access to potable water) are metrics most commonly used to measure ES (ibid.).

16.6 RESPONSE TO DESIGN CHALLENGES

To maximize the potential of offsets to achieve NNL/NG, there needs to be clarity about the objective of offsets; rigour in the choice of approach and use of metrics to ensure a fair and equivalent exchange; explicit and careful checking that the offset would satisfy key offset principles such as limits to what can be offset, additionality, landscape context and long-term outcomes; and avoidance of leakage. Tools being used to address these challenges comprise:

- 'multipliers' and time discounting;
- exchange rules; and
- independent peer review or validation.

16.6.1 Multipliers

Multipliers, sometimes referred to as 'mitigation ratios' or 'offset ratios', are applied to increase the size of an offset, and/or the number and type of offset activities, in order to be confident of achieving NNL/NG. They may be used to address uncertainties and/or risks associated with the success of restoration; 'out of kind' or 'trading up' exchanges; or location differences between the impact and offset sites (e.g., in or out of a specified watershed). They may be designed to ensure that conservation goals or targets would be met and/or that conservation of highly threatened biodiversity components would be assured (e.g., McKenney and Kiesecker, 2010; Quétier and Lavorel, 2011; Morandeau and Vilaysack, 2012; BBOP, 2012a). Multipliers are also used to address risks associated with time lags; the size of the offset is increased proportional to the risks of delays in achieving the offset (e.g., Bendor, 2009).

Use of multipliers has limitations. According to Bekessy et al. (2010, p. 153), '[u]ncertainty multiplier systems are at best complicated and unworkable in the eyes of proponents and are at worst token and not commensurate with the inherent uncertainty about restoration success'. The uncertainties associated with restoration are large; the multipliers are likely to be unworkably large (Moilanen et al., 2009), and far exceed what is currently applied in practice (Curran et al., 2014). Despite using large (up to 19-fold increases in pond habitat) multipliers to account for time lags between habitat creation for, and habitation by, a threatened frog species, Pickett et al. (2013) found only a 1.2–3.5-fold increase in population size. Time discounting, where future gains are valued in present-time units, is advocated on the grounds that it is unfair to compensate for a guaranteed immediate loss with a hypothetical and less certain future gain. Although this multiplier makes sense in terms of equity, the use of time discounting may do nothing to address the fact that delays in providing an offset can lead to critical shortages in ecological resources over time, no matter how large the offset. These critical shortages or 'ecological bottlenecks' could make it impossible to achieve offset gains comparable to losses from development (Gardner et al., 2013).

In some countries (e.g., South Africa), ratios are used as a mechanism to promote preservation of highly valued ecosystems and/or to ensure that biodiversity targets would be met; ratios of up to 20:1 for highly threatened biodiversity are required (Brownlie and Botha, 2009). In Colombia (Saenz et al., 2013), requiring higher offset ratios (up to 10:1) for ecosystems with few remaining patches relative to their historic distribution, experiencing a high rate of loss over the past six years, and/or with a large, intact proportion of their historic distribution remaining, is deemed appropriate.

While they are a common feature of offset policies, multipliers are a recent phenomenon and more research, societal debate and practical experience are needed to establish best practice (BBOP, 2012a). Importantly, increasing the size of an offset may not remedy the source of risk or uncertainty; it is crucial to relate the effect of the proposed multiplier and the risk/uncertainty. As noted by Gardner and others (2013), a lack of data and technical understanding means that multipliers are often generic and determined by the conservation significance of affected biodiversity, rather than being linked to specific risks and mitigation measures. Multipliers are not appropriate where the feasibility of the offset method is in question, such as untried restoration techniques: if a technique is methodologically uncertain and its outcomes unknown, attempting to implement it over twice the area is unlikely to reap twice the benefit. Conversely, multipliers are appropriate where uncertainty relates to the probable degree of success of a method. For example, where the probability of seedlings growing into mature trees has been proven in trials to be 50 per cent, it would be necessary to plant at least twice as many seedlings to achieve the desired outcome (BBOP, 2012a). Multipliers are inappropriate for situations where there is a risk that the offset intervention may fail entirely (Gardner et al., 2013).

According to Moilanen et al. (2009), very high offset ratios may be needed to guarantee a robustly fair exchange. They advocate using a 'bet-hedging' strategy in offset design to minimize risks; one where presently valuable offset areas are protected, and rehabilitation/restoration effort is split among a set of spatially dispersed and environmentally different sites.

16.6.2 Exchange Rules

Exchange rules are used to stipulate boundaries and limit risks in the impact–offset exchange. They require that the extent and ecological condition of habitat either do not change substantially or can only increase between impact and offset sites (e.g., Kiesecker et al., 2009). They may prohibit exchanges of different biodiversity types (e.g., grassland for forest). Efforts to ensure compositional similarity between losses and gains for unmeasured components of biodiversity can be assisted by 'rule-of-thumb' spatial restrictions such as the maximum distance between development and offset sites, or by restricting exchanges to within the same watershed, centre of endemism, vegetation type, or area in which people who may be affected by the impacts of development on biodiversity live (Salzman and Ruhl, 2000).

In the UK, Treweek et al. (2010) suggest introducing exchange rules to balance offsets that improve habitat condition through rehabilitation/restoration with protection offsets that increase 'biodiversity distinctiveness', to ensure that there is no loss of extent of priority habitat. In Victoria, Australia, similar exchange rules are in place (BBOP, 2009c); for example, vegetation proposed as an offset for impacts on 'very high' conservation significance habitat must be at least 90 per cent of the quality in the area being lost, and the re-vegetation component of the offset is limited to 10 per cent, whereas for 'medium' conservation significance offsets the quality requirement and re-vegetation limit is 50 per cent. This rule avoids exposing very high conservation values to the risks of exchanging small areas of high-quality habitat with large areas of immature or low-quality habitat.

16.6.3 Independent Review or Validation

Peer review or validation of a proposed offset design and implementation plan by independent experts provides assurance to all stakeholders of the offset's adequacy. Use of internationally accepted principles and 'good practice' standards (e.g., BBOP, 2012b) provides a benchmark against which to evaluate proposals.

Non-governmental, academic and research organizations involved in biodiversity conservation and wider 'sustainable development' arenas, as well as independent experts, can play an important role in ensuring the quality of offset design. In Australia, brokers are accredited by the government to bring together private landowners who generate compensation credits and the project developers that require them (Morandeau and Vilaysack, 2012).

16.7 ADDRESSING CHALLENGES IN OFFSET IMPLEMENTATION

Implementation must meet the requirements of laws, any financial institutions, and/or internal company policies. It can be carried out on a project-by-project basis by the proponent or appointed third party, including through habitat banks, auction and brokering schemes, and in lieu fee arrangements.

Conservation outcomes from offsets depend only partly on the scientific rigour applied in offset design. They are also strongly determined (e.g., Darbi et al., 2009; Doswald et al., 2012; Gardner et al., 2013) by effective monitoring and evaluation, and political and social acceptability and stakeholder 'buy-in'. These issues are discussed below.

16.7.1 Monitoring and Evaluation

Offset activities must be checked to ensure they are achieving the intended results, and reported to key audiences; monitoring and evaluation are thus crucial components of implementation (Villarroya et al., 2014). Hayes and

Morrison-Saunders (2007) found that the offset goal of NNL/NG was not being achieved partly because of a lack of oversight and enforcement, and problems with record keeping. Poor compliance has also been found in Australia and Canada (Bull et al., 2013). There is a need to increase follow-up and enforcement for IA and compensation/offset agreements (Bovarnick et al., 2010).

Many of the problems associated with offsets go undetected and thus not remedied because they occur beyond the time period of monitoring (Maron et al., 2012). A review of offset projects in the UK found that 49 per cent of projects included a monitoring period, and the average length of this period was less than two years; a period that would not account for impacts of succession on populations (Pickett et al., 2013).

Formal agreement defining the respective roles and responsibilities of parties involved in implementing offsets, as well as the duration of those responsibilities, provides a robust and auditable framework for implementation. The choice of indicators in monitoring key performance areas of the offset is crucial; monitoring must continue until the offset has been achieved, and in the long term. 'Leakage' must also be tracked as it would undermine results.

Adopting an adaptive management approach in offset implementation offers numerous advantages (BBOP, 2009d). It involves checking the outcomes of actions against performance targets, and adjusting management to improve that performance or ensure targets are achieved.

Independent verification of successful offset implementation, evaluated against internationally accepted principles and 'good practice' standards (e.g., BBOP, 2012b) and the project-specific BOMP or EMP/BMP, would provide assurance to stakeholders of the offset's outcome.

It can be advantageous to involve non-governmental or civil society organizations, representatives of local communities and/or academic or research institutions in monitoring. They may help in measuring and recording outcomes in the field, as input to independent verification and/or audit.

16.7.2 Political and Social Acceptability

Biodiversity offsets are unlikely to succeed in the long term unless they are supported by affected communities and political structures. Transparency and openness are key ingredients of acceptability, necessitating engagement with persons or groups who own/occupy, use, manage or regulate the area affected by the development and the offset areas; who could be affected by the project and/or offset activities; and whose involvement is needed to make the offset a success. They can include indigenous peoples and local communities, government authorities, civil society and non-governmental organizations, and scientific, academic or research institutions (BBOP, 2009d).

Partnerships with local communities, non-governmental organizations or civil society organizations help to build trust and acceptability of offsets, particularly where tangible benefits would flow to local communities; opportunities for partnerships have been used as a criterion for selecting offset sites (BBOP, 2009b). Establishing a group of technical specialists and/or involving research/academic institutions in the preparation of a BOMP, and with offset monitoring, may contribute to the offset's acceptability.

16.8 EMERGING TRENDS AND CONCLUSIONS

The following trends in offset policy and practice are evident:

1. The need to plan offsets more strategically through landscape-level interventions (e.g., Kiesecker et al., 2009). While IA is currently seen as the main vehicle for planning biodiversity offsets, its application globally is limited to specific activities. Taking a project-by-project approach to offsets overlooks cumulative effects and efficiencies that could be exploited when resources are allocated at broader scales (Sonter et al., 2014).
2. Increasing emphasis on the importance of regional and/or systematic conservation planning approaches, and corresponding metrics, to ensure that project-level offsets contribute to national conservation goals (e.g., Brownlie and Botha, 2009).
3. Rigid adherence to 'like for like' exchanges in offsets is questioned in light of both their conservation significance and constraints of land availability in land-hungry sectors. The challenge of determining appropriate metrics, multipliers or exchange rules for these 'out of kind' exchanges remains (e.g., Bull et al., 2014a).
4. Growing use of habitat banking as a favoured offset mechanism (BBOP, 2009c).
5. 'Offsets for uncertainty' have emerged in Australia (Middle and Middle, 2010) as a way to deal with 'acceptable' residual negative impacts of a project where there is significant uncertainty about these impacts; potential residual risk determines the level of offset.
6. Increasing attention to offset implementation, monitoring and management to improve outcomes for conservation.

Biodiversity offsets have obvious theoretical advantages as a means to address impacts on biodiversity and ES that otherwise remain due to partial application of the mitigation hierarchy. They should result in better outcomes for biodiversity, but only if sufficient effort is invested in avoiding impacts at source, enforcing implementation of minimization measures and monitoring biodiversity outcomes throughout project timelines. It is important to recognize that IA without the addition of offsets to the mitigation hierarchy has not had a good track record in arresting biodiversity decline. Adding an offsetting step has the potential to improve application of the mitigation hierarchy as a whole by forcing more explicit 'accounting' for biodiversity losses and gains throughout the IA process. However, there remains a need for 'proof of concept', and successful examples of offset implementation and long-term management, particularly in countries where regulatory capacity is limited.

NOTE

1. See http://www.equator-principles.com/index.php/about-ep/38-about/about; accessed 3 August 2014.

REFERENCES

BBOP (2009a), *The Relationship between Biodiversity Offsets and Impact Assessment: A BBOP Resource Paper*, Washington, DC: Business and Biodiversity Offsets Programme.
BBOP (2009b), *Compensatory Conservation Case Studies*, Washington, DC: Business and Biodiversity Offsets Programme.
BBOP (2009c), *Biodiversity Offset Design Handbook: Appendices*, Washington, DC: Business and Biodiversity Offsets Programme.
BBOP (2009d), *Biodiversity Offset Implementation Handbook*, Washington, DC: Business and Biodiversity Offsets Programme.
BBOP (2012a), *Biodiversity Offset Design Handbook – Updated*, Washington, DC: Business and Biodiversity Offsets Programme.
BBOP (2012b), *Standard on Biodiversity Offsets*, Washington, DC: Business and Biodiversity Offsets Programme.
BBOP (2012c), *Limits to What Can be Offset: A BBOP Resource Paper*, Washington, DC: Business and Biodiversity Offsets Programme.
BBOP (2012d), *Resource Paper: No Net Loss and Loss-Gain Calculations in Biodiversity Offsets*, Washington, DC: Business and Biodiversity Offsets Programme.
Bekessy, S.A., B.A. Wintle and D.B. Lindenmayer et al. (2010), 'The biodiversity bank cannot be a lending bank', *Conservation Letters*, **3**(3), 151–8.
Bendor, T. (2009), 'A dynamic analysis of the wetland mitigation process and its effects on no net loss policy', *Landscape and Urban Planning*, **89**(1), 17–27.
Bovarnick, A., C. Knight and J. Stephenson (2010), *Habitat Banking in Latin America and Caribbean: A Feasibility Assessment*, New York: United Nations Development Programme.
Brownlie, S. and M. Botha (2009), 'Biodiversity offsets: adding to the conservation estate, or "no net loss"?', *Impact Assessment and Project Appraisal*, **27**(3), 227–31.
Brownlie, S., N. King and J. Treweek (2013), 'Biodiversity tradeoffs and offsets in impact

assessment and decision making: can we stop the loss?', *Impact Assessment and Project Appraisal*, **31**(1), 24–33.

Bull, J.W., A. Gordon and E.A. Law et al. (2014a), 'Importance of baseline specification in evaluating conservation interventions and achieving no net loss of biodiversity', *Conservation Biology*, **28**(3), 799–809.

Bull, J.W., E.J. Milner-Gulland and K.B. Suttle et al. (2014b), 'Comparing biodiversity offset calculation methods with a case study in Uzbekistan', *Biological Conservation*, **178**, 2–10.

Bull, J.W., K.B. Suttle and A. Gordon et al. (2013), 'Biodiversity offsets in theory and practice', *Oryx*, **47**(3), 369–80.

Clare, S., N. Krogman and L. Foote et al. (2011), 'Where is the avoidance in the implementation of wetland law and policy?', *Wetlands Ecological Management*, **19**(2), 165–82.

Convention on Biological Diversity (CBD) (2010), *Global Biodiversity Outlook 3*, Montreal: CBD.

Convention on Biological Diversity (CBD) and Netherlands Commission for Environmental Assessment (NCEIA) (2006), *Biodiversity in Impact Assessment, Background Document to CBD Decision VIII/28: Voluntary Guidelines on Biodiversity-Inclusive Impact Assessment*, Technical Series No. 26, Montreal: CBD, accessed 8 August 2014 at https://www.cbd.int/doc/publications/cbd-ts-26-en.pdf.

Curran, M., S. Hellweg and J. Beck (2014), 'Is there any empirical support for biodiversity offset policy?', *Ecological Applications*, **24**(4), 617–32.

Darbi, M., H. Ohlenburg and A. Herberg et al. (2009), *International Approaches to Compensation for Impacts on Biological Diversity*, Berlin: Berlin University of Technology.

Doswald, N., M. Barcellos Harris and M. Jones et al. (2012), *Biodiversity Offsets: Voluntary and Compliance Regimes. A Review of Existing Schemes, Initiatives and Guidance for Financial Institutions*, Cambridge, UK/Geneva: UNEP-WCMC/UNEP FI.

Eftec, IEEP et al. (2010), *The Use of Market-based Instruments for Biodiversity Protection – The Case of Habitat Banking – Summary Report for the European Commission DG Environment*.

Gardner, T., A. von Hase and S. Brownlie et al. (2013), 'Biodiversity offsets and the challenge of achieving no net loss', *Conservation Biology*, **27**(6), 1254–64.

Habib, T.J., D.R. Farr and R.R. Schneider et al. (2013), 'Economic and ecological outcomes of flexible biodiversity offset systems', *Conservation Biology*, **27**(6), 1–11.

Hayes, N. and A. Morrison-Saunders (2007), 'Effectiveness of environmental offsets in environmental impact assessment: practitioner perspectives from Western Australia', *Impact Assessment and Project Appraisal*, **25**(3), 209–18.

Hughes, J., L. Ahuja and S. Brownlie et al. (2015), 'Using biodiversity plans to guide mitigation and offsets for a zinc mine in Northern Cape, South Africa'. Paper presented at the 35th Annual Conference of IAIA. 20–23 April 2015. Florence, Italy.

ICMM IUCN (2013), *Independent Report on Biodiversity Offsets*, prepared by The Biodiversity Consultancy, accessed 26 January 2016 at http://www.icmm.com/document/4934.

International Finance Corporation (IFC) (2012), *Performance Standard 6: Biodiversity Conservation and Sustainable Management of Living Natural Resources*, 1 January, Washington, DC: IFC.

Kiesecker, J.M., H. Copeland and A. Pocewicz et al. (2009), 'Development by design: blending landscape level planning with the mitigation hierarchy', *Frontiers in Ecology and the Environment*, **8**(5), 261–6.

Landers, D.H. and A.M. Nahlik (2013), *Final Ecosystem Goods and Services Classification System (FEGS-CS)*, EPA/600/R-13/ORD-004914, Washington, DC: US Environmental Protection Agency, Office of Research and Development.

Landsberg, F., J. Treweek and M.M. Stickler et al. (2013), *Weaving Ecosystem Services into Impact Assessment, A Step-By-Step Method*, Washington, DC: World Resources Institute.

Lawrence, D.P. (2007), 'Impact significance determination – back to basics', *Environmental Impact Assessment Review*, **27**(4), 755–69.

Madsen, B., N. Carroll and K. Brands et al. (2011), *Update: State of Biodiversity Markets*, Washington, DC: Forest Trends.

Maron, M., P.K. Dunn and C.A. McAlpine et al. (2010), 'Can offsets really compensate for habitat removal? The case of the endangered red-tailed black-cockatoo', *Journal of Applied Ecology*, **47**(2), 348–55.

Maron, M., R.J. Hobbs and A. Moilanen at al. (2012), 'Faustian bargains? Restoration realities in the context of biodiversity offset policies', *Biological Conservation*, **155**, 141–8.

Maron, M., J.R. Rhodes and P. Gibbons (2013), 'Calculating the benefit of conservation actions', *Conservation Letters*, **6**(5), 359–67.

McKenney, B. (2005), *Environmental Offset Policies, Principles, and Methods: A Review of Selected Legislative Frameworks*, Radolfzell, Germany: Global Nature Fund Biodiversity Neutral Initiative.

McKenney, B.A. and J. Kiesecker (2010), 'Policy development for biodiversity offsets: a review of offset frameworks', *Environmental Management*, **45**(1), 165–76.

Middle, G. and I. Middle (2010), 'A review of the use of environmental offset as a policy mechanism in the environmental impact assessment process (EIA) in Western Australia', *Impact Assessment and Project Appraisal*, **28**(4), 313–22.

Millennium Ecosystem Assessment (MA) (2005), *Ecosystems and Human Well-being: Synthesis*, accessed 26 January 2016 at http://www.millenniumassessment.org/documents/document.356.aspx.pdf.

Moilanen, A., A.J. van Teeffelen and Y. Ben-Haim et al. (2009), 'How much compensation is enough? A framework for incorporating uncertainty and time discounting when calculating offset ratios for impacted habitat', *Restoration Ecology*, **17**(4), 470–78.

Morandeau, A. and D. Vilaysack (2012), *Compensating for Damage to Biodiversity: An International Benchmarking Study*, Commissariat Général au Développement Durable, Etudes & Documents, No. 68, August, accessed 26 January 2016 at http://www.forest-trends.org/documents/files/doc_3209.pdf.

Moreno-Mateos, D., M.E. Power and F.A. Comín et al. (2012), 'Structural and functional loss in restored wetland ecosystems', *PLoS Biology*, **10**(1), e1001247.

NSW (2014), *Draft Framework for Biodiversity Assessment for Assessing and Offsetting State Significant Development and State Significant Infrastructure*, Sydney: NSW Biodiversity Offsets Policy for Major Projects, Office of Environment and Heritage for the NSW Government (Australia).

Pickett, E.J., M.P. Stockwell and D.S. Bower et al. (2013), 'Achieving no net loss in habitat offset of a threatened frog required high offset ratio and intensive monitoring', *Biological Conservation*, **157**, 156–62.

Pilgrim, J., S. Brownlie and J. Ekstrom et al. (2013), 'A process for assessing offsetability of biodiversity impacts', *Conservation Letters*, **6**(5), 376–84.

Quétier, F. and S. Lavorel (2011), 'Assessing ecological equivalence in biodiversity offset schemes: key issues and solutions', *Biological Conservation*, **144**(12), 2991–9.

Quigley, J.T. and D.J. Harper (2006), 'Compliance with Canada's Fisheries Act: a field audit of habitat compensation projects', *Environmental Management*, **37**(3), 336–50.

Rainey, H.J., E.H.B. Pollard and G. Dutson et al. (2014), 'A review of corporate goals of no net loss and net positive impact on biodiversity', *Oryx*, **49**(2), 232–8.

Rega, C. (2013), 'Ecological compensation in spatial planning in Italy', *Impact Assessment and Project Appraisal*, **31**(1), 45–51.

Rio Tinto (2004), *Rio Tinto and Biodiversity*, accessed 26 January 2016 at http://www.riotinto.com/documents/ReportsPublications/RTBidoversitystrategyfinal.pdf.

Rockström, J., W. Steffen and K. Noone et al. (2009), 'A safe operating space for humanity', *Nature*, **461**(7263), 472–5.

Saenz, S., T. Walschburger and J.C. González et al. (2013), 'A framework for implementing and valuing biodiversity offsets in Colombia: a landscape scale perspective', *Sustainability*, **5**(12), 4961–87.

Salzman, J. and J.B. Ruhl (2000), 'Currencies and the commodification of environmental law', *Stanford Law Review*, **53**, 607–94.

Sonter, L.J., D.J. Barrett and B.S. Soares-Filho (2014), 'Offsetting the impacts of mining to achieve no net loss of native vegetation', *Conservation Biology*, **28**(4), 1068–76.

Temple, H.J., S. Anstee and J. Ekstrom et al. (2012), *Forecasting the Path Towards a Net Positive Impact on Biodiversity for Rio Tinto QMM*, Gland, Switzerland: IUCN.

Ten Kate, K. and M.L.A. Crowe (2014), 'Biodiversity offsets: policy options for governments', an input paper for the IUCN Technical Study Group on Biodiversity Offsets, Gland, Switzerland: IUCN.

Ten Kate, K., J. Bishop and R. Bayon (2004), *Biodiversity Offsets: Views, Experience, and the Business Case*, Gland, Switzerland/Cambridge, UK: International Union for Conservation of Nature and Natural Resources (IUCN)/Insight Investment Management.

Treweek, J. (2009), *Scoping Study for the Design and use of Biodiversity Offsets in an English Context, Final Report to Defra*, accessed 26 January 2016 at https://www.gov.uk/government/uploads/system/uploads/attachment_data/file/218689/BiodiversityOffsets12May2009.pdf.

Treweek, J., B. Butcher and H.J. Temple (2010), 'Biodiversity offsets: possible methods for measuring biodiversity losses and gains for use in the UK', *In Practice*, **69**(18), 29–32.

Villarroya, A., A.C. Barros and J. Kiesecker (2014), 'Policy development for environmental licensing and biodiversity offsets in Latin America', *PLoS ONE*, **9**(9), e107144.

Walker, S., A.L. Brower and R.T. Stephens et al. (2009), 'Why bartering biodiversity fails', *Conservation Letters*, **2**, 149–57.

World Bank (2004), *Striking a Better Balance: The World Bank Group and Extractive Industries – The Final Report of the Extractive Industries Review*, Washington, DC: World Bank.

World Business Council for Sustainable Development (WBCSD) (2013), 'Eco4Biz: ecosystem services and biodiversity tools to support business decision making, Version 1', April.

17. Mitigation for the people: an ecosystem services framework*

Heather Tallis, Christina M. Kennedy, Mary Ruckelshaus, Joshua Goldstein and Joseph M. Kiesecker

17.1 INTRODUCTION

Research has made enormous advances in establishing the mitigation hierarchy framework, and developing clear methods for identifying biodiversity impacts, allowing partial satisfaction of the intent of these laws and regulations (Race and Fonseca, 1996; Geneletti, 2002; Landis, 2003; Lawrence, 2003; BenDor et al., 2008; Canter and Ross, 2010; McKenney and Kiesecker, 2010; BBOP, 2012a). However, ecosystem services (ES) and associated human impacts have largely been ignored in environmental impact mitigation (Brownlie et al., 2013; Bos et al., 2014). This presents a clear gap as many laws that establish mitigation requirements were largely designed to safeguard people from damages that can occur through environmental degradation: in other words, the laws were designed to protect people from the loss of ES (Villarroya et al., 2014). For example, some environmental laws generally state a desire to ensure future enjoyment of nature's benefits, such as Australia's National Strategy for Ecologically Sustainable Development. This law states its intent to, in part, 'enable development that improves the total quality of life, both now and in the future, in a way that maintains the ecological processes on which life depends'. Some countries are even more explicit about certain ES they aim to protect, such as the US Clean Water Act §404 that requires 'management programs shall conserve such [clean] waters for the protection and propagation of fish and aquatic life and wildlife, recreational purposes, and the withdrawal of such waters for public water supply, agricultural, industrial and other purposes'.

Beyond existing government regulations, new calls for inclusion of ES in Environmental Impact Assessment are being made by multi-lateral lending agencies, and other international organizations. Strategic Environmental Assessment (SEA) is being expanded through calls like that made by the United Nations Environment Programme (UNEP) (UNEP, 2014; also see

Chapter 3 in this volume for further discussion). Examination of current programs makes clear both the gap to be filled in meeting these new requirements, and the potential benefits that can come from doing so. For example, review of impacts from over 30 years of implementation of the US Clean Water Act has shown that failure to include ES in impact assessment and offset design has resulted in the transfer of wetlands and their associated benefits from high population urban centers to low population rural areas (BenDor et al., 2008).

Elements of ES mitigation have been advanced for specific types of assessments (e.g., SEAs; Geneletti, 2011) and in particular contexts (e.g., Kiesecker et al., 2010; Tallis and Wolny, 2010). However, we lack a consistent approach for integrating ES alongside biodiversity in the mitigation hierarchy, which underpins many countries' mitigation policies. We aim to fill this gap by presenting an integrated framework that applies important lessons and best practices from biodiversity in a tailored approach that recognizes and addresses the particular challenges presented by ES. This framework addresses the iterative steps of development siting, impact estimation and offset assessment (Figure 17.1). We discuss each step of the mitigation hierarchy, noting when advances from biodiversity mitigation can be applied to ES and when specialized approaches are needed. This framework is discussed in detail and complemented by deeper discussion of parallel advances needed for integrating biodiversity and ES mitigation elsewhere (Tallis et al., 2015).

17.2 SITING

In the initial phase of ES impact mitigation, development options for individual or multiple projects are best placed in the context of a larger landscape or seascape to guide appropriate siting. In this phase, assessment targets are chosen, the full spatial extent is identified and existing or new assessments can be used to determine ideal areas to both promote and avoid development.

17.2.1 Target Selection

Many ES are likely to be affected by new development actions, and it would be ideal to conduct a comprehensive analysis that includes all impacted services (Geneletti, 2011; IFC, 2012). However, constraints from limited data, capacity, or policy context will usually limit the number of ES that can be included in impact mitigation assessments. While the tendency will be to limit the number of services assessed, it is critical to

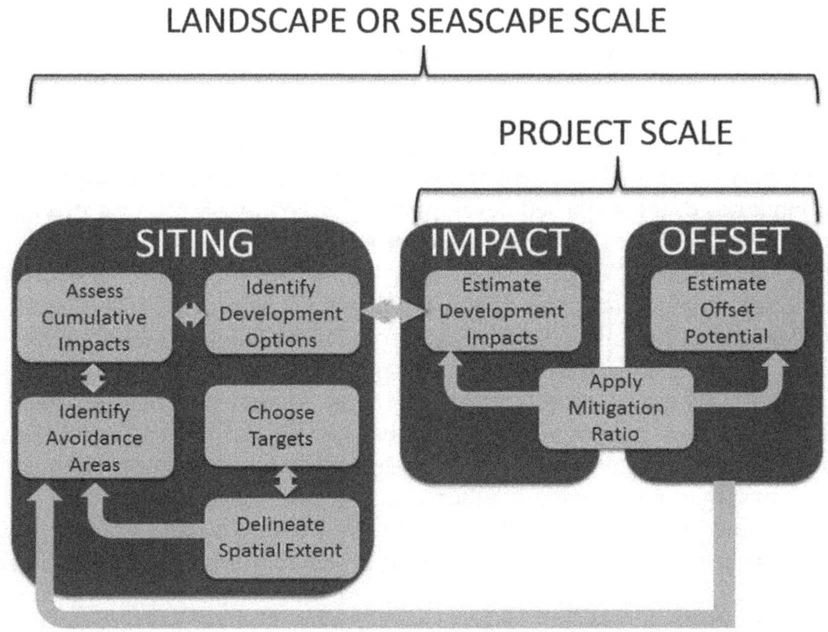

Note: ES and biodiversity mitigation will benefit from an interplay between landscape- or seascape-scale and project-scale approaches. In this context, siting, avoidance and offset decisions would be made on the basis of landscape- or seascape-scale assessments. This would ensure that areas critical for ES priorities are avoided and cumulative impacts are considered.

Figure 17.1 Integrated framework for biodiversity and ES mitigation for development impacts

recognize that services are seldom interchangeable with each other, and when assessed alongside biodiversity, are not interchangeable with species or habitats. Hummingbirds do not play the same ecological roles as polar bears, nor do they provide the same social or economic values to people. While this statement may be obvious, ES can often show dramatically different patterns in spatial and temporal variation, and so will not often act as useful surrogates for each other (e.g., Egoh et al., 2008; Naidoo et al., 2008; Cardinale et al., 2012). The selection of targets can strongly influence the outcomes of mitigation (Eiswerth and Haney, 2001). Given this sensitivity of the assessment process to this important step, there is a clear need for a systematic way to select ES targets that are as representative as possible.

Services may be selected because they provide a large amount of benefit in the proposed development area either by impacting a large number of people (beneficiaries) or by providing a relatively large benefit to a smaller or more vulnerable population (e.g., older people, under-represented groups). Services may also be selected as targets if they are expected to suffer disproportionately large losses from the proposed development. This may occur if a large proportion of their supply region falls within the boundaries of the proposed development (Luck et al., 2012), or if production of the service is especially sensitive to the kind of development being proposed. Finally, target services may be chosen to include services of especially high commercial or social value, such as provision of a locally produced traditional medicine plant, production of high-value commodity crops, or access to a spiritually significant place (ibid.).

Target ES may also be selected to act as 'umbrella services', expanding on the concept originated in Biodiversity Impact Assessment (Daily, 2000). Such umbrella services are presumed to respond to development in ways that mirror many other non-assessed services, capturing the trends and potential losses of these other services well. Existing studies or primary research can be used to identify which services may act as appropriate 'umbrella services' in each context. Close consideration is needed, as ES show highly variable relationships with each other on a case-specific basis (Anderson et al., 2009; Nelson et al., 2009; Cardinale et al., 2012), suggesting there are not universally appropriate umbrella services, but that such services may be identified using locally appropriate explorations. For example, regulating services show promising potential as umbrella services in the region of Quebec, Canada, as local studies found them to be most highly correlated with the most diverse set of services (Raudsepp-Hearne et al., 2010).

However, regulating services are not always good representative target services, as shown in another study comparing ES relationships in Southwest Australia and Argentina. In Southwest Australia, water quality regulation and carbon sequestration were good proxies for each other, showing highly correlated spatial and temporal patterns. These patterns broke down in the Argentinian pampas, where there was a trade-off between the same services, making them poor proxies for each other (Jackson et al., 2005). Variance in ES relationships will deserve even more attention in the selection of representative targets as climate and other global changes are likely to alter underlying ecosystem functions and ES inter-relationships.

Finally, when ES production is tightly dependent on specific species or habitat types, ES target selection can be aided by concepts used directly in biodiversity target selection. For example, just a few species commonly dominate commercial and recreational fishing (Holmlund and

Hammer, 1999), so recreational or economic benefits flowing from these services are tightly tied to the dynamics of the fish species of interest. Similarly, climate regulation through carbon sequestration is strongly associated with vegetation community characteristics, so may be represented in impact assessment by those habitat or land cover types (Gibbs et al., 2007). In a few cases, ES provision is strongly correlated with species diversity within a system, and so biodiversity proxies themselves may be used as ES targets (Cardinale et al., 2012). However, species diversity proxies should only be used as ES proxies in the few cases where the relationship between diversity and ES is strong and well described.

17.2.2 Spatial Extent

Traditionally, impact mitigation has focused on small spatial extents defined by the local impact site or sites proposed for development, and reference and control sites, if included (Atkinson et al., 2000; Gontier et al., 2006). However, as the science of landscape ecology has matured, work in the biodiversity arena has led to a call for an expansion in assessment scope to a larger landscape or seascape scale (Margules and Pressey, 2000; Geneletti, 2002; Groves et al., 2002; Landis, 2003; Landis and Wiegers, 2007). This evolution in conservation planning has in turn led to a similar shift from small site-focused mitigation to assessments that proactively consider cumulative impacts of development at broader landscape scales (Kiesecker et al., 2010). The same arguments for this expansion in biodiversity impact mitigation apply to ES assessment, so we emphasize the need to capture a spatial area large enough to reflect these forces, as well as the influence of projected future development (as done in Copeland et al., 2009).

Over the relevant landscape or seascape scale, Biodiversity Impact Assessment delineates an area that captures the spatial extent of all biodiversity targets and the ecological processes that support them. We build on this concept and extend it for the ES context. Servicesheds provide a spatial unit that captures the area needed for the ecological production of an ES benefit, as well as the area needed to reflect the physical and institutional connections between that ecological supply and human benefit. A serviceshed is defined as the area that provides a specific ES to a specific individual or group of people, referred to more generally as 'beneficiaries' (Figure 17.2) (Tallis and Polasky, 2009).

There are three key elements that define the spatial extent of a serviceshed. First, the serviceshed reflects the area of biophysical supply of a service. Without the production of a service (e.g., no water purification taking place, no fish that can be caught, no forest views to enjoy), there

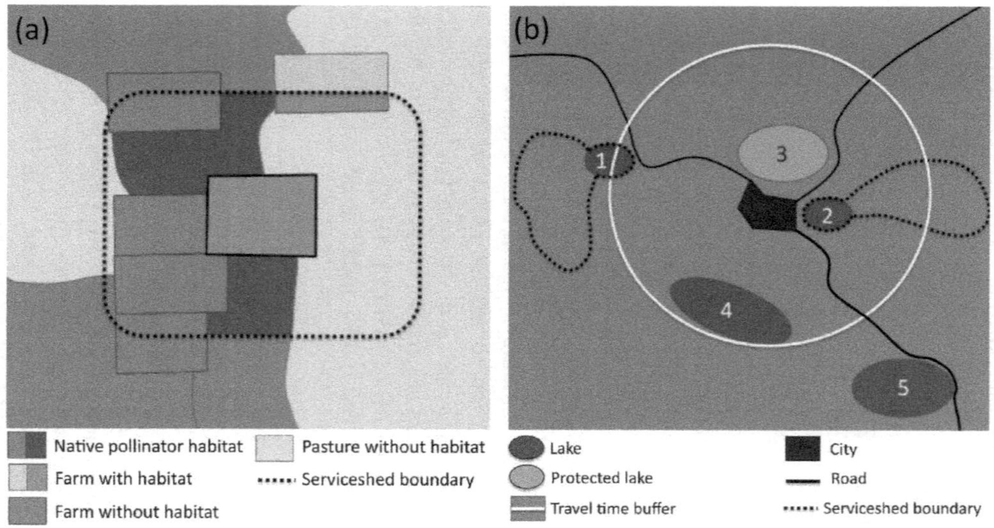

Note: (a) Darker-colored farm and native habitat are within the pollination serviceshed area, while lighter habitats are not. Pollinators may live outside the serviceshed boundary but will not provide pollination service to the focal farm. (b) Lakes 4 and 5 are outside the example recreational fishing serviceshed because potential beneficiaries lack physical access or are too far away, respectively. Lake 3 is within the potential serviceshed area but is protected, so access is denied due to institutional restrictions.

Source: Reproduced from Tallis et al. (2015).

Figure 17.2 Hypothetical serviceshed boundaries. Serviceshed boundaries for crop pollination (a) are determined by the flight distance(s) of pollinators in relation to crop fields and habitat areas. The serviceshed for recreational fisheries (b) is determined by the lakes (or rivers) with harvestable recreational fish species that are within an acceptable travel time and that allow physical access (public access, boat launches, etc.)

is no benefit to assess. Second, the serviceshed reflects the space needed to represent the physical connection between the supply area and beneficiaries, when services require physical interaction with nature. For some services, this element is not relevant. For example, existence value is a benefit some people receive from knowing that a species, such as a bird of paradise, exists. Since this benefit derives from a person's knowledge of a species existing, whether or not they ever see that species or physically experience it in any way, we do not need to consider any additional spatial boundaries in the drawing of the serviceshed for existence value. However, for other services, physical proximity and access is integral to the delivery of the service. For example, the provision of recreational fishing opportunities will only happen if people can physically access areas that are producing recreationally desirable fish. In this case, physical access through features such as walking paths, roads, or boat launches is essential for ES provision, and these features need to be considered in serviceshed delineation (Figure 17.2b).

The third factor informing the delineation of servicesheds is the influence of formal (laws, regulations) or informal (norms, cultural taboos) institutions on beneficiary access to ES supply regions. As with physical access, institutions can limit people's ability to benefit from an ES. Institutions that may be relevant include protected area boundaries, water use rights, land tenure, traditional access rights, hunting or fishing regulations such as seasonal or area closures, among others. Both physical and institutional access may affect different groups of people differently, emphasizing the need to specify beneficiary groups of interest, and delineate servicesheds for those specific groups. For example, consider an ES impact assessment in Sub-Saharan Africa where wildlife hunting is an ES of interest. The serviceshed for this ES would include the area needed for hunted species populations to be ecologically viable. But the extent of the serviceshed would vary based on who the hunters are, and what physical and institutional restrictions apply to their hunting activities. For example, an ES impact assessment on commercial trophy hunting would need to identify where trophy hunting is allowed, and where hunters are likely to be able to gain physical access by plane, car, boat or other means. Indigenous communities in the same region may be interested in wildlife hunting for subsistence meat provision. Including this service in an ES impact assessment would likely require the delineation of a different spatial extent that reflects the ranges of a different set of species, a different set of allowed access areas for these activities, and that reflects physical constraints on access such as footpaths.

Following the principles discussed here, serviceshed or other appropriate ES spatial boundaries can be drawn for target services and specified beneficiaries. When ES assessment is being done jointly with biodiversity impact assessment, the spatial extent will be the combined maximum extent of ES spatial units and the areas identified as important for biodiversity targets, such as species ranges or habitat extents, among others (Luck, Chan and Fay, 2009; Luck, Chan and Klien, 2012; Mandle and Tallis, 2012; Tallis, Polasky and Lozano et al., 2012; Arkema et al., 2013).

17.2.3 Avoidance Criteria

In an ideal mitigation process, it is critical to identify areas to avoid development activities because of the high likelihood that mitigation actions will not be sufficient to ensure no net loss, or ideally a net gain, of ES (Nichols and McElfish, 2009; BBOP, 2012b; Gardner et al., 2013; McKenney and Kiesecker, 2010). This precedent has been set through biodiversity mitigation, and while most countries refer to the need to first avoid impacts, avoidance actions and best practice for identifying biodiversity avoidance areas are seldom followed (Clare et al., 2011). Despite current challenges to implementation, the criteria underpinning biodiversity avoidance can inform ES avoidance area identification.

Biodiversity avoidance areas are often identified based on two general factors: irreplaceability (or uniqueness) and vulnerability (or threat) (Margules and Pressey, 2000; BBOP, 2012a; Pilgrim et al., 2013). Irreplaceability refers to the ability to find equivalent areas for mitigation that are similar enough to replace biodiversity impacted through development. This ties directly to the spatial extent of impact and the spatial extent of alternative or offset areas within the mitigation region that may be available for compensation. Separately, vulnerability addresses the likelihood that mitigation efforts can recover or restore the same biodiversity to the same state or quality as existed pre-impact (Wilson et al., 2005; BBOP, 2012a; IFC, 2012). As such, vulnerability addresses susceptibility to impact from proposed development actions, and from other drivers of change across the larger region. Within these general concepts, a wide array of specific factors can be used to specify avoidance areas (Figure 17.3). In any impact assessment, these factors and the conditions where avoidance may be needed can act as general guidelines for consideration, but landscape- and site-level quantitative assessments are needed to identify specific avoidance areas.

Some specific factors relevant to avoidance of biodiversity impacts are directly applicable to ES. For example, biodiversity areas containing species or habitats with few equivalent options for offsets within the larger landscape should be avoided. Similarly, it will be difficult to mitigate impacts to target services that have few alternative options within

General Factor	Ecosystem Services	AVOID if...
Landscape-level Goals	Options to protect/mitigate target services within servicesheds	Few
	Multiple values overlap	Yes
Irreplaceability (uniqueness)	ES link to specific place, habitat, or species	Strong
	Beneficiary dependence on ES	Strong
	Size of serviceshed area	Small
Vulnerability (threat)	Current rate & extent of loss	High
	Rate & likelihood of recovery	Low
	Beneficiary access to alternatives	Few
	Expected future ES demand	High

Note: Avoidance areas should be defined as those where impacts cannot be mitigated effectively and equivalently elsewhere. Conditions where mitigation will be challenging are identified in the 'AVOID if. . .' box. Factors relevant to both biodiversity and ES are shown in light gray, while those unique to ES are in dark gray.

Figure 17.3 Factors influencing avoidance for mitigation of ES impacts

the serviceshed where mitigation could replace lost benefits. Biodiversity impacts should be avoided when a proposed development area is home to many target species because finding equivalent offsets for all of them will be challenging. In the same regard, areas that provide a long list of target services will be challenging to replace, and so may be avoided. In terms of vulnerability, targets that have currently high rates and extents of loss from other drivers in the land or seascape should be avoided, as alternative areas for mitigation have a high likelihood of being lost or degraded in the near future (Langhammer, 2007; Pilgrim et al., 2013). Both biodiversity and ES targets that have a low likelihood of recovering from disturbance or of being established through restoration activities may be avoided, given the high likelihood of offset failure. Several other biodiversity avoidance criteria are not relevant to ES, or are only relevant when ES provision is tightly tied to specific species or habitats. For example, endemic, rare and unique species or habitats are recommended for avoidance because options for offsets will be few.

Finally, there are some avoidance criteria that are unique to ES. These are largely associated with the demand side of ES. For example, areas providing ES that many beneficiaries depend on for well-being will be more difficult to replace than areas that are providing ES to only a few people. The size of the serviceshed will also influence how replaceable an ES is, as a small serviceshed will provide few offset options to return the same benefits to the same people. In terms of vulnerability, beneficiaries with few alternatives to replace lost services will be more vulnerable than those with many options. For example, if drinking water quality regulation is lost in a rural area where no water source alternatives exist, avoidance may be necessary. If, however, multiple wells are present nearby and people can access them, or bagged or bottled water is abundant and cheap, then avoidance is less necessary. Avoidance may also be important in places where future demand for a target ES is expected to rise greatly.

In an ideal ES impact assessment, key factors related to irreplaceability and vulnerability would be assessed in the context of both proposed and cumulative development pressures at a landscape or seascape scale, as is recommended in Biodiversity Impact Assessment (Gardner et al., 2013). Often, practical limitations may hamper detailed, site-specific assessments for key avoidance thresholds. In such cases, decision criteria for avoidance decisions that have begun to emerge can pose a useful starting place. While in their early stages, these ideas include ways to identify ES thresholds such as 'safe minimum standards' that reflect the amount of ES required to maintain sustainable supplies of services (Fisher et al., 2008). This idea is similar to that promoted in the biodiversity field for minimum viable population size as a proxy for sustainable species populations (Kontogianni et al., 2010). Challenging work remains to determine how these thresholds can be defined in the face of uncertainty around future ES demands and relationships between biodiversity, ES and resilience (Bennett et al., 2009). If avoidance areas are identified up front through such a process, it can provide systematic guidance to individual projects as they are proposed and taken through the mitigation hierarchy.

17.3 IMPACT

During the impact stage, projected impacts are estimated and opportunities to minimize, restore and offset impacts are explored.

17.3.1 Minimize and Reduce Impacts

In biodiversity mitigation, minimization is used to reduce the impact of proposed development by altering the extent, intensity or duration of impacts through reductions in the area impacted, focusing of activities on less sensitive areas or using less damaging alternative activities in the development process. These options are all relevant to the minimization of ES impacts, although effectiveness of each will vary from site to site and among services. While research has been extensive in some areas of biodiversity minimization approaches (The Energy and Biodiversity Initiative, 2004), very little focus has been placed on this issue for ES. Still, we can draw useful lessons from related fields of practice such as water quality regulation, water supply management, flood regulation and air quality control (Nichols et al., 2008; Nichols and McElfish, 2009). One area ES seem poised to benefit more so than biodiversity is in on-site engineered options that may reduce impacts or replace services directly. For example, water quality regulation impacts on-site may be reduced through clearing of vegetation, soil impaction and disturbance, or other development related activities. Filtration or treatment facilities or actions taken on-site could reduce the impacts of these activities on local water supplies. Other ES that may benefit from such engineered on-site activities include air quality regulation (filters, scrubbers), fish production (sustainable, local aquaculture or re-stocking), carbon sequestration (carbon capture, alternative fuels), soil fertility (humus, organic fertilizer use), and coastal protection (breakwalls, levees, berms).

17.3.2 Impact Assessment

After minimization has been explored, residual impacts of allowed development activities can be assessed. Best practice emerging from Biodiversity Impact Assessment points to the benefits of functional impact assessments that reflect important dynamics and processes, such as the condition and functioning of habitats and species or community persistence (Kusler, 2003; Gontier et al., 2006; McKenney and Kiesecker, 2010; Quétier and Lavorel, 2011). Such functional assessments are preferred over area-based assessments, which have been widely criticized for their inability to reflect variation in habitat quality or function (Ruhl and Salzman, 2006; Wilkinson et al., 2009). The timeframe chosen for impact assessment will influence how much impact is anticipated, so should be chosen carefully. Biodiversity experience suggests considering impact at least over the lifetime of the proposed development project, and including cumulative impacts from the larger landscape or seascape as they are projected to unfold over the same timeframe.

Basic principles for ecosystem service impact assessment
A consistent conceptual basis for ES impact assessment in the context of mitigation has not been established, but methods and best practice from other applications of ES assessment are relevant (Slootweg et al.,

2001; Olander et al., 2015). For example, there is general recognition that services flow to people along a 'supply chain' from nature to people, and that services can be quantified at several steps along that chain. Steps are named differently by different authors, but one suggestion is to differentiate between the full potential of an ecosystem to provide a benefit, termed supply, the actual amount of service used or enjoyed by people, termed delivery, and people's preference for receiving that amount of service, termed value (Tallis and Polasky, 2009; Granek et al., 2010).

For an ES impact assessment to reflect anticipated changes in services, it must focus at least on the delivery step, indicating how the amount of service flowing to people may be altered (Olander et al., 2015). Ideally, ES impact assessment also captures projected changes in value, a broad term used to capture any contribution of an ES to human well-being. Value is not to be interpreted narrowly here as monetary value, but instead reflects any expression of value that indicates how much people care about receiving a benefit. Impact assessments that use purely ecological variables or indicators may give some indication of how an ecosystem's ability to support a service will change, but without making the explicit link to service delivery, an ES impact has not been assessed. As with biodiversity, ES impact assessment best practice would use a functional assessment, reflecting how variation in extent and quality of ecological production varies in space, and how human location, activities and preferences influence service delivery and value (Slootweg et al., 2001; Olander et al., 2015). The timeframe would span that of the development project lifetime and reflect cumulative effects at the land or seascape scale from other drivers.

As with biodiversity, it is critical in ES impact assessment to identify the projected change that can be attributed to the proposed development action. In ES literature, this is generally referred to as marginal change (Tallis and Polasky, 2009), or the difference in ES delivery or value between a baseline condition and a post-development condition. It can be tempting to estimate the total amount or value of service provided by a proposed development area and count that as the ES impact. This will often be misleading, overestimating impact as the future condition seldom provides zero service.

Methods for ecosystem service impact assessment

Following these general principles, some Biodiversity Impact Assessment tools can contribute to ES impact assessment, while others cannot. When target ES are provided by one or few species, or can be defensibly linked to a specific habitat type, methods developed for functional Biodiversity Impact Assessment can be part of the ES impact calculation process, but they do not fulfill best practice. Additional methods will be needed to translate the biological estimates of impact to service delivery and/or value estimates. For example, pollination of commercial crops may be carried out by a few identifiable pollinator species. The first step of an impact assessment may use well-developed methods to estimate the impact of development on target pollinator population viability. However, that alone does not indicate likely change in service, only change in supply. Here, the service provider unit concept can help extend biological assessments to ES assessments (Luck, Harrington and Harrison et al., 2009). Examples of ES where a clear and consistent link between a species or habitat type and a service are few, and so this approach should be treated with caution. For example, air and water pollution regulation, carbon sequestration, and flood mitigation are among the services of common concern that can be supplied to varying degrees by a diverse set of habitats and species.

Given the need to reflect marginal ES impacts, and to capture spatial variation that influences the supply, delivery and value of ES, area- or habitat-based assessment methods are often insufficient (Olander et al., 2015). These methods tend to focus on total ecosystem value, reflecting an often opaque set of services that may or may not match targets chosen for a given impact assessment. These methods also tend to rely on consistent, per-area levels of service delivery or value that fail to capture key variation in ecosystem function, physical or institutional access to services, and social context that alters preference for benefits (Barbier et al., 2008; Ricketts et al., 2008; Koch et al., 2009). As such, the use of benefit transfer methods and associated tools or datasets (e.g., Natural Assets Information System, Wildlife Habitat Benefits Estimation Toolkit, parts of Toolkit for Ecosystem Service Site-based Assessment) should proceed with extreme caution and attention to the selection of data sources that reflect only marginal changes from similar development changes, isolate the value of target services, and represent comparable ecological and socio-economic conditions (Plummer, 2009).

Many tools and models exist that can be used for functional ES assessments (e.g., Diaz et al., 2011). These approaches commonly use ecological production functions. These are often reflected as equations that capture key factors and relationships that allow for the calculation of marginal, service by service impacts. As such, these equations also usually effectively capture relevant variation in socio-ecological conditions and key connections to beneficiaries, allowing impact assessment that reflects how they vary in space and through time (Slootweg et al., 2001; Boyd and Wainger, 2003; NRC, 2005; Olander et al., 2015).

Some ES production functions have been translated into software tools that allow replicable ES impact assessment in a consistent framework

(ARIES, InVEST, OPAL, MEASURES, MIMES). As these tools are general ES assessment tools, others have reviewed their general strengths and weaknesses (Vigerstol and Aukema, 2011; Waage et al., 2011; Bagstad et al., 2013). Data availability presents a real challenge to the use of production functions in many contexts. To overcome this challenge, some production function models have been designed to use largely available land use/land cover data as one of the inputs to ES models (Nelson et al., 2009; Raudsepp-Hearne et al., 2010). While this shares some features with a habitat-based assessment approach, production functions overcome many of the area-based shortcomings by including additional factors in equations that represent important variability within habitat types (e.g., soils, elevation, climate, management data). In addition, production function models represent the necessary connections between ecological production and human benefit, allowing estimation of ES delivery or value.

Even with the incorporation of widely available land use and land cover (LULC) data, data limitations can inhibit the widespread use of production functions. Models that require readily available data have only been developed for a subset of ES (but see Kareiva et al., 2011; Guerry et al., 2015), and even these may be challenging for managers to use in a decision context. Such constraints can make the use of ES indicators and more simplified approaches attractive (e.g., De Groot et al., 2010 for land use planning), and some syntheses of such indicators have been done. We emphasize the challenges in using such approaches to achieve a robust ES impact assessment, as the relationship between ES and proxies is poorly understood for most services, and few proxies reflect marginal change for individual target services of interest (Naidoo et al., 2008). The development of ES impact assessment tools that are accessible to managers and decision-makers will be critical for advancing ES mitigation. It is clear from the biodiversity mitigation world that the primary reason detailed functional assessments are not widely utilized is lack of capacity (Clare et al., 2011).

17.4 OFFSET

Even under ideal planning processes, some development impacts cannot be avoided. When impacts do occur, offsets, in the form of restoration, protection or other management actions can be used to compensate for projected ES losses (BBOP, 2012b).

Biodiversity offsets are designed to create replacement actions that are additional (e.g., would not have happened without the action), produce measureable, equivalent improvements in targets, and do so effectively and efficiently. These relevant intents for ES offsets can be achieved through a two-step process. First, a large landscape- or seascape-scale assessment can be used to identify the potential of all areas to produce benefits through mitigation actions. Second, an efficient set of sites can be selected to meet identified offset requirements (Kiesecker et al., 2009; Tallis and Wolny, 2010).

17.4.1 Estimate Offset Potential

The amount of benefit – or additionality – that can be expected from plausible mitigation actions can be estimated using the same principles as those identified for impact assessment. Using the same methods to calculate both impacts and offsets will ensure that estimates are comparable, based on the same sets of assumptions, and have the highest possible likelihood of leading to actual additionality.

Offset design

In practice, mitigation for biodiversity impacts has largely been focused on on-site activities (Ruhl and Salzman, 2006; McKenney and Kiesecker, 2010). However, best practice would focus at a larger scale, using well-developed methods to identify efficient options for biodiversity offset design (e.g., ecoregional assessments, watershed management plans and other conservation planning exercises) (Kiesecker et al., 2009, Wilkinson et al., 2009; Kiesecker et al., 2010; Weber and Allan, 2010). Similar approaches can be tailored to the ES context and used to identify efficient ES offsets, though such methods are much less developed than those used for biodiversity. In general, areas identified in earlier steps for avoidance may provide prime options for offsets through protective management actions (Kiesecker et al., 2009, 2010).

In-kind vs out-of-kind offsets

Guidance on ES offsets is not well developed (but see Brownlie et al., 2012), so we offer three initial considerations: whether interchange among services will be allowed (equivalent to 'out-of-kind' biodiversity offsets), whether interchange between beneficiaries will be allowed (e.g., impacts to one city replaced by benefits to another city, or impacts to indigenous people offset by benefits to non-indigenous people), and whether monetary or other compensation for lost services will be allowed (another form of 'out-of-kind' offsets such as payments for lost services or technological provision of an alternative such as bottled water for lost water filtration services). Most decisions about what is allowed will be determined by political preference. However, we caution that any ES offset that does not return the same benefit to the same group of people is implicitly creating inequality, and therefore associated ES 'winners' and 'losers'. Such choices are very likely to create or worsen social inequities through ES delivery. Again, the US Clean Water Act outcomes in the United States demonstrate this risk (BenDor et al., 2008).

Spatially combining biodiversity and ecosystem service offsets

Once a landscape view of potential ES benefits has been produced, and information is gathered as to whether in-kind or out-of-kind offsets are allowed, several methods can be used to choose offset sites that will meet requirements efficiently. Prioritization approaches widely used for biodiversity targets (Possingham et al., 2000; Groves et al., 2002; Kiesecker et al., 2009, 2010) can be applied to ES information (Chan et al., 2006, 2011). When an ES impact assessment is being done separately from a biodiversity assessment, such prioritization methods can be applied directly to ES data.

When biodiversity and ES impact assessments are done jointly, decisions will have to be made about the order in which offsets are chosen. Given wide variation in the degree of spatial overlap between biodiversity and ES targets, the choice of which approach to use could dramatically affect the outcome. Offsets could be initially prioritized based on ES benefit potential, then assessed for biodiversity outcomes, and any remaining biodiversity requirement filled through a second biodiversity-driven prioritization. Conversely, an initial round of offset selection could be driven by biodiversity data, those sites assessed for ES offset potential, and any remaining ES offset requirement filled through a second ES-based prioritization (e.g., as done in a mine impact assessment case study in Colombia; Tallis and Wolny, 2010). In the third option, selection is truly integrated, and offsets are chosen through a joint optimization where sites are chosen first based on high potential to fulfill both biodiversity and ES offsets (e.g., as in a case study for maximizing agricultural yield while achieving Brazilian Forest Code compliance).

Comparisons of such selection exercises have not been done in enough contexts to identify whether one is generally more efficient than another, so exploration of these modes will be worthwhile until such generalities can be defined. The few cases where these options have been explored emphasize the sensitivity of offset efficiency to prioritization approach. For example, an impact assessment of a proposed road through the Peruvian Amazon showed that ES-driven prioritization was much more efficient at meeting ES offset needs than a biodiversity-driven approach (Mandle and Tallis, 2012). Driving the offset selection process with ES data reduced residual ES impacts to drinking water quality more than four-fold for sediment loads, 16-fold for nitrogen pollution and nearly 40-fold for phosphorous pollution.

The spatial prioritization method used for offset selection may also be directed by regulatory guidelines determining the degree to which overlap between biodiversity and ES offsets (termed 'stacking') is desired or allowed (Robertson et al., 2014). Stacking provides the potential to condense offsets into smaller areas that produce multiple benefits, but presents the risk that stacked offsets may not deliver a credible environmental benefit without unintended loss of unassessed ES.

Mitigation replacement ratios

While offset calculations aim to accurately estimate the potential benefit an offset can provide, some factors will often be left out of offset calculation because of knowledge, data or capacity constraints. To ensure that the common no-net-loss or net-gain outcomes desired by mitigation policies (McKenney and Kiesecker, 2010; Quétier and Lavorel, 2011; Gardner et al., 2013) are achieved under these constraints, practitioners commonly use mitigation ratios. These ratios define the number of credit units that need to be created by an offset to compensate or replace an equivalent single unit of loss at the development site.

Standard methods do not exist for the calculation of biodiversity or ES offset ratios. As a result, biodiversity ratios are negotiated rather than established scientifically, fail to represent many factors that determine equivalency of impact and offset sites consistently, and ratios vary dramatically (Moilanen et al., 2009; McKenney and Kiesecker, 2010; Saenz et al., 2013). ES offset ratio determination would benefit from a more standardized approach that considers a consistent set of factors, includes or excludes factors based on methods used for impact and offset calculation, and develops equations that can be used to reliably combine factors into an offset ratio. Drawing from and expanding beyond the literature on biodiversity offsets, we present a set of factors that can serve as a starting point for a more standardized approach to ES ratio calculation (Table 17.1). Factors address three major categories that may or may not need to be included, based on the methods for offset calculation used. These categories are (1) the magnitude of impact, (2) the quality of both impact and offset sites and (3) the mitigation method used to create the offset.

A large set of commonly used factors will not be needed in ratio calculation if the preferred functional assessment methods are used for both impact and offset calculation. Alternatively, these factors would be needed if the less effective area- or habitat-based estimation methods are used, as these methods fail to capture many important features of impact sites

Table 17.1 Biodiversity mitigation ratio factors and their applicability for ES offsets

Concept	Factor	Example Metric	Condition Associated with a Higher Offset	Relevance to Biodiversity		Relevance to Ecosystem Services			Ecosystem Service Considera-tions
				Area or habitat based	Functional assessment	Area or habitat based	Production function, delivery	Production function, value	
Magnitude of Impact	Type of development	Mine, road, agriculture	More damaging type of development	Yes	No	Yes	No	No	
	Impact site condition	Wetland function at site prior to impact, Value of service *x* at site prior to impact	Higher function	Yes	No	Yes	No	No	
	Vulnerability	Target loss rate(s) from other drivers, sensitivity of target(s) to type of development	More vulnerable targets	Yes	No	Yes	No	No	
	Duration of impact	Time frame of proposed development	Longer development term	Yes	No	Yes	No	No	

	Rarity of target(s)	Number of rare species at impact site	More rare species or habitats	Yes	No	Yes	Yes	No	Rarity is conceptually equivalent to service scarcity, where a service that is more scarce is more valuable. Valuation impact estimates would capture any stronger social preference for a more scarce service

Table 17.1 (continued)

Concept	Factor	Example Metric	Condition Associated with a Higher Offset	Relevance to Biodiversity		Relevance to Ecosystem Services			Ecosystem Service Considera-tions
				Area or habitat based	Functional assessment	Area or habitat based	Production function, delivery	Production function, value	
Magnitude of Impact	Intergenerational equity	Strength of desire to serve future generations	Stronger desire	Yes	Yes	Yes	Yes	No	Monetary impact assessments that included a future time frame of impact and a discount rate would directly capture this
	Uncertainty of impact estimate	Variance in impact estimates	Higher uncertainty	Yes	Yes	Yes	Yes	Yes	Most ecosystem service impact assessments are likely to have associated uncertainty

									that should be considered when setting an offset requirement
Quality of offset	Similarity of site conditions (in-kind vs out-of-kind)	Habitat type, level of function	Poorer match between sites	Yes	No	Yes	No	No	
	Distance from impact site	Distance, within impact watershed	Farther from impact site	Yes	Yes	Yes	No	No	Production function approaches generally link services to beneficiaries in space, so would capture important serviceshed context. Area- or habitat-based assessments likely would not

Table 17.1 (continued)

Concept	Factor	Example Metric	Condition Associated with a Higher Offset	Relevance to Biodiversity		Relevance to Ecosystem Services			Ecosystem Service Considera-tions
				Area or habitat based	Functional assessment	Area or habitat based	Production function, delivery	Production function, value	
Quality of offset	Leakage	Probability of development shifting to another area	Higher probability of leakage	Yes	Yes	Yes	Yes	Yes	This is not currently applied to biodiversity, but is relevant. Most assessments do not capture probability of leakage
	Risk of natural disturbance	Frequency of fire, frequency of floods	Higher frequency of disturbance	Yes	Maybe	Yes	Maybe	Maybe	Even functional assessments of biodiversity and ecosystem services may not include

									natural disturbances in impact and offset estimates
Additionality	Probability of development, deforestation rate	Lower risk of development	Yes	Maybe	Yes	Maybe	Maybe		Functional assessments of offsets may not include future scenarios of cumulative impacts or losses due to other drivers of change. If these are captured in the offset estimate, then they do not need to be included in a ratio

Table 17.1 (continued)

Concept	Factor	Example Metric	Condition Associated with a Higher Offset	Relevance to Biodiversity		Relevance to Ecosystem Services			Ecosystem Service Considera-tions
				Area or habitat based	Functional assessment	Area or habitat based	Production function, delivery	Production function, value	
Mitigation method	Risk of failure	Historical success rate of method	Lower rate of success	Yes	No	Yes	No	No	
	Time lag of recovery	Time period to recover target(s) to desired level	Longer recovery time	Yes	No	Yes	No	No	
	Perpetuity	Historical average duration of method success	Shorter period of outcome persistence	Yes	No	Yes	No	No	

Source: Modified with permission from Mandle et al. (2015).

and the surrounding landscape and so need to be accounted for through inclusion in the ratio. For example, functional assessment approaches commonly estimate the marginal impact of proposed development, while area-based approaches estimate total impact. To adjust area-based estimates to a more realistic view of impact, the ratio may be used to reflect more or less impactful types of development. Such adjustment is not needed if functional assessment captures likely impact from the specific kind of development accurately. As another example, area-based estimates of impact and offset generally assume that all possible offset areas have equivalent possible benefit. However, service provision areas farther from the impact site may not provide as much benefit, so ratios are often adjusted to reflect this. Again, if functional assessment captures spatial variation in ES delivery or value, no such adjustment is needed via a ratio.

The issue of discounting has been raised in biodiversity offsets (Denne and Bond Smith, 2012; Overton et al., 2013) and is especially relevant to ES offsets. The further into the future impacts are expected to persist before benefits are recovered, the greater the loss of ES value. Economists are in general agreement that people value a benefit more in the near term than the same benefit enjoyed further in the future. Discount rates are the most commonly used means to reflect this time-related preference. ES offset calculations that apply a discount rate reflect this issue and do not need to account for it in ratio calculation.

Even when robust functional assessment methods are used, not all factors relevant to achieving no net loss can be directly accounted for. For example, uncertainty associated with input data or assessment models will always be present, and best practice suggests increasing the offset ratio in areas where data or models are known to be highly uncertain (Bull et al., 2013).

In general, a consistent approach to ratio development could start by choosing factors for ratio calculation from a core set, adding or leaving out factors based on what is captured through the impact and offset calculation methods (e.g., Table 17.1). We provide an initial set of factors to consider and encourage further work to develop a method for consistently combining relevant variables into ratio equations in a replicable and transparent way.

17.5 CONCLUSIONS

We have outlined a general framework that starts to identify how best practice from biodiversity mitigation can be brought to bear on ES mitigation, and where alternative methods or considerations are needed.

We propose a means for selecting representative ES targets, identifying appropriate spatial assessment units, and consistently calculating ES offset ratios. In both impact and offset estimation, we emphasize the need to move away from area- or habitat-based methods to comprehensive functional assessments.

Implementation of ES in the mitigation hierarchy faces several scientific and political challenges. Research is needed to advance understanding of critical ES thresholds that can be used in identifying avoidance areas (Huggett, 2005; Chaplin-Kramer et al., 2015). Especially challenging here will be to identify these thresholds in the context of future ES needs that are likely to change as the human population grows and people's preferences evolve. Lessons from biodiversity impact mitigation emphasize both the shortcomings of area- and habitat-based assessment methods and the capacity shortcomings that prevent the widespread use of functional assessments, so rapid advances are needed to reduce capacity, data and model limitations for functional ES assessment. Finally, rigorous comparison of different offset prioritization approaches (e.g., biodiversity offsets first, ES offsets first or joint prioritization) is needed to help inform which method is appropriate under what circumstances.

On the policy side, there remains a clear need to translate this general framework for specific kinds of assessments. Several advances in this direction have been made in other chapters, such as Strategic Environmental Assessment (see Chapter 3), and Health Impact Assessments (Chapter 4). These exercises will be most useful if done in collaboration with standard-setting entities such as regulatory agencies, so as to ensure that analytical choices follow broader requirements of the regulating entity (Polasky et al., 2015). Regulatory context will also identify who has the authority to set a mitigation ratio and identify which prioritization methods may be allowed. Finally, many countries do not have regulations that require compensatory mitigation, nor have regulations that integrate site-scale and landscape- or seascape-scale assessment to help steer avoidance and offset decisions.

Even with these remaining challenges, impact mitigation processes provide one of the richest and fastest-growing opportunities to include ES information in major land use decisions. By continuing to advance ES methods beyond the conceptual and into the practical, impact assessments and mitigation programs will have more promising opportunities to achieve the joint environmental and social objectives of the laws and regulations that request them.

NOTE

* This work has been funded by an anonymous foundation, The Gordon and Betty Moore Foundation #2996 and the 3M Foundation.

REFERENCES

Anderson, B.J., P.R. Armsworth and F. Eigenbrod et al. (2009), 'Spatial covariance between biodiversity and other ecosystem service priorities', *Journal of Applied Ecology*, **46**(4), 888–96.

Arkema, K.K., G. Guannel and G. Verutes et al. (2013), 'Coastal habitats shield people and property from sea-level rise and storms', *Nature Climate Change*, **3**(10), 913–18.

Atkinson, S.F., S. Bhatia and F.A. Schoolmaster et al. (2000), 'Treatment of biodiversity impacts in a sample of US environmental impact statements', *Impact Assessment and Project Appraisal*, **18**(4), 271–82.

Bagstad, K.J., D.J. Semmens and S. Waage et al. (2013), 'A comparative assessment of decision-support tools for ecosystem services quantification and valuation', *Ecosystem Services*, **5**, 27–39.

Barbier, E.B., E.W. Koch and B.R. Silliman et al. (2008), 'Coastal ecosystem-based management with nonlinear ecological functions and values', *Science*, **319**(5861), 321–3.

BBOP (2012a), *Resource Paper: Limits to What Can Be Offset*, Washington, DC: Business and Biodiversity Offsets Programme.

BBOP (2012b), *Standard on Biodiversity Offsets*, Washington, DC: Business and Biodiversity Offsets Programme, accessed 27 January 2016 at http://bbop.forest-trends.org/guidelines/Standard.pdf.

BenDor, T., N. Brozović and V.G. Pallathucheril (2008), 'The social impacts of wetland mitigation policies in the United States', *Journal of Planning Literature*, **22**(2), 341–57.

Bennett, E.M., G.D. Peterson and L.J. Gordon (2009), 'Understanding relationships among multiple ecosystem services', *Ecology Letters*, **12**, 1394–404.

Bos, M., R.L. Pressey and N. Stoeckl (2014), 'Effective marine offsets for the Great Barrier Reef World Heritage Area', *Environmental Science and Policy*, **42**, 1–15.

Boyd, J. and L. Wainger (2003), *Measuring Ecosystem Service Benefits: The Use of Landscape Analysis to Evaluate Environmental Trades and Compensation*, Washington, DC: Resources for the Future.

Brownlie, S., N. King and J. Treweek (2013), 'Biodiversity trade-offs and offsets in impact assessment and decision making: can we stop the loss?', *Impact Assessment and Project Appraisal*, **31**(1), 24–33.

Bull, J.W., K.B. Suttle and A. Gordon et al. (2013), 'Biodiversity offsets in theory and practice', *Oryx*, **47**(3), 369–80.

Canter, L. and B. Ross (2010), 'State of practice of cumulative effects assessment and management: the good, the bad and the ugly', *Impact Assessment and Project Appraisal*, **28**(4), 261–8.

Cardinale, B.J., J.E. Duffy and A. Gonzalez et al. (2012), 'Biodiversity loss and its impact on humanity', *Nature*, **486**(7401), 59–67.

Chan, K.M.A., L. Hoshizaki and B. Klinkenberg (2011), 'Ecosystem services in conservation planning: targeted benefits vs. co-benefits or costs?', *PLoS ONE*, **6**(9), e24378.

Chan, K.M.A., M.R. Shaw and D.R. Cameron et al. (2006), 'Conservation planning for ecosystem services', *PLoS Biology*, **4**(11), 2138–52.

Chaplin-Kramer, R., R. Sharp and L. Mandle et al. (2015), 'Where matters: understanding how spatial patterns of agricultural expansion impact biodiversity and carbon storage at a landscape level', *Proceedings of the National Academy of Sciences of the United States of America*, **112**(24), 7402–7.

Clare, S., N. Krogman and L. Foote et al. (2011), 'Where is the avoidance in the implementation of wetland law and policy?', *Wetlands Ecology and Management*, **19**(2), 165–82.
Copeland, H.E., K.E. Doherty and D.E. Naugle et al. (2009), 'Mapping oil and gas development potential in the US Intermountain West and estimating impacts to species', *PLoS ONE*, **4**(10), e7400.
Daily, G.C. (2000), 'Management objectives for the protection of ecosystem services', *Environmental Science & Policy*, **3**(6), 333–9.
De Groot, R.S., R. Alkemade and L. Braat et al. (2010), 'Challenges in integrating the concept of ecosystem services and values in landscape planning, management and decision making', *Ecological Complexity*, **7**(3), 260–72.
Denne, T. and S. Bond-Smith (2012), *Discounting for Biodiversity Offsets*, report for the New Zealand Department of Conservation.
Diaz, S., F. Quetier and D.M. Caceres et al. (2011), 'Linking functional diversity and social actor strategies in a framework for interdisciplinary analysis of nature's benefits to society', *Proceedings of the National Academy of Sciences of the United States of America*, **108**(3), 895–902.
Egoh, B., B. Reyers and M. Rouget et al. (2008), 'Mapping ecosystem services for planning and management', *Agriculture Ecosystems & Environment*, **127**(1–2), 135–40.
Eiswerth, M.E. and J.C. Haney (2001), 'Maximizing conserved biodiversity: why ecosystem indicators and thresholds matter', *Ecological Economics*, **38**(2), 259–74.
Fisher, B., K. Turner and M. Zylstra et al. (2008), 'Ecosystem services and economic theory: integration for policy-relevant research', *Ecological Applications*, **18**(8), 2050–67.
Gardner, T.A., A. von Hase and S. Brownlie et al. (2013), 'Biodiversity offsets and the challenge of achieving no net loss', *Conservation Biology*, **27**(6), 1254–64.
Geneletti, D. (2002), *Ecological Evaluation for Environmental Impact Assessment*, Utrecht: Netherlands Geographical Studies.
Geneletti, D. (2011), 'Reasons and options for integrating ecosystem services in strategic environmental assessment of spatial planning', *International Journal of Biodiversity Science, Ecosystem Services & Management*, **7**(3), 143–9.
Gibbs, H.K., S. Brown and J.O. Niles et al. (2007), 'Monitoring and estimating tropical forest carbon stocks: making REDD a reality', *Environmental Research Letters*, **2**, 045023.
Gontier, M., B. Balfors and U. Mörtberg (2006), 'Biodiversity in environmental assessment – current practice and tools for prediction', *Environmental Impact Assessment Review*, **26**(3), 268–86.
Granek, E.F., P. Polasky and C.V. Kappel et al. (2010), 'Ecosystem services as a common language for coastal ecosystem-based management', *Conservation Biology*, **24**(1), 207–16.
Groves, C.R., D.B. Jensen and L.L. Valutis et al. (2002), 'Planning for biodiversity conservation: putting conservation science into practice: a seven-step framework for developing regional plans to conserve biological diversity, based upon principles of conservation biology and ecology, is being used extensively by the nature conservancy to identify priority areas for conservation', *BioScience*, **52**(6), 499–512.
Guerry, A., J. Lubchenco and S. Polasky et al. (2015), 'National capital and ecosystem services informing decisions: from promise to practice', *Proceedings of the National Academy of Sciences*, **112**(24), 7348–55.
Holmlund, C. and M. Hammer (1999), 'Ecosystem services generated by fish populations', *Ecological Economics*, **29**(2), 253–68.
Huggett, A.J. (2005), 'The concept and utility of "ecological thresholds" in biodiversity conservation', *Biological Conservation*, **124**(3), 301–10.
International Finance Corporation (IFC) (2012), *Performance Standards on Environmental and Social Sustainability*, IFC, accessed 27 January 2016 at http://www.ifc.org/wps/wcm/connect/115482804a0255db96fbffd1a5d13d27/PS_English_2012_Full-Document.pdf?MOD=AJPERES.
Jackson, R.B., E.G. Jobbagy and R. Avissar et al. (2005), 'Trading water for carbon with biological carbon sequestration', *Science*, **310**(5756), 1944–7.

Kareiva, P.M., H. Tallis and T. Ricketts et al. (2011), *Natural Capital: Theory and Practice of Mapping Ecosystem Services*, Oxford, UK: Oxford University Press.
Kiesecker, J.M., H. Copeland and A. Pocewicz et al. (2009), 'A framework for implementing biodiversity offsets: selecting sites and determining scale', *Bioscience*, **59**(1), 77–84.
Kiesecker, J.M., H. Copeland and A. Pocewicz et al. (2010), 'Development by design: blending landscape-level planning with the mitigation hierarchy', *Frontiers in Ecology and the Environment*, **8**(5), 261–6.
Koch, E.W., E.B. Barbier and B.R. Silliman et al. (2009), 'Non-linearity in ecosystem services: temporal and spatial variability in coastal protection', *Frontiers in Ecology and the Environment*, **7**(1), 29–37.
Kontogianni, A., G.W. Luck and M. Skourtos (2010), 'Valuing ecosystem services on the basis of service-providing units: a potential approach to address the "endpoint problem" and improve stated preference methods', *Ecological Economics*, **69**(7), 1479–87.
Kusler, J. (2003), *Reconciling Wetland Assessment Techniques*, Berne: Institute for Wetland Science and Public Policy, Association of State Wetland Managers.
Landis, W.G. (2003), 'Twenty years before and hence: ecological risk assessment at multiple scales with multiple stressors and multiple endpoints', *Human and Ecological Risk Assessment*, **9**, 1317–26.
Landis, W.G. and J.K. Wiegers (2007), 'Ten years of the relative risk model and regional scale ecological risk assessment', *Human and Ecological Risk Assessment*, **13**(1), 25–38.
Langhammer, P.F. (2007), *Identification and Gap Analysis of Key Biodiversity Areas: Targets for Comprehensive Protected Area Systems*, Gland, Switzerland: IUCN, accessed 29 January 2016 at https://portals.iucn.org/library/efiles/documents/PAG-015.pdf.
Lawrence, D. (2003), *Environmental Impact Assessment: Practical Solution to Reoccurring Problems*, Chichester, UK: John Wiley and Sons.
Luck, G.W., K.M.A. Chan and J.P. Fay (2009), 'Protecting ecosystem services and biodiversity in the world's watersheds', *Conservation Letters*, **2**(4), 179–88.
Luck, G.W., K.M. Chan and C.J. Klien (2012), 'Identifying spatial priorities for protecting ecosystem services', *F1000Research*, **1**, 17.
Luck, G.W., R. Harrington and P.A. Harrison et al. (2009), 'Quantifying the contribution of organisms to the provision of ecosystem services', *BioScience*, **59**(3), 223–35.
Mandle, L. and H. Tallis (2012), *Can the Pucallpa-Cruziero do Sul Road be Developed with No Net Loss of Natural Capital in Peru?*, report to the Latin America Conservation Council, The Nature Conservancy, Arlington, TX.
Mandle, L., H. Tallis and L. Sotomayor et al. (2015). 'Who loses? Tracking ecosystem service redistribution from road development and mitigation in the Peruvian Amazon', *Frontiers in Ecology and the Environment*, **13**(6), 309–15.
Margules, C.R. and R.L. Pressey (2000), 'Systematic conservation planning', *Nature*, **405**(6783), 243–53.
McKenney, B.A. and J.M. Kiesecker (2010), 'Policy development for biodiversity offsets: a review of offset frameworks', *Environmental Management*, **45**(1), 165–76.
Moilanen, A., A.J.A. van Teeffelen and Y. Ben-Haim et al. (2009), 'How much compensation is enough? A framework for incorporating uncertainty and time discounting when calculating offset ratios for impacted habitat', *Restoration Ecology*, **17**(4), 470–78.
Naidoo, R., A. Balmford and R. Costanza et al. (2008), 'Global mapping of ecosystem services and conservation priorities', *Proceedings of the National Academy of Sciences of the United States of America*, **105**(28), 9495–500.
National Research Council (NRC) (2005), *Valuing Ecosystem Services: Toward Better Environmental Decision-making*, Washington, DC: National Academy Press.
Nelson, E., G. Mendoza and J. Regetz et al. (2009), 'Modeling multiple ecosystem services, biodiversity conservation, commodity production, and trade-offs at landscape scales', *Frontiers in Ecology and the Environment*, **7**(1), 4–11.
Nichols, S.S. and J.M. McElfish (2009), *Wetland Avoidance and Minimization in Action: Perspectives from Experience*, Washington, DC: Environmental Law Institute.

Nichols, S.S., J. McElfish and J. Wilkinson (2008), *Environmental Law Institute, State Wetlands Protection Programs: Avoidance and Minimization Requirements*, accessed 6 February 2016 at https://www.eli.org/sites/default/files/eli-pubs/d18__05.pdf.

Olander, L., R. Johnston and H. Tallis et al. (2015), *Best Practices for Integrating Ecosystem Services into Federal Decision Making*, Durham, NC: National Ecosystem Services Partnership, Duke University.

Overton, J.M., R.T. Stephens and S. Ferrier (2013), 'Net present biodiversity value and the design of biodiversity offsets', *Ambio*, **42**(1), 100–110.

Pilgrim, J.D., S. Brownlie and J.M. Ekstrom et al. (2013), 'A process for assessing the offsetability of biodiversity impacts', *Conservation Letters*, **6**, 376–84.

Plummer, M.L. (2009), 'Assessing benefit transfer for the valuation of ecosystem services', *Frontiers in Ecology and the Environment*, **7**(1), 38–45.

Polasky, S., H. Tallis and B. Reyers (2015), 'Setting the bar: standards for ecosystem services', *Proceedings of the National Academy of Sciences of the United States of America*, **112**(24), 7356–61.

Possingham, H., I. Ball and S. Andelman (2000), 'Mathematical methods for identifying representative reserve networks', in S. Ferson and M. Burman (eds), *Quantitative Methods for Conservation Biology*, New York: Springer-Verlag, pp. 291–306.

Quétier, F. and S. Lavorel (2011), 'Assessing ecological equivalence in biodiversity offset schemes: key issues and solutions', *Biological Conservation*, **144**(12), 2991–9.

Race, M.S. and M.S. Fonseca (1996), 'Fixing compensatory mitigation: what will it take?', *Ecological Applications*, **6**(1), 94–101.

Raudsepp-Hearne, C., G.D. Peterson and E.M. Bennett (2010), 'Ecosystem service bundles for analyzing trade-offs in diverse landscapes', *Proceedings of the National Academy of Sciences of the United States of America*, **107**(11), 5242–7.

Ricketts, T.H., J. Regetz and I. Steffan-Dewenter et al. (2008), 'Landscape effects on crop pollination services: are there general patterns?', *Ecology Letters*, **11**(5), 499–515.

Robertson, M., T.K. BenDor and R. Lave et al. (2014), 'Stacking ecosystem services', *Frontiers in Ecology and the Environment*, **12**(3), 186–93.

Ruhl, J.B. and J. Salzman (2006), 'The effects of wetland mitigation banking on people', *National Wetlands Newsletter*, **28**(2), 1.

Saenz, S., T. Walschburger and J. González et al. (2013), 'A framework for implementing and valuing biodiversity offsets in Colombia: a landscape scale perspective', *Sustainability*, **5**(12), 4961–87.

Slootweg, R., F. Vanclay and M. van Schooten (2001), 'Function evaluation as a framework for the integration of social and environmental impact assessment', *Impact Assessment and Project Appraisal*, **19**(1), 19–28.

Tallis, H. and S. Polasky (2009), 'Mapping and valuing ecosystem services as an approach for conservation and natural resource management', *Annals of the New York Academy of Sciences*, **1162**, 265–83.

Tallis, H. and S. Wolny (2010), *Including Ecosystem Services in Mitigation. Report to Colombian Ministry of the Environment, Mines and Territorial Development*, Stanford, CA: Natural Capital Project.

Tallis, H., S. Polasky and J.S. Lozano and S. Wolny (2012), 'Inclusive wealth accounting for regulating ecosystem services', in *Inclusive Wealth Report 2012: Measuring Progress Towards Sustainability*, Cambridge, UK: Cambridge University Press.

Tallis, H., C.M. Kennedy and M. Ruckelshaus et al. (2015), 'Mitigation for one & all: an integrated framework for mitigation of development impacts on biodiversity and ecosystem services', *Environmental Impact Assessment Review*, **55**, 21–34.

The Energy and Biodiversity Initiative (2004), *Good Practice in the Prevention and Mitigation of Primary and Secondary Biodiversity Impacts*, Washington DC: Conservation International Publications.

UNEP (2014), *Integrating Ecosystem Services in Strategic Environmental Assessment: A Guide for Practitioners*, a ProEcoServ report by D. Geneletti, Nairobi: United Nations Environment Programme.

Vigerstol, K. and J.E. Aukema (2011), 'A comparison of tools for modeling freshwater ecosystem services', *Journal of Environmental Management*, **92**(10), 2403–9.
Villarroya, A., A.C. Barros and J.M. Kiesecker (2014), 'Policy development for environmental licensing and biodiversity offsets in Latin America', *PLoS ONE*, **9**(9), e107144.
Waage, S., K. Armstrong and L. Hwang (2011), *New Business Decision-making Aids in an Era of Complexity, Scrutiny and Uncertainty: Tools for Identifying, Assessing and Valuing Ecosystem Services*, San Francisco, CA: Business for Social Responsibility.
Weber, T.C. and W.L. Allen (2010), 'Beyond on-site mitigation: an integrated, multi-scale approach to environmental mitigation and stewardship for transportation projects', *Landscape and Urban Planning*, **96**(4), 240–56.
Wilkinson, J.B., J.M. McElfish and R. Kihslinger et al. (2009), *The Next Generation of Mitigation: Linking Current and Future Mitigation Programs with State Wildlife Action Plans and Other State and Regional Plans*, Washington, DC: Environmental Law Institute & The Nature Conservancy.
Wilson, K., R.L. Pressey and A. Newton et al. (2005), 'Measuring and incorporating vulnerability into conservation planning', *Environmental Management*, **35**(5), 527–43.

18. Promoting nature-based solutions for climate adaptation in cities through impact assessment

Davide Geneletti, Linda Zardo and Chiara Cortinovis

18.1 INTRODUCTION

18.1.1 Climate Change and Biodiversity: Challenges for Impact Assessment

Climate change is expected to be an increasingly important source of impacts on biodiversity, both directly and in combination with other activities and phenomena, such as land use change, habitat fragmentation, biological disruptions (e.g., introduction of non-native species, pests), and environmental pollution (Staudt et al., 2013). The effects of climatic shifts on biodiversity include changes in the ranges of species and habitat types, as well as in ecological dynamics and ecosystem functions (e.g., timing and quantity of water cycling, soil formation, nutrient cycling) (Chen et al., 2013; Bellard et al., 2012). These effects, in turn, may affect the provision of ecosystem services in terms of quality, quantity and spatial and temporal distribution. Nelson et al. (2013) illustrate how valuable ecosystem services in the USA will be impacted (positively or negatively) by the predicted change in climate variables. For example, large forest fires are likely to become more and more common (due to elevated spring and summer temperature), negatively affecting key forest services, such as recreation opportunities and timber. Water supply is expected to become more variable, both temporally and spatially, exacerbating water scarcity in some parts of the USA. Among the possible positive effects, the study mentions the enhancement of cultural services such as recreational fishing, due to more favorable habitat conditions for relevant alien fish species.

In response to the expected impacts of climate change on biodiversity and ecosystem services, the scientific literature has made several recommendations. Heller and Zavaleta (2009) synthesized these recommendations, by breaking them down into different sectors of action: conservation, science and technology, policy, and individual/communities. Recommendations include: increase connectivity, mitigate other threats (invasive species, pollution), maintain basic monitoring programs, protect many small reserves rather than single large ones, initiate long-term studies of species response to climate, re-assess conservation goals (i.e., move away from concepts of natural, embrace processes over patterns). In particular, the authors stress the need for extending conservation efforts beyond reserves and into human-dominated landscapes. These recommendations should help to move 'from what constitutes good biodiversity conservation practice to that which is good biodiversity practice in a changing climate' (Burch et al., 2014, p. 88).

Impact assessment potentially plays a key role in fostering this transition, and mainstreaming the recommendations into projects, plans and policies (PPP). The impact assessment community has been increasingly aware of the need to properly address climate change issues, as shown, for example, by recent guidance documents (EC, 2013a, 2013b). Addressing climate change in impact assessment involves understanding both the impact of PPP on future climate (e.g., in terms of changes of greenhouse gas emissions), and the effects of climate variability on the functioning and effectiveness of PPP (Posas, 2011). Despite the increasing interest, more efforts need to be devoted by the literature on the improvement of methods to predict and assess effects. For example, the results of a quick review of the scientific papers addressing Strategic Environmental Assessment (SEA) and climate change presented in Geneletti (2015a) show that 75 percent of the papers focus on reviews of current practices, and only the remaining 25 percent are devoted to the proposal of analytical methods that practitioners can apply to strengthen climate issues in SEA studies (e.g., indicator sets). Even though most review papers also include recommendations to improve future practice, these are mainly conceptual suggestions on the content and structure of the SEA, as opposed to operational methods to improve the understanding of the effects of policies and plans on climate (and vice versa).

Improved analytical methods are needed to address the complexity associated with the combination of climate change impacts and the impacts of the stressors determined by the PPP under analysis. Research has shown that often climate shifts exacerbate the effects of other environmental stressors, such as land use changes and pollution (chemicals, nutrients) (Grimm et al., 2013). In addition, there are many pathways through which climate change can affect biodiversity, and these also include the pathways determined by societal adaptation and mitigation responses. For example, Staudt et al. (2013) describe four different pathways through which temperature increase can affect native salmon population in a stream:

- Increase in stream water temperature negatively impacts salmon habitat.
- Increase in stream water temperature favors a competitive species, which in turn affects salmon population.
- Increase in irrigation (adaptation action by farmers), which determines a decrease in stream flow, affecting salmon habitat.
- Decrease in stream flow (determined by the increase in irrigation), which causes the remaining water to warm more rapidly, further favoring competitive species.

The complexity of Biodiversity Impact Assessment under climate change is increased by the fact that the effects are difficult to generalize, given that the timing, direction and magnitude of the expected shifts in habitat differ among species, regions and habitat types (Grimm et al., 2013). In addition, the prediction of climate change is inherently uncertain, given that it depends upon future patterns of greenhouse gas emissions, and the effects of these on the climate system, both of which are affected by uncertainties (IPCC, 2007). The uncertainty level is then further increased by the down-scaling of climate models that is often required to fit the scale of the impact assessment process. This may prove challenging, particularly for project-oriented processes, such as Environmental Impact Assessment (EIA), which operate at finer scales.

In the light of all these uncertainty factors, an important role that impact assessment processes can play is the promotion of 'nature-based solutions' for climate change adaptation. Nature-based approaches are no-regret options that offer the flexibility required for dealing with a changing climate, and the associated uncertainties.

18.1.2 Nature-based Solutions for Climate Adaptation

Nature-based solutions for climate adaptation (or 'ecosystem-based adaptation') are defined as the use of biodiversity and ecosystem services to help people to adapt to the adverse effects of climate change (Secretariat of the Convention on Biological Diversity, 2009). These approaches include management, conservation and restoration of ecosystems that deliver services that can help to reduce climate change exposures (Munang et al., 2013a). For example, restoring mangrove forest can contribute to dissipating the energy of storm surges, buffering human communities from floods and erosion (Erwin, 2009). Protecting groundwater recharge areas and floodplains can help to secure water resources and to cope with droughts (Jones et al., 2012). Enhancing green infrastructures in urban areas can reduce the heat island effect, and the associated health risks (Lafortezza et al., 2013). As opposed to more traditional infrastructure-based approaches (e.g., levees, sea walls, irrigation systems), nature-based solutions offer the advantage of promoting 'no-regrets' interventions, and potentially delivering multiple economic, social and environmental co-benefits that go beyond climate adaptation (Jones et al., 2012). These co-benefits include, among others, biodiversity conservation through enhanced habitat conditions; climate mitigation through increased carbon sequestration; conservation of traditional knowledge, livelihood and practices of local communities; improved recreation and tourism opportunities; enhanced food security (Vignola et al., 2009; Naumann et al., 2011; Munang et al., 2013b, 2013c; Demuzere et al., 2014). There is increasing evidence that nature-based solutions can provide flexible, cost-effective and broadly applicable alternatives to cope with the magnitude, speed and uncertainty of climate change (Munang et al., 2013a).

Achieving climate adaptation in urban areas is pivotal for sustainable development, as shown by growing actions undertaken by cities to pursue adaptation (Rosenzweig et al., 2010), as well as guidance documents produced to assist in this endeavor (e.g., ICLEI, 2010). Nature-based solutions can play an important role in urban contexts (EC, 2015) through interventions such as the design and improvement of green and blue infrastructures (e.g., urban parks, green roofs and facades, tree planting, rivers, ponds). Recently, Demuzere et al. (2014) presented a comprehensive analysis of the available empirical evidence about the contribution of green infrastructures to climate change adaptation in urban areas. Nevertheless, little evidence is available on the inclusion of nature-based approaches in urban planning (Geneletti and Zardo, 2016).

Impact assessment processes that support urban planning, such as in particular Strategic Environmental Assessment (SEA), can help to promote the proposal and implementation of nature-based solutions for climate adaptation in cities, as well as to compare the expected effectiveness of these solutions with regard to more traditional approaches (Geneletti, 2015b). This chapter aims at providing some insights and examples related to this topic, particularly by:

- presenting the findings of a review of the extent to which nature-based solutions for climate adaptation are currently included in planning documents at the urban scale (Section 18.2);
- illustrating a method to increase the evidence base that planners have at their disposal to develop and compare nature-based solutions, specifically concerning the analysis of the cooling capacity of green urban infrastructures (Section 18.3);

- testing the method in two applications related to impact assessment of urban planning for a selected case study (Section 18.4);
- drawing some conclusions and recommendations for future practice and research in this area (Section 18.5).

18.2 THE INCLUSION OF NATURE-BASED SOLUTIONS IN URBAN PLANNING

This section summarizes the findings of a review aimed at assessing the extent to which nature-based solutions are currently included in urban planning (Geneletti and Zardo, 2016). First, seven nature-based solutions that are relevant for urban contexts were identified and described (Table 18.1). Afterwards, a sample of European cities was selected, namely Amsterdam, Athens, Barcelona, Berlin, Copenhagen, Heidelberg, London, Madrid, Milan, Paris, Rome, Rotterdam, Stockholm, Venice and Warsaw. The urban climate adaptation plans (defined as plans that include strategies to reduce vulnerability to climate change in urban areas)

Table 18.1 Nature-based solutions for climate adaptation in urban areas

Code	Description
a	Ensuring ventilation from cooler areas outside the city through waterways and green areas to create air circulation and provide downwind cooling effect
b	Promoting green walls and roofs to improve the thermal comfort of buildings, particularly in hot and dry climate
c	Maintaining/enhancing urban green spaces (e.g., ecological corridors, trees, gardens) to reduce air and surface temperature by providing shading and enhancing evapotranspiration
d	Avoiding/reducing impervious surfaces to slow down water runoff and enhance water infiltration, reducing peak discharge and offering protection against extreme precipitation events
e	Re-naturalizing river systems to create space for floodwater to support higher base flows, reducing flood risk
f	Maintaining and managing green areas for flood retention and water storage to increase infiltration and induce the replenishment of groundwater
g	Promoting the use of vegetation adapted to local climate and drought conditions and ensuring sustainable watering of green space

Source: Modified after Geneletti and Zardo (2016) building on the list proposed by EEA (2012).

of these cities approved by the relevant municipal authority, and available on the Internet were collected. A direct content analysis (Hsieh and Shannon, 2005) was then performed, by reading all the documents associated to the selected plans, and identifying content related to nature-based solutions in all the components of the plans (i.e., baseline documents, description of vision and objectives, description of actions, and implementation proposals). Finally, a scoring system was applied to assess how well the different possible solutions were addressed in the different components (all methodological details can be found in Geneletti and Zardo, 2016).

A snapshot of the results is presented in Figure 18.1. As can be seen, solutions (c) and (f) are the most common ones, showing that there is strong awareness of the role that green areas play in addressing climate change challenges, both in terms of mitigating heat waves (c) and preventing floods (f). The frequency of these measures is perhaps not surprising giving that they result in the enhancement of green areas, which is a typical objective that planners pursue in order to improve the urban space for a variety of purposes that go beyond climate change adaptation (e.g., providing recreation opportunities, improving air quality) (Tzoulas et al., 2007). So, their frequency could be explained by the fact that these solutions rely on actions that are part of the standard portfolio that planners

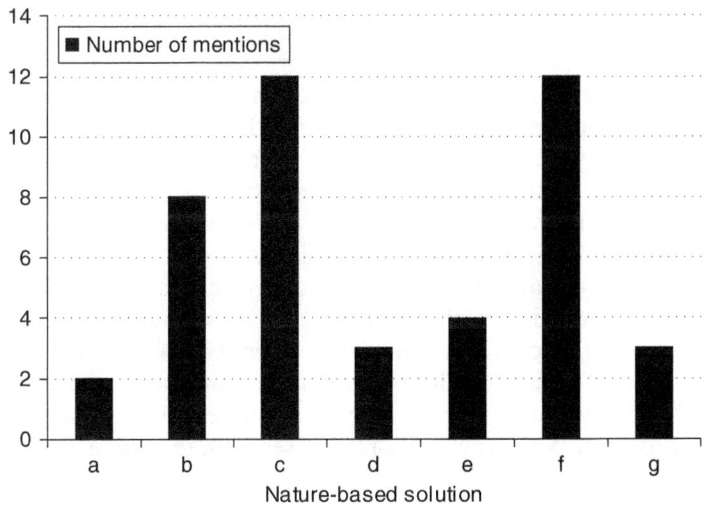

Source: Geneletti and Zardo (2016).

Figure 18.1 Number of mentions of the seven types of nature-based solutions (see Table 18.1) in the sample of plans

have been employing for decades. Green walls and green roofs (b) are found in more than half of the cities. These measures are well covered by the literature, which offers ample debate on the effectiveness of vegetated roofs and facades to improve the thermal comfort of buildings, providing data for different climate zones and recommendations for implementation (Cook-Patton and Bauerle, 2012; Santamouris, 2014). Solution (d) is rarely proposed by plans, despite the fact that ecosystem-based practices to reduce soil sealing and mitigate its negative effects on soil are often relatively cheap and quite straightforward to include in climate adaptation plans. Solution (a) is the least frequently encountered. One reason may be that the effectiveness of this measure is related more in general to the urban morphology. Hence, the design of urban waterways and open green areas that create air circulation needs to be undertaken jointly with other actions related to the built environment that go beyond the content of climate adaptation plans. Solution (g) is also rarely found in plans, but this may be explained by the fact that it encompasses a more limited set of actions, which may be relevant only in specific climate conditions.

Concerning how well nature-based solutions are reflected in the different plan components, the analysis revealed that a common shortcoming is the shortage of details provided in the information baseline. What is surprising in particular is the lack of climate vulnerability analysis, and the scarcity of information on the expected effects of the proposed solutions. The outcome of vulnerability analyses would allow better targeting of measures, in terms of design, location, priority level, and so on (Lankao and Qin, 2011; Holsten and Kropp, 2012). In no plan is there an attempt to quantify the expected effects, or to provide spatial distribution of the benefits that the measures will produce. For example, the enhancement of green areas to reduce heat or to prevent floods is typically proposed as a general measure that will do some good, without providing details and justification for critical decisions, such as the design and the location of these interventions, and the distribution and vulnerability of the expected beneficiaries. These issues play a key role in determining the effectiveness of the solutions (Kleerekoper et al., 2012; Kazmierczak, 2012). This speaks to the need for enhancing the baseline data that planners have at their disposal, including mapping of the potential of existing green/blue infrastructure assets to provide climate-related services, following, for example, the approach of EEA (2014).

The results of the review suggest that nature-based solutions are finding their way in climate adaptation plans, in response to a broad range of climate change challenges. However, the baseline information upon which these solutions are proposed and designed needs to be enhanced. Methods to assess the existing stock of green/blue infrastructures, and their potential to provide climate adaptation services must be mainstreamed in planning practice. Particularly, assessment of the flow of ecosystem services at local scales is often missing, given that many climate change impact and vulnerability studies provide results at larger scales, limiting their usefulness for developing local adaptation strategies (Vignola et al., 2009). An enhanced knowledge basis would allow better targeting of the design and implementation of nature-based solutions. The next two sections present a method to improve the knowledge basis on nature-based solutions, and its application to a case study in urban planning.

18.3 A METHOD TO ASSESS THE COOLING EFFECT OF GREEN URBAN INFRASTRUCTURES

Information on successful and cost-effective nature-based solutions such as green urban infrastructure (GUI) is lacking, especially at the local scale (Norton, 2015; Larondelle and Haase, 2013). Methodologies and tools for the assessment of the ecosystem services provided by GUI tend to be too data demanding for being routinely adopted by urban planning processes. More generally, research effort is needed to understand and to quantify at a proper scale how ecosystems provide services, and to spatially define the relationship between ecosystem structure, functions, services, and the associated benefits (Braat and De Groot, 2012). We addressed this problem by developing a method to assess the cooling effect provided by GUI that could be employed by planners to support the design and management of these infrastructures. An overview of this method is provided here. Full details can be found in Zardo et al. (under review).

The method is based on the 'cascade model' (Braat and De Groot, 2012), which describes the links between structure, functions and services provided by an ecosystem. We first identified the functions involved in the cooling capacity of GUI, and the associated components, that is, the physical attributes, such as shape, size, and so on. Afterwards, we performed an analysis of the literature to quantify the cooling capacity of a GUI associated to any given combination of attributes. The overall purpose was to provide planners with a relatively simple model to predict the cooling capacity of GUI, and to support their design and inclusion in urban plans.

The functions mainly involved in the cooling capacity of GUI are shading and evapotranspiration (McPhearson et al., 1997; Bolund and Hunhammar, 1999; Gomez-Baggethun and Barton, 2013; Larondelle and Haase, 2013). The two functions were first assessed individually, and then combined, in order to estimate the overall cooling capacity of GUI (Figure 18.2). The assessment of the shading function was based on the analysis

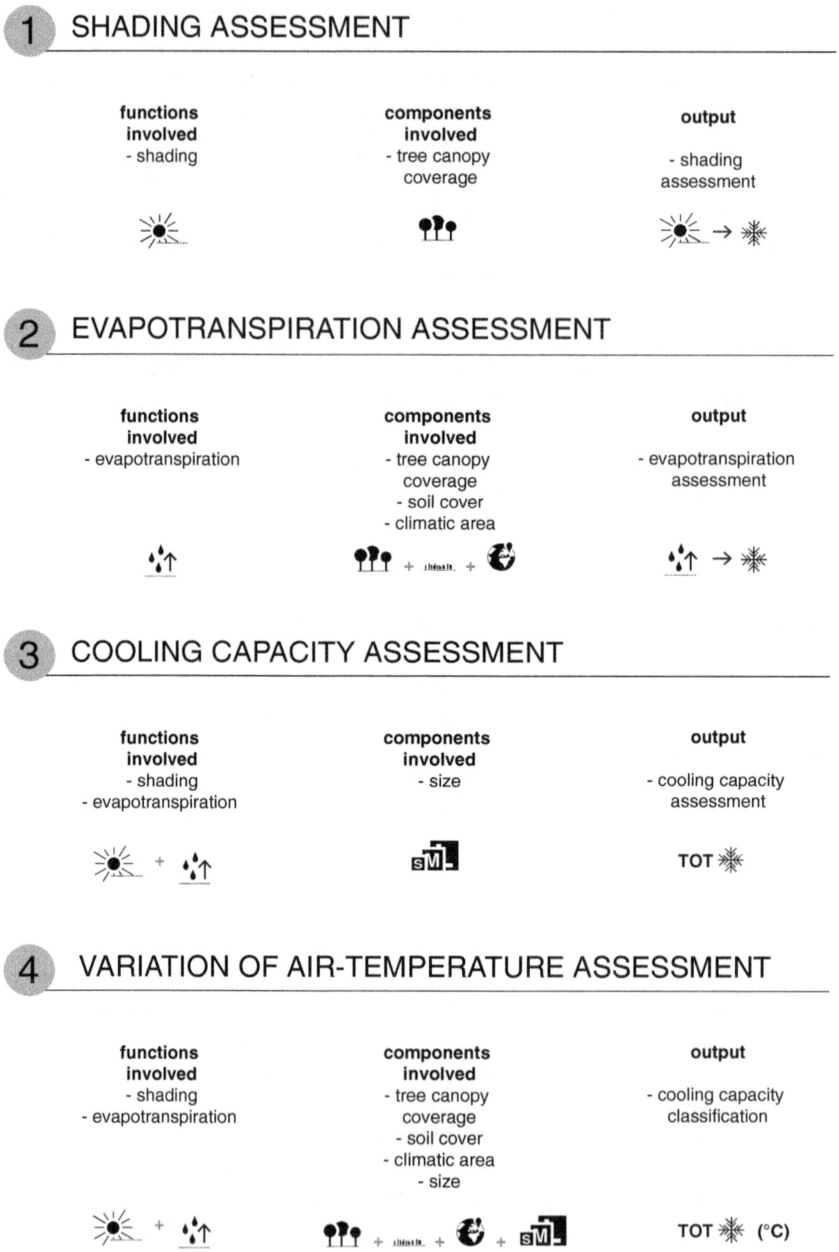

Figure 18.2 Overview of the method to assess the cooling capacity of GUI

of the tree canopy coverage, which is one of the key components that influences shading (Stronbach and Haase, 2012). Tree canopy coverage was classified into five categories: 0–20 percent, 20–40 percent, 40–60 percent, 60–80 percent and 80–100 percent. The assessment of evapotranspiration considered soil cover, tree canopy coverage and climatic area of the GUI, which are the main components involved in providing this function (Akbari, 1992; Souch and Souch, 1993; Chang et al., 2007; Bowler et al., 2010; Schwarz et al., 2011; Larondelle and Haase, 2013). Each of these components was classified into categories, and the categories were then combined into an overall evapotranspiration value, using approaches proposed in the literature (FAO, 1998). This value ranges between 0 and 100, where 0 represents the worst evapotranspiration performance.

The role played by shading and evapotranspiration to determine the overall cooling capacity depends, among other things, upon the size of the GUI (Chang et al., 2007), in ways that are not linear or easy to predict (Chang et al., 2007; Bowler et al., 2010). However, the literature suggests the existence of a 2 ha threshold. In areas smaller than 2 ha, the shares of contribution of evapotranspiration and shading to the overall cooling are 20 and 80 percent respectively (Shashua-Bar and Hoffman, 2000). In areas larger than 2 ha, the ratio becomes 60 and 40 percent (Cao, 2010). Using this rule, we combined the shading and evapotranspiration functions and assessed the overall cooling capacity of GUI. This capacity was then classified into six classes from 'E' to 'A+', adopting the European Union Energy Label classification (EU, 1992), where A+ represents the highest cooling performance. Finally, using data collected from empirical studies (Kjelgren and Clark, 1992; Souch and Souch, 1993; Shashua-Bar and Hoffman, 2000; Watkins, 2002; Potcher et al., 2006; Chang et al., 2007; Larondelle and Haase, 2013), the classes were linked to ranges of temperature decrease in different climatic areas (Mediterranean area, Continental area and Atlantic area). This method allows us to assess the cooling capacity of a GUI within the boundaries of the GUI itself. However, the cooling effect extends beyond these boundaries (Bowler et al., 2010). Decay functions can be applied to assess also this effect (Shashua-Bar and Hoffman, 2000), as shown in the case study presented next.

The method was applied to the city of Trento, located in an alpine region in Northeastern Italy. Trento is a city of 100 000 people that sprawls about 10 km across the Adige Valley floor (Figure 18.3). The GUI of Trento were mapped and classified according to the key components described earlier, namely tree canopy coverage, soil cover and size. This information was then aggregated using the cooling capacity values for the Mediterranean climatic area (see details in Zardo et al., under review), obtaining the map presented in Figure 18.4 (left). The cooling effect that

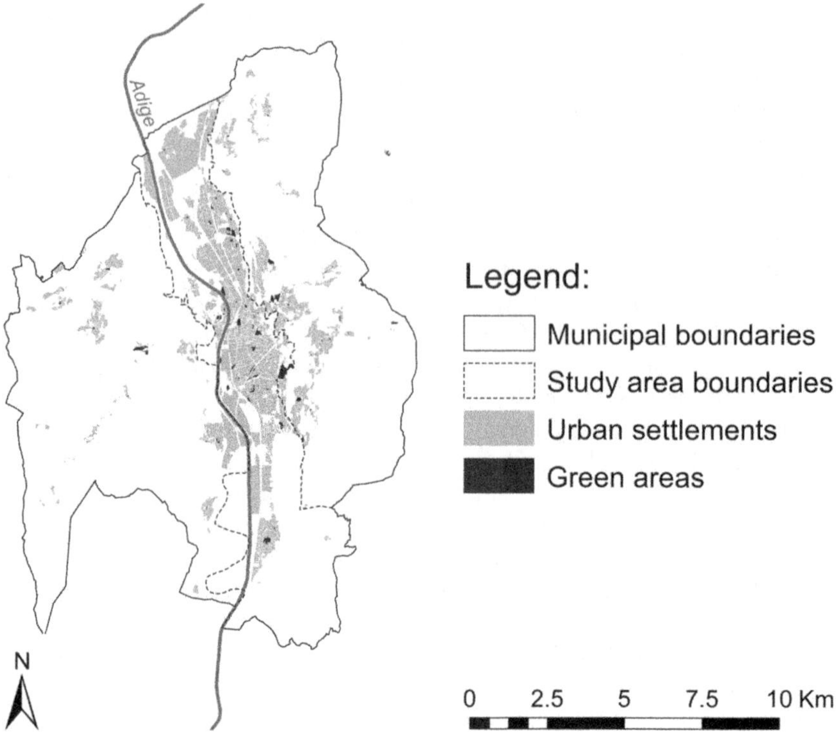

Figure 18.3 The city of Trento

extends beyond the GUI was considered by applying a 5m buffer to clusters of trees (Taha et al., 1991) to account for the shading function, and the decay function proposed by Shashua-Bar and Hoffmann (2000) to account for the evapotranspiration function. The results are shown in Figure 18.4 (right).

18.4 APPLICATIONS TO IMPACT ASSESSMENT FOR URBAN PLANNING

This section shows how the results of the cooling capacity assessment performed on the GUI of Trento can be applied to support urban planning, and the associated impact assessment processes. In particular, we performed two different applications related to urban development opportunities regulated by the existing Urban Plan. We simulated a situation where an impact assessment is carried out to define possible

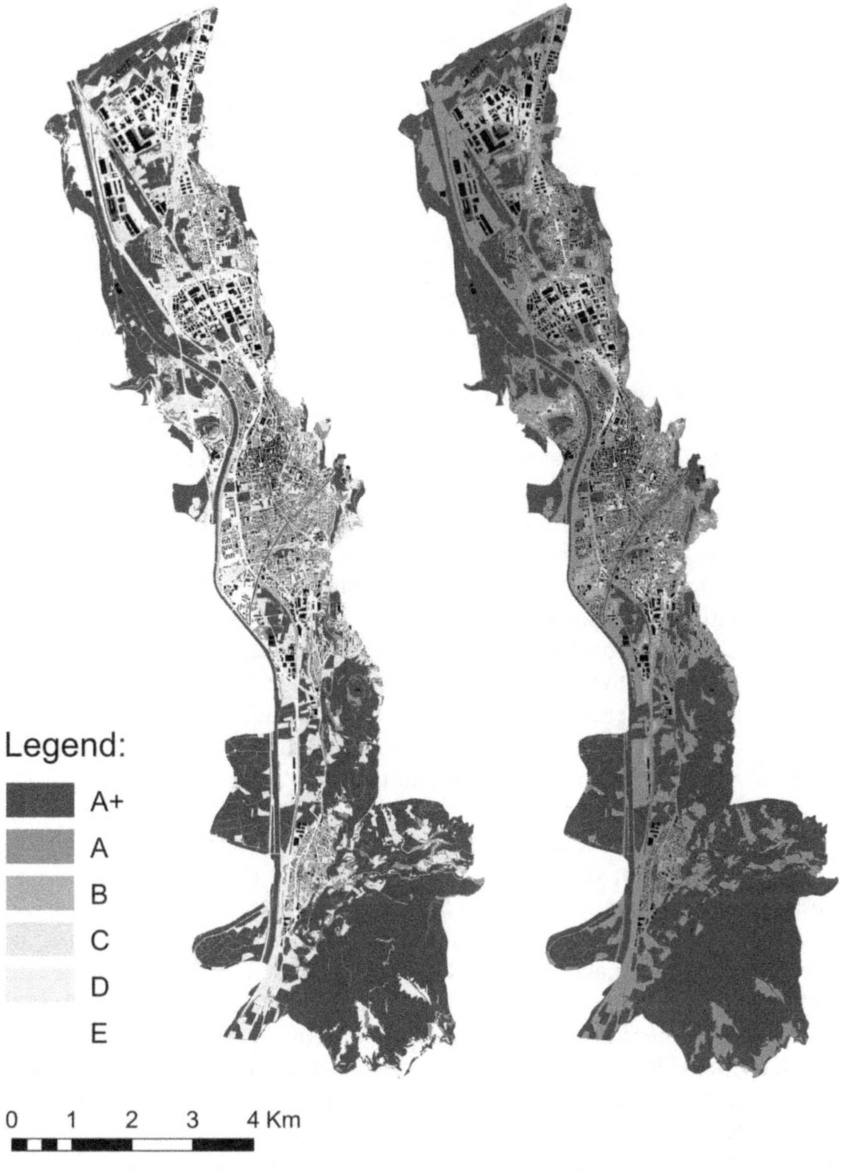

Figure 18.4 *Classification of the cooling capacity within GUI (left) and of the resulting cooling effect (right) in the city of Trento*

implementation strategies of an existing plan, or to suggest revisions to the plan itself. The two applications refer to the completion of vacant lots inside the existing urban fabric, and the re-development of brownfields, respectively. In the first case, the selected lots are seen as receptors of the thermal benefit provided by the surrounding GUI. The analysis aims at understanding which lots should receive priority for development, by virtue of the expected thermal benefits from existing GUI (Section 18.4.1). In the second case, different greening interventions in brownfields, and their associated cooling capacity, are simulated, in order to identify the sites, and the types of intervention, that provide the highest benefits to citizens, considering also their vulnerability to heat waves (Section 18.4.2).

18.4.1 Identify Priority Sites for Urban Development

The first application is based on a biophysical analysis. The goal is to rank the residential zoning vacant lots according to the thermal benefits that they are expected to receive from the existing GUI. This thermal benefit, provided to the future residents, is used as a criterion to assess the suitability of each area for residential development. This allows, other things being equal, us to prioritize interventions in order to take the maximum advantage of the microclimate regulation provided by GUI. Also, the analysis provides the basis from which to define different nature-based actions to be implemented by the developers as measures to compensate for the different starting conditions and to reach an adequate level of thermal comfort for the future inhabitants.

Seventy-three vacant lots with a surface that varies from around 500 m^2 to around 5000 m^2 are identified in the study area by the existing Urban Plan. Almost all of them are located inside or at the borders of existing residential settlements (Figure 18.5, left). We modeled the thermal benefit provided to these sites by the surrounding GUI (see previous section). The internal cooling capacity was set to zero, in order to decouple the benefit determined by the internal conditions from the external contribution that we wanted to measure. Then, for each vacant lot we calculated the cooling capacity provided by the surrounding GUI classified into six classes and computed the percentage of the lots area that falls in each class. A weighting method was applied in order to draw an average score for each vacant lot: a value from 5 to 1 was assigned to the cooling classes from A+ to D, and then weighted by the share of the total area covered by each class. The resulting score allowed to compare the 73 vacant lots and to rank them according to the thermal benefit that they receive, as shown in Figure 18.6.

The morphology of the valley floor where the city is settled (characterized by the closeness to the surrounding wooded hills and by the presence

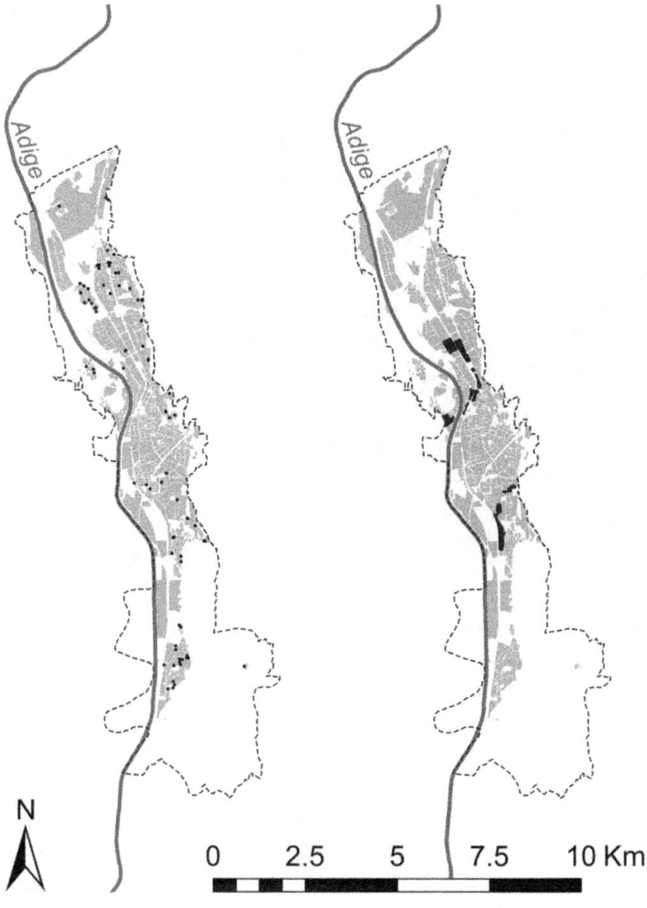

Figure 18.5 The sites classified by the Trento Urban Plan as vacant lots for residential development (left) and as brownfields for future re-development (right) that have been considered in this study

of the Adige River and its tributaries) ensures that the majority of the vacant lots are in good thermal condition. Seventy-five percent of the analyzed lots receive a score equal or higher than 3, which corresponds to an average B class of cooling capacity. The worst-performing lots are mostly located at the edge of the city or within the northern suburb, where the mix of residential and industrial areas scarcely equipped with green infrastructures, as well as the presence of major transportation infrastructures, have a negative impact on the cooling capacity. Vacant lots with the

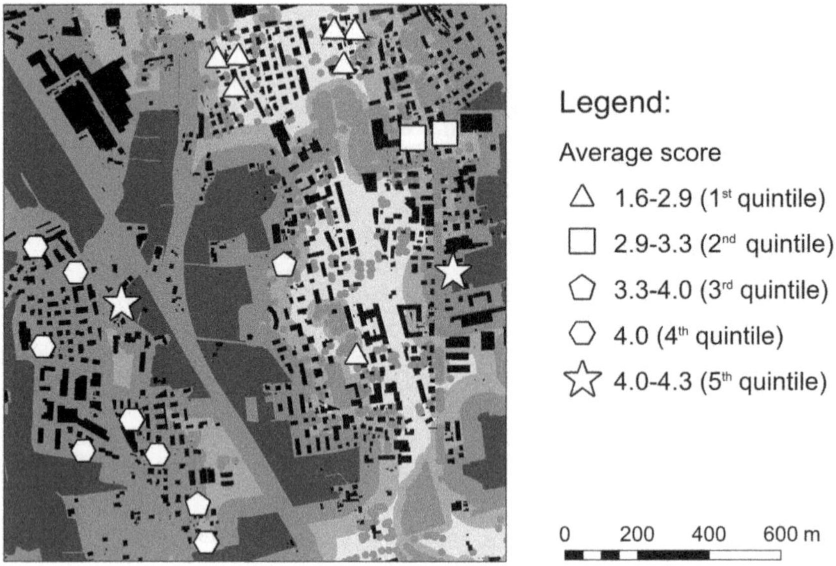

Note: The lots are classified according to the quintile of the score they belong to (the first quintile includes the lots that receive the lowest benefits).

Figure 18.6 Sub-window of the central sector of the study area showing the classification of the thermal benefits received by the vacant lots

highest scores are in general the most peripheral and can be found both in the northern part of the study area, at the borders of a green wedge that penetrates the built spaces, and in the southern part, next to the surrounding wooded slopes. Nevertheless, some vacant lots inside the city center also reach the highest level of thermal benefit provided by the surrounding green infrastructure due to the proximity to large green areas (urban parks) and to water bodies.

18.4.2 Assessing the Impact of Brownfield Re-development on GUI Beneficiaries

The second application adds the quantification of the beneficiaries of the cooling effect to the biophysical analysis, in order to compare and assess different brownfield re-development scenarios. The goal is to understand how the surrounding population, and vulnerable groups in particular, are affected by land cover changes occurring in the re-development sites envisioned by the Urban Plan. More specifically, we simulated two different greening interventions resulting in different cooling performances, and we measured the number of people that would be affected by the change. This enables us to understand how and where it is more cost-effective to intervene through nature-based solutions, answering questions such as: Which area can be more effectively transformed through greening actions? Which level of performance is required to increase the well-being of the surrounding inhabitants? In which area is the same investment expected to obtain the biggest gain?

To define vulnerable groups we referred to the analyses by Kabisch and Haase (2014) and Kazmierczak (2012) and identified elderly people (above 65 years old), children (under 5 years old), and foreigners as the most sensitive and less adaptive to heat stresses. Census data provided by the municipality about total number of residents, age group distribution, and presence of foreigners for each census block have been collected and linked to the spatial map. To be as accurate as possible in the spatial definition of the population, we identified the residential buildings inside each census block and distributed the residents only in the surface covered by their footprint.

The analysis was performed on 13 brownfields identified by the Urban Plan as areas for future re-development. Most of them are former industrial sites or partially abandoned residential areas (see Figure 18.5, right). Their dimension ranges from 0.5 to 9.9 ha. Among them, seven are larger than 2 ha, thus the maximum distance that can be reached by their cooling effect on the surroundings is around 250 m. For the smaller ones the maximum buffer of effect is 100 m from the boundary (see previous section).

For each area we simulated two transformation scenarios (through greening interventions), and modeled the expected cooling capacity. Scenario A refers to the 'best cooling performance' that can be reached according to the size of the brownfield. That is, sites larger than 2 ha were converted into class A+, and sites smaller than 2 ha into class A. Scenario B simulates a 'medium-level cooling performance'. That is, sites larger than 2 ha are converted into class B, and sites smaller than 2 ha into class C (see Zardo et al., under review, for details on this modeling). The same class of internal cooling capacity can be obtained through different interventions corresponding to different combinations of land cover and tree coverage. In scenario A, we simulated the presence of a homogeneous grassy area with tree coverage higher than 80 percent (e.g., an urban forest or an intensely planted urban park). In scenario B, we simulated a greening intervention able to reach a high tree coverage over a sealed surface (> 40 percent for the bigger areas and > 60 percent for the smaller ones).

We considered all the transformations independently and calculated for each scenario the map of the cooling capacity inside the maximum area of influence of each re-development site. Through an overlay between this map and the census data, we determined the number of people living within each cooling class, for the different scenarios and broken down by different vulnerable groups (foreigners, children, elderly people). Figure 18.7 and Tables 18.2, 18.3 and 18.4 present the results obtained for one of the re-development sites.

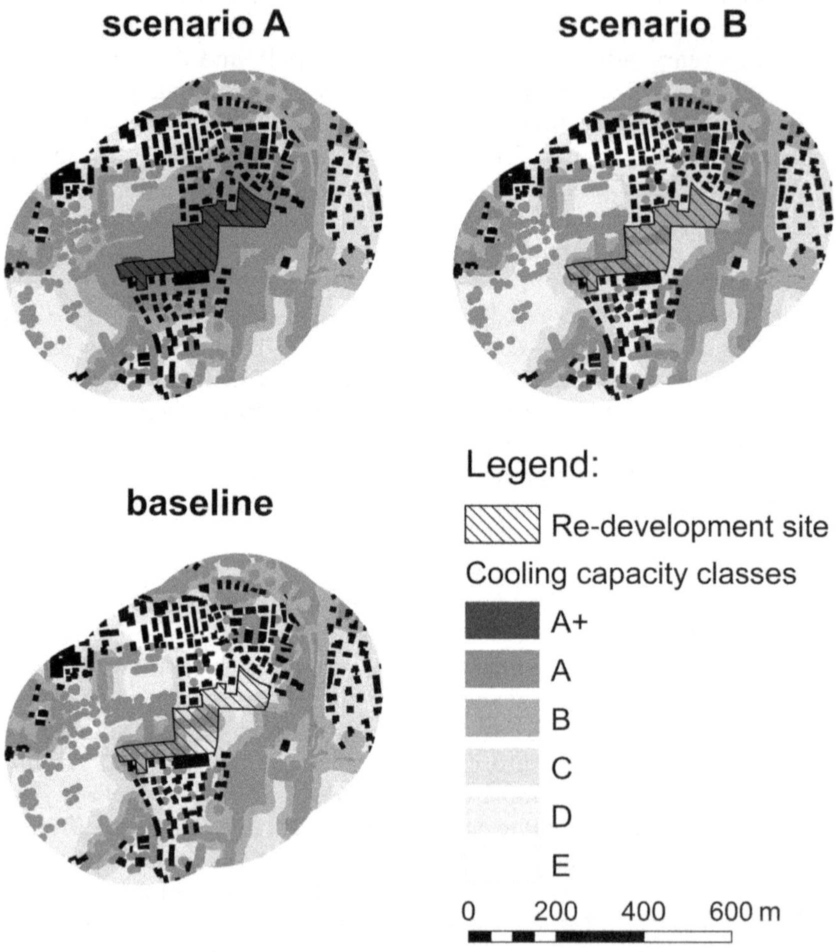

Figure 18.7 Cooling capacity classes modeled for one of the re-development sites (Site 11) in the baseline conditions and under the two transformation scenarios

Table 18.2 Number of beneficiaries within each cooling capacity class (Site 11, baseline conditions)

Cooling Capacity Class	Overall Population	Vulnerable Population		
		Foreigners	Children	Elderly people
A+	0	0	0	0
A	255	40	11	73
B	376	53	16	108
C	725	104	29	201
D	938	127	38	263
E	250	31	11	72

Table 18.3 Number of beneficiaries within each cooling capacity class (Site 11, scenario A)

Cooling Capacity Class	Overall Population	Vulnerable Population		
		Foreigners	Children	Elderly people
A+	0	0	0	0
A	788	126	30	225
B	582	82	23	165
C	758	107	32	211
D	402	38	19	113
E	14	2	1	4

Table 18.4 Number of beneficiaries within each cooling capacity class (Site 11, scenario B)

Cooling Capacity Class	Overall Population	Vulnerable Population		
		Foreigners	Children	Elderly people
A+	0	0	0	0
A	255	40	11	73
B	337	45	14	97
C	1031	149	40	287
D	765	98	33	215
E	157	22	8	45

In order to assess the benefits provided by the different transformation scenarios, we performed a spatial analysis of the changes between each scenario and the baseline conditions. First, we assigned a numerical value to each class (from 6 for class 'A+', to 1 for class 'E') in the maps of the cooling capacity, and then subtracted the two maps representing the condition after and before the transformation. The result shows the spatial distribution of the expected differences in class: positive values stand for changes from lower to upper classes, negative values for changes from upper to lower classes. The overlay between this map and the census data allowed us to estimate the number of people, for each group, affected by positive or negative changes.

Figure 18.8 shows the results of this analysis. As an overall indicator, we calculated the population and the vulnerable population (i.e., elderly + children + foreigners) affected by the transformation: positive values indicate a net positive change, negative values indicate a net negative change. Finally, the results were normalized by the area of the brownfield, in order to allow for a comparison across sites. Hence, this final analysis provides an estimation of the expected number of beneficiaries per unit of area of intervention.

The results show that medium-level interventions (scenario B) are not expected to produce remarkable changes, except for specific re-development sites located inside urban areas with low cooling capacity. Indeed, most of the transformations modeled in scenario B do not generate any benefit for the surrounding inhabitants, whose thermal conditions are mostly determined by other factors. Only two areas over 2 ha, when transformed to reach a class B internal cooling capacity, bring some limited thermal benefits to the surrounding residential areas. This is a consequence of the general good conditions of the study area (see Figure 18.4).

The results of scenario A, which simulates a green area almost totally covered by trees, demonstrate that even a land cover change in a limited area can bring significant benefit to the citizens. Moreover, the comparison permits to identify what interventions are more cost-effective. For example, considering scenario A, re-development site no. 11 produces the best results both in absolute and relative terms, being a potentially large green area, inside a heavily built-up and densely populated part of the city. On the other hand, the second rank depends on the indicator: the greening intervention on re-development site no. 12 (resulting in a class A+ of cooling capacity) has a positive effect on the highest number of residents, but the same intervention on re-development site no. 5 (resulting in a class A due to the small dimension) performs relatively better, in particular when the number of vulnerable people is

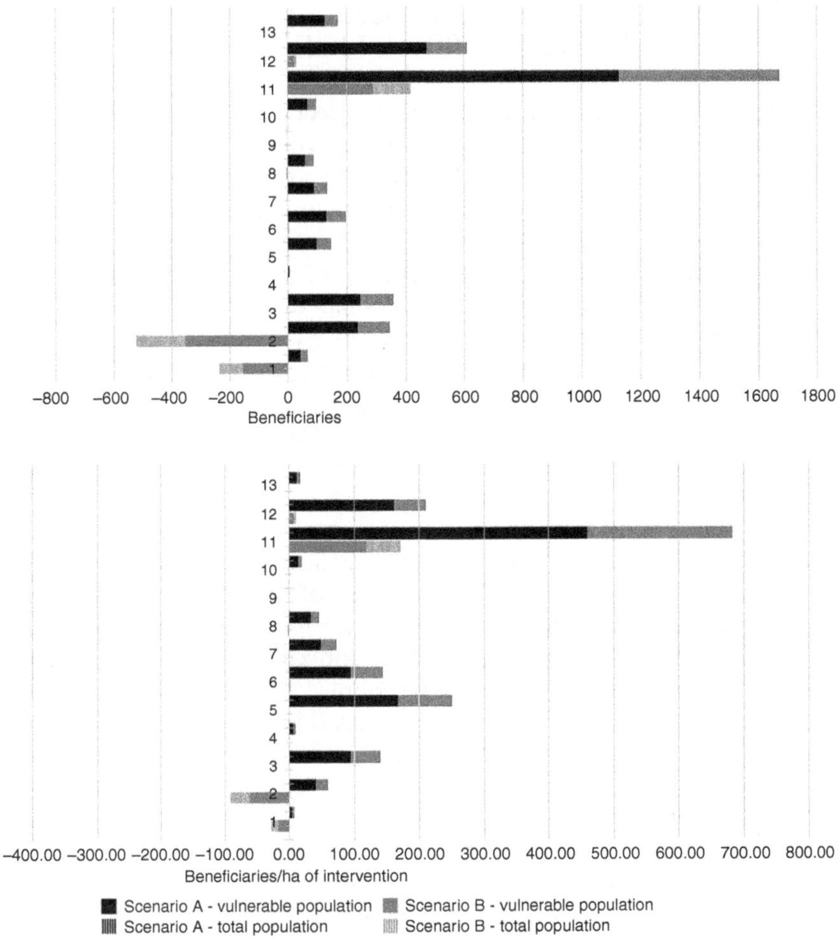

Note: Positive values indicate a net positive change, negative values indicate a net negative change in cooling capacity class.

Figure 18.8 Population and vulnerable population affected by the transformation scenarios in each re-development site

considered. This means that, depending on the present conditions of the urban environment and the density and vulnerability characteristics of the resident population, the same investment for greening interventions can be more cost-effective when directed at a small area, even if it results in a lower cooling capacity compared to the one that can be obtained in larger sites.

18.5 CONCLUDING REMARKS FOR FUTURE RESEARCH AND PRACTICE

The review of the planning documents (Section 18.2) revealed that a common shortcoming in practice is the shortage of information about climate change that can be used to support urban planning. There is a need to enhance these baseline data, including mapping of the potential of existing green/blue infrastructure assets to provide climate-related services, following, for example, the approach of EEA (2014). The method presented in this chapter to assess the cooling capacity of GUI represents a step in the direction of enhancing the knowledge basis that planners have at disposal to design and implement nature-based solutions for climate adaptation in cities. Impact assessment processes can contribute to this by proposing solutions that maximize the benefits provided by GUI, and by supporting them with an explicit analysis of the expected benefits and beneficiaries.

The explicit consideration of ecosystem service beneficiaries increases the added value provided by impact assessment processes to decision-making, as discussed also in Chapter 17 of this book. Additionally, accounting for beneficiaries promotes more integrated forms of impact assessment, given that – as shown by the application presented in Section 18.4 – biophysical analyses need to be coupled with socio-economic ones. This integration of impact assessment types is instrumental to identifying a shared set of alternatives that are responsive to the defined objectives, and that can be compared and iteratively refined along the decision-making process (Geneletti, 2014).

Another key area that requires improvement with respect to current practice concerns the analysis of the co-benefits associated to nature-based solutions. One of the strongest motivations for promoting nature-based solutions is that they bring environmental and socio-economic benefits, beyond climate adaptation. These co-benefits range from biodiversity conservation to climate mitigation, and from improved recreation opportunities to enhanced food security. In the sample of plans reviewed in Section 18.2, co-benefits are never really analyzed, but just mentioned, hampering the possibility to understand their actual magnitude. A more formal analysis of the magnitude of the co-benefits needs to be promoted in impact assessment, in order to provide a stronger rationale for decisions involving the design and implementation of nature-based solutions. Ideally, comparisons between nature-based and more traditional adaptation solutions should be performed, as advocated by Jones et al. (2012). These analyses can take advantage of the methodologies and findings presented in the growing literature on the assessment and evaluation of ecosystem services (Kareiva et al., 2011), including its emerging streams focused on spatial planning (McKenzie et al., 2014) and impact assessment (Geneletti, 2011, 2013a, 2013b).

Finally, a key remark concerns the handling of uncertainty factors. Impact assessment processes that include climate change considerations should be performed consistently with three principles suggested by EC (2013a): acknowledge all assumptions and limitations of current knowledge; base analysis and recommendations on the precautionary principles, and propose alternatives and monitoring schemes that prepare for adaptive management of PPP. However, current impact assessment practice does not seem to properly account, both in the analytical and in the communication components, for the problems associated with climate change uncertainty (Larsen et al., 2013). Improving the consideration of climate change-related uncertainties should become a priority for future impact assessment practice, as well as research. Concerning the latter, there is a more general need to develop analytical methods that study the interactions between climate and other stressors, given that the understanding of precise pathways, types and character of interactions is still very limited (Staudt et al., 2013).

REFERENCES

Akbari, H., S. Davis and S. Dorsano et al. (1992), *Cooling our Communities. A Guidebook on Tree Planting and Light-Colored Surfacing*, Washington, DC: Environmental Protection Agency.

Bellard, C., C. Bertelsmeier and P. Leadley et al. (2012), 'Impacts of climate change on the future of biodiversity', *Ecology Letters*, **15**(4), 365–77.

Bolund, P. and S. Hunhammar (1999), 'Ecosystem services in urban areas', *Ecological Economics*, **29**(2), 293–301.

Bowler, D.E., L. Buyung-Ali and T.M. Knight et al. (2010), 'Urban greening to cool towns and cities: a systematic review of the empirical evidence', *Landscape and Urban Planning*, **97**(3), 147–55.

Braat, L.C. and R.S. de Groot (2012), 'The ecosystem services agenda: bridging the worlds of natural science and economics, conservation and development, and public and private policy', *Ecosystem Services*, **1**(1), 4–15.

Burch, S., P. Berry and M. Sanders (2014), 'Embedding climate change adaptation in biodiversity conservation: a case study of England', *Environmental Science & Policy*, **37**, 79–90.

Cao, X., A. Onishi and J. Chen et al. (2010), 'Quantifying the cool island intensity of urban parks using ASTER and IKONOS data', *Landscape and Urban Planning*, **96**(4), 224–31.

Chang, C.R., M.H. Li and S.D. Chang (2007), 'A preliminary study on the local cool-island intensity of Taipei city parks', *Landscape and Urban Planning*, **80**(4), 386–95.

Chen, I.C., J.K. Hill and R. Ohlemuller et al. (2013), 'Rapid range shifts of species associated with high levels of climate warming', *Science*, **333**(6045), 1024–6.

Cook-Patton, S. and T. Bauerle (2012), 'Potential benefits of plant diversity on vegetated roofs: a literature review', *Journal of Environmental Management*, **106**, 85–92.

Demuzere, M., K. Orru and O. Heidrich et al. (2014), 'Mitigating and adapting to climate change: multi-functional and multi-scale assessment of green urban infrastructure', *Journal of Environmental Management*, **146**, 107–15.
Erwin, K.L. (2009), 'Wetlands and global climate change: the role of wetland restoration in a changing world', *Wetlands Ecology and Management*, **17**(1), 71–84.
European Commission (EC) (2013a), *Guidance on Integrating Climate Change and Biodiversity into Strategic Environmental Assessment*, Luxembourg: Publications Office of the European Union, accessed 29 January 2016 at http://ec.europa.eu/environment/eia/pdf/SEA%20Guidance.pdf.
European Commission (EC) (2013b), *Guidance on Integrating Climate Change and Biodiversity into Environmental Impact Assessment*, Luxembourg: Publications Office of the European Union, accessed 29 January 2016 at http://ec.europa.eu/environment/eia/pdf/EIA%20Guidance.pdf.
European Commission (EC) (2015), *Towards an EU Research and Innovation Policy Agenda for Nature-Based Solutions & Re-Naturing Cities. Final Report of the Horizon 2020 Expert Group on 'Nature-Based Solutions and Re-Naturing Cities'*, Luxembourg: Publications Office of the European Union, accessed 28 January 2016 at http://ec.europa.eu/programmes/horizon2020/en/news/towards-eu-research-and-innovation-policy-agenda-nature-based-solutions-re-naturing-cities.
European Environment Agency (EEA) (2012), *Urban Adaptation to Climate Change in Europe. Challenges and Opportunities for Cities Together with Supportive National and European Policies*, EEA Report No. 2/2012, Luxembourg: Office for Official Publications of the European Union, accessed 28 January 2016 at http://www.eea.europa.eu/publications/urban-adaptation-to-climate-change.
European Environment Agency (EEA) (2014), *Spatial Analysis of Green Infrastructure in Europe*, EEA Technical Report No. 2/2014, Luxembourg: Publications Office of the European Union, accessed 28 January 2016 at http://www.eea.europa.eu/publications/spatial-analysis-of-green-infrastructure.
European Union (EU) (1992), *Council Directive 92/75/EEC of 22 September 1992 on the indication by labelling and standard product information of the consumption of energy and other resources by household appliances*, accessed 28 January 2016 at http://eur-lex.europa.eu/legal-content/EN/TXT/?uri=CELEX%3A31992L0075.
Food and Agriculture Organization of the United Nations (FAO) (1998), 'Crop evapotranspiration – guidelines for computing crop water requirements', *FAO Irrigation and Drainage Paper No. 56*, Rome: FAO.
Geneletti, D. (2011), 'Reasons and options for integrating ecosystem services in strategic environmental assessment of spatial planning', *International Journal of Biodiversity Science, Ecosystem Services & Management*, **7**(3), 143–9.
Geneletti, D. (2013a), 'Ecosystem services in environmental impact assessment and strategic environmental assessment', *Environmental Impact Assessment Review*, **40**, 1–2.
Geneletti, D. (2013b), 'Assessing the impact of alternative land-use zoning policies on future ecosystem services', *Environmental Impact Assessment Review*, **40**, 25–35.
Geneletti, D. (2014), 'Integration of impact assessment types improves consideration of alternatives', *Impact Assessment and Project Appraisal*, **32**(1), 17–18.
Geneletti, D. (2015a), 'Research in strategic environmental assessment needs to better address analytical methods', *Journal of Environmental Assessment, Policy and Management*, **17**(1), 1550014.
Geneletti, D. (2015b), 'A conceptual approach to promote the integration of ecosystem services in strategic environmental assessment', *Journal of Environmental Assessment Policy and Management*, **17**(4), 15550035.
Geneletti, D. and L. Zardo (2016), 'Ecosystem-based adaptation in cities: an analysis of European climate adaptation plans', *Land Use Policy*, **50**, 38–47.
Gomez-Baggethun, E. and D.N. Barton (2013), 'Classifying and valuing ecosystem services for urban planning', *Ecological Economics*, **86**, 235–45.
Grimm, N.B., M.D. Staudinger and A. Staudt et al. (2013), 'Climate-change impacts

on ecological systems: introduction to a US assessment', *Frontiers in Ecology and the Environment*, **11**(9), 456–64.
Heller, N.E. and E.S. Zavaleta (2009), 'Biodiversity management in the face of climate change: a review of 22 years of recommendations', *Biological Conservation*, **142**(1), 14–32.
Holsten, A. and J.P. Kropp (2012), 'An integrated and transferable climate change vulnerability assessment for regional application', *Natural Hazards*, **64**(3), 1977–99.
Hsieh, H.F. and S.E. Shannon (2005), 'Three approaches to qualitative content analysis', *Qualitative Health Research*, **15**(9), 1277–88.
Intergovernmental Panel on Climate Change (IPCC) (2007), *Climate Change 2007: Synthesis Report. Contribution of Working Groups I, II and III to the Fourth Assessment Report of the Intergovernmental Panel on Climate Change*, Geneva: IPCC.
International Council for Local Environmental Initiatives (ICLEI) (2010), *Changing Climate, Changing Communities: Guide and Workbook for Municipal Climate Adaptation*, Toronto: ICLEI.
Jones, H.P., D.G. Hole and E.S. Zavaleta (2012), 'Harnessing nature to help people adapt to climate change', *Nature Climate Change*, **2**(7), 504–9.
Kabisch, N. and D. Haase (2014), 'Green justice or just green? Provision of urban green spaces in Berlin, Germany', *Landscape and Urban Planning*, **122**, 129–39.
Kareiva, P., H. Tallis and T.H. Ricketts et al. (2011), *Natural Capital: Theory and Practice of Mapping Ecosystem Services*, Oxford, UK: Oxford University Press.
Kazmierczak, A. (2012), *Heat and Social Vulnerability in Greater Manchester: A Risk–Response Case Study*, Manchester, UK: EcoCities project, University of Manchester.
Kjelgren, R.K. and J.R. Clark (1992), 'Microclimates and tree growth in three urban spaces', *Journal of Environmental Horticulture*, **10**(3), 139–45.
Kleerekoper, L., M. van Esch and T.B. Salcedo (2012), 'How to make a city climate-proof, addressing the urban heat island effect', *Resources, Conservation and Recycling*, **64**, 30–38.
Lafortezza, R., C. Davies and G. Sanesi et al. (2013), 'Green infrastructure as a tool to support spatial planning in European urban regions', *iForest – Biogeosciences & Forestry*, **6**(3), 102–8.
Lankao, P.R. and H. Qin (2011), 'Conceptualizing urban vulnerability to global climate and environmental change', *Current Opinion in Environmental Sustainability*, **3**(3), 142–9.
Larondelle, N. and D. Haase (2013), 'Urban ecosystem services assessment along a rural–urban gradient: a cross-analysis of European cities', *Ecological Indicators*, **29**, 179–90.
Larsen, S.V., L. Kørnøv and P. Driscoll (2013), 'Avoiding climate change uncertainties in strategic environmental assessment', *Environmental Impact Assessment Review*, **43**, 144–50.
McKenzie, E., S. Posner and P. Tillmann et al. (2014), 'Understanding the use of ecosystem service knowledge in decision making: lessons from international experiences of spatial planning', *Environment and Planning C: Government and Policy*, **32**(2), 320–40.
McPhearson, E.G., D. Nowak and G. Heisler et al. (1997), 'Quantifying urban forest structure, function, and value: the Chicago Urban Forest Climate Project', *Urban Ecosystems*, **1**(1), 49–61.
Munang, R., I. Thiaw and K. Alverson et al. (2013a), 'Using ecosystem-based adaptation actions to tackle food insecurity', *Environment: Science and Policy for Sustainable Development*, **55**(1), 29–35.
Munang, R., I. Thiaw and K. Alverson et al. (2013b), 'The role of ecosystem services in climate change adaptation and disaster risk reduction', *Current Opinion in Environmental Sustainability*, **5**(1), 47–52.
Munang, R., I. Thiaw and K. Alverson et al. (2013c), 'Climate change and ecosystem-based adaptation: a new pragmatic approach to buffering climate change impacts', *Current Opinion in Environmental Sustainability*, **5**(1), 67–71.
Naumann, S., G. Anzaldua and P. Berry et al. (2011), *Assessment of the Potential of Ecosystem-based Approaches to Climate Change Adaptation and Mitigation in Europe. Final Report to the European Commission*, Oxford, UK: Ecologic Institute and Environmental Change Institute, Oxford University Centre for the Environment.

Nelson, E.J., P. Kareiva and M. Ruckelshaus et al. (2013), 'Climate change's impact on key ecosystem services and the human well-being they support in the US', *Frontiers in Ecology and the Environment*, **11**(9), 483–93.
Norton, B.A., A.M. Coutts and S.J. Livesley et al. (2015), 'Planning for cooler cities: a framework to prioritise green infrastructure to mitigate high temperatures in urban landscapes', *Landscape and Urban Planning*, **134**, 127–38.
Posas, P.J. (2011), 'Exploring climate change criteria for strategic environmental assessments', *Progress in Planning*, **75**(3), 109–54.
Potcher, O., P. Cohen and A. Bitan (2006), 'Climatic behaviour of various urban parks during hot and humid summer in the Mediterranean city of Tel Aviv, Israel', *International Journal of Climatology*, **26**(12), 1695–711.
Rosenzweig, C., W. Solecki and S.A. Hammer et al. (2010), 'Cities lead the way in climate-change action', *Nature*, **467**(7318), 909–11.
Santamouris, M. (2014), 'Cooling the cities – a review of reflective and green roof mitigation technologies to fight heat island and improve comfort in urban environments', *Solar Energy*, **103**, 682–703.
Schwarz, N., A. Bauer and D. Haase (2011), 'Assessing climate impacts of planning policies – an estimation for the urban region of Leipzig (Germany)', *Environmental Impact Assessment Review*, **31**(2), 97–111.
Secretariat of the Convention on Biological Diversity (2009), *Connecting Biodiversity and Climate Change Mitigation and Adaptation: Report of the Second Ad Hoc Technical Expert Group on Biodiversity and Climate Change*, Technical Series No. 41, Montreal: Secretariat of the Convention on Biological Diversity, accessed 29 January 2016 at http://www.unep.org/roe/Portals/139/documents/BiodiveConf/Main/STRA-CO_2009_10_CBDbiod%20_clim_chg_report_KEY_adv.pdf.
Shashua-Bar, L. and M.E. Hoffman (2000), 'Vegetation as a climatic component in the design of an urban street', *Energy and Buildings*, **31**(3), 221–35.
Souch, C.A. and C. Souch (1993), 'The effect of trees on summertime below canopy urban climates: a case study', Bloomington, IN, *Journal of Arboriculture*, **19**(5), 303–12.
Staudt, A., A.K. Leidner and J. Howard et al. (2013), 'The added complications of climate change: understanding and managing biodiversity and ecosystems', *Frontiers in Ecology and the Environment*, **11**(9), 494–501.
Stronbach, M.W. and D. Haase (2012), 'Above-ground carbon storage by urban trees in Leipzig, Germany: analysis of patterns in a European city', *Landscape and Urban Planning*, **104**(1), 95–104.
Taha, H., H. Akbari and A. Rosenfeld (1991), 'Heat island and oasis effects of vegetative canopies: micro-meteorological field-measurements', *Theoretical and Applied Climatology*, **44**(2), 123–38.
Tzoulas, K., K. Korpela and S. Venn et al. (2007), 'Promoting ecosystem and human health in urban areas using green infrastructure: a literature review', *Landscape and Urban Planning*, **81**(3), 167–78.
Vignola, R., B. Locatelli and C. Martinez et al. (2009), 'Ecosystem-based adaptation to climate change: what role for policy-makers, society and scientists?', *Mitigation and Adaptation Strategies for Global Change*, **14**(8), 691–6.
Watkins, R. (2002), 'The impact of the urban environment on the energy used for cooling buildings', PhD thesis, Department of Mechanical Engineering, Brunel University, UK.
Zardo, L., D. Geneletti and M. Pérez-Soba et al. (under review), 'An ecosystem-based model for assessing the cooling capacity of different types of Green Urban Infrastructures'.

19. Where are the best places for the next billion people? Think globally, plan regionally*

Richard T.T. Forman and Jianguo (Jingle) Wu

19.1 FOUNDATIONS

19.1.1 Global Population Trends and the Next Billion

The United Nations recently projected that the world population will increase to 9.6 billion in 2050 and 10.9 billion in 2100, with a continuing upward trend (Figure 19.1) (Gerland et al., 2014; but see Lutz et al., 2008). This new forecast overturns the widely accepted projection in the literature during the past 20 years that the global population will stabilize at about 9 billion by the end of this century (e.g., Lutz et al., 2008). The world population increased from 6 billion in 1999 to 7 billion in 2012,

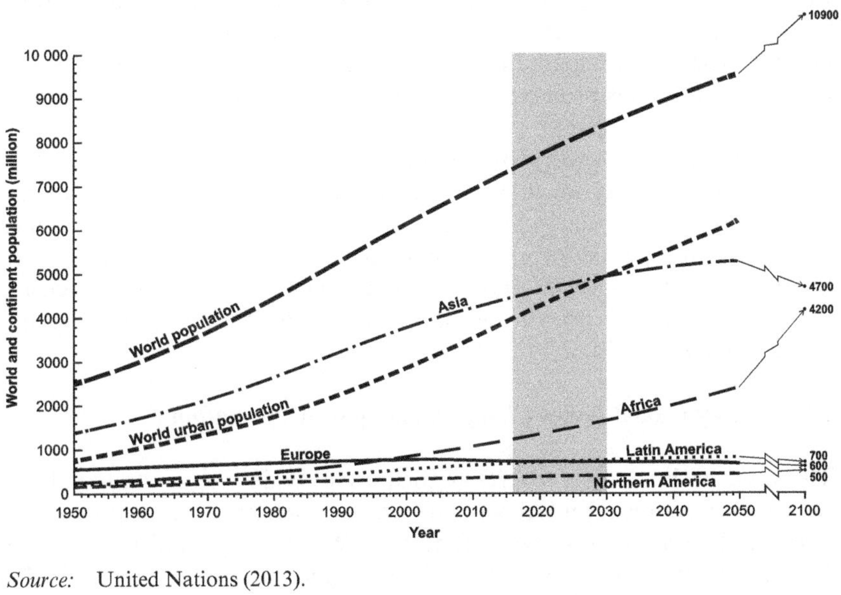

Source: United Nations (2013).

Figure 19.1 Population growth of world, urban areas, and continents

and is expected to reach 8.1 billion by 2025 (United Nations, 2013). Based on UN projections, there will be an increase of approximately 1 billion people in 14 years, for example, 1.02 billion from 2016 to 2030 (Figure 19.1).

Future population growth will be quite uneven geographically, with almost all future growth likely occurring in Africa and Asia (Figure 19.1). By 2050, the Asian population is expected to peak and then decline, whereas the African population will continue to soar. Excluding small nations (< 50 000 km^2), nearly all African and Middle Eastern nations currently have a high annual population growth (> 2 percent), and European and North American nations a low growth rate (< 1 percent) (World Bank, 2015). In contrast, many Latin American and Asian (including Oceania) nations have low growth rates, but medium and high growth rates are also present, producing a patchy pattern within these continents.

Future global trends are characterized by declining fertility and growth rates, increased aging populations (especially in developed nations), rapid urbanization, increasing transnational migration (from developing to developed countries), and accelerated environmental deterioration (air, water and soil pollution, plus land degradation) (Lutz et al., 2008; Harper, 2014). Almost all future population growth is expected to be urban (see Figure 19.1), though dispersed growth in rural areas may occur (Kotkin, 2010). In 2014, 54 percent of the world's population was urban, and 66 percent urban is anticipated in 2050 (United Nations, 2014). By 2050, the urbanization level is projected to reach 64 percent for Asia and 56 percent for Africa.

Imagine the entire world population urbanized by 2092 (Batty, 2011) with everyone living in an urban area. The rapid urbanization of the planet Earth, particularly in its dry and resource-poor regions, implies myriad environmental, economic, and social challenges to humankind in the coming decades. Cities must be a major focus of sustainability science and practice, if both people and nature are to thrive long term (Forman, 2008; Wu, 2013, 2014).

19.1.2 Key Global Issues Changing Along with Population

The area of land (and water) needed to provide energy and material resources consumed, and absorb the wastes discharged, for a population is referred to as the ecological footprint (Wackernagel and Rees, 1996). Human demands for ecosystem goods have apparently exceeded the Earth's regenerative capacity since the 1980s, overshooting the global biocapacity by 20 percent in 1999 (Wackernagel et al., 2002). Therefore, adding a billion people will certainly further degrade our finite planet. In effect, the best or most suitable places for major population growth are 'least bad'. To reduce the future environmental and human crises puts a premium on protecting and improving farmland, natural land and built areas.

Cities have been a major focus of footprint analysis because of their rapidly escalating environmental impacts near and far. For example, the ecological footprint of Vancouver in 1991 was estimated as 200 times the geographic area of the city (Rees, 1997). The footprint of the 29 cities in Baltic Europe was at least 565 to 1130 times the total area of the cities themselves (Folke et al., 1997). Large US cities also have an ecological footprint a few hundred times larger than their physical size (Luck et al., 2001). Shrinking the ecological footprint of a city population to the urban region makes sense as a policy goal. Meanwhile, localizing the area of resources consumed and wastes discharged near a city not only reduces transportation costs, but also enables people to see and live with their environmental effects.

The projected global population growth (see Figure 19.1) is associated with, and will catalyze, other major environmental, human and spatial changes. Consider shortages of clean freshwater, air and water pollution, food shortages, natural habitat loss and degradation, rapid outward urbanization, movements of people and climate change effects. Currently these are major problems in many of the world's cities and regions (Hardoy et al., 2001; McGranahan and Satterthwaite, 2003). Before identifying more suitable places for major population growth, we briefly highlight global freshwater availability, followed by urban region characteristics.

19.1.3 Freshwater Availability

Increasingly a global characteristic, water shortages are most severe in the world's vast drylands. These areas normally have low precipitation, high climatic variability, infertile soils and slow economic development. Drylands cover 41 percent of the ice-free land area of the Earth, and support > 38 percent of the global population (Millennium Ecosystem Assessment, 2005; Reynolds et al., 2007). About 90 percent of the dryland population resides in developing countries, with an average GDP per capita 30 percent lower than the global average. Making things worse, climate models predict that most drylands will become drier with more frequent extreme weather and climatic events (IPCC, 2013).

Water use has been increasing twice as fast as world population, and 'By 2025, 1.8 billion people will be living in countries or regions with absolute water scarcity, and two-thirds of the world population could live under water stress conditions' (UN, 2015). Providing acceptable-quality water for people long term, without devastating ecosystem functions and biodiversity, is a huge challenge. Thus freshwater supply is considered a primary constraint on where the best places are globally and regionally for the next billion people.

19.1.4 The Urban Region

Many big problems are especially visible to urban residents (Hardoy et al., 2001; McGranahan and Satterthwaite, 2003), including heat and flooding associated with widespread impervious surface; everything bathed in human-distributed chemicals and wastes; water systems massively disrupted, polluted and truncated; farmland altered and abandoned; and natural ecosystems extensively lost, degraded and fragmented. These correlate with population growth. Yet, without changing the amounts of these attributes, their negative effects can be significantly reduced by creating wise spatial arrangements around cities (Alberti and Marzluff, 2004; Forman, 2008, 2014; Ahern, 2013).

A functional urban region concept highlights the area of active two-way interactions between a city and its surroundings (Forman, 2008). Often with a radius of some 70–100 km or more, and unlike a city, an urban region is especially suitable for urban or land use planning to reduce the ecological footprint and promote sustainability.

Think globally, plan regionally, and then act locally. At the global scale, focus sustainability efforts on dryland regions, less-developed regions (especially Africa and Asia), and rapidly urbanizing regions. On local and regional scales of village, town, city and region, minimize the amount, and increase the efficiency, of resource use; better design and plan landscapes; and rediscover and invent more sustainable ways of living, working and playing.

We now sequentially highlight: (1) best places globally/continentally for the next billion people; (2) best places in urban regions for major population growth; and (3) strategies for improving farmland, natural land, and built areas.

19.2 BEST PLACES GLOBALLY FOR THE NEXT BILLION PEOPLE

For consistency in this global and continental analysis, we used information and maps for diverse variables from the *Atlas of Global Conservation* (Hoekstra et al., 2010). Highlighting the importance of environmental constraints and existing population density, we began by excluding water-stressed regions (similar to UNEP, 2008) as inappropriate for major population growth. Other arid areas (including deserts and desertified areas) were next excluded, followed by tundra-and-ice areas. Very high-density population areas not already marked out were then excluded. Similarly, centers of endemic species, high-density population areas, and boreal forest were sequentially excluded. Centers of endemic species highlight the importance of biodiversity (many other 'hotspots' of species exist). We have not added socio-economic and other variables such as availability of jobs, quality of government, conflict areas, or deforestation, since these often evolve with the environmental and population-density constraints and fluctuate at shorter time scales.

The best areas for major population growth on the global/continental scales are those remaining after excluding the preceding seven variables (Figure 19.2). Thus the most suitable areas for the next billion people are in well-watered warm regions: (1) large areas of South America; (2) areas of southern Canada and northern and eastern USA; (3) south-central Africa; (4) Asia north of the Himalayas and in a west-to-east strip from the Black Sea to North China; and (5) scattered areas in Oceania.

A second category of areas with somewhat less severe constraints may be considered slightly less suitable areas for growth. These are the high-density population and boreal forest areas (Figure 19.2): (1) central Canada; (2) West Africa; (3) wide west-to-east strip from Scandinavia to Kamchatka; (4) northeastern China; and (5) half of Southeast Asia.

A striking disconnect emerges. The most suitable areas globally (Figure 19.2), especially those characterized by relatively satisfactory freshwater availability, differ markedly from the areas (see Figure 19.1) projected to grow rapidly in population in the decades ahead. This seems to imply that transnational migration pressure from high-growth population to 'best place' high-resource areas (see Figures 19.1 and 19.2) will be evident in the decades ahead.

Migration of tens or hundreds of thousands (but normally not millions) of people has been driven by disease, religion, ideology, space, war and more, and may now be increasingly spurred by spreading water stress. Although migration between cities and regions within a nation is widespread, overwhelmingly people remain residents of their nation. Transnational migration, rather than promising a sustainability solution (Tanton, 1994), commonly poses significant problems for a source area, the travel route, and the recipient, usually urban, region.

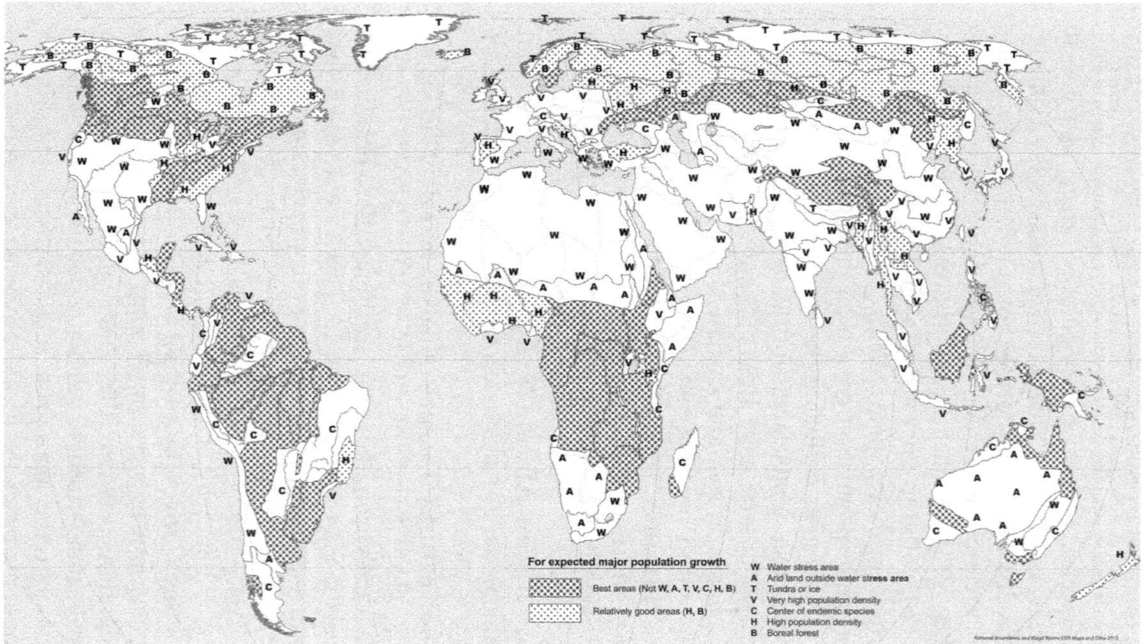

Note: Water stress indicates that freshwater withdrawal exceeds human requirements (UNEP, 2008); only extreme and high water-stress areas included. Very high-density population area = > 100 people/km^2; high-density population = 50–100/km^2. Center of endemic species = high concentration of bird, mammal, reptile and/or plant species unique to a region. See text.

Source: Based on Hoekstra et al. (2010); ESRI 2013 base map.

Figure 19.2 Proposed most suitable areas globally for major population growth based on environmental and population constraints

19.3 BEST PLACES IN URBAN REGIONS FOR THE NEXT BILLION PEOPLE

19.3.1 Spatial Options

Most cities began where good agricultural soil was near a water body and natural land. Thus over the years, typically outward urbanization has covered valuable soil, polluted local water bodies and eliminated the nearest natural lands.

Fifteen types of places in an urban region are identified where major population growth could be concentrated (Figure 19.3). For each, the effects on both total area and highest quality of (1) food-producing land, and of (2) natural land that would be eliminated and/or degraded are estimated. The results suggest that the best places (least damaging to farmland and/or natural land) for major population growth are (Figure 19.3): city center; ring immediately around city center; inner and outer suburbs; and within satellite cities in either farmland or natural land.

The worst places (most damaging) appear to be areas surrounding satellite cities, areas surrounding towns/villages and a newly built city/town, again in either farmland or natural land (Figure 19.3). Exurban sprawl areas and towns/villages in farmland emerge as intermediate in suitability. In farmland, any focused development, except within satellite cities and towns/villages, is likely to noticeably degrade food production capacity. In natural land, any major development outside satellite cities can be expected to seriously erode ecosystem values and services.

However, space available for major population growth must be considered. Overall, space availability for the best and intermediate suitability places (Figure 19.3) is:

City center	0	None
Ring around city center	X	Little
Inner suburbs	X	Little
Outer (almost all-built) suburbs	XX	Medium
Exurban sprawl area	XXX	Extensive
Satellite cities (in farmland or natural land)	XX	Medium
Towns/villages in farmland	XXX	Extensive

Thus, combining the Figure 19.3 results with space availability suggests that the best, or least problematic, places to focus population growth are: (1) outer suburbs; (2) exurban sprawl areas; (3) satellite cities; and (4) towns/villages in farmland. Importantly, this assumes urban region

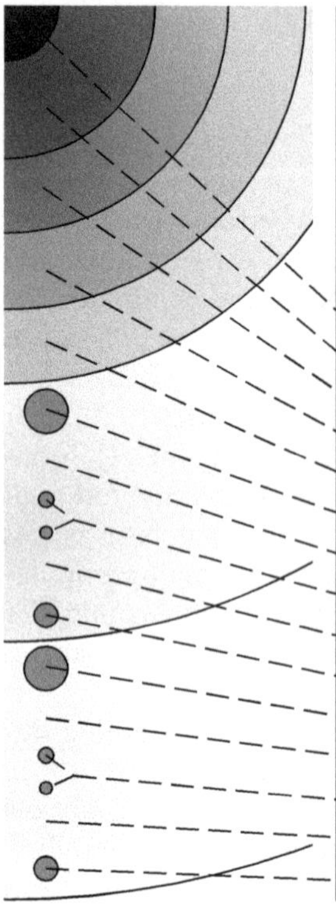

Place where major population growth might occur or be targeted (O = None; X = Little; XX = Medium; XXX = Extensive)		Amount of area eliminated and/or degraded: Food-producing land		Amount of area eliminated and/or degraded: Natural land	
		Total area	Highest quality	Total area	Highest quality
City center	(C)	O	O	O	O
Ring immediately around city center	(C)	O	O	O	O
Inner suburbs	(D)	O	O	O	O
Outer suburban area almost all-built	(D)	X	O	X	O
Exurban sprawl area	(D)	X	X	X	X
Satellite city in farmland	(C)	O	O	X	O
Area surrounding farmland satellite city	(D)	XX	XX	X	X
Towns/villages in farmland	(C)	X	X	X	O
Area surrounding farmland towns/villages	(D)	XXX	XXX	X	X
New city/town built in farmland	(C)	XX	XXX	X	X
Satellite city in natural land	(C)	O	O	X	O
Area surrounding natural-land satellite	(D)	X	X	XXX	XX
Towns/villages in natural land	(C)	O	O	X	XX
Area surrounding natural-land towns/villages	(D)	X	X	XXX	XX
New city/town built in natural land	(C)	O	O	XXX	XXX

Note: Farmland is mostly active and fallow cropland; natural land (forest, woodland, grassland, or desert) includes semi-natural land. Total area refers to food-producing or natural land in a city's urban region. Highest-quality food-producing land is mainly minimally polluted high-quality agricultural soil, or a large relatively intact agricultural area. Highest-quality natural lands are large natural areas (including large semi-natural patches in farmland), major water bodies, and vegetation strips alongside water bodies. The ring immediately around city center is often characterized by former or old industry, mixed use, scarcity of parkland and rapid change. We assume typical characteristics: villages, towns and cities developed where good agricultural soil and a surface water body are in proximity; farmland contains some surface water bodies and patches of semi-natural vegetation; and relatively contiguous natural land and surface water bodies are present in the outer portion of the urban region (Forman, 2008). C = concentrated in; D = dispersed across.

Figure 19.3 Possible places for major population growth in an urban region, with estimated effects on food-producing land and natural land

planning, and that growth locations fit effectively within the plan outlined. Below we look more closely at key issues to consider in the rapidly changing areas beyond the inner suburbs. Rapid change represents opportunity, a time to improve trajectories.

BOX 19.1 URBANIZATION/SETTLEMENT AROUND AFRICAN CITIES AND A RURAL AREA

Landscape transformation processes over 23 years (1986–2009) were studied in and around four cities of four nations in dry Sub-Saharan West Africa (Brinkmann et al., 2012). The 2009 populations of Kano, Niamey, Sikasso and Bobo-Dioulasso varied from 160 000 to 2.3 million, and all had intensive urban production of vegetables. During the period, woodland and shrubland were eliminated, cropland > 10 km from city edge expanded (associated with irrigation in two cases), and then cropland changed to barren land and was abandoned. Also, built area urbanization spread, but the population grew much faster, creating rapid densification. Urbanization expanded from the edge of cities, but also increased in small towns within 25 km of the city center. In two cases urbanization was concentrated by water and roads, while in two cases it was further away from them.

In Kenya, major settlement and progressive population growth occurred in a remote area by a national park with large wildlife populations (Muriuki et al., 2011). The villages began next to woodlands. Small woodland patches were removed first for firewood and home uses, so patch number rapidly decreased and average size increased. Larger wooded patches then became more open and shrank in size during charcoal production. Small, cultivated patches continued to increase in number (to approx. 300). They then coalesced into five large subsistence farming areas subject to wind/water erosion. In addition to farmland, wildlife, biodiversity and natural systems were severely degraded by this changing land mosaic spatial pattern associated with population growth.

19.3.2 Primary Outward Urbanization Patterns

Four basic models describe outward urbanization worldwide: bulges (from a city border), satellite cities growth, dispersed sites sprawl, and transportation corridor (strip/ribbon) development (Forman, 2008, 2014). The first two emerge as better environmentally, the last two worse. Each city or region portrays its own combination of the patterns. The greatest population growth is projected for Africa (see Figure 19.1), yet today's urbanization patterns there seem to seriously degrade both farmland and natural land (Box 19.1).

Elsewhere, bulge expansion, which only adds groups of buildings next to existing buildings, minimizes degradation of food-producing and natural lands, and may enhance communities. Outside Rome over 40 years (1960–2000), cropland, vineyard/orchard, and woodland each dropped by about a third, as low-density housing tripled in area (Salvati et al., 2012). Yet, in addition to sprawl, towns tripled in building density. Outside Dongguan (South China) over 18 years (1988–2006), initial urbanization was both expansion from the city edge and dispersed 'leapfrog' development (Liu et al., 2010).

Over time, leapfrog growth sharply dropped, infilling rapidly increased and edge expansion remained high throughout. Around diverse USA cities long-term settlement patterns seemed to occur in four patterns (Clark et al., 2009): clustered growth adjoining a city; clustered growth separated from the city; dispersed growth; and linear growth. Commonly a city showed two of the patterns. Finally, recent studies suggest that cities worldwide are becoming increasingly complex in their landscape structure with increasing diversity of land use types, edge density (total boundary length between land uses) and patchwork geometry (Jenerette and Potere, 2010; Wu, 2014).

Spatial urbanization patterns affect sustainability on the scale of cities and urban regions (Borrego et al., 2006; Jabareen, 2006; Bereitschaft and Debbage, 2013). Cities with compact forms, high population densities, interstitial green spaces, and houses with small yards are likely to have low per-capita ecological footprints (Jabareen, 2006; Sushinsky et al., 2013). At the regional scale, protecting farmland and natural areas, and maximizing landscape connectivity of green and blue spaces, minimizes the overall environmental impacts of urbanization (Forman, 2008; Lin and Fuller, 2013; Wu, 2014).

Now we turn to practical solutions for improving food-producing land, natural land and communities. Rather than mainly global prescriptions and policies, such as more crop genetics, harvesting marine plankton, doubling of protected land or building only multi-unit housing, we highlight a more place-based focus, where existing and future people will live and work.

19.4 IMPROVING FARMLAND

High-quality agricultural soils, normally with ample water, mineral nutrients and organic matter, are prime for food production, especially in large contiguous areas and with minimal pollutants added. Protection of these areas from urbanization, fragmentation and industrial pollution, plus improvement, remains a priority to feed the rapidly growing world population.

19.4.1 In Urban Regions

An array of urban region solutions will improve food-producing land despite a new billion people:

Prevent or minimize development on the best soils, thus providing food and stability for the future. Use these best soils relatively near cities mainly for vegetable and fruit production, supplemented by other crops, but not for livestock production. Clean water, biodiversity and recreational benefits are added by strategically locating bits of nature across farmland. Flooding is reduced with natural vegetation covering small headwater streams and intermittent channels. Outside Barcelona, avian diversity, especially rare species, is negatively related to both forest area (mostly fragmented) and built area, but positively correlated with cropland and with shrubland-grassland (Santos et al., 2008). Mosaic areas of woodland, cropland, and built area are especially rich in species.

Establish with a strong legal framework at least two large 'market gardening' (truck farming) areas of good soil, preferably on a floodplain or with irrigation (Figure 19.4). These strategic highly productive areas of mainly perishable vegetables and fruits readily transported at minimal cost, provide fresh produce to city markets and restaurants. With two or more market gardening areas providing flexibility and stability, they also help provide relatively clean moist cool air for the city. Urban social benefits are also evident, including farming families and employees providing diversity, urban community groups growing food, and food provided for needy low-income city residents.

To avoid contamination of these best soils, and hence the foods produced, nearby and upwind polluting industries, incinerators, and power plants are prohibited or moved. Analogously, to sustain quality soils long term, agricultural chemical use is limited. Modern industrial centers are established elsewhere with economically efficient power, water, transportation resources and waste disposal, where older and poorly located industries can relocate, and new industries can invest. Often this process helps clean up local water bodies. Also, research and pilot projects using urban wastewater aquaculture to produce clean food, consistent with public health and clean water constraints, warrant acceleration.

At a finer scale, urban agriculture within a city includes growing food in window boxes/balconies, home vegetable gardens, community gardens, rooftops and so forth. Impressive results, especially for vegetables in many African cities, are documented worldwide (Forman, 2014). Results can be quickly achieved as, when the Soviet Union terminated subsidies, Cuban urban agricultural production skyrocketed.

Street trees, parks and other public spaces can also produce a range of usable products such as fruits, nuts and fiber (Clark and Nicholas, 2013). Thus urban region food production, urban agriculture and urban food forestry are a key to urban sustainability or resilience. Not only securing food supply, they also decrease energy use and greenhouse gas emission, reduce the global spread of infectious diseases, enhance physical and mental health (Reeves, 2014), and help maintain the social-environmental identity and diversity in cities.

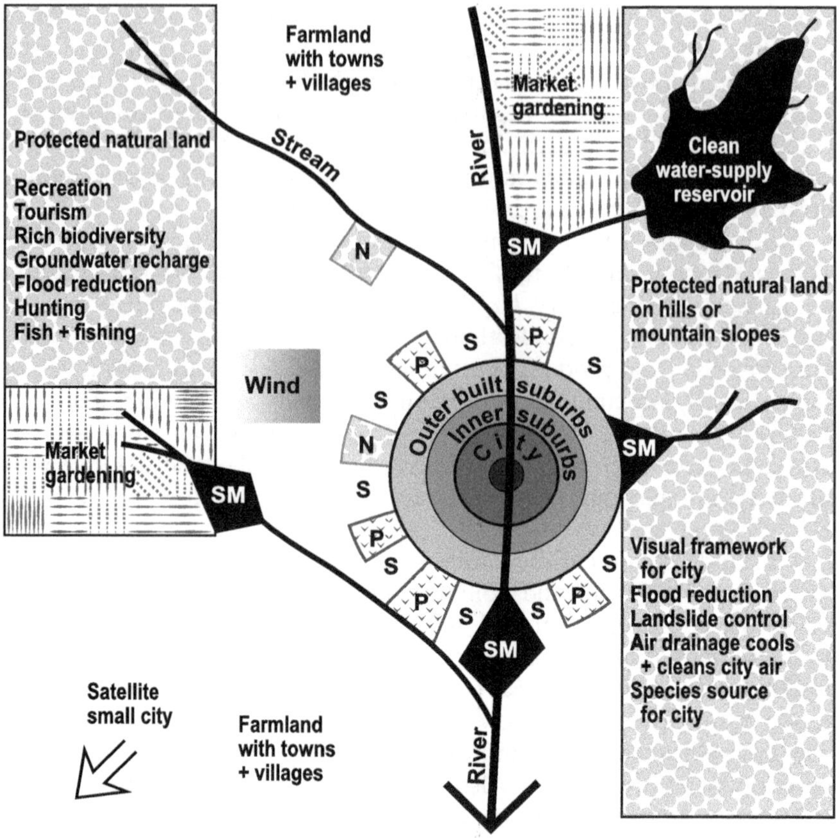

Note: An urban area (in moist climate, typically ringed by farmland, with woodland beyond) includes: city, with city center and ring immediately around it; suburbs, both inner and outer, and almost all-built; and sprawl area, with low population density and scattered developments. The highest-priority natural systems areas are highlighted; see text for their primary values and uses. P = park, which includes intensive recreational use; N = patch of semi-natural vegetation; S = sprawl area; SM = stormwater mitigation area.

Figure 19.4 Diagrammatic model to sustain ecosystem services for cities

19.4.2 Agricultural Land Worldwide

With noticeably more people, plus existing food shortages, greatly increasing the number of major food species and varieties makes good sense. This builds on the familiar ecological principle of each species using a somewhat different set of environmental resources. Drought-tolerant food species for expanding drylands are especially needed.

Many productive annual grains are potentially replaced by perennial grains, which greatly decreases soil degradation and erosion and minimizes the use of machinery and fossil fuel (Jackson, 1980; Cox et al., 2006; Pimentel et al., 2012). For instance, a mosaic with large areas of perennials and small areas of annuals may be effective in many semiarid and mesic regions.

Some 30 percent of the non-ice-covered land is used for livestock production. Consistent with the second law of thermodynamics, it takes a lot of plant food (e.g., 10 kg) to make a kilogram of meat. Noticeably reducing livestock production, or reducing the amount of animal products in future diets, would quickly free up extensive areas of land for other uses. Crops on better pasture soils would produce considerable plant food for the next billion people. Natural systems rapidly covering less suitable pasture soils, including many dry areas, would significantly reduce erosion, improve groundwater and surface water bodies, and improve wildlife habitat. Furthermore, the extensive area of crops currently grown for animal feed would drop. Also, crop areas planted for transportation fuel seem more appropriate for food production (Dirks et al., 2012). In moist areas, portions of today's cropland might be converted to fruit production.

Adding these solutions together, even some of them or a small number of all, would creatively provide ample food-producing land for the billion new people.

19.5 IMPROVING NATURAL LAND

Overall the most valuable natural areas are considered to be: (1) large contiguous natural and semi-natural areas; (2) major water bodies; and (3) vegetation strips alongside water bodies (Forman and Sperling, 2011). Large vegetation patches in farmland and in urban land are appropriately included in (1), though key functions differ somewhat. An 'emerald network' of large green areas connected by corridors or 'stepping stones' is a particularly valuable integrative pattern for sustaining nature. This provides multiple ecosystem functions, such as sustaining biodiversity, providing habitat connectivity, protecting water bodies and providing flexibility for the future (Forman, 1995, 2008; Lovell and Taylor, 2013).

19.5.1 In Urban Regions

Large natural or semi-natural land (especially woodland) adjoining or near an urban area provides a huge array of environmental, human and economic benefits (see Figure 19.4) (Forman, 2014), including: (1) reservoir protection and clean water supply; (2) flood reduction; (3) nature-based recreation opportunities; and (4) biodiversity protection and enrichment for the urban area. On vegetated hillslopes/mountain slopes by the built area, natural land provides major additional benefits: (a) erosion, mudslide, sedimentation and flood protection; (b) cool air drainage at night that both cools and cleans the urban air, especially important with climate warming; and (c) no damage to human structures from earthquakes and fire sweeping upslope. Large natural land upwind supplies clean moist cool air to the urban area. Two or more large protected woodlands provide flexibility and stability. Protected natural-ecosystem patches (see 'N' in Figure 19.4) between a large protected area and the urban area serve as key stepping stones for species movement.

A ring of relatively large parks (see Figure 19.4), each generally containing considerable semi-natural areas, immediately around a city or metro area also provides many societal values. Intensive recreation opportunities are close to most residential areas, and neighborhood groups help care for the parks when government maintenance budgets are down. The ring also provides stepping stones for species movement around and into the city.

Greenspaces within a city, some with semi-natural vegetation, help cool summer air, reduce flooding, sustain high biodiversity and provide recreational opportunity. Large-to-small parks and other greenspaces can be spatially arranged to achieve these values over the entire built area (Forman, 2014). In short, economic and social benefits of greenspaces in urban areas are manifold and increasingly recognized (Lovell and Taylor, 2013; Forman, 2014; Wolch et al., 2014; Wu, 2014).

19.5.2 Natural Lands Worldwide

Typically the most effective way to sustain a clean water source for people is to surround it with continuous natural vegetation. Vegetation covering the drainage basin of a reservoir minimizes surrounding soil erosion, and thus sedimentation and loss of reservoir capacity. Minimizing agricultural, industrial and other pollutants, and therefore water treatment costs, provides economic value. With a concentration of microhabitats around water sources, such protected natural land is particularly effective in sustaining biodiversity.

Stormwater and atmospheric pollutants from transportation, industry, agriculture and built areas can be effectively cleaned, especially by filtering through most natural soils and wetlands. In dry climates, considerable water often exists around built areas, but it is largely polluted and needs cleaning for diverse human uses and ecosystem functions.

Transportation's dense road networks subdivide nature's large populations of animals and plants into small populations (Forman et al., 2003; Laurance and Balmford, 2013; Perz, 2014). But small populations commonly have both demographic and genetic problems, which increase the probability of local species extinction. Perforating roads with wildlife crossings (especially when upgrading bridges and culverts), and constructing strategically located wildlife underpasses/overpasses, helps reconnect the land for critical species movement. Many approaches can reduce traffic noise that inhibits sensitive animals in wide areas near busy roads. Near built areas, such solutions can also increase scarce interconnected trail networks for relaxed walking and for hikers.

Looking further ahead, a 'netway-with-pods' system would transform much of today's extensive transportation imprint back to natural ecosystems and other uses for society, and reconnect all surrounding land (Forman and Sperling, 2011). Further benefits of the netway system include increased safety and efficiency, no fossil fuel use, no greenhouse emissions, and increased market gardening and trail networks near communities.

19.6 IMPROVING BUILT COMMUNITIES

19.6.1 Sprawl and Outer Suburbs

Distinct diverse vibrant town and neighborhood centers are the crown jewels here. Mixed-use patterns at both sprawl/suburb and residential area scales provide walkable/bikeable employment opportunities near homes. An efficient transportation system has industrial and commercial centers connected to the regional highway system; primary routes between commercial and middle-income residential areas; secondary routes connecting low-income residential with commercial and industrial areas; and tertiary routes connecting middle-income residential with industrial center and low-income residential areas.

Major population growth might be targeted to degraded natural and farmland spaces (e.g., abandoned fields), or to communities losing population. Or concentrate growth near industrial areas, highway intersections, or railway facilities. Thus jobs and housing are close together. Many informal squatter settlements across urban areas grow in such locations. With inadequate infrastructure investment, serious environmental problems are characteristic. Yet these residential areas also seem to have disconnected impervious surfaces, plus considerable bare soil, native and non-native woody plants, spontaneous vegetation, and somewhat rich biodiversity (Muller et al., 2010; Biggs et al., 2010; Forman, 2014).

Alternatively, concentrate population growth in low-density residential areas with attached or detached single-unit houses on large, medium or even small plots. Densify near existing and proposed wastewater sewer lines, near public transport routes and near convenient shopping areas. Use novel approaches. Change property lines so houses are built between houses, reduce building setback regulations, greatly increase floor area ratios, convert large houses into apartments/condominiums, grow buildings on lawns and yards, create narrow wildlife corridors through blocks, and so forth. But for each housing unit, retain a tiny green private space for children's use, vegetable growing, aesthetics, and physical and mental health (Kellert et al., 2008; Reeves, 2014). Do single-family housing densities of 5, or 25, or 40 units per hectare (Davies et al., 2009; Cilliers et al., 2012) provide an appropriate minimum size for private spaces?

19.6.2 City and Inner Suburb

Major population growth in densely populated areas seems counterintuitive if overall quality of life and links with nature are important. Existing high-rise, low-rise and courtyard/patio residential areas are commonly short on green spaces, local employment, local shops, and public transport, while suffering from excess urban heat, air pollution, traffic and noise. Urban infill commonly exacerbates the problem (Forman, 2014).

Is the urban status quo improvable near term to permit suitable living for major population growth? For instance, redesign wide, dangerous, noisy traffic-jammed roads to create quiet, vegetation-embroidered neighborhood spaces, easily connected by walking/biking, plus taxi-like vehicles and service vehicles. Create cultural, shopping, civic meeting places, as valued centers of distinctive communities for residents in the urban sea of buildings and roads. Shrink our home ranges, with jobs, residences, parks, schools and so forth arranged closer together, thus reducing transportation time and cost. Increase public transportation of varied modes, with walking/biking-only routes alternating with personal vehicle and public transportation routes. Test more visionary approaches.

Across the entire urban area, create a tree-anchored 'green net', tying together an abundance of parks, greenspaces, neighborhood centers, playgrounds, ballfields, urban agriculture, flower gardens and planters (Rosenzweig, 2003; Forman, 2014). Locate trees to clean the air, and also trees to cool the air. Emphasize plants with low water usage and high bird/pollinator usage. Enhance urban neighborhoods and a sense of place for residents who readily care for the green net (Ndubisi, 2014).

To decrease urban heat, impervious surfaces are reduced, whitened or covered with plants. To decrease flooding and polluted local water bodies, impervious surface areas are fragmented into small sections and disconnected from water bodies. Stormwater runoff is readily funneled to detention and retention basins/ponds/wetlands, as well as diverse greenspaces. Stormwater mitigation for streams leaving an agricultural area, plus entering and leaving a built area, is strategically valuable (see Figure 19.4). So, reduce, fragment and disconnect impervious surfaces, while increasing and connecting greenspaces and waterways.

Yet, before densifying city and inner suburb areas, much more needs improvement – water supply, human wastewater, food systems, solid waste, pests and disease vectors. Furthermore, big disturbances happen, often becoming 'disasters' in built areas. Solutions providing both resistance and resilience, and largely spatial, are reasonably well known (Forman, 2008; Ahern, 2013; Wu and Wu, 2013), but alas, rarely implemented in advance.

To mold the land so nature and people both thrive long term (Forman, 2008) future sustainability will require better design and planning of our land – backyard, neighborhood, city and urban region. Before adding a billion people, what actions would provide multiple values for farmland, natural land and built communities, and noticeably improve each region of the globe (Table 19.1)?

19.7 CONCLUSION

Based on environmental constraints (especially water stress; and including key biodiversity areas) plus existing population density, the best places for major human population growth at the global scale seem to be large areas of South America; across central and eastern Northern America; scattered areas of Oceania; Asia north of the Himalayas and in a central east–west strip; and south-central Africa. This pattern sharply contrasts with population growth projected to overwhelmingly occur in Asia and Africa. Global/regional scale migration from high growth rate areas toward more suitable freshwater and other resource areas is likely to continue, and increase.

At the urban region scale, based on expected degradation or loss of food-producing land and natural land, today's outer suburbs, exurban sprawl areas, satellite cities, and towns/villages in farmland seem best suited for adding the next billion people. This conclusion also differs from projected population growth mainly in cities.

Global-scale land planning and human migration regulation currently seem impractical for this enormous population growth challenge. Thus urban region-scale planning and implementable solutions seem urgent.

Table 19.1 Encapsulated actions to improve lands during major population growth

Farmland	Natural Land	Built Communities
Protect best soils and enrich others	*Protect/expand* the integrity of large green areas	*Protect/enhance* water supply with vegetation
Establish market gardening areas by all cities	*Remove* remote roads and buildings degrading nature	*Disconnect* impervious surface areas
Accelerate urban agriculture	*Connect* large green areas with corridors and/or stepping stones	*Shrink* human home ranges
Close/relocate industries polluting key farmland	*Perforate* roads to facilitate wildlife crossing/ movement	*Transform* busy noisy roads to quiet neighborhood spaces
Reduce livestock and meat to gain land for edible crops	*Expand* vegetation protecting water bodies, and along headwaters and intermittent channels	*Arrange* urban greenspaces to cool air, reduce flooding, support biodiversity and enhance recreation
Reduce cropland used for fuel and animal feed	*Increase* bits of nature across farmland and built areas	*Convert* sprawl areas to denser mixed-use areas with vibrant neighborhoods
Use more food species and varieties		*Grow* a tree-anchored green net
Replace annual grains with perennial grains		

Unplanned urbanization, notably its sequential spatial arrangement, has major impacts on food-producing areas, natural land and built areas. A wide range of actions to improve, or minimize negative effects on, areas for food, nature and community are outlined. No single overarching solution emerges. Yet many individual solutions address a number of problems. Therefore, a package of specific solutions, most strongly spatial and each addressing two or more issues, seems promising to mold a better land ahead. With the human ecological footprint apparently exceeding the Earth's biocapacity, unless the land worldwide is effectively improved, another billion people is a bad omen.

NOTE

* The authors warmly thank Taco I. Matthews for excellent illustrations, and Andrew F. Bennett, Michael W. Binford, Gilbert Metcalf, and Lawrence Buell for valuable insights during the preparation of this chapter.

REFERENCES

Ahern, J. (2013), 'Urban landscape sustainability and resilience: the promise and challenges of integrating ecology with urban planning and design', *Landscape Ecology*, **28**(6), 1203–12.

Alberti, M. and J.M. Marzluff (2004), 'Ecological resilience in urban ecosystems: linking urban patterns to human and ecological functions', *Urban Ecosystems*, **7**(3), 241–65.

Batty, M. (2011), 'When all the world's a city', *Environment and Planning A*, **43**(4), 765–72.

Bereitschaft, B. and K. Debbage (2013), 'Urban form, air pollution, and CO_2 emissions in large U.S. metropolitan areas', *The Professional Geographer*, **65**(4), 612–35.

Biggs, T.W., E. Atkinson and R. Powell et al. (2010), 'Land cover following rapid urbanization on the US–Mexico border: implications for conceptual models of urban watershed processes', *Landscape and Urban Planning*, **96**(2), 78–87.

Borrego, C., H. Martins and O. Tchepel et al. (2006), 'How urban structure can affect city sustainability from an air quality perspective', *Environmental Modeling & Software*, **21**(4), 461–7.

Brinkmann, K., J. Schumacher and A. Dittrich et al. (2012), 'Analysis of landscape transformation processes in and around four West African cities over the last 50 years', *Landscape and Urban Planning*, **105**, 94–105.

Cilliers, S., S. Siebert and E. Davoren et al. (2012), 'Social aspects of urban ecology in developing countries, with an emphasis on urban domestic gardens', in M. Richter and U. Weiland (eds), *Applied Urban Ecology: A Global Framework*, Chichester, UK: John Wiley, pp. 123–38.

Clark, K. and K. Nicholas (2013), 'Introducing urban food forestry: a multifunctional approach to increase food security and provide ecosystem services', *Landscape Ecology*, **28**(9), 1649–69.

Clark, J.K., R. McChesney and D.K. Munroe et al. (2009), 'Spatial characteristics of exurban settlement pattern in the United States', *Landscape and Urban Planning*, **90**(3–4), 178–88.

Cox, T.S., J.D. Glover and D.L. van Tassel et al. (2006), 'Prospects for developing perennial-grain crops', *BioScience*, **56**(8), 649–59.

Davies, Z.G., R.A. Fuller and A. Loram et al. (2009), 'A national scale inventory of resource provision for biodiversity within domestic gardens', *Biological Conservation*, **142**(4), 761–71.

Dirks, L.C., G.W. Dirks and J.G. Wu (2012), 'Evolving perspectives on biofuels in the United States', *Frontiers in Energy*, **6**(4), 379–93.

Folke, C., A. Jansson and J. Larsson et al. (1997), 'Ecosystem appropriation by cities', *Ambio*, **26**(3), 167–72.

Forman, R.T.T. (1995), *Land Mosaics: Ecology of Landscapes and Regions*, New York: Cambridge University Press.

Forman, R.T.T. (2008), *Urban Regions: Ecology and Planning Beyond the City*, New York: Cambridge University Press.

Forman, R.T.T. (2014), *Urban Ecology: Science of Cities*, New York: Cambridge University Press.

Forman, R.T.T. and D. Sperling (2011), 'The future of roads: no driving, no emissions, nature reconnected', *Solutions*, **2**(5), 10–23.

Forman, R.T.T., D. Sperling and J.A. Bissonette et al. (2003), *Road Ecology: Science and Solutions*, Washington, DC: Island Press.

Gerland, P., A.E. Raftery and H. Sevcikova et al. (2014), 'World population stabilization unlikely this century', *Science*, **346**(6206), 234–7.

Hardoy, J.E., D. Mitlin and D. Satterthwaite (2001), *Environmental Problems in an Urbanizing World: Finding Solutions for Cities in Africa, Asia and Latin America*, London: Earthscan.

Harper, S. (2014), 'Economic and social implications of aging societies', *Science*, **346**(6209), 587–91.

Hoekstra, J.M., J.M. Molnar and M. Jennings et al. (2010), *The Atlas of Global Conservation: Changes, Challenges, and Opportunities to Make a Difference*, Berkeley, CA: University of California Press and The Nature Conservancy.

Intergovernmental Panel on Climate Change (IPCC) (2013), *Climate Change 2013: The Physical Science Basis*, IPCC WGI Fifth Assessment Report, Geneva: IPCC.

Jabareen, Y.R. (2006), 'Sustainable urban forms: their typologies, models and concepts', *Journal of Planning Education and Research*, **26**(1), 38–52.

Jackson, W. (1980), *New Roots for Agriculture*, Lincoln, NE: University of Nebraska Press.

Jenerette, G.D. and D. Potere (2010), 'Global analysis and simulation of land-use change associated with urbanization', *Landscape Ecology*, **25**(5), 657–70.

Kellert, S.R., J.H. Heerwagen and M.L. Mador (eds) (2008), *Biophilic Design: The Theory, Science and Practice of Bringing Buildings to Life*, New York: John Wiley.

Kotkin, J. (2010), *The Next Hundred Million: America in 2050*, New York: Penguin.

Laurance, W.F. and A. Balmford (2013), 'A global map for road building', *Nature*, **495**(7441), 308–9.

Lin, B.B. and R.A. Fuller (2013), 'Sharing or sparing? How should we grow the world's cities?', *Journal of Applied Ecology*, **50**(5), 1161–8.

Liu, X., X. Li and Y. Chen et al. (2010), 'A new landscape index for quantifying urban expansion using multi-temporal remotely sensed data', *Landscape Ecology*, **25**(5), 671–82.

Lovell, S. and J. Taylor (2013), 'Supplying urban ecosystem services through multifunctional green infrastructure in the United States', *Landscape Ecology*, **28**(8), 1447–63.

Luck, M., G.D. Jenerette and J. Wu et al. (2001), 'The urban funnel model and the spatially heterogeneous ecological footprint', *Ecosystems*, **4**(8), 782–96.

Lutz, W., W. Sanderson and S. Scherbov (2008), 'The coming acceleration of global population ageing', *Nature*, **451**(7179), 716–19.

McGranahan, G. and D. Satterthwaite (2003), 'Urban centers: an assessment of sustainability', *Annual Review of Environmental Resources*, **28**, 243–74.

Millennium Ecosystem Assessment (MA) (2005), *Ecosystems and Human Well-Being: Current State and Trends*, Washington, DC: Island Press.

Muller, N., P. Werner and J.G. Kelcey (2010), *Urban Biodiversity and Design*, New York: Wiley-Blackwell.

Muriuki, G., L. Seabrook and C. McAlpine et al. (2011), 'Land cover change under unplanned human settlements: a study of the Chyulu Hills squatters, Kenya', *Landscape and Urban Planning*, **99**(2), 154–65.

Ndubisi, F.O. (ed.) (2014), *The Ecological Design and Planning Reader*, Washington, DC: Island Press.

Perz, S.G. (2014), 'The promise and perils of roads', *Nature*, **513**(7517), 178–9.

Pimentel, D., D. Cerasale and R.C. Stanley et al. (2012), 'Annual vs. perennial grain production', *Agriculture, Ecosystems and Environment*, **161**, 1–9.

Rees, W.E. (1997), 'Urban ecosystems: the human dimension', *Urban Ecosystems*, **1**(1), 63–75.

Reeves, F. (2014), *Planet Earth: How an Unhealthy Environment Leads to Heart Disease*, Vancouver: Greystone Books.

Reynolds, J.F., D.M. Stafford Smith and E.F. Lambin et al. (2007), 'Global desertification: building a science for dryland development', *Science*, **316**(5826), 847–51.

Rosenzweig, M.L. (2003), *Win-Win Ecology: How the Earth's Species Can Survive in the Midst of Human Enterprise*, Oxford, UK: Oxford University Press.

Salvati, L., M. Munafo and V.G. Morelli et al. (2012), 'Low-density settlements and land use changes in a Mediterranean urban region', *Landscape and Urban Planning*, **105**(1–2), 43–52.

Santos, K.C., J. Pino and F. Roda et al. (2008), 'Beyond the reserves: the role of non-protected rural areas for avifauna conservation in the area of Barcelona (NE of Spain)', *Landscape and Urban Planning*, **84**(2), 140–51.

Sushinsky, J.R., J.R. Rhodes and H. Possingham et al. (2013), 'How should we grow cities to minimize their biodiversity impacts?', *Global Change Biology*, **19**(2), 401–10.

Tanton, J.H. (1994), 'End of the migration epoch?', *The Social Contract*, **4**(3), 162–73.

United Nations (UN) (2013), *World Population Prospects: The 2012 Revision*, New York: UN, accessed 29 January 2016 at http://esa.un.org/unpd/wpp/index.htm.

United Nations (UN) (2014), *World Urbanization Prospects: The 2014 Revision*, New York: UN.

United Nations (UN) (2015), 'International decade for action "Water for Life" 2005–2015', accessed 29 January 2016 at http://www.un.org/waterforlifedecade/scarcity.shtml.

United Nations Environment Programme (UNEP) (2008), *Vital Water Graphics: An Overview of the State of the World's Fresh and Marine Waters*, Nairobi: UNEP.

Wackernagel, M. and W.E. Rees (1996), *Our Ecological Footprint: Reducing Human Impact on the Earth*, Gabriola Island, BC: New Society Publishers.

Wackernagel, M., N.B. Schulz and D. Deumling et al. (2002), 'Tracking the ecological overshoot of the human economy', *Proceedings of the National Academy of* Sciences (USA), **99**(14), 9266–71.

Wolch, J.R., J. Byrne and J.P. Newell (2014), 'Urban green space, public health, and environmental justice: the challenge of making cities "just green enough"', *Landscape and Urban Planning*, **125**, 234–44.

World Bank (2015), 'Data – indicators', accessed 29 January 2016 at http://data.worldbank.org/indicator.

Wu, J.G. (2013), 'Landscape sustainability science: ecosystem services and human well-being in changing landscapes', *Landscape Ecology*, **28**(6), 999–1023.

Wu, J.G. (2014), 'Urban ecology and sustainability: the state-of-the-science and future directions', *Landscape and Urban Planning*, **125**, 209–21.

Wu, J.G. and T. Wu (2013), 'Ecological resilience as a foundation for urban design and sustainability', in S.T.A. Pickett, M.L. Cadenasso and B.P. McGrath (eds), *Resilience in Urban Ecology and Design: Linking Theory and Practice for Sustainable Cities*, New York: Springer, pp. 211–30.

PART IV

CONCLUSIONS

20. Strengthening biodiversity and ecosystem services in impact assessment for better decisions

Davide Geneletti

The contributions to this *Handbook* illustrated the different ways in which information on biodiversity and ecosystem services (ES) can permeate different types of impact assessment across many planning and policy sectors. The variety of case studies described in the book provided examples of how this information is used to improve decision-making at all levels, from strategic choices to individual projects. This final chapter aims at summarizing some of the overarching lessons drawn from the book. These lessons are articulated into two sections, addressing analysis and frameworks, respectively.

Specifically, Section 20.1 presents an overview of the innovative methods and the associated challenges that have been proposed in the book to improve impact analysis. Section 20.2 expands the perspective and discusses the new – or enriched – frameworks that emerge from the different chapters, and the way in which they may contribute to better decisions, by enhancing the consideration of biodiversity and ES in impact assessment.

20.1 STRENGTHENING IMPACT ANALYSIS

Better decisions require, among other things, an improved understanding of impacts, which can be achieved through state-of-the-art methods that take advantage of scientific advances and the increasing availability of data and technology (Geneletti, 2015a). In this *Handbook*, innovative methods have been proposed to enhance the treatment of biodiversity and ES in impact assessment. Particularly, several chapters described advanced spatial analyses.

Lisa Mandle and Heather Tallis, in Chapter 2, presented a spatially explicit assessment of impacts to ES, based on recently developed tools, such as the software platform InVEST Version 2.3 (Tallis et al., 2012). These tools allow using the simplest possible approach to modeling complex processes as a way to provide sound scientific estimates, while requiring minimal data, time and resources. InVEST models were applied, for example, to a case study on mining permitting in Colombia to estimate marginal impacts on target ES. Ultimately, these impact estimates enable us to understand which mines are likely to contribute the most to downstream population level impacts.

Spatial approaches with similar level of sophistication, but applied to biodiversity modeling, are described by Berit Balfors and colleagues in Chapter 8. This contribution shows the advantages of adopting a landscape perspective (even for site-level development proposals), and applying spatial ecological analysis to model different biodiversity components, assess the quantity, quality and connectivity of related habitat types, evaluate planning alternatives, and communicate the evaluation outcomes. The case studies from the Stockholm region presented in this chapter show that a landscape approach provides opportunities to address different scales of ecological processes, strengthening important landscape structures and functions in urban, regional and infrastructure planning. In addition, a landscape approach can facilitate multi-scalar and multi-sectoral studies, by providing a link between different planning actions at the local, regional and national levels. This promotes the application of the integrated frameworks discussed in the next section.

Spatial analysis can improve also surveys that aim at collecting citizens' values and perceptions. Leena Kopperoinen and colleagues presented an online participatory GIS survey that allows to examine the inhabitants' perception of cultural ES, with a great level of thematic and spatial detail (Chapter 10). Respondents pinpointed targets on a map corresponding, among others, to scenic and symbolic places, or good places to relax and enjoy nature. This information can provide significant input to impact assessment processes of spatial/urban planning and policy-making, particularly if associated with data that show the current accessibility to the identified places, and model how this will change under different development options and future scenarios.

Advanced methods for spatial scenarios generation and assessment are proposed in Chapter 9, by Christine Fürst and colleagues. Their integrative modeling tool ('GISCAME') combines land use and land management change scenarios, with the assessment of their effects on regional targets, using an ES approach. Both analytical (rule-based) and participatory design instruments can be used to generate scenarios. The latter ones are particularly interesting for impact assessment, as they allow for a strong interaction with stakeholders. For example, scenarios can be generated that answer questions such as 'What kind of changes on ES could happen if. . .' or 'What would need to be changed to enhance a specific ES?'. Another strength of the platform, which makes it particularly suitable to support impact assessment in complex settings, is its visual, qualitative, and quantitative feedback, which includes key elements such as ES provision maps, risk maps (water erosion, mass movement), habitat fragmentation maps, diagrams on the overall balance of ES bundles, including temporal variability. Clearly, this tool can potentially satisfy most of the needs in terms of analytical support for ES analysis in impact assessment.

The innovative analyses and tools described above focus on a central stage of any impact assessment: the prediction of the consequences of decisions on target recipients. However, there are other relevant stages where information on biodiversity and ES can be mainstreamed into impact assessment. These stages, and the associated analyses, are illustrated by Davide Geneletti in Chapter 3, with specific reference to Strategic Environmental Assessment (SEA). The analyses include: establishing an ES context, by identifying ES and beneficiaries and reviewing relevant regulations and policies (see also Geneletti, 2015b); determining priority ES and assessing their baseline conditions and trends; and developing possible alternatives that enhance opportunities and reduce risks for ES. The overall purpose of the proposed analyses is to ensure that all relevant information of ES is collected, processed and used to support decision-making effectively throughout the impact assessment process.

A critical issue that influences the effectiveness of ES information in impact assessment is the choice of the evaluation methods. ES can be evaluated using monetary measures, or non-monetary ones, such as, for example, biophysical or social parameters (De Groot et al., 2002). Both approaches offer advantages, but have also limitations (Spangenberg and Settele, 2010; Sherrouse et al., 2011). The specific objectives and scope of the impact assessment (including the foreseen interaction with stakeholders), as well as the availability of data and resources, should play a key role in guiding the selection of the most appropriate way to evaluate ES.

The contributions to this *Handbook* acknowledge the complementarity of different evaluation approaches, and discuss their merits and challenges. Particularly, Léa Tardieu presented in Chapter 6 the application of economic valuation of ES loss to a transportation infrastructure case study. She concludes that the main advantages of the economic valuation reside in the enhanced guidance provided to the design of avoidance measures during the preliminary environmental studies, and in the possibility of testing the cost-effectiveness of mitigation measures. Davide Geneletti and colleagues illustrated how the biophysical evaluation of the climate regulation service provided by green urban infrastructures can be used in impact assessment (Chapter 18). In their case study, the outcome of the evaluation is used to support urban planning, by suggesting nature-based solutions that maximize the expected cooling effect for the most vulnerable population.

Finally, many of the chapters pinpointed specific analytical issues that are still challenging and worthy of additional research. The assessment of cumulative effects is one such issue. In Chapter 14, Asha Rajvanshi provided a detailed account of how cumulative effects associated with dams can be addressed and modeled. These impacts are characterized by complex pathways, involving multiple root causes and lower- and higher-order effects, interlinked by cause–effect relationships. She concludes that key challenges are, among others, the development of measurable biodiversity standards against which the value of a site can be assessed, and the delineation of appropriate boundaries of the study area.

The assessment of cumulative effects on biodiversity also appears to be critical for another energy sector: wind energy (Chapter 13). Lea Bulling and Johann Köppel maintained that assessing cumulative and long-term effects is bound to become even more crucial in the future, with further wind energy development, posing a great challenge to Environmental Impact Assessment. Additionally, the magnitude of cumulative effects also depends on the remaining habitat availability and quality outside the impacted area, as well as on other anthropogenic activities, such as other forms of energy production. Their approach based on different siting levels, and management interventions, represents a promising path towards the limitation of cumulative impacts on biodiversity of wind farms.

Cumulative effects also play a role in the design of biodiversity and ES offsets. Susie Brownlie and Jo Treweek concluded their contribution 'Biodiversity offsets for "no net loss" through impact assessment' (Chapter 16) by stressing the need for more strategic and landscape-level offset planning. Designing offsets, and planning their implementation, requires clearly defining objectives, choosing the most appropriate type and location, and measuring biodiversity losses and gains. A project-by-project approach to offsets overlooks cumulative effects, and reduces resource allocation efficiency.

Another analytical issue that emerged as challenging in Chapter 16 is the setting of the baseline, that is, the background rate of biodiversity loss against which offsets must be determined. The perception of whether offsets interventions are successful may change significantly when outcomes are measured against different baselines. The same challenge is described by Assumpció Antón and colleagues in their contribution on Life Cycle Assessment (LCA) (Chapter 7). Impact assessment in LCA requires comparing the environmental quality linked to the studied system to a reference state. The authors described the implications of adopting the three main alternatives for this reference state that have been proposed in the literature, namely natural or quasi-natural vegetation, historic land use and current land quality state. The selection of a suitable reference state poses a challenge because it has a strong influence on the results of the assessment, and their interpretation.

20.2 STRENGTHENING IMPACT ASSESSMENT FRAMEWORKS

Morrison-Saunders et al. (2014) put forward a call for more integration in impact assessment frameworks to move beyond silos of expertise, and achieve progress towards sustainable development. Canter and Ross (2014) and Greig and Duinker (2014) also acknowledged the benefits of more integrated impact assessments. Among these benefits, Geneletti (2014) highlighted the enhanced consideration of project (or plan, policy) alternatives, hence the increased support provided to decision-making. Along the same lines, many chapters of this *Handbook* advanced proposals to strengthen existing impact assessment frameworks that address biodiversity and ES, by expanding their traditional content and promoting integration across sectors

This is explicit in Chapter 2, where Lisa Mandle and Heather Tallis describe a framework for integrating development impacts to the environment and to people through ES. Heather Tallis and co-authors, then, further expand this framework (Chapter 17), by including the design of measures to mitigate and offset impacts on ES. The framework acknowledges that ES provision depends not just on how much of a particular ecosystem is present and where, but also on where the people who have the potential to benefit from its functions are located. The authors introduce the concept of 'servicesheds', defined as those areas that provide a specific ES to a specific beneficiary. Servicesheds are used to track benefits and negative impacts of development to specific groups of people, and to identify suitable alternatives and mitigation measures. For example, in their case study on a proposed highway in Peru, the authors conclude that the loss of ES is not distributed equally across population centers, or between indigenous and non-indigenous communities. Hence, the concept of servicesheds is a promising way to foster consideration of equity, a central sustainability issue that is often overlooked by impact assessments (Lamorgese and Geneletti, 2015).

Similarly, Tara Hooper and co-authors promote the use of a broad framework based on ES to assess and monitor impacts of marine planning (Chapter 11). As opposed to traditional biodiversity-centered frameworks, capturing information also about the flow of ES and their beneficiaries can support comprehensive assessments of the dynamic interactions between marine planning and the socio-ecological systems. Their Bristol Channel case study illustrates how this kind of assessment may prompt, for example, the identification of key areas of service provision, informing decisions about the future protection of these areas.

Another contribution that highlights the benefits of cross-sector integration is Chapter 5, in which Leena Karrasch provides a conceptual framework to include ES approaches in Social Impact Assessment. The application to a case study in regional development demonstrates how the framework helps to develop an appropriate communication strategy and supports awareness-raising among stakeholders. In particular, it proved useful to clarify the different demands of the stakeholders, and associate them to the supply of natural resources. By doing an evaluation of causalities and characteristic values described by the sector representatives, it was possible to make an assessment of their interests, preferences, and needs in relation to natural resources and, therefore, ES.

Health Impact Assessment (HIA) too may benefit from a stronger integrative framework, where ecosystems are seen as settings for health, and ES allow us to make explicit the ecological and economic conditions that underpin health consequences. Pierre Horwitz and Margot W. Parkes (Chapter 4) propose this framework to orient the scoping phase of HIA, where the process is planned and the required information and resources are determined. The framework is based on four key steps: documentation of health issues (e.g., exposure to infectious diseases), association of relevant ES to each issue (e.g., water purification, biological control agents for pathogens), identification of health benefits if the services are maintained/ enhanced (e.g., confidence in the quality and safety of interacting with local ecosystems), and identification of health consequences of disruption to the services (e.g., exposure to water-borne disease). The case study on coal seam gas production illustrated in the chapter demonstrates the utility of framing ES to ensure that scoping embraces a systemic understanding of interrelated health impacts, as well as to link health to other sectors for better integrated decision-making.

All these proposals to promote more integrated impact assessment frameworks center on the concept of ES, showing their relevance across different environmental and socio-economic topics and decision sectors. The case studies that accompany the different proposals provide an account of the merits and the potential of the 'ES philosophy' as a framing device for the impact assessment of projects, plans and policies, as recently discussed by Baker et al. (2013). In different forms, all the proposals acknowledge the advantages of making explicit the interactions between changes in biodiversity (and more generally the biophysical environment), effects on ES, and consequences for people's well-being.

Dilys Roe and Davide Geneletti focus on one specific sub-set of well-being consequences, namely those associated to poverty alleviation (Chapter 15). They advance a proposal for an improved conceptual framework to guide thinking about the interactions between different components of biodiversity and different dimensions of poverty. The framework illustrates how the actual impact that biodiversity has on poverty is mediated by the environmental, political and institutional context that regulates the nature and severity of poverty, the natural availability of biodiversity, and the ability of poor people to access and use biodiversity. The impact could be positive, neutral or negative (e.g., if human wildlife conflict or vector-borne disease exacerbates poverty). The use of biodiversity by poor people will also have an impact on biodiversity itself. The nature of this impact (biodiversity enhancement or depletion) will determine the long-term sustainability of use and ongoing ability of biodiversity to provide poverty alleviation services. Understanding how poor people interact with biodiversity and how this interaction contributes to their livelihoods can provide a more nuanced understanding of the social consequences of biodiversity impacts and allow more insights to be derived from impact assessments.

Better consideration of the broader socio-political context to improve the understanding of biodiversity impacts is advocated also by David Newsome and Michael Hughes in Chapter 12, 'Understanding the impacts of ecotourism on biodiversity: a multi-scale, cumulative issue influenced by perceptions and politics'. Their case studies in Bwindi and National Parks in Uganda illustrate that tourist interest in a specific iconic species, such as gorillas or chimpanzees, can have wider conservation implications, as conservation of habitat is vital to conservation of species of major tourist interest. Moreover, by conserving the habitat of the iconic species other ecotourism values can be realized by virtue of tourism-related biodiversity conservation outcomes. Hence, ecotourism can function by adding economic and employment values to local communities and can be a positive activity for the conservation of biodiversity. However, these implications can be properly captured by Environmental (and Social) Impact Assessments only if they are able to progress toward consideration of ecotourism impacts within the wider socio-political and geographical context.

Finally, Richard T.T. Forman and Jianguo Wu, in Chapter 19, conceptualize a broader framework for land planning, under the motto 'think globally, plan regionally'. The framework moves from predicted population growth rate, and includes other global changes that are relevant to progress towards sustainability. This enables to identify spatial options for the best places in urban regions (least damaging to farmland and/or natural land) where major population growth could be concentrated in the future. The proposed best places are: city centers, rings immediately around city centers, inner and outer suburbs, and within satellite cities in either farmland or natural land. The worst places appear to be areas surrounding satellite cities, areas surrounding towns/villages and newly built cities and towns. The authors then outline a wide range of actions to improve, or minimize negative effects on farmland, natural land and built communities, acknowledging that no single overarching solutions emerge.

It is the role of impact assessment to identify and propose packages of solutions that better fit a specific context, and decision-making level. The 'big picture' provided by Richard T.T. Forman and Jianguo Wu can stimulate practitioners and decision-makers to understand, and address, the implications for the scale of their specific impact assessment exercise. The chapters of this book highlighted the diversity of approaches that are being developed to strengthen the consideration of biodiversity and ES in impact assessment. They also demonstrated their applicability in different contexts and sectors the world over. It is my hope that this *Handbook* will help practitioners and researchers to improve the practice of this evolving field, and contribute to better decisions about the use of our lands and waters.

REFERENCES

Baker, J., W.R. Sheate and P. Phillips et al. (2013), 'Ecosystem services in environmental assessment – help or hindrance?', *Environmental Impact Assessment Review*, **40**(1), 3–13.

Canter, L. and B. Ross (2014), 'A basic need for integration – bringing focus to the scoping process', *Impact Assessment and Project Appraisal*, **32**(1), 21–2.

De Groot, R.S., M.A. Wilson and R.M. Boumans (2002), 'A typology for the classification, description and valuation of ecosystem functions, goods and services', *Ecological Economics*, **41**(3), 393–408.

Geneletti, D. (2014), 'Integration of impact assessment types improves consideration of alternatives', *Impact Assessment and Project Appraisal*, **32**(1), 17–18.

Geneletti, D. (2015a), 'Research in strategic environmental assessment needs to better address analytical methods', *Journal of Environmental Assessment Policy and Management*, **17**(1), 1550014.

Geneletti, D. (2015b), 'A conceptual approach to promote the integration of ecosystem services in strategic environmental assessment', *Journal of Environmental Assessment Policy and Management*, **17**(4), 1550035.

Greig, L. and P. Duinker (2014), 'Strengthening impact assessment: what problems do integration and focus fix?', *Impact Assessment and Project Appraisal*, **32**(1), 23–4.

Lamorgese, L. and D. Geneletti (2015), 'Equity in sustainability assessment. A conceptual framework', in A. Morrison-Saunders, J. Pope and A. Bond (eds), *Handbook of Sustainability Assessment*, Cheltenham, UK and Northampton, MA, USA: Edward Elgar Publishing.

Morrison-Saunders, A., J. Pope and J.A.E. Gunn et al. (2014), 'Strengthening impact assessment: a call for integration and focus', *Impact Assessment and Project Appraisal*, **32**(1), 2–8.

Sherrouse, B.C., J.M. Clement and D.J. Semmens (2011), 'A GIS application for assessing, mapping, and quantifying the social values of ecosystem services', *Applied Geography*, **31**(2), 748–60.

Spangenberg, J.H. and J. Settele (2010), 'Precisely incorrect? Monetising the value of ecosystem services', *Ecological Complexity*, **7**(3), 327–37.

Tallis, H.T., T. Ricketts and A.D. Guerry et al. (2012), *InVEST 2.3 Beta User's Guide*, Stanford, CA: Natural Capital Project.

Index